BEEF PRODUCTION AND THE BEEF INDUSTRY:
A Beef Producer's Perspective

Robert E. Taylor
Department of Animal Sciences
Colorado State University
Fort Collins, Colorado

Burgess Publishing Company
Minneapolis, Minnesota

Discipline Editor: Jim Ruen
Developmental Editor: Nancy Crochiere
Production: Pat Barnes, Judy Vicars, Morris Lundin
Cover Photo: Courtesy of Norden Laboratories
Composition and Design: Kingsport Press

Printed in the United States of America

Library of Congress Cataloging in Publication Data

Taylor, Robert E. (Robert Ellis), 1934–
Beef production and the beef industry.

Bibliography: p.
Includes index.
1. Beef industry—United States. 2. Beef industry.
3. Beef cattle. 4. Beef. I. Title.
HD9433.U4T39 1984 338.1′762′0830973 83–26192
ISBN 0–8087–3652–3

Burgess Publishing Company
7108 Ohms Lane
Minneapolis, MN 55435

J I H G F E D C

CREDITS FOR COLOR SECTION

We would like to acknowledge the following sources of the photographs in the color plate section: American Angus Association, BBU, American Brahman Breeders Association, International Brangus Breeders Association, Charolais Journal, American Breeders Service, North American Gelbvieh Association, North American Limousin Foundation, American Polled Hereford Association, Dickinson Ranch, Santa Gertrudis Breeders International, Leachman Cattle Co., American Scotch Highland Breeders Association, American Shorthorn Association, and American Simmental Association. Plates E-H are courtesy of Iowa State University and Colorado State University.

Dedication

This book is dedicated to my family—the most important focal point of my life. My father and mother (Wallace and Etta Taylor) instilled in me an intense interest in beef cattle and an understanding of the importance of work. My wife, Carole, and our children Scott, Lori, Mitchell, Alan and Aaron have been patient, understanding and supportive as I have pursued my vocation—beef cattle.

Contents

Figure i. The slaughter steer is the focal point of the U.S. beef industry, although slaughter heifers, cows, and bulls contribute to the beef supply. Time is a major factor in the beef production process from "sperm to steak" or from "the range to the range." Each beef industry segment uses many biological and economic relationships to efficiently produce a desirable beef product. Drawn by Dennis Giddings.

Preface

The cattle industry has been evolving and changing for centuries. Its emergence gained significance when man recognized that domesticating cattle would result in a consistent supply of milk, meat, raiment, and power.

The organizational structure of the cattle industry started to become complex when agricultural production efficiency made rapid strides through increased mechanization. This allowed people to pursue vocations other than raising their own supply of food and other basic necessities. The release of time from the land, an accumulation of capital, and man's creative ability generated the Industrial Revolution. In turn, industrial development has provided an abundant supply of goods and services.

Increased emphasis on education for all people and recognition of research needs, with time and dollars to support it, have generated an age of involved technology and specialization. Rapid changes in technology continue to allow farmers and ranchers to increase their productivity. This change has been phenomenal over the past century. In 1900 each farmer in the United States produced enough food for 8 consumers, while in 1940 the number was 11 and currently it is 78. In contrast, each Russian farmer presently produces food for 4 consumers. The United States has become an urban society with less than 4% of the approximate 230 million people on farms and ranches. This change in itself has created communication challenges for beef producers and beef consumers.

These and other changes have brought the beef industry to its present status of several loosely interlocking segments operating rather independently of each other. Each segment operates with different economic parameters although there is some overlap between segments.

Beef cattle exist for the following reasons: (1) they are a resource to use land which has been inherited or purchased for investment (land in the United States currently has a market value that usually exceeds its agricultural productive value), (2) beef cattle are owned, in numerous cases, to reflect status and prestige and (3) to provide a source of livelihood and to produce a product of high consumer preference.

This book is written for two primary purposes: (1) to identify and clarify the primary biological principles which contribute to the raising of productive and profitable beef cattle and (2) to describe and assess the several beef industry segments which hopefully will contribute to a better understanding and effective communication by those involved in the beef industry.

The book is written primarily from the viewpoint of commercial beef producers. The material identifies the essential ingredients needed by commercial producers to raise productive and profitable cattle which can meet the specifications and requirements of the beef industry segments.

The first ten chapters of the book cover the segments of the beef industry, while most of the remaining chapters cover biological principles needed to understand beef cattle productivity. Beef producers must understand how cattle economically perform under different environmental conditions and how cattle or the products produced contribute to the total beef industry. The intent of this book is to tie these needs together in a single volume (Figure i).

Because the beef industry is dynamic and constantly changing, some of the material in this book is outdated the moment it is printed. Therefore it is my intent to revise the book every two years so students and producers can relate to a changing industry.

Acknowledgments

Many individuals, too numerous to mention, have reviewed several chapters or the entire manuscript and made helpful suggestions. Photographs and illustrations have been willingly supplied by people associated with the beef industry.

Pat Beebe and Jackie Nelson have spent endless hours typing and organizing the material for this book. To them I am most appreciative.

1

An Overview of the Present Beef Industry

INTERNATIONAL

Cattle, including the domestic water buffalo, contribute food, fiber, fuel, and work for many of the four billion people in the world.

The beef producer in the United States is influenced significantly by other countries. Some of these influences are reflected in international trade, drought, population changes, political pressures, business opportunities, new sources of breeding stock, and disease problems.

World Cattle Numbers

In 1974, the USDA Foreign Agricultural Service reported that there were 1.34 billion head of cattle in the world, of which approximately 150 million were the domestic water buffalo. The rate of increase in world cattle during recent years has been approximately 2% per year, although the growth in numbers during 1980 was 0.5%. Since 1975, the USDA has reported cattle numbers in only major, selected countries; therefore, estimates of total world cattle numbers have not been made. The leading countries in cattle numbers are shown in Table 1.1. India far exceeds any other country in cattle numbers. Cattle numbers per person in India are above average; however, per capita beef consumption (Table 1.9) is very low. The primary reason for this large discrepancy in numbers and consumption is that cattle are sacred to Hindus and they do not eat beef.

Table 1.1 also shows cattle numbers on a per capita and square mile basis. Countries having ratios of cattle to people higher than 1:1 are typically beef exporting countries (Table 1.3). Cattle density (number of cattle per square mile) can be misleading as cattle numbers are typically concentrated in areas of high forage and feed grain production. Figure 1.1 shows cattle densities in various regions of the world.

Table 1.1. Leading Countries in Cattle and Buffalo Numbers (1980)

Country	Mil. Head	No. Cattle Per Capita	No. Cattle Per Sq. Mile
1. India	241	0.36	196
2. Soviet Union	115	0.43	13
3. United States	111	0.50	31
4. Brazil	93	0.76	28
5. China	71	0.10	19
6. Argentina	59	2.18	30
7. Mexico	29	0.40	38
8. Columbia	27	0.98	59
9. Australia	26	1.78	9
10. France	24	0.45	114
Other countries important in beef consumption and world trade with the United States			
Uruguay	10	3.45	143
New Zealand	8	2.50	80
Canada	12	0.50	3
Japan	4	0.03	29

Source: USDA Foreign Agriculture Circular, February 1981.

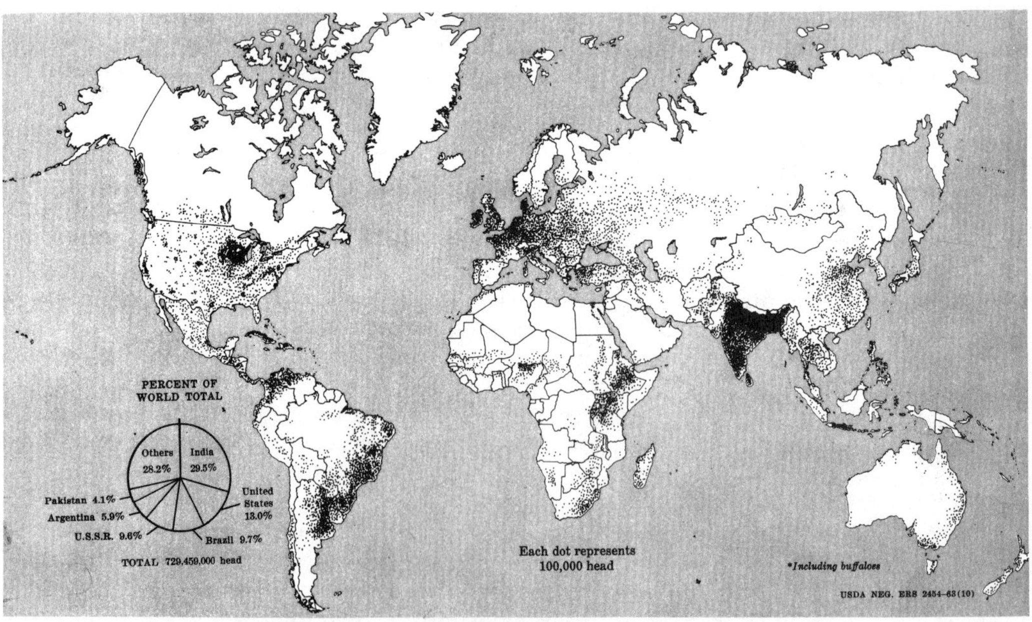

Figure 1.1. World cattle and buffalo numbers, average 1957–1961. *Source:* USDA.

International Trade

Tables 1.2, 1.3, 1.4 and 1.5 reflect the importance of beef and beef products in international trade. The United States imports a rather large amount of beef (approximately 10% of annual production when computed on a carcass weight basis). While total beef imported is over 2 billion lbs, approximately three-fourths of that amount is subject to the meat import law. Meat imports subjected to the law totaled 1.422 billion lbs in 1980. Under the counter-cyclical formula in the new import law, the amount of beef imported is controlled by beef produced in the United States. When domestic output

Table 1.2. Value of United States Exports of Beef and Beef Products

Item	1980 Exports (mil dollars)	1980 Imports (mil dollars)
Beef and veal	249	1780
Variety meats	300	8
Tallow	727	1
Live cattle	55	237
Cattle hides	682	83
Total (cattle)	2014	2109
Total (all livestock and meat)	3043	3219

Source: USDA.

Table 1.3. Leading Countries in Boneless Beef Imports to the United States, 1980

Country	Mil lbs
1. Australia	358
2. New Zealand	144
3. Canada	34
4. Mexico	26
5. Costa Rica	21

Source: USDA.

Table 1.4. Leading Countries for Inedible Tallow Exported from the United States, 1980

Country	Mil lbs
1. Egypt	187
2. Korea	114
3. Japan	106
4. Netherlands	95
5. Mexico	78

Source: USDA.

Table 1.5. Leading Countries for Hides Exported from the United States, 1980

Country	Mil lbs
1. Japan	7.5
2. Korea	2.7
3. Mexico	2.0
4. Taiwan	1.3
5. Romania	1.0

Source: USDA.

is large, imports will be reduced; when domestic production is down, imports will be increased. While beef imports in the United States become a fervent issue with cattlemen, they should realize the dependency of the U.S. beef industry on the export market for hides and tallow. Income generated from these two major export items basically offsets the dollar cost of the imported beef. On the other hand, a 680 million dollar hide export market returns to the United States as a 3 billion dollar import cost of leather, shoes, and leather products.

World Implications of Beef as a Food

Tables 1.6, 1.7, 1.8 and 1.9 show the countries where red meats (beef, veal, lamb, pork, and horse) are primarily produced and consumed. The United States produces over four times the amount of red meat (38.9 versus 9.5 billion lbs) and beef (21.5 versus 6.4 billion lbs) when compared to the countries in second place. Consumption patterns are similar to those of production; for example, total red meat consumed (United States, 41.1 versus West Germany, 10.1 billion lbs) and beef consumed (United States, 24.0 versus Argentina, 5.3 billion lbs). The United States is the unquestioned leader in total pounds of beef produced and consumed.

Japan's eleven pounds per capita consumption of beef is rather low. The per capita consumption in that country, however, has doubled during the past ten years due to rapid economic developments and increased standard of living. These changes in living standards have increased their demand for fed beef. Their level of production is below

Table 1.6. Leading Countries in Red Meat Production

Red Meat Produced (Carcass Weight Basis, 1980)	
Country	***Bil lbs***
1. United States	38.9
2. W. Germany	9.5
3. France	8.1
4. Argentina	7.3
5. Brazil	7.3
World total—approx.	175.0

Source: USDA.

Table 1.7. Leading Countries in Red Meat Consumption

	Per Capita and Total Red Meat Consumption (Carcass Weight Basis, 1980)	
Country	***Lbs Per Capita***	***Total (bil lbs)***
1. New Zealand	250	0.8
2. Argentina	229	6.2
3. Uruguay	207	0.6
4. United States	185	41.1
5. Australia	178	2.6

Source: USDA.

Table 1.8. Leading Countries in Beef Production

Carcass Beef Produced (1980)	
Country	***Bil lbs***
1. United States	21.5
2. Argentina	6.4
3. Brazil	4.8
4. France	4.0
5. Australia	3.3
6. W. Germany	3.3
World total—approx.	87.3

Source: USDA.

Table 1.9. Leading Countries in Beef Consumption

	Per Capita and Total Beef Consumption (Carcass Weight Basis, 1980)	
Country	***Lbs Per Capita***	***Total (bil lbs)***
1. Argentina	196	5.3
2. Uruguay	172	0.5
3. New Zealand	125	0.4
4. United States	106	24.0
5. Australia	103	1.5
	Other Countries of Interest	
India—less than	1	0.4
Japan	11	1.3

Source: USDA.

Figure 1.2. Water buffalo are important sources of power in preparing rice fields. Courtesy of Dr. Budi S. Nara.

their consumption level which identifies Japan as an excellent export market for other countries. Price of beef is high in Japan compared to most other countries. Politics, high middleman mark-ups, and time-delayed product handling, however, pose hurdles to increasing beef exports from the United States to Japan.

In many countries, cattle and buffalo, along with other animals, provide 75 to 90% of the power for agriculture (McDowell 1978). This power indirectly provides food (rice and other cereal grains) for some heavily populated areas of the world (Figure 1.2).

Annual cattle slaughter provides nearly 90 billion lbs of meat for the world's population. Assuming an equal distribution throughout the world, this would be approximately 20 lbs of beef per person. If all people in the world consumed as much beef as those in Argentina, a ten-fold increase in pounds of beef produced would be needed. The restricted land areas with marked population increases will likely not allow such an expansion in cattle numbers.

Some individuals strongly feel that beef cattle have little, if any, place in feeding a hungry world. They point to a rapidly expanding world population (Figure 1.3) which they feel will soon demand only the most efficiently produced foods.

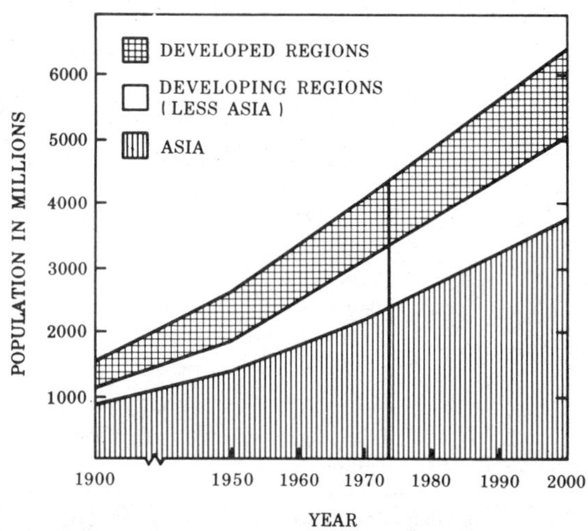

Figure 1.3. Past, present, and projected world population, 1900–2000. Courtesy of USDA.

The greatest pressure for food appears to be in Asia and some African countries where population concentration is heaviest, population growth is rapid, and present food supply is limited. There are areas of the world where people are undernourished (Figure 1.4).

Beef and other animal products are nearly always consumed in quantity when they are available. Availability of animal products in most countries is highly related to the economic status of the people and their agricultural technology. Vegetarianism in countries such as India may be the long-term result of intense population pressures and scarcity of feed for animals because of competition between humans and animals for food. Rising population pressures, particularly in developing countries (Southeast Asia, Africa, and Latin America) force people to consume foods primarily of plant origin.

In many of these developing countries, the standard of living is low because large numbers of people must work to produce food if their society is to avoid widespread starvation; in this situation, very few people are available to provide other goods and services. In developed nations (United States, Canada, Europe, USSR, Japan, Republic of South Africa, Australia, and New Zealand) by contrast, relatively few people are involved in producing food needed for their society, and many people are engaged in occupations that provide other goods and services.

Cereal grains are the most important source of energy in world diets. Energy derived from cereal grains, however, is twice as important in developing countries (as a group,

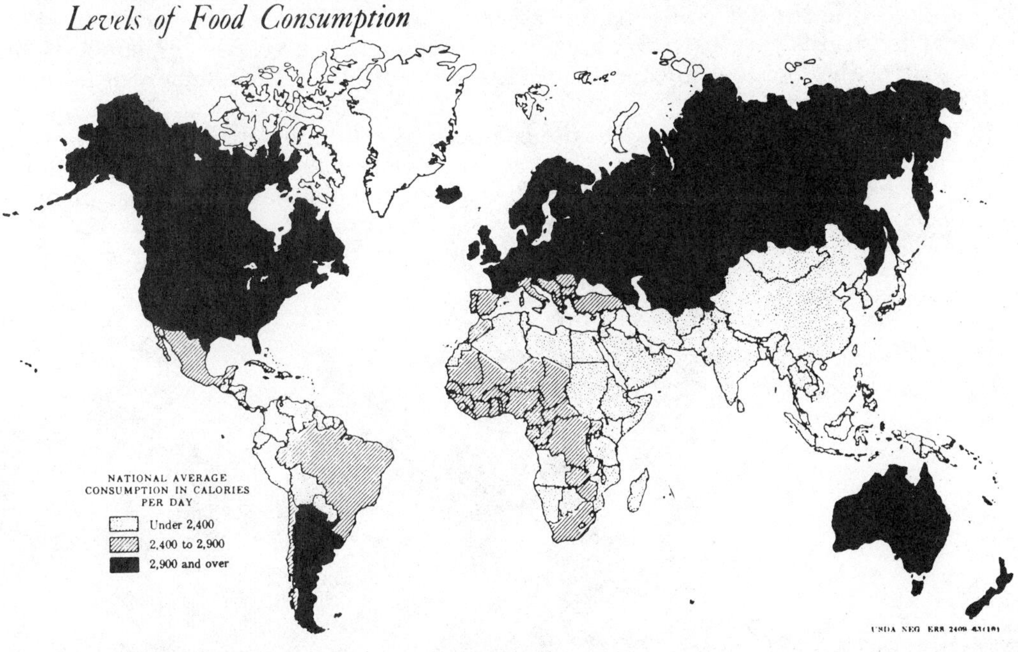

Figure 1.4. Level of food consumption in the world based on calories per person per day. Daily requirements in the United States are 2600 calories and 40 grams of protein. For tropical zone developing countries the level is 2300 calories and 38 grams of protein. Courtesy of USDA.

but there are exceptions) as in developed countries. Beef and other animal products provide approximately 25% of the dietary energy supply for people in developed countries as contrasted to approximately 7% for developing countries. Protein supplied in human diets from animal products is approximately 55% for developed countries, while only 21% for developing countries.

There are two scientifically based reasons why beef and other meats are important as a food. The first is that the assortment of amino acids in animal protein more closely matches the needs of the human body than does the assortment of amino acids in plant protein. The second is that vitamin B_{12}, which is required in human nutrition, may be obtained in adequate quantities from consumption of meat or other animal products but not from consumption of plants.

Another major reason for human use of animals for food is that most countries have land areas unsuitable for growing cultivated crops. Approximately two-thirds of the world's agricultural land is permanent pasture, range, and meadow; of this, about 60% is not suitable for producing cultivated crops that would be consumed directly by humans. This land, however, can produce roughage in the form of grass and other vegetation that is digestible by grazing ruminant animals (of which cattle and sheep are the most important). These "walking food factories," traveling without fossil fuel, can harvest and convert the vegetation, which is undigestible for the most part by humans, to high-quality protein food (Figure 1.5). Every 26 lbs of usable range forage eaten by a cow–calf combination will yield approximately one pound of calf. Each 10 lbs of range forage eaten by a calf directly will produce a pound of calf in a cow–calf operation. In the United States, about 385 million acres of rangeland and forest, representing 44% of the total land area, are used for grazing. Although this acreage now supports only about 40% of the total cattle population, it could carry twice this amount if developed and managed intensively.

Animal agriculture therefore does not compete with humans for the produce of most of the land used as permanent pasture, range, and meadow. On the contrary, the use of animals as intermediaries provides a means by which land that is otherwise unproductive for humans can be made productive.

There are current concerns about energy, protein, population pressures, and land

Figure 1.5. Cattle forage many untillable acres in the United States and throughout the world. Courtesy of American Simmental Association.

resources as they relate to animal agriculture. The quantities of energy and protein present in foods produced by animals are smaller than the quantities of the same nutrients consumed by animals in their feed because animals are inefficient in the ratio of nutrients utilized to nutrients produced. More acres of cropland are required per person for a diet high in foods from animals than for a diet including only plant products. As a consequence, animal agriculture has been criticized for wasting food and land resources that could otherwise be used to feed persons with inadequate diets. Consideration must be given to economic systems and consumer preferences to understand why agriculture perpetuates what critics perceive as resource-inefficient practices. These practices relate primarily to providing food-producing animals with feed that could be eaten by humans and using land resources to produce crops specifically for animals instead of producing crops that could be consumed by humans.

Agriculture producers produce what consumers want to eat as reflected in the prices consumers can and are willing to pay. Eighty-five percent of the world's population desires food of animal origin in its diet, perhaps because of a higher palatability preference for foods of animal origin. In most countries, as per capita income rises, consumers tend to increase their consumption of meat and animal products, which are generally more expensive pound for pound than products derived from cereal grains.

If many consumers in countries in which animal products are consumed at a high rate were to decide to eat only food of plant origin, consumption and price of foods from plants would increase, and consumption and price of foods from animals would decrease. Agriculture would then adjust to produce greater quantities of food from plants and lesser quantities of food from animals. Ruminant animals can produce large amounts of meat without grain feeding. The amount of grain feeding in the future will be dictated by cost of grain and the price consumers are willing to pay for beef.

Some people in the United States are proponents of a shift from consumption of food from animals to foods from plants. They see this primarily as a moral issue, stating that it is not ethical to let people elsewhere in the world starve when people in the United States could meet their own food needs by eating foods from plants rather than feeding plants to animals. The United States could then send the balance of the plant-derived foods abroad. These people believe that grain should be provided at minimum cost to needy countries because we have a surplus in the United States and we can ship it with comparative ease. Providing free food to other countries has met with limited success in the past. It puts economic pressure on the American farmer who then cannot make an adequate living, which eventually raises taxes. Free food has also not been distributed adequately in developing countries because transportation and marketing distribution systems are poor, and it has upset their agricultural production. Many people in the United States feel they have a moral obligation to share their abundance with other people in the world, particularly those in the developing countries. It appears that this can best be done by sharing our time and our basic, but not necessarily advanced, technology.

Citizens of developed countries who feel committed to feeding a hungry world should give consideration to all areas of unproved food production instead of singling out the inefficiency of cattle and other grain-fed livestock. The following need careful consideration: (1) Reducing food waste. A survey of more than 350,000 residents of Tucson, Arizona showed that they annually discard some 950 tons of feed, most of it edible, valued at 9 to 11 million dollars. From this survey and other studies, it is estimated that in the United States approximately 20% of the calories purchased at retail are discarded. (2) Better land utilization by locating new housing and industrial developments on less productive land (Figure 1.6). (3) More food could be raised in residential areas by producing food instead of lawns and fruit trees instead of ornamental trees. (4) In considering priority of shifting the crops that are produced, which should have the lesser priority—grain for cattle, grain for alcohol, or land producing tobacco? (5) An analysis of a 17+ million dollar pet industry for which feed grains provide more than 3 million tons of pet food valued at more than one million dollars. (6) Moderation in calorie consumption by people in developed countries where a high percentage of people have been reported as overweight.

There is evidence that development of animal agricultural systems in developing countries could initiate a self-sustaining, more productive agriculture requiring only small inputs of fossil-fuel energy. Efficient utilization of grazing animals in specific integrated farming systems could provide energy inputs from draft and transportation and also increase food and other useful by-products from renewable sources of energy.

People need self-motivation to improve. They need to be shown how to help themselves step by step, and need to fully develop agricultural resources in their own countries. Increases in efficiency of animal production in the United States has been remarkable during the past half century. For example, pounds of liveweight beef marketed per breeding female has increased from 220 in 1925 to 482 in 1975 (CAST Report No. 82). This increase in efficiency has occurred primarily because people had an incentive to progress under a free enterprise system. They learned how to improve their standard of living using the available resources. This progress has taken time under the environmental conditions that motivated people to be successful.

Remarkable changes in agriculture efficiency in the United States have occurred in the stepwise development of a stable and efficient political system, compulsory and free education, followed by industrial development.

Not only should advanced technology be shared with other people throughout the world, but United States citizens should recognize how these achievements have been made. The achievements have been built on knowledge developed through meaningful experience and research; extension of knowledge to producers; and development of an industry to provide inputs, transportation, processing, and marketing. Dwindling dollars presently being put into agricultural research and extension of knowledge may not provide technology needed for future food demands.

Population growth and agricultural development will be challenges in future years if people in the world are to be adequately fed. These challenges will be greater in developing countries.

About 30% of the world human population, and 32% of the ruminant animal population live in developed regions of the world, but ruminants of these same regions produce two-thirds of the world's meat and 80% of the world's milk. In developed regions, a higher percentage of animals are used as food producers, and these animals are higher in productivity on a per animal basis than in developing regions. This is the primary reason for the higher level of human nutrition in developed countries.

There is reason to believe that many so-called developing regions of the world can achieve levels of plant, animal, and eventually human food productivity similar to those of developed regions. Except perhaps in India, there are abundant supplies of animal feed resources, not competitive with human food production, to support expansion of

Figure 1.6. Each year thousands of acres of highly productive farmland are converted to urban development. Courtesy of UAES Public., USU, Logan, Utah.

animal populations and production. It has been estimated that through changes in resource allocation, an additional 8 billion acres of arable land (twice what is presently being used) and 9.2 billion acres of permanent pasture and meadow (23% more than presently being used) could be put into production in the world. This, plus the potential increase in productivity per acre and per animal presently reflected in the developed countries, demonstrates the magnitude of world food production potential. This potential, however, cannot be realized without proper government planning and increased incentive to individual producers.

In the long run, each nation must assume the responsibility of more of its own food supply by efficient production, barter, or purchase, and by keeping future food production technology ahead of population increases and demand. The extensive untapped resources that exist throughout the world can greatly enhance food production, including an ample supply of animal products. The greatest resource is each individual human being who can become self-motivated to become more productive and self-reliant.

UNITED STATES

The term *beef industry* implies that the total beef production system is a fairly unified operation subject to an overall management program. Actually, the beef industry is a series of several loosely interlocking segments. While these various segments are generally connected in series, they operate quite independently from each other. Each segment has different economic parameters and management problems, and they market different products. In many cases they are in direct competition with one another. In a real sense, the different beef industry segments are considered separate industries because of these distinctly different characteristics.

Table 1.10 gives an overview of the beef industry with the primary purpose of showing the animal and product flow between the various segments.

The beef industry in the United States can best be described in numbers, pounds, and dollars (Table 1.11) as related to product produced and product consumed.

Figure 1.7 shows that cattle numbers in the United States have increased dramatically since the turn of the century. There are, however, marked peaks and valleys in cattle numbers dictated by supply, demand, drought, and general economic conditions.

The January 1983 cattle inventory of 115.2 million head was considerably less than the peak numbers of 131.8 million head in 1975. This was the most marked decline in cattle numbers that has been experienced in the past century. Cattle cycles reflect the boom and bust periods of the cattle industry. The influence of the cattle cycle in marketing cattle is discussed in Chapter 18.

Changes in cow numbers over the years reflect an increase in beef cows and a decrease in dairy cows (Figure 1.8). Total milk production from dairy cows has remained at approximately the same level due to a marked increase in productivity for each dairy cow. The beef producer should take a lesson from this marked improvement on a per animal basis.

Distribution of beef cows by states is shown in Figure 1.9. Note the concentration of beef cows in the Great Plains area. This area, covering the states of Texas north through North Dakota and the eastern parts of New Mexico, Colorado, Wyoming, and Montana accounts for over 50% of the U.S. beef cow population. Some of the Corn Belt states and the southeast states have significant numbers of cows.

Table 1.10. An Overview of the Beef Industry Segments in the United States

Segments of the Beef Industry	Product Produced	Traits of Economic Importance	Approximate Number	Billion Dollars Invested
Seedstock Producers	Breeding stock—primarily bulls (14 to 30 mos old) and some heifers and cows. Some steers and heifers for feeding. Slaughter cows and bulls.	1. Same traits as commercial cow–calf producer. 2. All other genetically influenced traits (primarily weaning weight, yearling weight, and carcass composition.	150,000 breeders	15.0
* ↓ ↑ **				
Commercial Cow–Calf Producers	Calves (6 to 10 mos old), weighing 300 to 600 lbs. Slaughter cows and bulls with majority over 5 years of age weighing 800 to 1500 lbs (cows) and 1000 to 2500 lbs (bulls).	1. Percent calf crop. 2. Weaning weight. 3. Longevity of cows.	1.5 million producers	180.0
↓ ↑				
Yearling or Stocker Operator	Feeder steers and heifers (most of them 12 to 20 mos old weighing 500 to 900 lbs).	1. Gaining ability. 2. Livability and adaptability.		
↓ ↑				
Feeders	Slaughter steers, heifers, cows, and bulls (most steers and heifers 16 to 30 mos old, weighting from 900 to 1400 lbs).	1. Average daily gain. 2. Feed efficiency—(lbs feed per 100 lbs gain).	3,000 feedlots (feed 70% of cattle)	7.5
↓ ↑				
Packers	Carcasses (approximately 600 to 800 lbs). Boxed beef (carcasses into subprimal cuts).	1. Carcass yield grade. 2. Carcass quality grade. 3. Shelf-life of product.	750 packers	3.8
↓ ↑				
Purveyors	Carcasses (approx. 600 to 800 lbs) and resulting retail cuts.	1. Carcass quality grade. 2. Carcass yield grade. 3. Shelf-life of product.	?	?
↓ ↑				
Retailers	Retail cuts	1. Carcass yield grade. 2. Carcass quality grade. 3. Shelf-life of product.	250 food chains	50.0
↓ ↑				
Consumers	Retail cuts. Cooked portions.	1. Tenderness. 2. High ratio of lean to fat. 3. High ratio of lean to bone. 4. Flavor and juiciness. 5. Consistency.	230 million consumers	

* Animal and/or product flow.

** Demands and expectations.

Table 1.11. Cattle Numbers, Prices, Population, and Consumption in the United States from 1920 to Present Time

Year	Human Population (mil)	No. Cattle (mil)	No. Beef Cows (mil)	Carcass Beef Produced (bil lbs)	Per Capita Carcass Beef Consumption (lbs)	Disposable Per Capita Income (dollars)		Percent Income Spent for Beef
						Current Dollars	*1972 Dollars*	
1920	105.7	70.4	12.5	6.3	59.1			
1925	115.0	63.4	11.2	6.9	59.5			
1930	122.8	61.0	9.1	5.9	48.9	598	1709	
1935	126.9	68.8	11.1	6.6	53.2	458	1544	
1940	131.8	68.3	10.7	7.2	54.9	569	1849	
1945	139.2	85.6	16.5	10.3	49.4	1065	2420	
1950	151.1	78.0	16.7	9.5	63.4	1355	2386	
1955	164.0	96.6	25.7	13.6	82.0	1654	2577	2.65
1960	179.3	96.2	26.3	14.8	85.1	1937	2697	2.72
1965	193.0	109.0	33.4	18.7	99.5	2436	3152	2.48
1970	201.9	112.4	36.7	21.7	113.7	3376	—	2.55
1971	204.9	114.6	37.9	21.9	113.0	3587	3714	2.52
1972	206.7	117.9	38.8	22.4	116.1	3838	3837	2.66
1973	209.6	121.5	40.9	21.3	109.6	4315	4083	2.69
1974	211.6	127.7	43.0	23.1	116.8	4667	4013	2.72
1975	213.8	131.8	45.4	23.8	120.1	5075	4055	2.71
1976	215.9	128.0	43.8	25.7	129.2	5477	4161	2.54
1977	218.1	122.8	41.4	25.0	125.9	5954	4266	2.29
1978	220.5	116.3	38.8	24.0	120.1	6571	4409	2.41
1979	223.0	110.9	37.0	21.3	107.6	7293	4493	2.42
1980	224.4	111.0	37.1	21.5	105.8	8002	4473	2.27
1981	226.8	114.3	38.7	22.4	104.4	8827	4526	2.09
1982	229.1	115.7	39.4	22.4	104.4	9377	4567	2.00
1983	231.1	115.2	38.1					

Sources: 1. Cattle, January 1983. Crop Reporting Board, ESCS, USDA.
2. Livestock and Meat Situation, ESCS, USDA, 1983.

Changes in beef cow numbers by states over the past decade are reflected in Figure 1.10. These changes are primarily influenced by changes in economic conditions and forage supply. The latter can be dramatically influenced by drought, renovation of previously unproductive land, water development, and conversion of cropland to pasture and other forage for cattle.

Figure 1.11 shows the increased number of fed cattle marketed over the past 25 years. The feedlot industry which has had its major growth after the Second World War had a marked decline in 1973–1975. This decline was due primarily to a one- to three-fold increase in feed grain prices in addition to low fed-cattle prices.

Distribution of fed-cattle marketings by states is portrayed in Figure 1.12. Only 23 states are identified, primarily because other states are relatively unimportant in their contribution to the total cattle fed in the United States. The primary reason for this

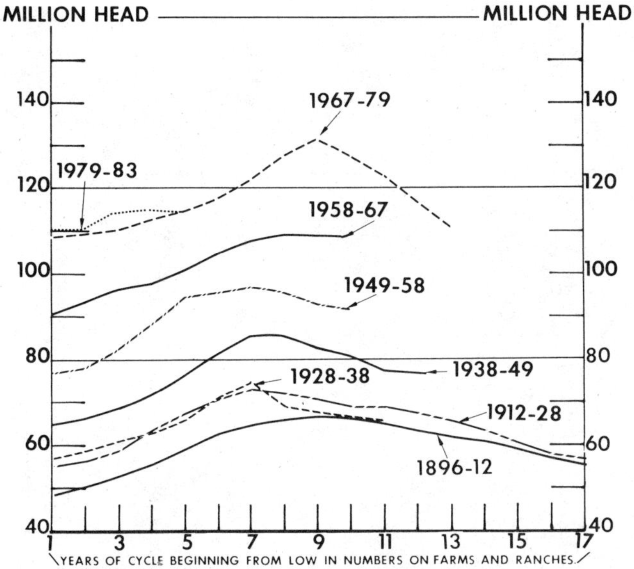

Figure 1.7. Cattle on farms by cycles. Courtesy of Western Livestock Round-Up.

distribution of fed cattle is availability of feed grains and environmental conditions that favor cattle feeding.

Figure 1.13 depicts changes in fed-cattle numbers over the past ten years. The greatest growth in fed-cattle numbers has occurred in several Plains States while significant reductions have occurred in the Corn Belt and California.

Comparison of Figures 1.12 and 1.14 identifies major cattle feeding states with large commerial feedlots as contrasted with states where smaller farm feedlots are important. Iowa is an example of a farm feedlot state where 2.5 million head of fed cattle are

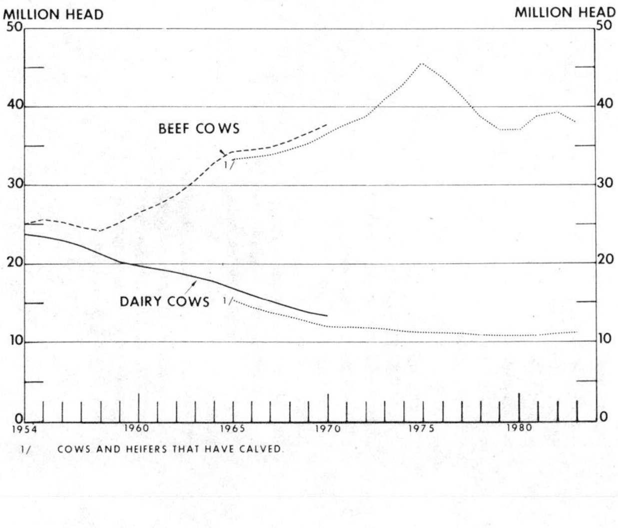

Figure 1.8. Cow numbers 1 January 1954 to 1983. Courtesy of Western Livestock Round-Up.

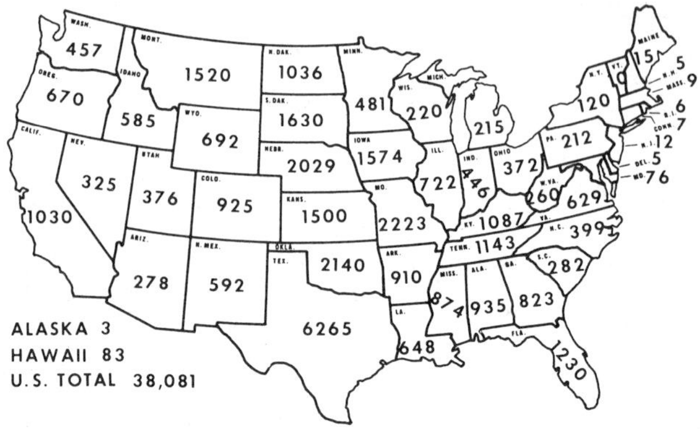

Figure 1.9. Beef cows that have calved 1 January 1983 (1000 head). Courtesy of Western Livestock Round-Up.

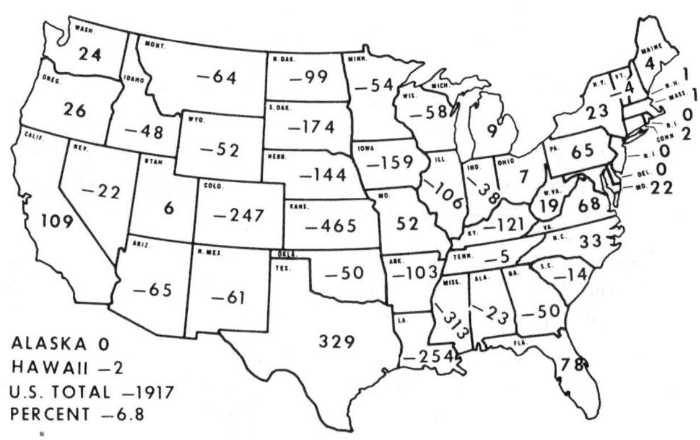

Figure 1.10. Change in beef calf crop 1972 to 1982 (1000 head). Courtesy of Western Livestock Round-Up.

Figure 1.11. Fed cattle marketings, 1982 (1000 head). Courtesy of Western Livestock Round-Up.

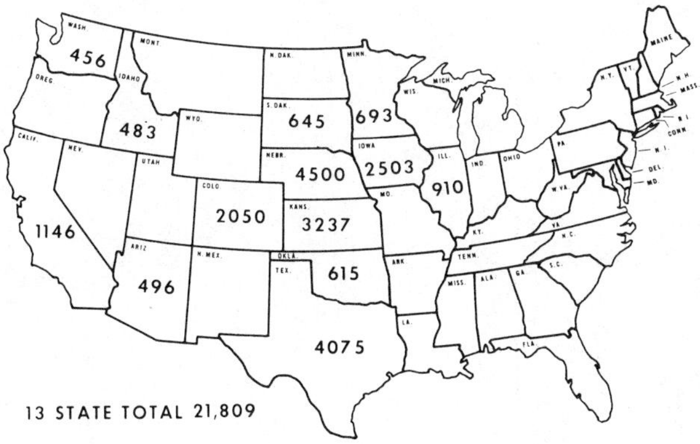

Figure 1.12. Fed cattle marketings (13 states), 1960–1982. Courtesy of Western Livestock Round-Up.

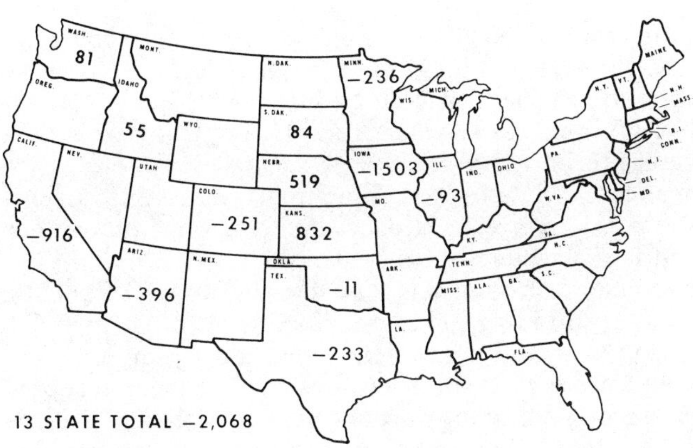

Figure 1.13. Change in fed cattle marketings 1971–1981 (1000 head). Courtesy of Western Livestock Round-Up.

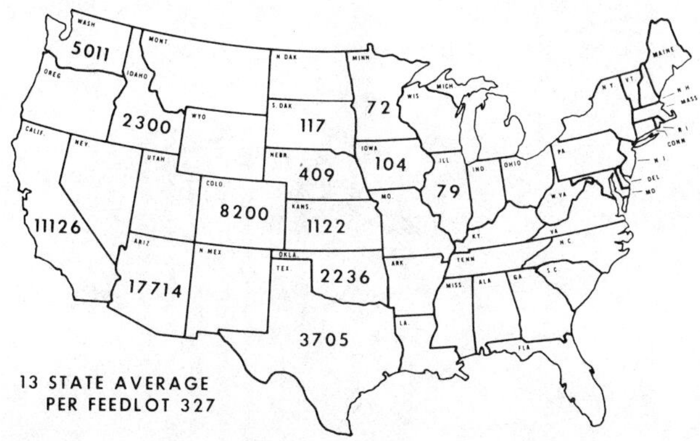

Figure 1.14. Average number of cattle marketed per feedlot, 1981. Courtesy of Western Livestock Round-Up.

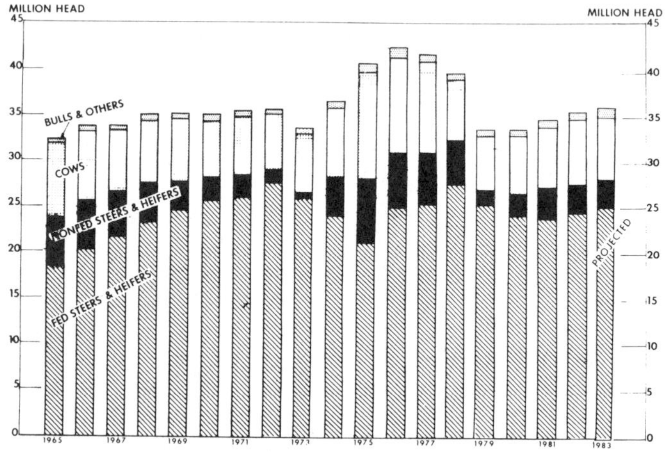

Figure 1.15. Commercial cattle slaughter, 1965–1981 (mil head). Courtesy of Western Livestock Round-Up.

marketed annually from feedlots selling approximately 100 head per feedlot. In contrast, Arizona cattle feeders marketed more than 19,000 head per feedlot in 1981.

The magnitude of the beef packing industry as reflected by number of cattle slaughtered is shown in Figure 1.15. Fed steers and heifers comprise the highest percentage of the more than 30 million head slaughtered annually.

Figure 1.16 shows the distribution of cattle slaughtered. The importance of the various states in cattle slaughter is similar to where the cattle are fed. Packing plants are typically located where cattle feeding is most heavily concentrated.

Tonnage of carcass beef and other red meats produced is shown in Figure 1.17. Packers, purveyors, and retailers have a big undertaking in slaughtering, processing, and distributing approximately 20 billion lbs of beef and veal to the consuming public.

Figure 1.18 shows how the major end product of the beef industry has been accepted by consumers. Per capita consumption of pounds of carcass beef has shown a large and consistent growth over the past 25 years. The marked decline in per capita beef consumption from 1976–1979 is due primarily to smaller supplies of available beef.

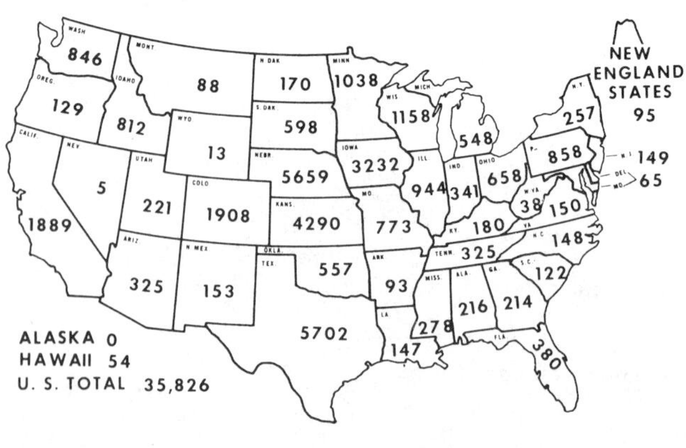

Figure 1.16. Commercial cattle slaughter, 1981. (1,000 head). Courtesy of Western Livestock Round-Up.

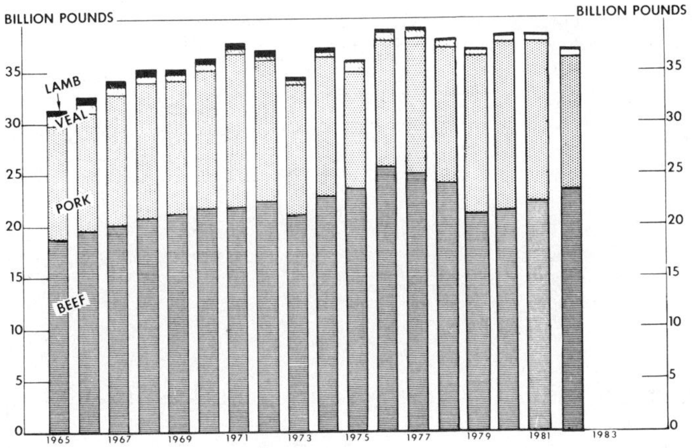

Figure 1.17. Production of red meat. Courtesy of Western Livestock Round-Up.

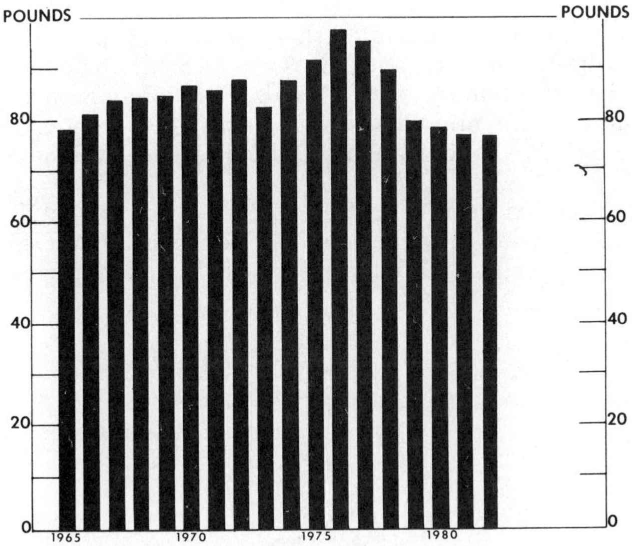

Figure 1.18. Carcass beef consumption per person (retail weight basis). Courtesy of Western Livestock Round-Up.

Beef Industry Profitability

Table 1.12 represents an attempt to account for all costs of producing beef in the United States during 1981. The data used are from government and industry sources. Total cost of producing a typical 1100 lb Choice steer from conception to consumer was about \$1070 in 1981. Gross returns from the sale of the edible portion as well as by-products totalled about \$1025, resulting in a net loss to the industry of approximately \$45. In 1981, this loss was largely absorbed by the cattle feeding segment of the industry.

Several beef industry leaders have commented that during the past 15 to 20 years there has not been one year when all segments have made a profit. If this is true,

Table 1.12. Estimated Cost of Producing Beef in the United States, 1981

Item	$
Producing 450 lb feeder calf	385
Finishing to 1100 lb slaughter wt	430
Killing, processing and shipping 430 lb boxed beef	75
Wholesale and retail costs	180
Total costs	1,070

Source: Ritchie 1982.

then it means profit in one segment of the beef industry comes as a loss in another segment.

Contribution of Beef to U.S. Economy

Figure 1.19 shows cash receipts from meat animals and other agricultural products received by farmers and ranchers. Receipts from cattle and calves was 34.8 billion dollars of the 44.3 billion dollars received from the sale of meat animals. States having over 1 billion dollars in receipts from sale of cattle and calves were Texas (4.9), Nebraska (3.1), Kansas (3.0), Iowa (2.8), Colorado (2.2), California (1.9), Oklahoma (1.8), Missouri (1.4), and South Dakota (1.2).

The January 1983 inventory of 115 million head of cattle and calves was valued at approximately 50 billion dollars. These cattle will generate from between 30 and 45 billion dollars in cash receipts each year depending on price of cattle and numbers

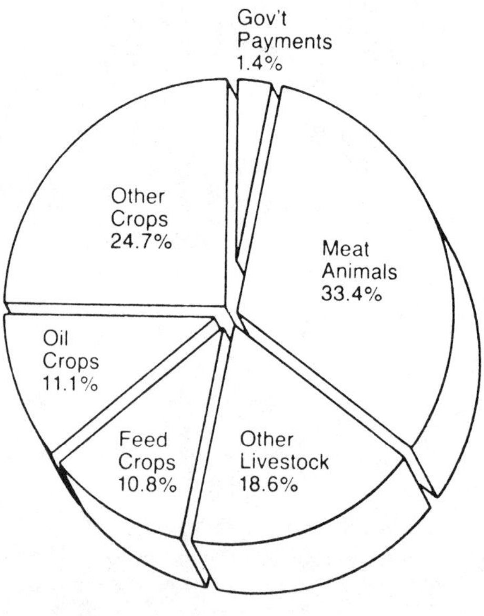

Figure 1.19. Cash receipts (total 132.6 billion dollars) from farming and ranching, 1979. Courtesy of USDA.

Table 1.13. U.S. Industries and Their Resulting Sales Totally or Partially Supported by Beef Cattle

Industry or Association Organizations	Total Sales (Mil $)	Estimated Percent of Sales Generated by Cattle
Animal health	1,900	
biologicals	200	
pharmaceuticals	600	
feed additives	1,100	
Finance (credit)		
Livestock publications	40	88
Transportation		
Market outlets		
Artificial insemination		
Embryo transfer	20	
Feed manufacturing	20,000	20
Veterinary equipment		

sold. Obviously the existence of cattle and the resulting cattle support many other industries and multiply billions of additional dollars in the U.S. economy. Some industries that function because of the existence of cattle are shown in Table 1.13.

SELECTED REFERENCES

Publications

Baldwin, R. L., 1980. *Animals, Feed, Food and People. An analysis of the Role of Animals in Food Production.* Boulder: Westview Press.

Council for Agricultural Science and Technology (CAST), 1975. *Ruminants as Food Producers.* Spec. Public. No. 4.

Fitzhugh, H. A., Hodgson, H. J., Nguyen, T. H., Scoville, O. J., and Byerly, T. C., 1978. *The Role of Ruminants in Support of Man.* Morrilton, Ark.: Winrock International Livestock Research and Training Center.

Food From Animals: Quantity, Quality and Safety, 1980. Council for Agric. Sci. and Tech. (CAST). Report No. 82.

Foreign Agricultural Circulars, 1980, 1981. (U.S. Livestock Product Exports–Imports, World Livestock Numbers, World Meat Consumption, World Meat Production) USDA, Foreign Agricultural Service, Washington, D.C.

Foreign Agricultural Circulars, 1980, 1981. (U.S. Livestock Product Exports–Imports, World Livestock Numbers, World Meat Consumption, World Meat Production) USDA, Foreign Agricultural Service, Washington, D.C.

Koch, R. M. and Alego, J. W., 1983. *The Beef Industry: Changes and Challenges* J. Anim. Sci. 57:28.

Livestock and Meat Situation, 1981. USDA, Washington, D.C. 20250.

McDowell, R. E., 1978. *Contributions of Animals to Human Welfare.* New York's Food and Life Sciences 11:17.

Plant and Animal Products in the U.S. Food System, 1978. (Symposium Proceedings.) National Academy of Sciences. Washington, D.C.

Reid, J. T., White, O. D., Anrique, R., and Fortin, A., 1980. *Nutritional energetics of livestock: Some present boundaries of knowledge and future research needs.* J. of Anim. Sci. 51:1393.

Ritchie, H. D., 1982. *Future and direction of the U.S. beef industry.* Proceed. Ann. Conf. on AI and Embryo Transfer in Beef Cattle. Denver, Co.

Simpson, J. R. and Farris, D. E., 1982. *The World's Beef Business.* Ames: Iowa State Univ. Press.

Smith, A. J. and Gunn, R. G., 1980. *Intensive animal production in developing countries.* The British Soc. of Anim. Prod. Occas. Public No. 4. Thames Ditton, England: Milk Marketing Board.

U.S. meat exports and imports, 1981. NCA Beef Business Bulletin. May 1.

Ward, G. M., Sutherland, T. M., and Sutherland, J. M., 1980. *Animals as an energy source in third world agriculture.* Science 208:570.

The Water Buffalo: New Prospects for an Underutilized Animal, 1981. Washington, D.C.: National Academy Press.

Western Livestock Round-Up, 1982. (Published monthly by the Cooperative Extension Services of the 17 Western Land Grant Universities.)

Visuals

"What Makes a Cattleman?" 16 mm film (approx. 10 min). 1982. Ralston Purina, Checkerboard Square, St. Louis, Mo. 63164.

"Sunchanger." 16 mm film (sound and color). 10 minutes. Produced by Elanco Products Co., P.O. Box 1705, Indianapolis, IN 46206.

2

The Consumer and the Beef Industry

Individuals involved with production, processing, and retailing of beef realize that the demand for beef is established by the consumer. The consumer may be a customer in a retail food store purchasing meat for the family, a customer in one of the many fast-food outlets or people seated in an elite restaurant contemplating the choice of steak or prime rib. In these situations and others, the consumer's choice of the kind of beef and how much, gives eventual direction and momentum to the supply of beef to be produced.

CONSUMPTION PATTERNS

Most Americans have considered beef the *King of the Meats.* They have shown a preference for beef not only over other red meats (Table 2.1) but over most other animal protein foods as well. Beef continues to be the number 1 meat in the minds of most consumers primarily because of beef's flavor, eating satisfaction, and prestige factor.

Table 2.1. Per Capita Consumption of Red Meats on a Carcass, Retail and Cooked Weight Basis, 1982

Meat	Carcass Weight (lbs)	Retail Weight (lbs)	Cooked Weight (lbs)
Beef	104.4	77.3	37.1
Veal	2.0	1.6	0.5
Pork	62.7	58.5	20.9
Lamb	1.7	1.5	0.5
Total	170.9	138.9	60.7

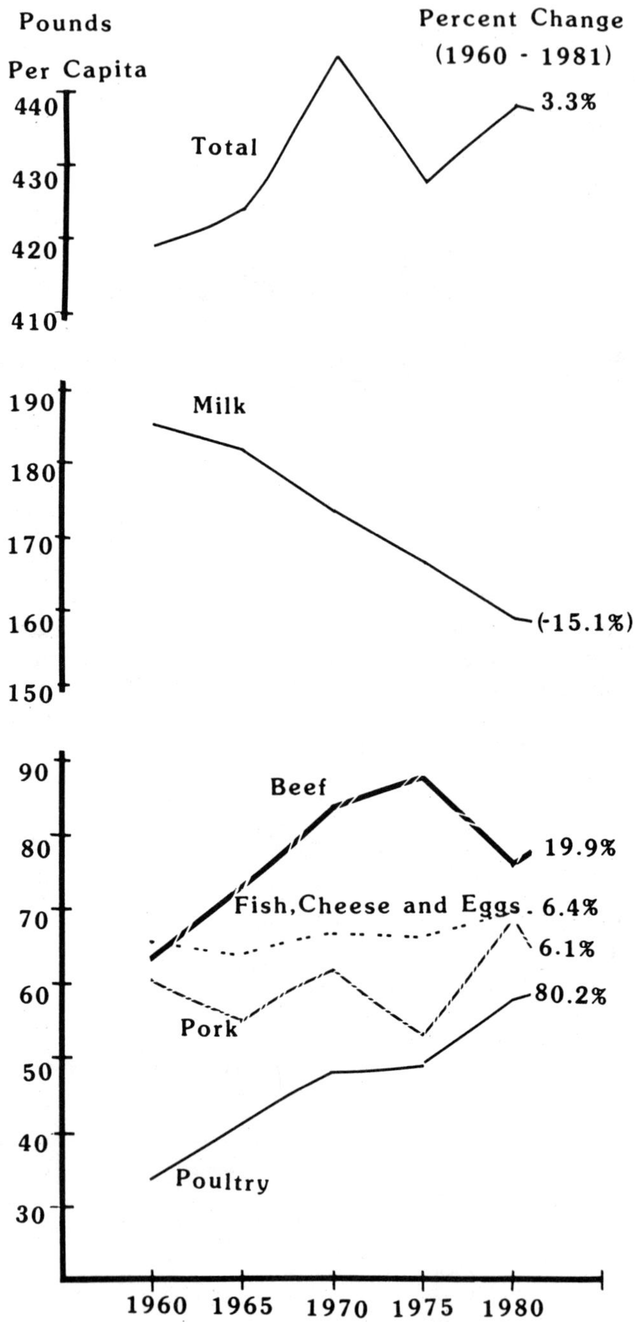

Figure 2.1. Per capita consumption of animal protein foods (retail weight, 1960–1981). Courtesy of National Cattlemen's Association (NCA).

Consumption patterns for different food items can have significant changes depending on availability, cost, disposable income for food, technological advances in food processing, and new information on nutrition and food safety. Most animal food products are preferred to plant food products if they are readily available and the standard of living allows consumers to purchase them.

Beef consumption has had a dramatic increase over the last several decades. During the past few years, beef consumption has declined (Figure 2.1) primarily because the supply of beef has been greatly reduced due to economic pressures which have been on consumers. Pork and poultry consumption has increased significantly in recent months because of an abundant supply and a cost per pound much less than beef (Figures 2.2 and 2.3).

There has been disagreement as to the primary causes of a lower per capita beef consumption in recent years. Some analysts believe consumers are losing their taste for beef. Others feel it is primarily economic reasons—relatively higher beef prices and reduced purchasing power (Figure 2.4).

Consumers having the highest beef consumption are blue-collar workers in average- to high-income levels for their type of work and teenage boys. The low beef consumers are children under 12. Some primary factors affecting beef consumption patterns in the United States besides income level are region, family size, and age of household. Obviously several of these factors are interrelated. Per capita beef consumption is highest in the northeast with the south being the lowest. Pork consumption is the reverse of this relationship. Per capita beef expenditures decline as family size increases. Blacks spend considerably fewer dollars for beef than whites. Less beef is consumed where

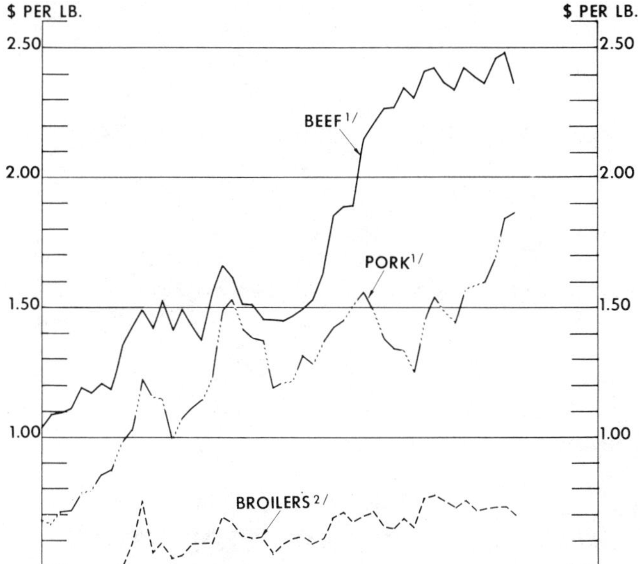

Figure 2.2. Retail beef, pork, and poultry prices, 1971–1982. Courtesy of Western Livestock Round-Up.

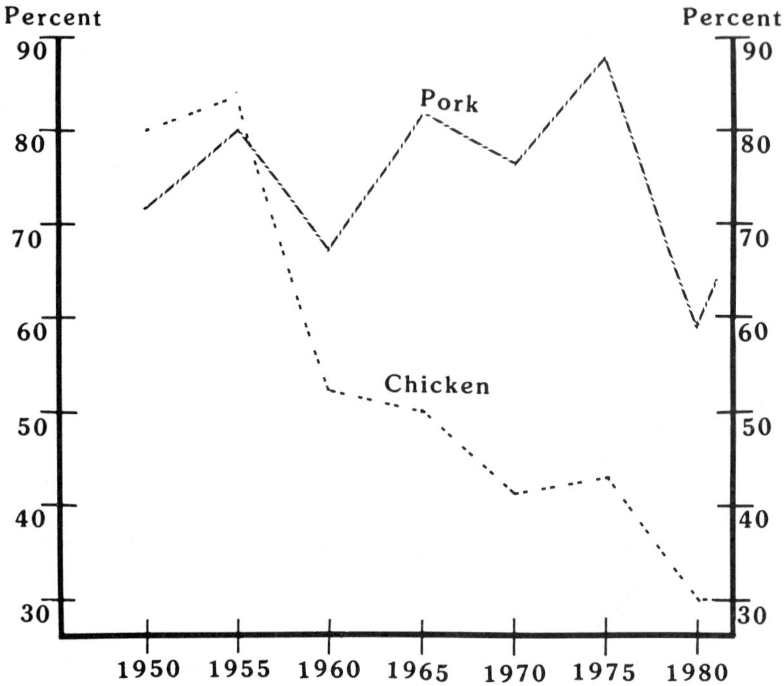

Figure 2.3. Pork and chicken prices as a percent of beef prices. Courtesy of National Cattlemen's Association (NCA).

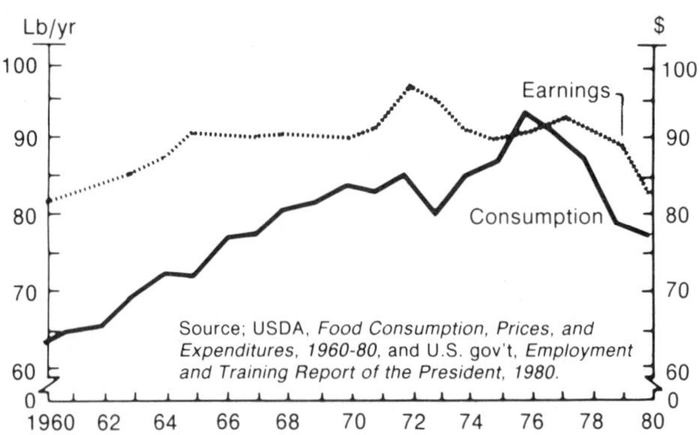

Figure 2.4. Per capita beef consumption and average real spendable weekly earnings. Courtesy of USDA.

households are headed by persons 65 years of age or older, which no doubt reflects income level.

CONSUMER EXPENDITURES FOR BEEF AND OTHER FOOD

The amount of beef consumed in the United States appears to be most highly influenced by the amount of money consumers have available to spend for goods and services (Table 2.2).

Table 2.2 Beef and Food Expenditures as Related to Per Capita Disposable Income

Year	Disposable Income Per Person		Percent of Income Spent for Food	Percent of Income Spent for Beef	Per Capita Beef Consumption Carcass wt (lbs)
	Current Dollars	1972 Dollars			
1955	1654	2577	25.3	2.65	81
1960	1937	2697	20.2	2.72	85
1965	2436	3152	18.2	2.48	100
1970	3376	—	17.3	2.55	113
1975	5075	4055	17.0	2.71	118
1976	5477	4161	16.8	2.54	129
1977	5954	4266	16.7	2.29	126
1978	6571	4409	16.5	2.41	120
1979	7293	4493	16.4	2.42	108
1980	8002	4473	16.6	2.27	106
1981	8827	4526	16.7	2.09	104
1982	9377	4567	16.1	2.00	104

Source: USDA.

Today more food is being purchased away from home, primarily in some 33,000 fast-food establishments.

Higher income families purchase more of their food away from home and have a higher per capita consumption of beef. This is further evidence that beef consumption is closely associated with standard of living. Future beef consumption patterns will be greatly influenced by the economic well-being of the country. Food purchased away from home accounts for more than $87 billion, which is approximately 35% of the American food dollar. In most cases cost of food, particularly meat items, is higher than the same food prepared at home. About one-third of the total hamburger is consumed away from home, primarily in the fast-food outlets.

The higher the disposable income, the greater the per capita consumption of beef. This is true as long as there is an ample supply of beef to be purchased. This is best demonstrated by the income elasticity of beef. Income elasticity for all beef has been estimated at 0.47. This means that for each one percent increase (decrease) in income, the amount of beef demanded increases (decreases) by 0.47 percent. At the present level of beef consumption (104 lbs), an increase (decrease) by one percent in per capita disposable income would increase (decrease) consumption by approximately one-half pound.

Food prices have risen significantly over the past few years as a result of inflation. Figure 2.5 shows that percent of per capita disposable income for food is lower than most previous years. Figure 2.6 shows that costs of other goods and services have been greater than that for food.

There continues to be a general feeling among U.S. consumers, and even Congress, that food in this country should be provided at minimum cost. This "free food" policy stifles the profit incentive of producers and lowers priority of private and public research dollars needed to meet future food demands.

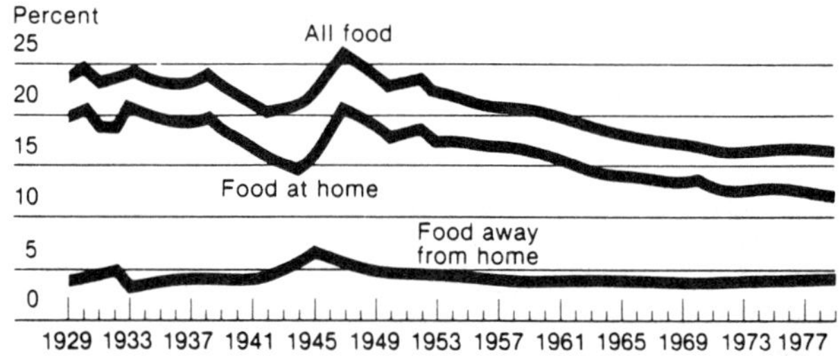

Figure 2.5. Expenditures for food as a percent of disposable personal income, 1929–1979. Courtesy of U.S. Dept. of Commerce.

Food price increases in the United States have been modest when compared to increases in some other countries (Figure 2.7). Table 2.3 shows that supply and demand greatly influence beef prices in different parts of the world.

Consumers having more economic pressures will more critically assess the cost of meat per serving. Table 2.4 shows that hamburger is very competitive with chicken and that there is a need for more intermediate beef products between hamburger, steaks, and roasts. Most steaks and roasts are less competitive on a price basis, and they obviously are purchased at higher costs per serving because of beef's taste appeal.

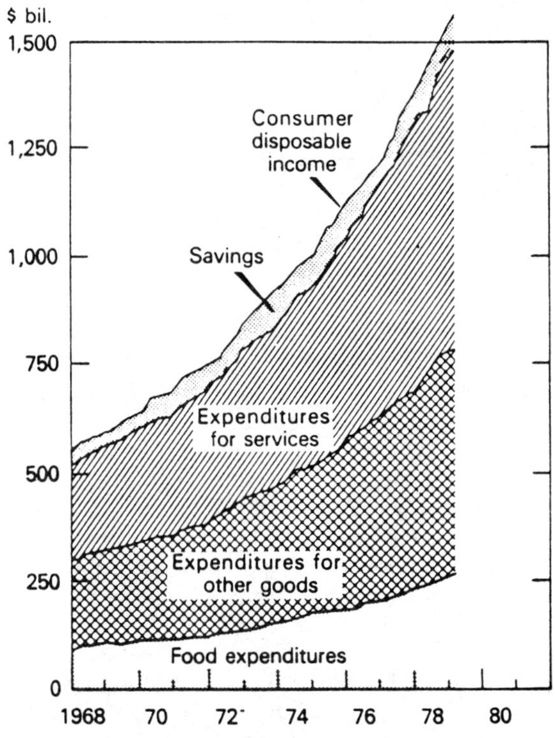

Figure 2.6. Income and expenditures. Courtesy of the USDA Handbook of Agricultural Charts, 1979.

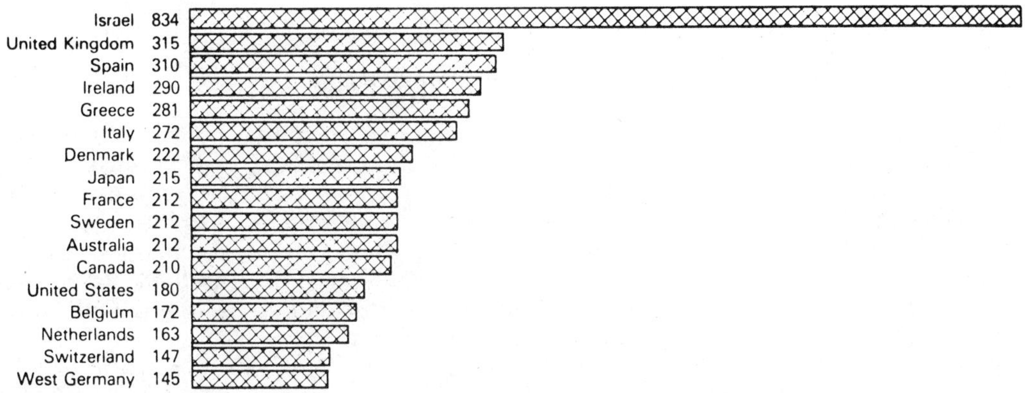

Figure 2.7. Percentage increases in food prices in several countries, 1970–1978. Courtesy of the USDA Handbook of Agricultural Charts, 1979.

Table 2.3. Retail Beef Prices and Hours Worked to Buy One Pound of Steak in Selected World Capitals (July 1980)

Country	Boneless Sirloin (U.S. $s)	Hours Worked (hrs:min)
Bonn	$ 6.50	0:58
London	6.44	1:31
Mexico City	1.85	1:47
Tokoyo	15.26	2:47
Wash. D.C.	3.69	0:23

Table 2.4. Costs of 3 Ounces of Lean Meat (Denver, Colorado, Mid-February 1979)

	Retail Price Per Pound	Percent of Cooked Lean Meat Per Retail Pound	Cost of 3 Ounces Cooked
Beef liver	$0.69	69.4	$0.19
Hamburger	1.29	72.1	0.34
Chicken, whole, cut-up	0.73	39.1	0.35
Turkey	0.89	46.9	0.36
Chicken breasts	1.15	53.6	0.40
Pork, picnic	0.99	40.8	0.46
Ham, whole	1.35	53.6	0.47
Ocean perch, fillet	1.95	64.6	0.64
Ham, canned	2.34	75.0	0.59
Haddock, fillet	2.19	64.6	0.64
Round steak	2.09	55.1	0.71
Chuck beef roast, (bone in)	1.65	41.7	0.74
Rib roast	1.99	41.7	0.90
Sirloin steak	2.15	43.6	0.92
Pork loin roast	1.89	37.5	0.95
Pork chops, center cut	2.15	41.7	0.97

Source: USDA, NCA.

CONSUMER PREFERENCES FOR BEEF

Consumers obviously eat food for the life-sustaining nutrients, but they primarily have preference for foods which provide eating satisfaction and convenience. Beans, peanut butter, eggs, milk, and chicken are typically cheaper sources of protein than beef on a per unit cost basis. Most consumers, however, give preference to the palatability characteristics of beef and will pay the higher price unless extreme economic pressures dictate otherwise. Recent surveys have shown that 91% of U.S. households serve beef compared to 68%, 80%, and 4% for poultry, pork, and lamb respectively. Only milk and eggs rank above beef in total household acceptability.

Consumers prefer beef that is tender, flavorful, and juicy, with a high ratio of lean to fat, and a high ratio of lean to bone. They also have preference for a consistent product, and obviously, price is a major consideration with many consumers.

Consumers prefer to buy fresh beef as attempts to market frozen beef has met with little success. Consumers indicate that the inability to judge the freshness of frozen beef is their primary reason for not purchasing it in the frozen condition.

Several national surveys have shown that most consumers are not familiar with the USDA grades. Most consumers confuse the grade with the inspection marks. They believe all beef is graded and frequently report certain grades when no such grade designation exists (e.g., "Grade A," "First Cut," and "Grade AA").

Color

Color of meat in retail cases ranks as a very important selection criterion for most consumers. Their preference is a bright, cherry red color in contrast to a dark, less bright, red color. In past years this was a useful criterion to evaluate beef since color identified meat from younger and older cattle. Meat from older cattle is a darker color and usually less tender than meat from a younger animal. Currently however, most meat from older cattle goes into processed meats, where by mechanical means the tenderness problem is largely eliminated. Most darker-colored beef in retail cases results from meat staying in the retail case for a period of time. After meat is cooked, the darker color disappears and eating qualities are similar to brighter-colored meat.

Marbling

Fat within the muscle is used by many consumers as an important criterion in selecting beef. Consumers in certain areas of the United States will prefer more marbling than consumers in other areas. Most consumers feel that increased marbling in meat indicates increased tenderness and other palatability characteristics. A summary of research work (Texas Agric. Expt. Sta. Rpt. 22, 1970) shows that marbling accounts for only 10% to 15% of the differences in palatability. Current research work continues to support this earlier summary. Therefore, the relationship between marbling and tenderness and other palatability characteristics is low.

There is increasing evidence that feeding of concentrates to cattle greatly contributes to consumer palatability characteristics. It appears that marbling merely reflects the fact that cattle have been fed concentrates. The problem arises in the fact that all cattle fed concentrates will not marble, and yet their meat is usually excellent in palatability. Some totally forage-fed cattle may have adequate levels of marbling but be relatively poor in palatability. Some forage-fed cattle are quite satisfactory from a palatability standpoint, although in general they are less tender and juicy. They will not give a uniform product unless consumers change their preference from fed beef.

Beef which is too lean will not be as juicy as meat with at least a minimum fat level. It has been shown that hamburger containing less than 18 percent fat is not as palatable as hamburger containing 18% to 24% fat.

Tenderness

Consumers rate tenderness as the most important eating characteristic of beef. Tenderness is primarily a function of type of cut and animal age, with 14- to 26-month-old fed slaughter steers being the most tender.

There is an increasing amount of evidence that cooking and serving methods significantly affect tenderness of meat. Studies show that when internal cooking temperatures of beef exceed 63 °C (145 °F), meat is less tender (*J. Anim. Sci.* 43:114, *J. Anim. Sci.* 46:965). Tenderness and flavor are most desirable if meat is not allowed to cool quickly to room temperature (22 °C or 72 °F).

Other factors influencing tenderness are discussed in Chapter 15.

Lean to Fat

The ratio of lean to fat and lean to bone can be most effectively controlled by proper breeding and feeding programs with limited fat trimming at the packing plant or retail level. Concentrate feeding, without overfeeding, will keep lean to fat ratio at the desired level resulting in beef having desired consumer palatability characteristics if beef has been processed correctly.

Consumer surveys give evidence that consumers use fat as a selection criterion. They avoid excess fat because of sensory characteristics (e.g., too greasy, undesirable taste), concern about relationship of fat to health problems and weight control as it is affected by caloric intake.

Leanness and Price

Since 1979 several leading retailers have featured their own name brands of beef (e.g., Safeway Quality, King Soopers Generic Beef) with less emphasis on USDA Choice. These "house" grades or name brands are high yield grades (primarily 2's) but are not quality graded. This "no-roll" beef would be essentially grade Good under USDA standards. These retailers have been selling their own name brand beef for 2 to 20 cents less per pound than choice grade, depending on the particular cut involved. The featured lean product selling for less per pound has been widely accepted by consumers in certain areas. Safeway has been selling lean beef in 75% to 80% of their stores and a combination of lean beef and Choice grade in the remaining 20% to 25% of their stores. These retailers have evidence from consumer acceptability that beef "fed like Choice," will "eat like Choice" even if the marbling is not the same.

COMPETITION FROM OTHER MEATS AND SYNTHETICS

Other Red and White Meats

The primary substitutes for beef are pork and poultry, and to a smaller extent, lamb. As the prices of these substitutes go up relative to the price of beef, consumers will purchase more beef. Cross-price elasticity is the best measure of how prices of substitutes affect the amount of beef demanded. Cross-price elasticity is the ratio of percentage change in quantity of beef demanded divided by associated change in price of the beef substitute (Table 2.5).

An example of using cross-price elasticities would be if the price of pork should increase by one percent, the amount of beef demanded would increase by about 0.10%.

During 1980 and 1981, considerable discussion occurred on the question, "Is there a decreased consumer demand for beef?" Increased consumption of pork and poultry, a decreased percentage of income spent for beef, and beef-health issues have been cited as evidence for a decreased demand for beef. Others have stated that decreased per capita consumption of beef is merely a reflection of decreased supply which is an integral part of traditional cattle cycles. Caution should be exercised in using percent of income spent for beef as the sole indicator of demand for beef. If real incomes of individuals increase, percent of disposable income spent for beef could decline even when demand for beef was high.

One way to evaluate demand for beef and other competitive meats is to look at total consumer expenditures for these products. Figure 2.8 shows the total consumer expenditures for beef, pork, and broilers in "real" terms with all prices adjusted for inflation by using the Consumer Price Index.

Real consumer expenditures for all three meats, including beef, peaked in 1973. That year the average consumer spent a total of $152.07 in terms of 1967 dollars for all three meats. This was broken down as follows: beef, $86.61; pork, $47.23; broilers, $18.23. In terms of constant dollars, expenditures for the three meats have generally declined since 1973, reaching a low of $129.38 in 1980. During 1980, in terms of 1967 dollars, the average consumer spent $75.21 for beef, $39.27 for pork, and $14.90 for poultry.

It is very difficult to quantify consumer demand. Beef remains the preferred protein source of most people. It will have to price itself out of the market before consumers will turn it down. The demand for beef, however, appears to be at the crossroads. The direction, no doubt, will be influenced by economic conditions as reflected by disposable income available for food purchases. The evidence and publicity in beef-health issues will also give direction to change in consumer demand for beef.

Figure 2.9 shows that total per capita beef, pork, poultry, and fish consumption has changed very little over the past 15 years. The part that each of these four protein foods comprise of the total does change, therefore beef has to be competitive for consumers' protein dollar. Estimated per capita consumption for 1982, on a retail weight (lbs)

Table 2.5. Cross-Price Elasticities

Pork	0.100
Chicken	0.075
Lamb	0.040

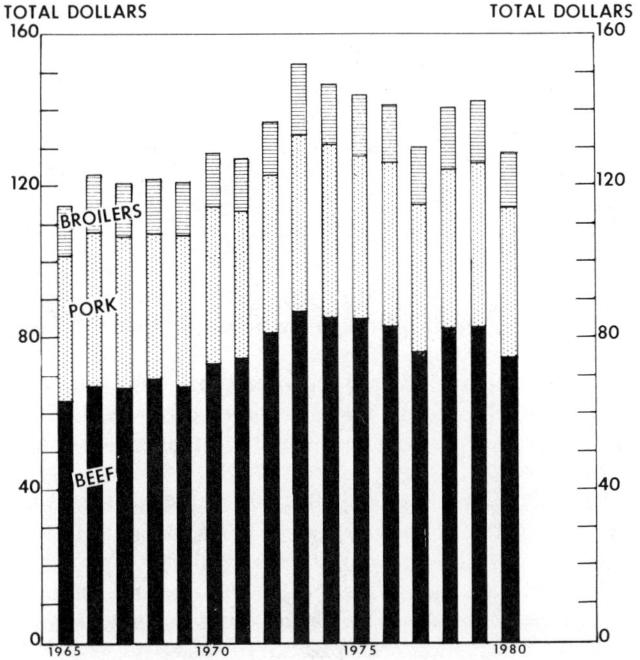

Figure 2.8. Real per capita expenditures for beef, pork, and broilers, 1965–1980 (adjusted for inflation). Courtesy of The Western Livestock Round-Up.

basis is: beef, 76.2; pork, 55.1; chicken, 55.3; turkey, 10.4; total red meat and poultry, 196.1. This is less than 200 lbs for the first time since 1975. Poultry consumption in 1982 is expected to exceed pork for the first time in history.

Synthetic Meats

Synthetics or analogs are products produced by artificial means rather than occurring in natural form. Soybean proteins have penetrated the meat market as analogs which

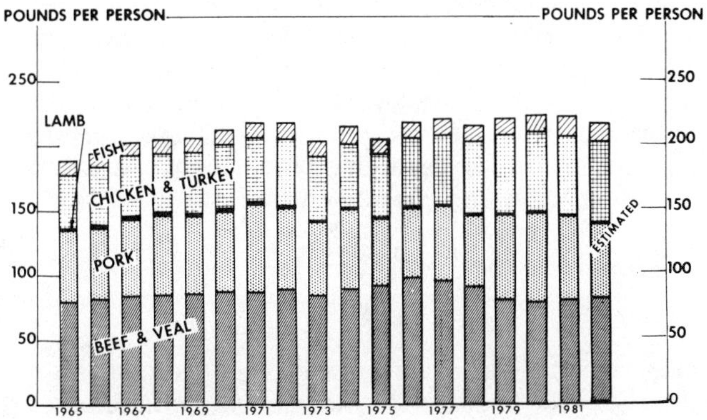

Figure 2.9. Consumption of meat, poultry, and fish, 1965–1982. Retail weight equivalent. Courtesy of Western Livestock Round-Up.

Table 2.6. Protein Costs from Beef and Soy Products

Protein Source	Dollar Cost Per Pound of Net Usable Protein
Beef	$3.26
Soy flour and grits (40–45% protein)	0.05–0.11
Soy concentrates (60–70% protein)	0.18–0.25
Soy isolates (90–97% protein)	0.35–0.40

resemble specific meats in color, taste, and texture and as substitutes for meat in processed items (patties, chili, casserole-type dishes, and others).

Certain individuals feel that meat-type extenders have the greatest market potential along with some analogs of processed meats (bologna, weiners, and so forth). There have been projections that soy analogs could replace as much as eight percent of the red meat producing animals in the 1980s.

The primary marketplace could well be institutions (such as schools, hospitals, and so forth) who look carefully at costs of the nutrient supply. Some costs, based on 1972 data, are included in Table 2.6.

Soybeans were introduced into the American diet on a large commercial scale in 1973 as "meat extenders" when meat prices climbed. The premixed soybean product was combined with fresh hamburger. This beef blend was prevalent until meat prices dropped. A 1974 USDA study found that when the price differential for the extended product versus ordinary hamburger dropped below 20 cents a pound, most people stopped buying it.

At the present time, the analogs do not appear to have any major effect on beef consumption. The USDA estimates that approximately 50 million bushels, or 2% of the yearly U.S. production of soybeans, goes into protein products for human consumption. Only about 1% of all U.S. households consume meat substitutes. Large companies such as Miles Laboratory, General Mills, Pillsbury, and General Foods have been investing millions of dollars into development of meat analogs. Although meat product extension and replacement was predicted to be the primary market for textured plant protein, the growth area has been in pet foods, bakery ingredients, and dairy products. Success of meat analogs will likely be dependent on cost of production, price of various competitive meats (particularly ground beef), and the economic status of the consuming public.

Another consideration may be a deterrent to expanded use of soybeans in human diets. In 1981, a renowned team of hematologists at the University of Kansas Medical Center announced results of a research project where soy protein products from processed soybeans dramatically inhibited absorption of iron in humans. No doubt there will be several other research projects in the future verifying these results.

OTHER USABLE PRODUCTS FROM CATTLE

Figure 2.10 shows the major usable products that are obtained from beef cattle.

The hide is the best known by-product and usually the highest valued. The hide

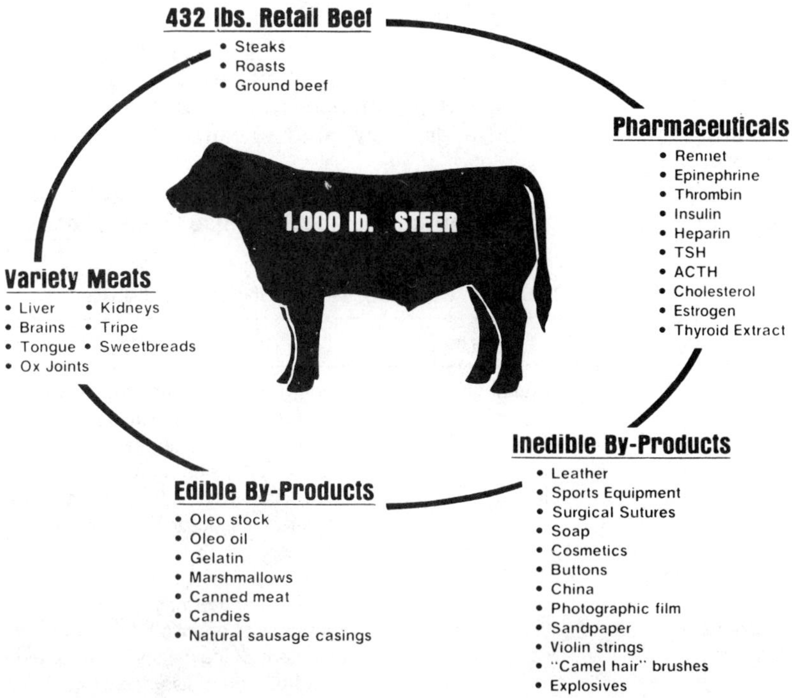

Figure 2.10. The major products from beef cattle. Courtesy of Live Stock and Meat Board.

gives three types of leather (latigo, suede, tooling), felt, certain textiles, a base for many ointments and insulation, and is a binder for plaster and asphalt. The hair from the hide is used in production of insulation and rug pads. Fine hair from the ear is used to make artist's brushes, so-called "camel hair" brushes. Gelatin from hides is used in foods, film, and glues.

During the 1981 convention of the American Surgical Association, there was a report on the successful use of artificial skin used on severely burned patients. The artificial skin was made from cowhide, shark cartilage, and plastic.

The primary edible by-products are liver, heart, kidneys, brains, tripe (walls of the stomach), sweetbreads (pancreas), and tongue. These products have long been known for their high nutritive value and are considered gourmet items by some consumers. Other edible by-products come from fats (oleo stock and oil for margarine and baker's shortening, while oleo sterine is used for making chewing gum and certain candies); bones, horns, and skins (gelatin for making marshmallows, ice cream, canned meats, and gelatin dessert); intestines (natural sausage casings and surgical sutures).

Other inedible by-products, besides the hide and hair, come from inedible fats (industrial oils, lubricants, tallow for tanning, soap and glycerine which is used in explosives, lipsticks, face and hand creams, and some medicines); fatty acids (production of chemicals, biodegradable detergents, pesticides, and flotation agents); bones, horns, and hooves (buttons, bone china, piano keys, glues, adhesives, animal feeds, fertilizers, neatsfoot oil, nonedible gelatin for photographic film, paper, wall paper, emery cloth, sandpaper, combs, tooth brushes, violin strings, and bone charcoal which is used in production of high-grade steel ball bearings); arteries to replace human arteries that are defective.

Table 2.7 shows the source of the primary pharmaceuticals from beef cattle and their value to humans.

There are some unusual uses of cattle by-products. The Chinese and Japanese take gallstones from old cows, grind them up in a powder and use them in a similar manner

Table 2.7. The Pharmaceuticals from Beef Cattle—Their Source and Utilization

Pharmaceutical	Source	Utilization
Epinephrine	adrenal gland	relief of hay fever, asthma, and other allergies; heart stimulation.
Thrombin	blood	assists in blood coagulation; treatment of wounds; skin grafting.
Fibrinolysin	blood	dead tissue removal; wound cleansing agent; healing of skin from ulcers or burns.
Desoxycholic acid	bile	used in synthesis of cortisone for asthma and arthritis.
Liver extract	liver	treatment of anemia.
Ox bile extract	liver	treatment of indigestion, constipation, and bile tract disorders.
Heparin	lungs	anti-coagulant; prevention of gangrene.
Insulin	pancreas	treatment of diabetes for 10 million Americans.
Chymotrypsin	pancreas	remove dead tissue; treatment of localized inflammation and swelling.
Glucagon	pancreas	counteracts insulin shock; treatment of some psychiatric disorders.
Trypsin	pancreas	cleansing of wounds.
Rennet	stomach	assists infants in digesting milk; cheese making.
Ovarian hormone	ovaries	to treat painful menstruation and prevent abortion.
Parathyroid hormone	parathyroid gland	treatment of human parathyroid deficiency.
Corticotrophin (ACTH)	pituitary gland	diagnostic assessment of adrenal gland function; treatment of psoriasis, allergies, mononucleosis, and leukemia.
Hyaluranidase	testes	enzyme that aids drug penetration into cells.
Thyrotropin (TSH)	pituitary gland	stimulates functions of thyroid gland.
Vasopressin	pituitary gland	control of renal function.
Cholesterol	nervous system	male sex hormone synthesis.
Thyroid extract	thyroid gland	treatment of cretinisin.
Amfetin (trade name)	amniotic fluid	reduces postoperative pain and nausea and enhances intestinal peristalsis.

as our grandparents used castor oil—as a cure-all for everything. The market price for them is $1400 per pound. And cattle hair balls are utilized by the Chinese as bandaids. They feel there is healing power in a hair ball.

This summary of beef by-products is not complete. Research has and no doubt will continue to identify useful by-products from beef cattle.

DIET–HEALTH CONSIDERATIONS

Recent consumer surveys indicate that Americans are sensitive to health and nutrition relationships with the food they purchase. Nearly two-thirds of the people surveyed said they had changed their diets during the past three years (1977–1980), primarily for health or nutrition reasons.

The relationship of human health to diet is a very controversial subject at the present time. It is important to understand the true nature of this controversy and clarify as many known relationships as possible.

Nutrients in Beef

Beef is an excellent source of protein, iron, and B vitamins. Table 2.8 shows the major nutrients supplied from three ounces of cooked beef.

The annual meat supply provides a substantial percentage of the total food nutrient supply. Table 2.9 shows how meat has contributed to the nutrient supply over the past decade. The beef industry supplies Americans with 12% of their total food energy and 23% of their protein intake. No other single agricultural product exceeds this level of importance in the United States food supply.

BEEF AND HUMAN DISEASE

Heart disease and cancer are currently two of the most dreaded human diseases. Over the past several years, millions of dollars have been utilized in research attempting to identify factors causing these diseases. The causitive factors presently appear to be many and complex. This has allowed only limited progress in identifying valid cures for the diseases. Medical scientists are not in agreement as to major factors in causing heart disease and cancer.

Sound, well-based research work appears to be often lost in an emotionally stimulated population. Many politicians, governmental regulating agencies, and other organized groups have occasionally based judgements and decisions on emotion rather than the best accumulated facts that research provides. Some individuals feel that emotionalism is stirred by a communication system that builds part of its followership on the sensational.

Heart Disease

The major known risk factors associated with coronary heart disease are high blood cholesterol concentrations, smoking, hypertension, physical inactivity, and obesity. It

Table 2.8. Percentages of Daily Recommended Nutrient Allowances.* (Based on Three Ounces of Cooked Beef)†

	Age Yrs.	Protein	Calories	Iron	Zinc	Phosphorus	Magnesium	Thiamin	Riboflavin	Niacin	Vit. B6	Vit. B12
Children	1–3	110	17	21	51	20	12	11	42	43	53	177
	4–6	85	13	32	51	20	9	9	30	33	36	117
	7–10	70	10	32	51	20	7	7	28	24	27	88
Males	11–14	57	8	18	34	14	5	6	22	22	20	59
	15–18	47	8	18	34	14	4	5	19	20	16	59
	19–22	47	8	32	34	20	5	5	19	20	16	59
	23–50	45	10	32	34	20	5	6	21	22	16	59
	51+	45	10	32	34	20	5	7	22	24	16	59
Females	11–14	57	10	18	34	14	6	7	26	24	20	59
	15–18	53	11	18	34	14	6	7	24	28	16	59
	19–22	55	11	18	34	20	6	7	24	28	16	59
	23–50	55	11	18	34	20	6	8	28	30	16	59
	51+	55	13	32	34	20	6	8	30	33	16	59

* Figures based on 1974 National Research Council Recommended Dietary Allowances and nutrient composition data from (a) Composition of Foods, Agriculture Handbook No. 8, USDA, 1963; (b) The Nutritive Value of Cooked Meat, Miscellaneous Publication MP-49. Oklahoma State University, 1958; (c) Murphy, E. W., Willis, B. W., and Watt, B. K., Provisional Tables on the Zinc Content of Foods, J. Am. Dietet. Assoc. 66:345–355, 1975.

† Courtesy of the National Live Stock and Meat Board

Table 2.9. Contribution of Meat to Nutrient Supply Available for Civilian Consumption, 1967 and 1980*

Food Group	Year	Food Energy	Protein	Fat	CHO	Ca	P	Fe	Mg	Vit. A	Thiamin	Riboflavin	Niacin	B_6	B_{12}
									Percent						
Meat (incl.	1967	21	40	36	0.1	4	26	31	13	23	29	22	47	39	68
poultry & fish)	1980	21	43	36	0.1	4	29	31	14	24	28	24	45	41	72

* Courtesy of National Food Review.

is generally accepted that consumption of animal fat by humans will cause an increase in the level of cholesterol in the blood. It is also accepted by many people that consumption of vegetable oils (polyunsaturated fats) causes a decrease in blood cholesterol concentration. These relationships lead to the theory that there is a relationship between consumption of animal fats and the incidence of atherosclerosis ("plugging" of the arteries with fatty tissue) which in turn would result in death from coronary heart disease.

Evidence supporting the proposed cholesterol–heart disease relationship is still theory. Studies in which dietary fat has been modified, either in kind or amount, has not reduced mortality. Changing diets from animal fats has not improved the heart disease record. There are data which give evidence to poorer health conditions where extended diets have been high in polyunsaturated fats.

Cholesterol is a natural occurring substance in the human body. Every cell manufactures cholesterol on a daily basis. The average human "turns over" (uses and replenishes) 2000 mg of cholesterol daily. Average dietary consumption of cholesterol is approximately 600 mg daily. Therefore the body makes 1400 mg each day to meet its needs.

Most consumers do not know the difference between saturated and polyunsaturated fats. Through many margarine and vegetable oil commercials, however, they are informed that saturated is "bad" and unsaturated is "good." In a recent review of the diet and heart disease relationship, Dr. Mann relates that the dietary-heart hypothesis became popular because a "combination of urgent needs of health agencies, oil–food companies, and ambitious scientists had transformed that fragile hypothesis into treatment dogma. Diet-heart enthusiasts began using a show of hands or mail polls to influence national food policy" (*The New England J. of Med.* 297:644).

Recent research work shows that when polyunsaturated fats are substituted for saturated fats in the diet, they do not always lower the blood cholesterol level. In most cases, consumption of polyunsaturated fat increases the concentration of cholesterol in tissue. Thus, animal fats may be necessary to keep down the accumulation of total body cholesterol.

Cholesterol cannot be used by the body unless it is joined with a water soluble protein, creating complexes known as lipoproteins. There are several different types of lipoproteins. Recently, research workers have become very interested in two of these lipoproteins, HDL (high density lipoprotein) and LDL (low density lipoprotein). LDLs have been generally associated with cardiovascular problems while some research data show that HDLs may reduce heart attacks by 20%. Additional research with laboratory animals has shown that those fed beef had HDL levels 33% higher than those animals fed soybean diets.

There is evidence of genetic differences in the proportion of HDLs and LDLs in individuals. People who are overweight, non-exercisers, and cigarette smokers have higher proportions of LDLs than individuals who are lean, exercisers, and non-smokers.

Although the exact role of cholesterol in development of coronary heart disease is not known, it would appear logical from existing information to use prudence in implementing drastic changes in dietary habits.

Cancer

Statements have been made that consumers eat too much meat, and the excessive meat consumption causes cancer. Evidence to support this statement is questionable, and a point can be made as to what is excessive. The approximate 65 lbs of cooked red meats consumed per person per year converts to 2.8 ounces per day (1.6 oz is beef). This provides approximately 25 grams of protein which is less than half the average recommended daily allowance for protein. The American Heart Association recommendation has been 3½ ounces of cooked meat per person on a daily basis. If this recommendation is valid, then the per capita consumption of beef and meat is low in the United States.

There have been research reports linking meat consumption to cancer of the colon. Also, it has been reported that frying hamburgers too long at temperatures above 300 °F may produce carcinogenic substances. Cause and effect relationships have not been clearly identified. One physician recently remarked, "The evidence is very lean and shaky to say that meat causes cancer. It is preposterous to advocate dietary changes as preventive measures."

There have been claims that estrogens are carcinogenic substances; therefore, no estrogens should be included in feed additives or growth implants to cattle. The major concern has been diethylstilbestrol (DES), a synthetic estrogen used to increase growth rate and feed efficiency in cattle. In recent times, DES has been banned for use as a growth stimulant in cattle production. This action was controversial because scientific evidence to support this action appeared to be very weak. This action was taken because of the Delaney Amendment which was passed by Congress in 1958. This amendment forbids issuance of a regulation permitting the use of any substance if it "is found to induce cancer when ingested by man or animal, or if it is found, after tests which are appropriate for the evaluation of the safety of food additives, to induce cancer in man or animal."

There is evidence that large doses of estrogens can cause cancer. Vaginal or cervical cancer has been reported in some daughters of women who were given massive doses of diethylstilbestrol (300 to 18,000 micrograms) in an attempt to prevent miscarriages. Large doses of estrogen administered to laboratory animals have produced tumors of the mammary gland, pituitary gland, uterus, and testes.

Estrogens are naturally occurring substances essential for life's processes, especially for reproduction. Small amounts of estrogen are beneficial, while large amounts of estrogen are harmful. Most substances produce harmful effects when given in excessive amounts. This is true of essential nutrients such as water and minerals (salt).

A considerable amount of knowledge is available on the impact of uses of hormonally active substances. There is no concrete evidence of a cancer hazard attributable to use of hormonally active substances in livestock production. Some evidences to support this statement are, first, only in a few instances (less than one percent) has DES been detected in beef liver. DES does not concentrate in other muscles except in rare cases of extreme illegal overdosage. If all liver consumed by the population of the United States was contaminated with DES at a concentration of 2.0 parts per billion (a considerable overestimate), the average daily intake would be 0.0038 micrograms per day. Intake of 0.0038 micrograms per day would compare to the dosage of up to 2500 micrograms of DES–equivalent per day for contraceptive purposes, or 500 to 1000 micrograms of estrogen for treatment of menopausal symptoms or osteoporosis. Another way of comparing these is that a person would have to daily consume 5500 lbs of liver containing trace levels of DES to equal the amount of the hormone used daily in human therapy.

Second, there would be 0.0001 micrograms of DES absorbed from the 0.0038 micrograms ingested. This would compare to the normal body production of estrogen which would be 400 to 1400 times greater in a 50 lb boy, 2500 to 25,000 times greater in a 120 lb woman, and with even a higher level during pregnancy.

Third, estrogens are produced in plants as well as animals. Wheat, soybeans, green leaves, and vegetable oils all contain far more estrogenic activity than has ever been found in liver from estrogen-treated cattle. Wheat germ, for example, contains 2000 times the DES-equivalent of an equal weight of beef liver containing DES at two parts per million.

Antibiotics

There are claims that using antibiotics in cattle is dangerous to human health because microorganisms, exposed to antibiotics, can develop resistant strains. It is assumed that these resistant strains will infect human beings, then antibiotic therapy will be unproductive.

There is no conclusive evidence that these antibiotic resistant strains can be easily transferred from livestock to man. Large research projects are currently underway to determine potential problems in this area and the seriousness of them.

DIETARY GUIDELINES

The United States Government has released several reports giving dietary guidelines and dietary goals for United States citizens. There are mixed reactions to these reports. Some say that it is the first step of government to control diets of people. Others feel the guidelines are good if they are scientifically based, not emotionally implemented, and where people are given freedom of choice.

Beef producers and other industry people have expressed concern because in 1977 the Senate Select Committee on Nutrition released a report entitled "Dietary Goals for the United States." One of the proposals was to decrease consumption of meat and increase consumption of poultry and fish. The idea behind the proposal was to reduce cholesterol, overall fat, and saturated fat consumption. A recent report (*Beef Business Bulletin,* 4 June 1982) indicated that if all Americans were to adopt Dietary Goals, beef and other red meat consumption would be reduced by approximately 48%.

On 4 February 1980, officials from the USDA and HEW (Health, Education, and Welfare) released new Federal Dietary Guidelines. These guidelines did moderate some of the initial ones made in 1977. Recommendations were as follows: eat a variety of foods, maintain ideal body weight, avoid too much fat, saturated fat and cholesterol, eat foods with adequate starch and fiber, avoid too much sugar, avoid too much sodium, and if you drink alcohol, do so in moderation.

These guidelines continue to cause much concern and feeling in the nutrition world. Questions are raised as to how much is "too much." Medical doctors, some with extensive nutrition education and others with little or none, continue to make conflicting statements. Nutritionists point out that beef has comparative levels of cholesterol when compared with equal serving sizes of other red meats, poultry, and fish; for example 3 oz cooked, lean beef has 73 mg cholesterol compared to shrimp (128 mg), turkey (dark meat, 86 mg), turkey (white meat, 65 mg), and veal (84 mg), chicken (74 mg), pork (77 mg), and flounder (52 mg). The recommended dietary cholesterol level per day by the American Heart Association is 300 mg.

Current legislation is pending which may change the existing food safety laws. The Food Safety Amendments of 1981 would revise several laws including The Federal Food, Drug, and Cosmetic Act and the Meat Inspection Act. The primary concern is the basis for assessing risk. There would be a change from "zero risk" to "significant risk" in determining danger to the consumer, a basis that many contend is practical in light of technology that can now detect drug residues in parts per trillion.

The amendments also call for a review of food safety decisions, made by a panel of scientific experts, based on the best scientific information available. Another provision provides for gradual phase out of a substance deemed to have questionable danger.

One of the sponsors of the bills containing the amendments has stated, "We need to build a system which gives our people continued assurance that their food is safe and free from substances which could do real harm. At the same time, we need to deal with some rigid inflexible provisions of existing law which could produce serious and costly disruptions without any offsetting benefits."

Regardless of the source of information consumers have used, there is substantial evidence that Americans have been changing their diets in recent years for health and nutrition reasons. Surveys have shown that 64% of the sampled households have made diet changes since the USDA Dietary Guidelines were published in 1977 (*National Food Review,* Fall 1980). This survey identified the following primary points with the percent showing the portion of households making changes:

1. Weight control is a prominent factor in food choice (43%).
2. Citing fat intake as their concern, some consumers were switching to leaner beef, skim milk, and reducing or discontinuing the use of bacon, sausage, fresh pork, and beef (28%).
3. Less consumption of beef, eggs, and shellfish to reduce intake of cholesterol (23%).
4. Fiber was not a prominent factor in food selection.
5. A reduction in sugar intake was cited by households most often for making a change in food use (52%).
6. Reduction in salt intake (20%).

A recent study has shown beef to be lower in calories, saturated fats, and cholesterol than previously thought. A 3-ounce serving of cooked, lean beef had only 192 calories, or 8% of the average daily calorie intake of adult males. Fried chicken had 10% of the average intake, while pork was equal to beef, roast chicken was 7%, and flounder was the lowest with 3% of the average daily calorie intake.

OTHER CONSUMER REACTIONS

There have been several occasions during the past several decades when the retail price of beef has risen rapidly and to high levels in a short period of time. These price changes have initiated strong consumer reaction to the point where beef boycotts occurred in 1969 and 1973. Consumer organizations came into existence which allowed verbal attacks and protest marches to spread across the United States. Beef producers angrily responded that they were not responsible for high prices and definitely were not gaining profit windfalls.

President Nixon put a ceiling on wholesale and retail beef prices on 29 March 1973. Consumers were doubtful that this would solve the problem, and beef producers were angry with government intervention. Some observers have felt that President Nixon's meat-freeze action was totally political in an attempt to get the heat of Watergate off his back.

There have been recent allegations that increased beef prices are one of the major causes of inflation. Figure 2.11 shows changes in beef prices relative to cost of other consumer items.

There have been marked increases and decreases in beef prices over time. These price changes appear to be highly related to the beef supply. In 1973 there was a temporary shortage of beef, primarily due to fewer cattle being fed because of high grain prices and other factors. Beef tonnage increased over the next few years because beef producers were rapidly increasing cow numbers.

Beef price changes and resulting consumer reactions appear to be tied to cattle cycles and some short-run influences affecting beef supply. Unless the beef industry can solve some erratic supply problems, producers may expect consumer reaction or government intervention in future years. In May 1979, *The National Enquirer* ran an article under the title, "Cattlemen Roping in Billions in . . . The Big Beef Rip-off," stating that cattlemen had systematically depleted their herds to force prices up.

The environmentalist movement has significantly affected the beef industry through consumer groups and newly formed governmental organizations, such as the EPA (Environmental Protection Agency) and others. Pollution control, especially with feedlots,

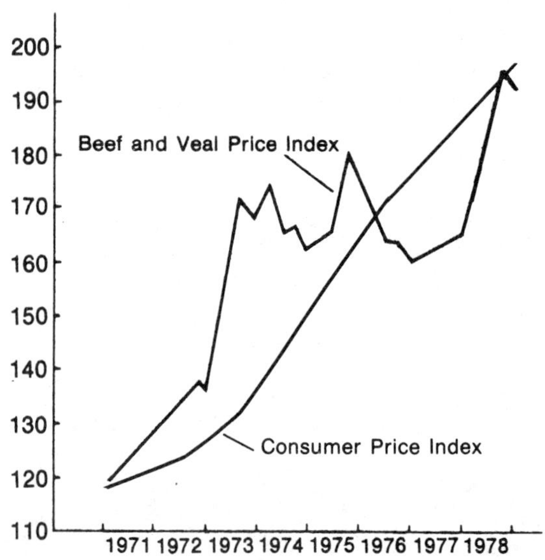

Figure 2.11. Indexes of retail beef prices and all consumer items (1967 = 100). Courtesy of USDA.

has been implemented. Concern is expressed by beef producers as attempts are being made to move from a multiple land use (grazing, recreation, hunting) concept to a single use concept.

There are other consumer reaction pressures being felt by the beef industry. Predictions have been made that the vegetarian movement will affect greater numbers in the future. In a press release from the president of Vegetarian Information Service at their 1982 conference, he was credited as commenting that "the annual slaughter of nearly five billion warm-blooded animals for food, the associated disruptions of the earths ecosystems and productive capacity, and the resulting deaths of 1.5 million people from chronic diseases as the most monstrous machine of death and destruction that the world has ever known." He also called on the conference participants "to become 100 times more effective in putting an early end to this evil." There is no known evidence to demonstrate that the vegetarian movement has had a significant effect on beef consumption in the United States.

Vegetarian and animal welfare groups have been planning books for preschool, kindergarten, and elementary grades designed to promote equality of humans and food-producing animals. The primary educational thrust is that there is no alternative to vegetarianism.

Animal Welfare

In recent months, those promoting animal's rights have been generating considerable publicity. These individuals see animal confinement and various other management practices as an infringement on the rights of animals. An increasing number of people appear to be dedicated to this crusade, and they plan to seek legislation to remedy what they term serious injustices imposed on animals by man.

Representatives of organizations espousing vegetarianism and animal rights met in July 1981, to apparently mobilize their diverse interests into various action programs. Their theme of the coalition conference was that the animal rights movement has come of age and is now ready to proceed with action. Task force committees have been formed which will focus on production and standards of treatment for farm animals, veal calf production methods, an evaluation of the Humane Slaughter Act, and several others.

ADVERTISING AND PROMOTION

Many consumer products are highly advertised and promoted to be competitive with other consumable products. Beef has enjoyed an enviable position of receiving increased consumer acceptance without a large advertising and promotion budget.

There is, however, a different consumer environment emerging. Beef producers have been challenged by consumer boycotts, rising production costs, the environmentalist movement, a consumer-oriented government, competition from meat analogs, limited dollars going into needed research projects, potential health problems, and no doubt several others. Within the past decade, the phrase, "It's time to reach consumers," has been spoken more frequently by beef producers.

Some states, aided by a check-off program, have collected funds for advertising and promotions. Some of these states have had frustrating experiences in attempting to advertise a state-labeled beef product. Beef, carrying such labels, usually loses its identity as it moves through the market place. Whose beef is it if it is raised in Alabama, fed in a Texas feedlot, and retailed in California? Some state beef promotion dollars have been used effectively in other areas such as education and research. The National Livestock and Meat Board has provided extensive educational and promotional material on beef with a rather limited budget.

The beef industry attempted a more realistic self-help program designed to collect funds (0.3% of animal value at each market transaction) for market development, consumer information, and research. The Beef Research and Information Act, which needed a two-thirds majority vote by qualified beef producers, however, failed to pass. Another vote was taken on this act (called Beeferendum) early in 1980. A simple majority vote was needed, but it failed again. The reasons for the failure are not totally evident. It does appear that some primary factors were (1) concern about any government involvement (Secretary of Agriculture would give approval to the elected Governing Board), (2) feeling of some loss of independence, (3) not really sure what was happening, coupled with a feeling that it was being forced on them, and (4) many votes coming from owners of small cow herds who make their livelihood elsewhere.

More and more beef producers are realizing that dollars are needed for promotion, education, and research. Table 2.10 shows a comparison of the beef industry with other industries on dollars spent for education and promotion. Figure 2.12 shows dollars spent for promotion by several agricultural commodity groups in 1980. The dairy industry estimates that for every dollar invested in advertising, dairy farmers get back over two dollars in increased sales. Citrus growers expect a 10 to 1 return.

During 1982 the Beef Industry Council of the Livestock and Meat Board and state beef councils launched the largest beef advertising and promotion campaign in history. Over six million dollars was spent on national TV commercials, commercial and consumer magazine ads, radio commercials, and billboards. The advertising focused on the eating satisfaction of beef with the primary theme, "Somehow, Nothing Satisfies Like Beef."

In 1983 some $7.5 million check-off dollars will be spent on TV and print advertising with the theme, "Beef Gives Strength." This theme, based on extensive consumer research, is expected to improve sales to light beef users (more concerned about nutrition,

Table 2.10. Advertising by Commercial Companies, 1979

Company	Investment as a Percent of Sales
Wrigley and Co.	12.100
General Foods	8.100
Proctor and Gamble	6.300
Ralston Purina	1.600
Beef Industry	0.015

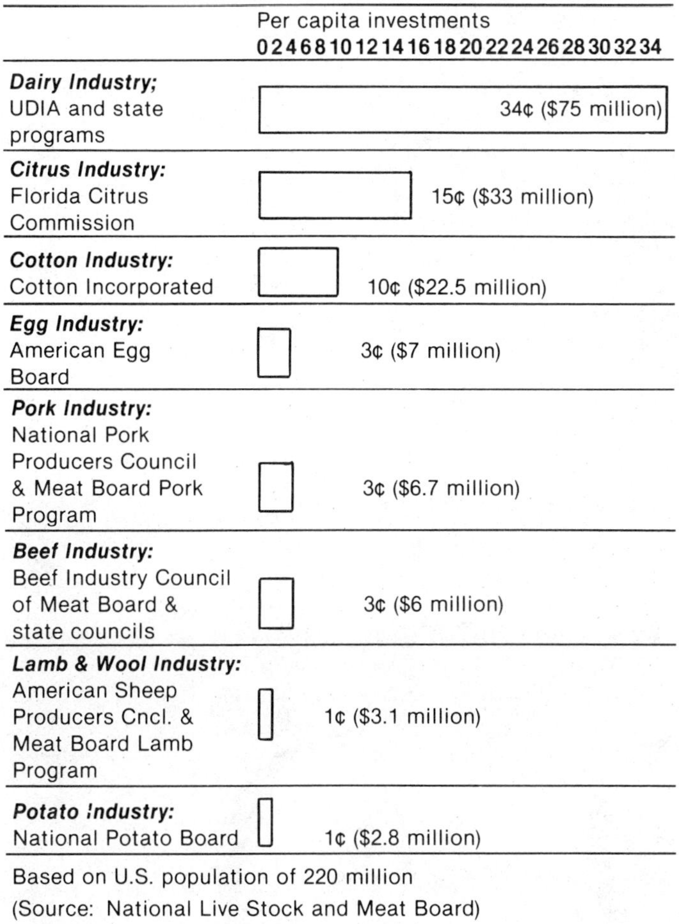

Figure 2.12. National commodity program budgets, 1980. Courtesy of National Live Stock and Meat Board.

fitness, and fashionability) as well as moderate and heavy users. Highlights will include new nutrient data on beef and a program to improve beef's image as a fashionable food.

Advertising dollars come from states with checkoff programs. Several states have a 25-cent checkoff, with the ultimate goal of $1.00 per head (Figure 2.13).

Beef producers are highly dependent on consumers. Verbal accusations between the producer and consumer segments have benefited neither group. Consumers like beef and will consume it at high levels as long as disposable income is reasonably high and the price is in line with other competitive foods. Producers should realize that their financial support of education, research, promotion, and advertising programs is essential. This support appears to be crucial in maintaining a desirable consumer image of beef, removing emotionally based misunderstandings, and communicating the need for a profit incentive for the survival of the beef industry.

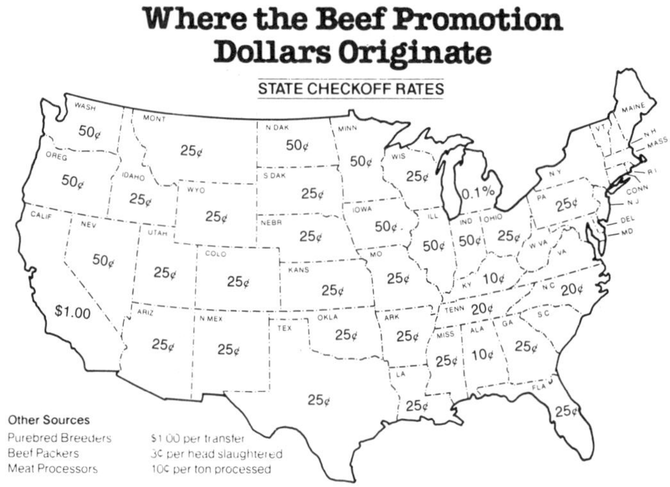

A

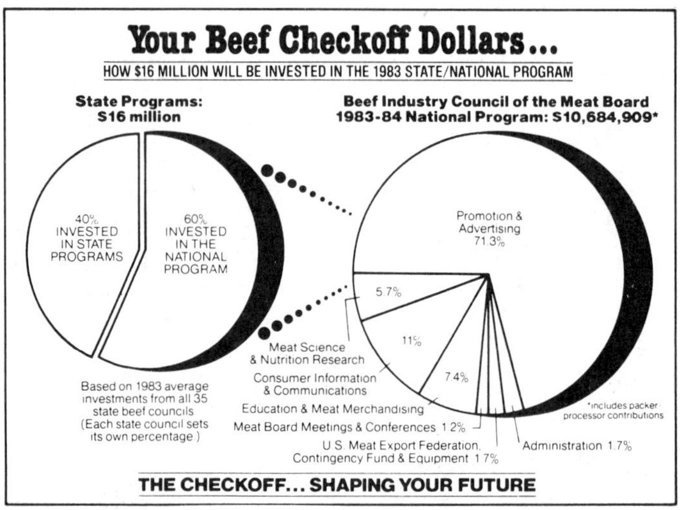

B

Figure 2.13. A. Checkoff amounts per head for the 35 states generating funds for the beef market development programs. B. Where dollars are invested in the 1983 beef program. Courtesy of National Live Stock and Meat Board.

SELECTED REFERENCES

Publications

Ahrens, R. A., 1979. *Red meat competitors: Their impact and nutritive value.* J. Anim. Sci. 48:401.

Baldwin, R. L., 1980. *Animals, Feed, Food, and People. An Analysis of the Role of Animals in Food Production.* Boulder: Westview Press.

Breidenstein, B. C. and Carpenter, Z. L., 1983. *The red meat industry: product and consumerism.* J. Anim. Sci. 57:119 (Suppl. 2).

Consumers' Knowledge and Use of Government Grades for Selected Food Items, 1970. USDA Marketing Research Report No. 876.

Cross, H. R., Stanfield, M.S., Koch, E. J., 1976. *Beef palatability as affected by cooking rate and final internal temperature.* J. Anim. Sci. 43:114.

Diet, Nutrition and Cancer: A critique, 1982. Council for Agric. Sci. and Tech. (CAST). Spec. Public. No. 13.

Fitzhugh, H. A., Hodgson, H. J., Scoville, O. J., Nguyen, T. D. and Byerly, T. C., 1978. *The Role of Ruminants in Support of Man.* Winrock International.

Food Questions and Answers: Environmental Issues, Production, Processing, Safety, Nutrition, Diet, Beverages, Drugs, Economic Issues, World Food Issues, 1981. Council for Agric. Sci. and Tech. (CAST). Special Report No. 7.

Foods from Animals: Quantity, Quality and Safety, 1980. Council for Agric. Sci. Tech. (CAST). Report 82.

The Future for Beef. National Cattleman's Association Beef Business Bulletin, 5 March 1982.

The Good Things We Get from Cattle besides Beef, 1975. National Live Stock and Meat Board.

Hegarty, P. V. J., 1979. *Nutrition and health factors in meat consumption.* J. Anim. Sci. 48:408.

Hormonally Active Substances in Foods: A Safety Evaluation, 1977. *Council for Agric. Sci. and Tech.* (CAST) Report No. 66.

Jeremiah, L. E., 1970. *Beef Quality I. Marbling as an Indicator of Palatability.* Texas Agric. Expt. Sta. Tech. Rpt. 22.

Livestock and Meat Situation, USDA, Washington, D.C.

Mann, G. V., 1977. *Diet–Heart: End of an Era.* The New Engl. J. of Med. 297:644.

Menzie, E. L., Gordin, Z., and Archer, T. F., 1976. *Consumer Attitudes, Knowledge, and Buying Habits Relative to Beef: An Arizona Case Study.* Ariz. Agric. Expt. Sta. Tech. Bul. 226.

National Food Review, USDA, Washington, D.C.

Nutrition and Your Health, Dietary Guidelines for Americans, 1980. Washington, D.C.: USDA and U.S. Dept. of Health, Education and Welfare.

Plant and Animal Products in the U.S. Food System, 1977. National Academy of Sciences Symposium.

Regulation of Potential Carcinogens in the Food Supply: The Delaney Cause, 1981. Council for Agric. Sci. and Tech. (CAST). Report No. 89.

Scientific Aspects of the Welfare of Food Animals, 1981. Council for Agric. Sci. and Tech. (CAST). Report No. 91.

Toward Healthful Diets, 1980. Washington, D.C.: National Academy of Sciences.

Visuals

"A Promise for Life." Nine minute film produced by the Calif. Beef Council (CBC). Shows beef cattle as a natural resource in producing food and inedibile by-products (produced for grades 4 through 12). CBC, 551 Foster City Blvd., Suite A, Foster City, CA 94404.

Slide and tape set on farm animal welfare and care. Univ. of Nebraska, Dept. of Ag. Communications, Lincoln, NE (68583–0918). Cost is $30.

Catalog of Literature and Audio-Visual Aids. National Live Stock and Meat Board, 444 N. Michigan Ave., Chicago, Il 60611.
Meat Photographs for Advertising, Editorial and Educational Use. National Live Stock and Meat Board, 444 N. Michigan Ave., Chicago, Il 60611.
"Sunchanger." 16 mm film (sound and color). 10 minutes. Produced by Elanco Products Co., P.O. Box 1705, Indianapolis, IN 46206.

3
The Retailer–Purveyor and the Beef Industry

Retailers provide the final link between beef producers and consumers. Figure 3.1 shows channels of meat production and processing prior to reaching some three-quarter million retail meat outlets. The 265,000 meat retailers are represented primarily by supermarket chains and independent grocers who provide consumers with ready-to-cook portions of meat. The 500,000 food service operations are represented primarily by hotels, restaurants, institutions, and fast-food establishments. Many of these operations are referred to as the HRI (hotel, restaurant and instituational). It is estimated that HRI trade handles approximately 40% of the final meat sales. Purveyors provide portion-control beef cuts, primarily as a processing link between the packer and several food service operations.

Meat Retailers

The supermarket and grocery store industry is the largest in retailing, with 1978 sales of more than 150 billion dollars, which represents 20% of total retail sales. Grocery stores handle about three-fifths of the food consumed in the United States and thus provide a critical link between consumers and the rest of the food delivery system. Most food retailing businesses are owned by United States citizens, however, 15 foreign firms own 23 United States grocery firms which accounted for 10% of the total grocery store sales in 1979. Foreign investment in the United States, in all areas, has doubled in the last five years, but the amount of foreign investment is still only about one-fourth as great as United States investment in foreign firms.

There are approximately 175,000 grocery stores today with less expansion in number of stores compared to several years ago. This is because of a slower population growth and people choosing to eat more meals away from home. While the grocery store industry has been experiencing some sales problems, new opportunities for growth have been occurring in the food service industry.

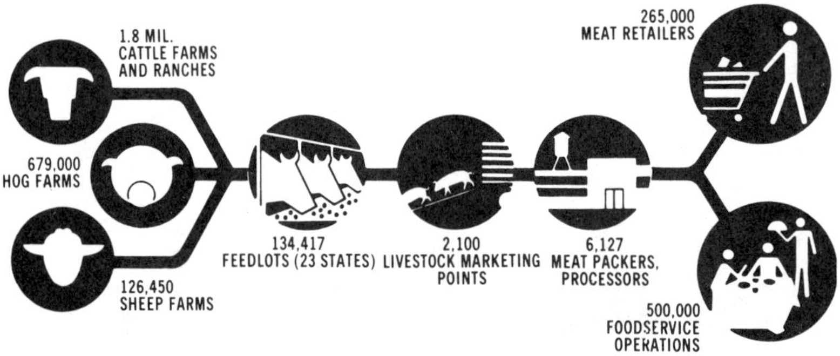

Figure 3.1. Channels of meat production and processing in reaching the retail segment. Courtesy of National Live Stock and Meat Board.

Supermarkets are attempting to counteract the slowdown in growth and sales by adopting scanning computerized checkouts, product diversification, and moving from meat cutting in the backroom of the store to centralized plants. The latter has been adopted in many areas, but union resistance has slowed it down in other areas.

Meat is the backbone of the nation's supermarkets. Meat accounts for approximately 25% of food sales, while meat, poultry, and fish represent approximately 30% of the total grocery and supermarket sales. Because of the magnitude of fresh beef sales (11.5 billion dollars), any problems which affect either the beef producer or the supermarket will eventually be felt by both (Table 3.1).

Foodservice Operations

Consumers spend approximately 30% of their food dollar for meals away from home. Restaurants, cafeterias, and fast-food places together account for about three-fifths of expenditures for food consumed away from home. Fast-food outlets are the fastest growing segment of the away-from-home market although these markets appear to have matured and less growth is anticipated in future years.

Food service operations attract many consumers who like the convenience and atmosphere of eating meals away from home. Almost half of the dollars spent for beef are expended away from home (Table 3.2). Since a service cost is included for the meat portion of a meal, the percent away from home in terms of tonnage would be less. The tonnage for beef consumed away from home is estimated from 25% to 35% of the total United States sales.

Sales at eating places, excluding sales tax, equaled about 56 billion dollars in 1977. Increased energy and labor costs, combined with reduced growth of real per capita incomes, have slowed the growth of the food service industry since 1979. Also, the hamburger fast-food chains have adjusted to a reduced hamburger supply and higher hamburger prices by increasing their menu offerings of poultry, pork, and seafood items.

Table 3.1. Consumer Expenditures for Selected Products Sold in Food Stores, 1977

Food and Non-food Products	Billion Dollars
Fresh and cured meat, fish and poultry	$30.3
Fresh Meat	15.5
Beef	11.5
Veal	0.7
Lamb	0.6
Pork	2.7
Provisions	10.3
Cured hams and picnics	1.8
Packaged bacon	1.8
Frankfurters	1.1
Sausage and products	2.6
Other provisions	3.1
Poultry	3.4
Fish and other seafood	1.1
Frozen foods	1.6
Meat	0.2
Poultry	1.1
Fish and other seafood	0.4
Canned foods	8.9
Meat	1.5
Poultry	0.1
Fish and other seafood	1.1
Vegetables	2.1
Fruits	1.2
All other	2.9
Produce	15.2
Fresh vegetables	10.1
Fresh fruits	5.9
Non-Food items	40.8
Household supplies	10.0
Pet foods	2.4
Tobacco products	5.3
Health and beauty aids	5.4
Beer	6.8

Source: National Food Review, Winter 1979 (taken from the September 1978 issue of *Supermarketing*).

McDonalds Corporation, the world's largest food service firm, operates 6700 restaurants in all 50 states, Washington, D.C., and 29 international markets including more than 250 in Japan. Its 1980 annual report told of the sale of the firm's 35 billionth hamburger.

Table 3.2 Consumer Expenditures for Beef and Other Meats, 1979

Meat	Total	Away From Home	Percent Away From Home
	(Billion dollars)		
Beef	45.7	21.1	46.0
Veal	0.9	0.3	
Lamb	0.8	0.3	
Pork	24.7	6.5	26.3
Red meat	72.0	28.2	
Chicken	10.1	4.3	
Turkey	2.6	0.9	
Poultry	12.7	5.2	41.0

Source: USDA.

RETAIL CUTS AND THEIR VALUE

Carcass Preferences

Table 3.3 shows the carcass weights, quality grades, and yield grades having the greatest preference by the various retail beef outlets. This table demonstrates very well how the large variation of weights, sexes and grades of cattle are utilized by the retail trade because of various consumer products and preferences.

Wholesale and Retail Cuts

Figure 3.2 shows wholesale and retail cuts used in the retail trade. In the past, most beef was acquired by retailers in half or quarter carcasses; however, at present approximately 75% of the beef is cut into smaller portions. This boxed beef is purchased as primal or subprimal (e.g., chuck, loin, rib or round, or 2 to 3 smaller cuts from each of these wholesale cuts).

Yield Grade Differences

Retail value differences, reflected in different yield grades, can be estimated by using percent boneless, closely trimmed, retail cuts (BCTRC) from the yield grade formula (Table 3.4).

The $76.80 difference in the two yield grades is a conservative estimate because the average retail price of $2.40 (July 1981) is the average for all retail cuts with the bone in. The 32 pounds of retail cuts (superiority of carcass A over carcass B) are from the round, loin, rib, and chuck and they are boneless, both of which would elevate the price per pound.

Table 3.5 shows actual carcass cutout by yield grade and value difference for each retail cut. Some of these yield grade value differences may not be entirely realistic for beef in the retail trade. For example, outside fat is trimmed to approximately 0.3

Table 3.3. Carcass Preferencess by Different Retail Markets

Retail Market	Quality Grade	Yield Grade	Carcass Class	Carcass Weight
A. Retail fresh beef				
1. National & local supermarkets				
a. Boxed beef	Choice, ungraded (Good)	1–3	Steer & heifer	6–800
b. Hanging quarters	Choice, ungraded (Good)	2&3	Steer & heifer	6–800
2. Neighborhood & economy grocers	Ungraded (Good)	1–3	Steer & heifer	6–800
3. Meat shops	Prime, Choice & ungraded (Good)	2&3	Steer & heifer	6–700
4. Locker and freezer beef	Ungraded (Good), Choice	2&3	Steer & heifer	5–650
B. Food service suppliers				
1. Fancy hotels, clubs & restaurants	Prime & Choice	3–5	Steer & heifer	6–800
2. Cafeterias, hotels restaurants & institutions	Choice & ungraded (Good)	2&3	Steer & heifer	6–800
3. Economy restaurants	Ungraded (Good, Standard & Commercial)	1&2	Steer, cow & heifer	6–800
4. Fast food (hamburger, roast beef)	Ungraded (Good, Standard & Utility)	1&2	Steer, cow & heifer	6–900
C. Processed meat				
1. Hot dogs, bologna, pizza, etc.	Ungraded (Commercial, Utility, Cutter and Canner)	1&2	Bull & cow	8–900

inch for merchandising purposes. It is obvious that a greater amount of fat is trimmed in going from a YG4 to a YG3 compared to a YG2 to a YG1, because a YG1 typically has less than 0.3 inch of fat prior to any trimming. YG4 and 5 carcasses are generally not preferred by meat retailers even though outside fat can be removed. Large amounts of fat remain between the muscles which cannot be conveniently removed.

Quality Grade Differences

Most beef going into steaks and roasts comes from the Prime, Choice, and Good grading carcasses with lower grades producing processed beef (hamburger and so forth). The difference between Choice and Good varies from $1 to $10 per hundredweight of carcass depending on supply and demand of beef. Table 3.6 shows whole carcass prices for various quality grades, yield grades, and sexes of cattle, while Figure 3.3 demonstrates some carcass value differentials for some of these same factors over the past several years.

	Saleable Beef—lbs	Other lbs
• CHUCK *164.8 lbs (26.8% of total carcass)*		
Blade pot-roast	59.3	
Stew or ground beef	32.1	
Arm pot-roast	22.3	
Cross rib pot-roast	10.7	
Boston cut	9.9	
Fat and bone		30.5
TOTAL	134.3 lbs	30.5 lbs
• BRISKET *23.4 lbs (3.8% of total carcass)*		
Boneless	9.4	
Fat and bone		14.0
TOTAL	9.4 lbs	14.0 lbs
• SHANK *19.1 lbs (3.1% of total carcass)*		

	Saleable Beef—lbs	Other lbs
• RIB *59.0 lbs (9.6% of total carcass)*		
Standing rib roast	24.2	
Rib steak	12.4	
Short ribs	4.7	
Braising beef	2.7	
Ground beef	3.5	
Fat and bone		11.5
TOTAL	47.5 lbs	11.5 lbs
• LOIN *105.8 lbs (17.2% of total carcass)*		
Porterhouse steak	18.7	
T-bone steak	9.5	
Club steak	5.2	
Sirloin steak	41.4	
Ground beef	2.9	
Fat and bone		28.1
TOTAL	77.7 lbs	28.1 lbs

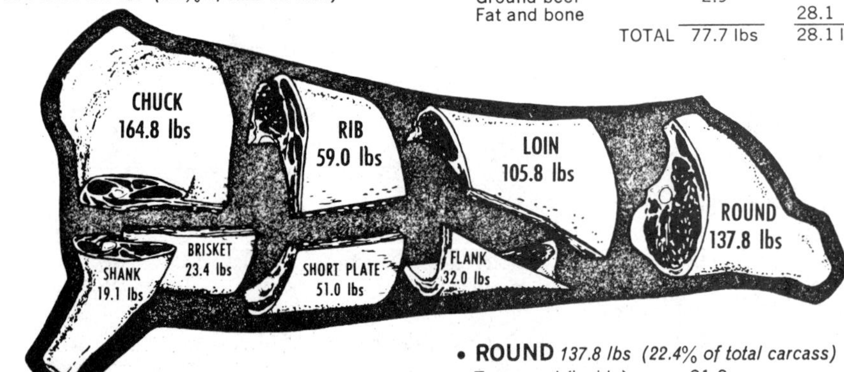

	Saleable Beef—lbs	Other lbs
• SHORT PLATE *51.0lbs (8.3% of total carcass)*		
Plate, stew, short ribs	40.8	
Fat and bone		10.2
TOTAL	40.8 lbs	10.2 lbs
• FLANK *32.0 lbs (5.2% of total carcass)*		
Flank	3.2	
Ground beef	12.6	
Fat		16.2
TOTAL	15.8 lbs	16.2 lbs
MISC. *22.1 lbs (3.6% of total carcass)*		
Kidney, hanging tender	3.6	
Fat, suet, cutting losses		18.5
TOTAL	3.6 lbs	18.5 lbs

	Saleable Beef—lbs	Other lbs
• ROUND *137.8 lbs (22.4% of total carcass)*		
Top round (inside)	21.0	
Bottom round (outside)	20.3	
Tip	13.1	
Stew	8.3	
Rump	4.8	
Kabobs or cubes	2.1	
Ground beef	14.2	
Fat and bone		54.0
TOTAL	83.8 lbs	54.0 lbs

SUMMARY
(1000 lb choice steer)

Dresses out 61.5%	615 lbs
Less fat, bone and loss	183 lbs
Saleable beef	432 lbs

NATIONAL LIVE STOCK AND MEAT BOARD

Figure 3.2. The location of the wholesale cuts of beef and the approximate percent of retail cuts from each wholesale cut. Courtesy of National Live Stock and Meat Board.

Table 3.4. Estimating Retail Value Differences Using the Carcass Yield Grade Formula

Carcass	Carcass wt (lbs)	Yield Grade	BCTRC, Lbs. (Car. wt Times % Cutability)	Retail Price per Pound	Dollar Value
A	700	2.5(51.2%)	358	$2.40	$859.20
B	700	4.5(46.6%)	326	2.40	782.40
				Difference	$ 76.80

Table 3.5 Comparison of Yields of Retail Cuts and Retail Sales Values of Choice Beef Carcasses, by Yield Grades, April 1981

Yield Grades		1		2		3		4		5	
Retail Cut	Price per Pound	% of Carcass	Value/ Cwt Carcass	% of Carcass	Value/ Cwt Carcass	% of Carcass	Value/ Cwt Carcass	% of Carcass	Value/ Cwt Carcass	% of Carcass	Value/ Cwt Carcass
Rump											
boneless	2.60	3.7%	9.62	3.5%	9.10	3.3%	8.58	3.1%	8.06	2.9%	7.54
Inside round	2.96	4.9	14.50	4.5	13.32	4.1	12.14	3.7	10.95	3.3	9.77
Outside round	2.70	4.8	12.96	4.6	12.42	4.4	11.88	4.2	11.34	4.0	10.80
Round tip	2.69	2.7	7.26	2.6	6.99	2.5	6.72	2.4	6.46	2.3	6.19
Sirloin,											
boneless	3.20	4.8	15.36	4.5	14.40	4.2	13.44	3.9	12.48	3.6	11.52
Short loin	3.52	5.3	18.66	5.2	18.30	5.1	17.95	5.0	17.60	4.9	17.25
Blade chuck	1.53	9.9	15.15	9.4	14.38	8.9	13.62	8.4	12.85	7.9	12.09
Rib, short cut (7")	2.56	6.3	16.13	6.2	15.87	6.1	15.62	6.0	15.36	5.9	15.10
Chuck, arm											
boneless	2.05	6.4	13.12	6.1	12.51	5.8	11.89	5.5	11.27	5.2	10.66
Brisket,											
boneless	2.42	2.5	6.05	2.3	5.57	2.1	5.08	1.9	4.60	1.7	4.11
Flank steak	3.63	0.5	1.82	0.5	1.82	0.5	1.82	0.5	1.82	0.5	1.82
Lean trim	1.86	14.9	27.71	13.8	25.67	12.7	23.62	11.6	21.58	10.5	19.53
Ground beef	1.50	13.3	19.95	12.2	18.30	11.1	16.65	10.0	15.00	8.9	13.35
Kidney	0.71	0.3	0.21	0.3	0.21	0.3	0.21	0.3	0.21	0.3	0.21
Fat	0.10	8.1	0.81	13.3	1.33	18.5	1.85	23.7	2.37	28.9	2.89
Bone	0.03	11.6	0.35	11.0	.33	10.4	0.31	9.8	0.29	9.2	0.27
Total		100.0	179.66	100.0	170.52	100.0	161.38	100.0	152.24	100.0	143.10

Difference in retail value between yield grades—$9.14 per cwt of carcass.

Source: USDA.

Table 3.6. Wholesale Dressed Beef Prices (Dollars per 100 Pounds)

Carlot Basis		CAF East Coast		Central U.S. FOB Omaha		Central U.S. FOB Amarillo		CAF Los Angeles	
		May 9 1981	May 8 1982	May 9 1981	May 8 1982	May 9 1981	May 8 1982	May 9 1981	May 8 1982
Carcass Steer Beef:									
Choice 3	600–700 lb	—	—	102.85	113.10	104.00	—	112.10	114.83
	700–800 lb	—	—	102.75	113.10	105.00	—	111.17	114.67
	800–900 lb	—	—	102.70	113.10	105.00	—	—	—
Good 3	500–600 lb	—	—	—	—	—	—	111.75	111.50
	600–700 lb	—	—	100.90	—	102.50	—	109.38	109.50
	700–800 lb	—	—	100.45	—	102.50	—	109.38	109.50
Choice 2	600–700 lb	—	—	107.50	112.50	104.25	113.75	113.50	116.62
	700–800 lb	—	—	107.00	—	—	—	112.50	115.52
Good 2	500–600 lb	—	—	—	—	—	—	111.75	111.50
	600–700 lb	—	—	100.90	—	102.50	—	110.12	109.67
	700–800 lb	—	—	100.45	—	102.50	—	110.12	109.67
Choice 2–3	600–700 lb	—	—	—	—	—	—	—	—
	700–800 lb	—	—	—	—	—	—	—	—
Good 2–3	600–700 lb	—	—	100.90	110.80	102.50	110.25	109.38	109.67
	700–800 lb	—	—	100.45	110.70	102.50	109.75	109.38	109.67
Choice 4	600–800 lb	—	—	93.20	103.50	—	105.00	96.38	108.00
	800–900 lb	—	—	93.20	103.50	—	105.00	96.38	108.00
Carcass Heifer Beef:									
Choice 3	500–600 lb	—	—	101.75	109.80	103.00	—	105.50	110.50
	600–700 lb	—	—	101.75	111.00	103.00	—	105.50	—
Good 3	500–600 lb	—	—	98.88	—	101.50	—	—	—
	600–700 lb	—	—	99.31	—	101.50	—	—	—
Choice 2	500–600 lb	—	—	103.38	—	—	—	—	—
	600–700 lb	—	—	103.38	—	—	—	—	116.50
Good 2	500–600 lb	—	—	98.88	—	101.50	—	—	—
	600–700 lb	—	—	99.31	—	101.50	—	—	—
Choice 4	500–700 lb	—	—	93.20	103.75	—	104.50	—	107.50
Carcass Cow Beef:									
Utility									
(Breaking) 2–4	450 lb up	—	—	75.00	75.75	76.00	—	79.00	78.50
(Boning) 1–2	400 lb up	—	—	77.50	78.83	78.50	—	82.00	80.50
Canner & Cutter 1–2	350 lb up	—	—	82.75	80.62	—	—	85.44	88.00
Carcass Bull Beef:									
Yield Grade 1–2	500 lb up	—	—	—	87.33	—	—	91.75	89.00
Choice 3 Primal Beef Cuts:									
Steer Hinds	120–165 lb	—	—	123.00	—	—	—	126.50	135.50
	165–210 lb	—	—	123.00	—	—	—	126.50	135.50
Heifer Hinds	120–165 lb	—	—	119.75	130.00	—	—	—	—
Steer & Heifer Fores	135–210 lb	—	—	87.20	93.38	—	—	93.00	102.50
Arm Chucks	80–130 lb	—	—	—	88.12	—	—	88.00	97.00
Ribs (7 Rib)	24–30 lb	—	—	—	—	—	—	152.50	162.50
	30–40 lb	—	—	—	—	—	—	152.50	162.50

Table 3.6. (*Continued*)

Carlot Basis		May 9 1981	May 8 1982	May 9 1981	May 8 1982	May 9 1981	May 8 1982	May 9 1981	May 8 1982
		CAF East Coast		Central U.S. FOB Omaha		Central U.S. FOB Amarillo		CAF Los Angeles	
Loins (Trimmed)	35–50 lb	—	—	—	—	—	—	—	182.50
	50–65 lb	—	—	—	—	—	—	180.00	182.50
Steer Rounds	60–95 lb	—	—	—	—	—	—	117.00	129.00
Heifer Rounds	60–85 lb	—	—	—	—	—	—	117.00	129.00
Full Plates		—	—	—	69.00	—	—	—	—
Short Plates		—	—	—	70.50	—	—	57.00	74.50
Briskets		—	—	—	68.50	—	—	60.50	70.00
Flanks (Untrimmed)		—	—	51.00	—	—	—	52.00	69.00
Boneless Beef	90% Fresh	120.00	115.12	—	—	—	113.50	121.50	113.75
(Chemical lean)	Frozen	—	—	118.00	—	—	—	—	—
	85% Fresh	110.38	109.50	—	—	—	—	113.50	108.50
	Frozen	—	—	—	—	—	—	—	—
	75% Fresh	96.00	95.00	92.00	—	—	—	—	—
	Frozen	—	—	90.50	92.58	—	—	—	—
	50% Fresh	61.25	73.00	59.62	67.33	—	—	58.00	71.00
	Frozen	61.00	—	58.00	68.67	—	68.00	—	—
Boneless Bull Beef Fresh 92%		—	—	—	—	—	—	—	119.00

Source: USDA, *Livestock Market News.*

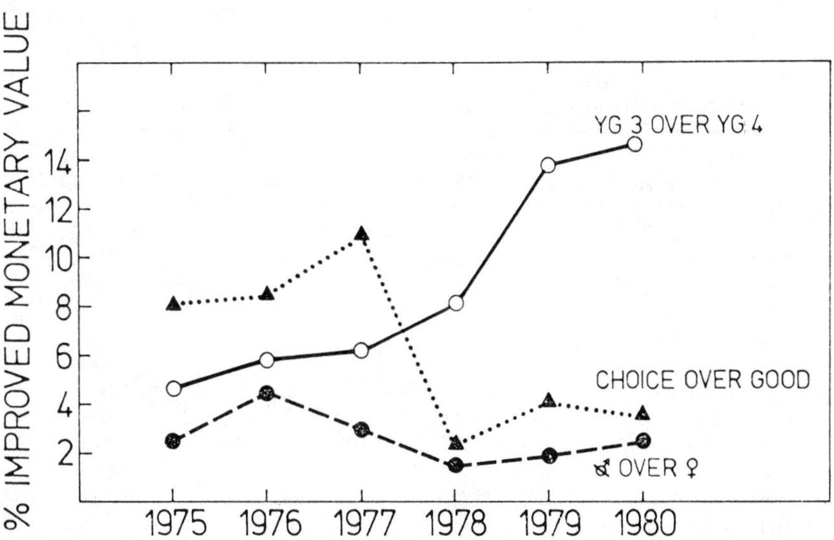

Figure 3.3. Carcass value differentials for YG3 versus YG4, Choice versus Good and steers versus heifers. *Source:* Kauffman, Proc. Natl. Beef Grading Conf., Iowa State Univ., Jan. 1981.

COSTS IN RETAILING BEEF

A component part of the retail price of beef is marketing cost. Figure 3.4 shows a breakdown of these costs as they apply to all foods. Table 3.7 gives evidence of changes in food marketing costs over the past ten years.

The increased retail price of beef during 1978 has been labeled as one major cause of inflation. The reality of this situation is that the price of beef resulted from an undersupply of beef, particularly processing beef. Most of the inflationary part of the higher retail cost of beef has come from increased costs in processing and marketing. Labor

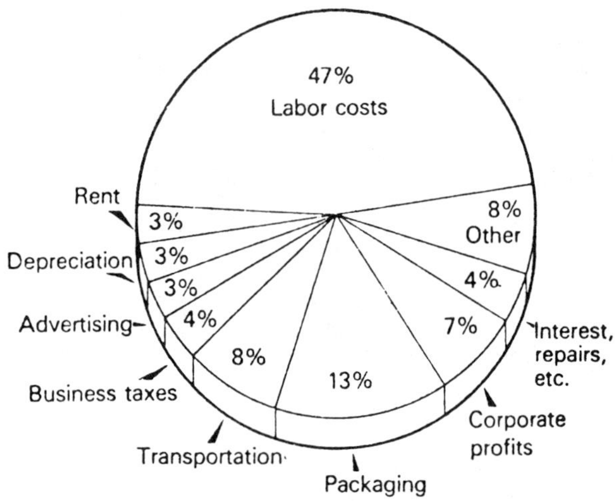

Figure 3.4. Costs of the food marketing Bill, 1976. *Source:* USDA.

Table 3.7. Marketing Bill and Components

Component	1967	1977	Increase 1967–1977
	Billion Dollars		Percent
Marketing bill	61.4	123.5	101
Labor	25.9	58.8	127
Packaging	7.2	16.0	122
Transportation	4.3	10.4	142
Profits	3.4	8.1	138
Minor components	8.5	19.0	124
Business taxes	2.4	5.4	129
Depreciation	1.8	3.9	119
Rent	1.6	3.3	113
Advertising	1.5	3.0	98
Repairs, bad debts, contributions	0.9	2.1	144
Interest	0.4	1.3	208

Source: National Food Review, USDA, June 1978.

unions have the leverage to increase costs through higher wage demands and in preventing some efficiencies to come into the beef marketing system. It is an interesting paradox to note that 500,000 members of the Amalgamated Meat Cutter and Butcher Workmen (an affiliate of AFL-CIO) have a locked-in, increasing, high hourly wage, while cattle producers wonder each year if their hourly labor will have a positive or negative return.

SELECTED REFERENCES

Publications

Aders, R. O. *Meat is the backbone of the nations supermarkets.* The National Provisioner, 24 March 1979.

Longen, K. *Grocery store sales by product class.* National Food Review, Winter 1979.

Meat in the Foodservice Industry, 1977. Chicago: National Live Stock and Meat Board.

Van Dress, M. G. and Grinnell, G. *Grocery stores and eating places: Industries in transition.* National Food Review, Spring 1980.

Visuals

Catalog of Literature and Audio-Visual Aids. National Live Stock and Meat Board, 444 Michigan Ave., Chicago, Il 60611.

Meat Photographs for Advertising, Editorial and Educational Use. National Live Stock and Meat Board, 444 N. Michigan Ave., Chicago, Il 60611.

4
The Packer and the Beef Industry

NUMBER AND DISTRIBUTION OF PACKING PLANTS

Cattle slaughtering plants can be divided into two types; (1) federally inspected and (2) non-federally inspected. Federally inspected plants are those transporting beef out of state. Federal officials inspect these plants, the cattle, and the products produced to assure wholesomeness, proper labeling, and freedom from disease and adulteration. Non-federally inspected plants utilize beef within the state and are subject to inspection criteria enforced by the state where the plant is located.

Approximately 94% of the commercial cattle slaughtered in 1979 was in 1477 federally inspected packing plants, while approximately 87% of the commercial calf slaughter was in 745 federally inspected packing plants. Figure 4.1 shows the states having 5 or more federally inspected plants, each state slaughtering over 500,000 head of cattle in 1979.

SIZE OF PACKING PLANTS

Table 4.1 shows the size of packing plants reflecting number of plants and number of cattle slaughtered. Large packing plants, slaughtering more than 50,000 head per year, are responsible for most of the cattle slaughtered in the United States. (Figure 4.2).

In the 1950s and 1960s, the packing industry was typified by four major meat packing firms—Armour, Swift, Wilson and Cudahy. The "Big Four," as they often were called, specialized in livestock slaughter and fresh meat merchandising, and in 1960 accounted for roughly one-half of the total federally inspected livestock slaughter in the United States.

Over the past two decades, the packing industry has evolved into a $65 billion per year, multifaceted agribusiness. Phenomenal developments in technology and production efficiency have transformed the "old-line" meat packers into highly diversified corporate

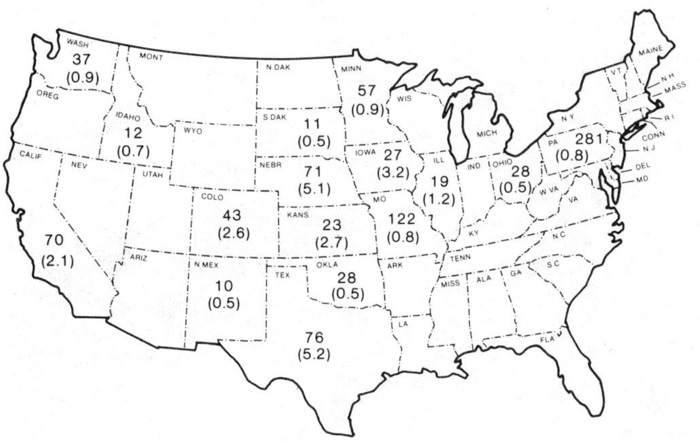

Figure 4.1. States having 5 or more federally inspected beef packing plants. Numbers slaughtered shown in ( ) in millions of head, 1979. Drawn by CSU Graphics.

giants, many of whom operate as subsidiaries to large conglomerate firms. A recent list of the top ten U.S. meat packing and processing companies reflects the structure and scale of today's meat industry (Table 4.2).

In 1980, the top ten companies had net sales of more than $22 billion and employed about 86,200 people. Their involvement in the industry encompasses nearly every aspect of meat production, processing, and merchandising.

It is not unusual to find some packing operations slaughtering more than 2000 cattle in an eight-hour day and some plants running two shifts per day.

PRODUCTS PRODUCED

The packer is involved in slaughtering beef animals and separating them into component parts of carcasses and the various by-products. It is important to understand the products and their respective economic values which are derived during slaughtering and processing of beef animals. During the slaughtering process, the head, hide, and shanks are removed from the animal. Internal organs (offal) are then removed and the carcass is split into two halves (Figure 4.3). The hide and offal are commonly referred

Table 4.1. Federally Inspected Cattle Slaughtering Plants, 1979

Size Group	No. Plants	No. Cattle Slaughtered (Thousand Head)
Under 1,000	822	304
1,000– 9,999	338	1,116
10,000–49,999	164	4,332
50,000+	153	26,140
Total	1477	31,892

Source: USDA.

Figure 4.2. Aerial view of large packing plant complex at Dakota City, Nebraska: ① corporate headquarters, ② processing plant, ③ load out area for carcasses and boxed beef, ④ cattle holding area, ⑤ truck washing compound, ⑥ hide plant, and ⑦ railroad cars for beef tallow. Courtesy of Iowa Beef Processing, Inc.

to in the trade as "the drop." Value of the drop may mean the difference between a profit or a loss for the packers (Figure 4.4).

Carcasses are moved from the slaughter floor to a cooler where the carcasses are stored 24 hours at 33°F. Heat removal is important to eliminate spoilage; however, extremely cold temperatures in the cooler may produce cold shortening of muscle fibers which reduces tenderness of meat.

After chilling, carcasses are shipped as halves, quarters, or processed in primal or subprimal cuts and shipped in boxes (Figure 4.5).

Carcass Beef

Commercial carcass beef production in 1981 was approximately 22 billion lbs on a carcass weight basis with more than 35 million head of cattle slaughtered. The carcass

Table 4.2. The Top Ten Meat Packing and Processing Companies in the United States

Firm	Parent Company	Net Sales (Billion Dollars)
1. Iowa Beef Processors, Inc.	Occidental Petroleum	4.6
2. Armour and Co.	Greyhound Corp.	3.2
3. Swift and Co.	Esmark	2.5
4. Wilson Foods Corp.	LTV Corp.	2.3
5. John Morrell and Co.	United Brands	2.2
6. Swift Ind. Packing Co.	Esmark	2.0
7. Oscar Mayer and Co.	General Foods	1.8
8. Excel	Cargill	1.3
9. Geo. A. Hormel and Co.	—	1.3
10. Spencer Foods Inc.	Land O'Lakes	1.2

Figure 4.3. Carcasses are eviserated (internal organs removed) as workers move on a conveyor belt along with the carcasses, which are moved on a chain. In the background, the organs are inspected by a federal inspector and other workers are washing and processing the viscera. Courtesy of Iowa Beef Processors, Inc.

Figure 4.4. Beef hide room where hides are processed, tied, and stored in preparation for shipping. Courtesy of Iowa Beef Processors, Inc.

Figure 4.5. Carcasses are temporarily stored in a cooler where rapid chilling prevents spoilage of the meat. Note that the carcasses have been ribbed between the twelfth and thirteenth ribs for grading purposes. Courtesy of Iowa Beef Processors, Inc.

beef eventually appears to the consumer in several categories: (1) "block" or table beef (primarily steaks and roasts), (2) processing beef going into ground beef, and (3) processing beef going into processed products which include cured, flaked, and formed (sausage, including frankfurters), cooked, canned, prepared dinner products, and others. Processed product ingredients also utilize beef head, cheek meat, and variety meats. Many processed products will also include pork trimmings and variety meats, textured vegetable proteins, cereals, dairy products, and poultry.

The amount of carcass beef going into block or processed beef depends on the phase of cattle cycle, grain prices, current livestock, and meat prices, factors affecting imports

Table 4.3. Beef Supply—Source and Average Yield of "Block" and Processing Beef, 1978

Source	Billion Pounds	Percent	
		Block	Processing
Domestic			
Slaughter			
Fed Steers & Heifers	17.3	80	20
Cull Cows	4.3	15	85
Bulls	0.6	—	100
Non-fed Steers & Heifers	1.8	68	32
Imported Beef	2.3	10	90
Total	26.3		

Source: Livestock and Meat Situation, USDA, April 1979.

Table 4.4. Processing Beef Disposition and Consumption, 1978

Item	Disposition (Billion Pounds)	Per Capita Consumption (Lbs, Carcass Wt, Equiv.)
Processed products	4.5	20.8
Ground beef	5.9	27.1

Source: Livestock and Meat Situation, USDA, April 1979.

and likely several others. Distribution of the 1978 beef supply into block beef and processing beef is shown in Table 4.3.

The proportion of processing beef going into processed products and ground beef and per capita consumption of each is shown in Table 4.4.

Boxed Beef

During the 1970s, an increased amount of carcass beef was fabricated into boxed, vacuum packed, primals, and subprimals at the packing plant rather than being shipped as half or quarter carcasses. The exact amount of boxed beef produced has not been accurately determined; however, it is estimated that nearly three-fourths of slaughtered steers and heifers go into boxed beef. The leading packers in boxed beef production in 1976 are shown in Table 4.5.

Figure 4.6 diagrams the stages of beef fabrication in producing boxed beef.

Boxed beef is produced at central cutting facilities owned by retailers in addition to that produced at the packing plant level. Boxed beef is stored at near freezing temperatures but is not frozen. It is sold at retail typically within 21 days of slaughter. Most boxed beef comes from yield grade 2 or 3 carcasses (Figures 4.7, 4.8, and 4.9).

A large percent of beef going to restaurant trade is fabricated into final retail cuts. The beef is then frozen so it can be stored for a longer period of time than fresh, boxed beef.

Table 4.5. Leading Packing Companies in Boxed Beef Production, 1976

Company	Million Lbs
IBP	1630
Missouri Beef Packers (MBPXL)	610
Monfort	243
Morrell	193
Spencer	180
Armour	158
Nat. Beef	136
Sterling	130
Swift	130
Flavorland	110

Source: The National Provisioner, 6 January 1979.

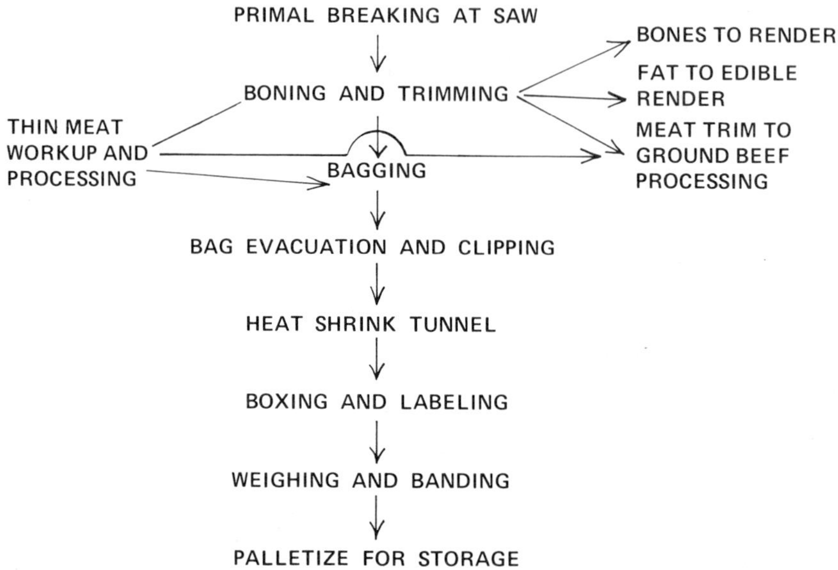

Figure 4.6. Schematic diagram showing the stages of boxed beef production. Courtesy of University of Nebraska.

Figure 4.7. Boning, trimming, and separation of thin meat cuts from primal cuts is done by workers who perform their specific tasks on stationary platforms on either side of the moving conveyor belt. Courtesy of Iowa Beef Processors, Inc.

Figure 4.8. Prior to boxing of the beef, each primal cut is bagged, then oxygen is removed from inside the bag. This bagging process extends shelf-life of the meat. Courtesy of Iowa Beef Processors, Inc.

Figure 4.9. Boxed beef in storage. The equipment in the background can automatically travel along the isle, adding and removing boxes. After the beef is boxed, there is minimal human labor involved in moving, storing, and loading the boxed beef. Courtesy of Iowa Beef Processors, Inc.

During recent years, concern has been expressed as to how boxed beef would influence the price of fed cattle. In the past, carcass beef volume and price has been used as the live cattle price indicator. Prices are being reported on boxed beef several ways; however, the volume is not well known. Prices are being quoted on various fabricated beef cuts as shown in Table 4.6.

Table 4.6. Boxed Fabricated Beef Cuts (Vacuum Packaged). Weekly Average of Daily Quotations in Dollars Per 100 Lbs

Carlot Basis		Aug. 23 1980	Aug. 22 1981	Aug. 23 1980	Aug. 22 1981	Aug. 23 1980	Aug. 22 1981
		East Coast		**Central U.S.**		**Los Angeles**	
IMPS Item No.		*CAF*		*FOB*		*CAF*	
107 Rib, Oven Prepared	17–28#	—	—	—	—	—	—
109 Rib, Roast Ready	14–24#	246.75	219.50	241.00	216.67	249.38	225.88
112 Ribeye Roll	6–11#	—	—	—	—	466.25	463.00
112A Ribeye Roll, Lip on	7–12#	—	349.00	362.92	345.50	375.50	360.00
114 Shoulder Clod	13–24#	151.25	145.25	147.08	139.88	150.25	144.75
116A Chuck Roll	13–25#	151.50	149.00	146.17	141.25	160.88	151.88
120 Brisket (Bnls), Deckle off	6–14#	132.00	111.33	127.38	106.50	131.50	116.50
126 Armbone Chuck (Bnls)	60–90#	126.25	117.25	121.69	112.40	132.00	121.33
160 Round, Shank off, (P Bnls)	45–80#	—	—	150.50	141.00	—	—
161 Round, Shank off (Bnls)	40–75#	167.00	—	164.58	153.50	—	—
167 Knuckle	8–13#	168.25	—	163.67	161.00	172.50	166.50
167A Knuckle (TRMD)	8–13#	—	—	—	—	—	—
168 Top (Inside) Round	14–23#	188.25	176.50	183.85	169.33	192.12	180.75
170 Bottom (Gooseneck) Round	18–29#	152.70	143.25	148.60	140.00	152.25	149.75
171 Bottom (Gooseneck) Round (Untrmd)	18–29#	—	—	—	—	—	—
171A Bottom (Gooseneck) Round (Untrmd, heel out)	17–28#	—	—	—	—	—	—
173 Short Loin (TRMD)	17–30#	—	—	—	—	—	—
175 Strip Loin	11–24#	242.50	—	240.50	226.00	—	—
179 Strip Loin, Short Cut	8–14#	—	—	—	—	—	—
180 Strip Loin, Short Cut (Bnls)	8–10#	—	—	—	295.00	350.00	—
	10–12#	327.50	—	343.50	325.00	352.50	321.88
	12# up	341.25	300.00	343.50	—	351.25	321.88
184 Top Sirloin Butt	7–10#	—	—	241.25	222.50	250.00	—
	10–12#	—	233.00	248.25	224.50	253.12	235.00
	12–14#	—	233.00	255.00	229.00	253.12	237.62
185 Bottom Sirloin Butt	5–7#	—	—	121.75	103.00	—	—
	7–9#	131.50	—	123.25	107.00	—	112.00
186 Bottom Sirloin Butt (TRMD)	2–7#	—	—	—	—	—	—
189 Full Tenderloin	5–7#	350.00	335.00	330.75	330.00	355.00	327.50
185A Bottom Sirloin, Flap	1–3#	—	—	—	160.00	205.00	187.88
185B Bottom Sirloin, Ball Tip	1–2#	—	—	—	—	185.00	—
	2# up	—	—	—	—	222.50	185.00
185C Bottom Sirloin, Triangle	1½–5#	—	—	—	—	205.00	185.00

Source: USDA Livestock Market News, 15 June 1982.

Also, the USDA estimates carcass value based on the value of fabricated beef cuts as quoted from the USDA *Livestock Market News,* 15 June 1982.

"The weekly average gross cut out value of Choice 600–700 lb carcasses, based on current prices of Choice fabricated beef cuts (FOB Midwest Basis) for the week ending *8–22–81* was *$109.29,* per cwt, compared with *$110.07* the previous week and *$121.05* a year ago."*

* *Note:* Based on average industry yields with no adjustment for packer processing costs.

Figure 4.10 shows how IBP uses formula pricing for CATTLE PAK® (their name brand of boxed beef). The formula for all twenty styles is based on the top half of the price quotation for 600 to 700 lb carcasses in the National Provisioner "Yellow Sheet," River Points, market report, plus a fee of $0.75 per cwt for selection of yield grade 3 or better. IBP allows full credit for all thin meats and trimmings at the current market price. Fat and bone is credited at the current edible price. A processing and packaging fee is charged to fully fabricate the entire carcass of beef. Cost of the carcass, plus processing and packaging fee, minus thin meat cuts, gives the product cost. Product cost is divided by thick meat yield to give price per cwt, F.O.B., IBP's plant for CATTLE PAK® Cattle.

By-products

Several by-products and their respective consumer uses were previously identified in Figure 2.10 in Chapter 2. Packers typically sell most of their by-products to other

DATE 8-29-76

CATTLE-PAK PRICING: CHOICE BONELESS CHUCK & BONELESS ROUND
CATTLE-PAK CATTLE

	YIELD/LBS	PRICE/CWT		VALUE/CWT
*Choice 6/7 Cattle	650.00	@ 80.75 =		$ 524.88
			PROCESSING & PACKAGING FEE	$ 34.84
			SUB TOTAL	$ 559.72

THIN MEAT CREDITS:		F.O.B. PLANT	
Boneless Brisket	16.71	@ 85.00 =	14.20
Shank Meat	7.93	@ 94.00 =	7.45
Inside Skirts	2.92	@ 120.00 =	3.50
Outside Skirts	2.92	@ 170.00 =	4.96
Flank Steaks	2.73	@ 235.00 =	6.42
Kidneys	1.30	@ 20.85 =	0.27
Hanging Tenders	1.63	@ 145.00 =	2.36
Short Ribs	4.52	@ 75.00 =	3.39
IBP Special Trim	2.72	@ 115.00 =	3.13
Regular Trim	86.69	@ 53.10 =	46.03
Fat & Bone	120.74	@ 11.66 =	14.08
Floor Scraps	3.25	@ 1.50 =	.05
Cut Loss	3.25	@ -0- =	-0-
	257.31		

TOTAL THIN MEAT CREDIT	($ 105.84
TOTAL PRODUCT COST	$ 453.88
DIVIDED BY YIELD	÷ 392.69
PRICE/CWT FOB PLANT	$ 115.58
FREIGHT	$ ______
SELLING PRICE C.A.F.	$ ______

*TOP HALF SHEET
RIVER POINTS
CHOICE STEERS PLUS $0.75/CWT

FORM DATED 7-3-78

Figure 4.10. Iowa Beef Processors method of pricing their boxed beef. Courtesy of Iowa Beef Processors, Inc. CATTLE-PAK® Boxed Beef brochure.

processors who produce the final consumable products. Table 4.7 shows the average yield of by-products and their comparative values.

The current (June 1982) by-product value per hundredweight (liveweight) for a 1000 pound steer is $5.85, for a total of $58.50. The by-product value a year ago was $6.00 per hundred pounds of liveweight. Packers' profit margins are heavily dependent on the value of by-products. These values have fluctuated quite widely over short periods of time.

Fluctuations in hide prices will usually have the most economic impact on the packer since hides comprise the largest part of total by-product value. During the past decade, hide prices have ranged from a high of $0.60 to a low of $0.10 per pound.

Figure 4.11 shows how hide prices have varied over the past several years. Currently (June 1982) the packer is receiving approximately $0.40 per pound for hides.

Table 4.7. By-Product Yield and Value from an Average 1100 Pound Steer

Product	Expected Yield Per Head (Lbs)	Market Value Per Cwt (7–23–83)	Total Value Per Head
Edible			
Select liver	13.00	$ 20.60	$ 2.68
Boneless heart	3.75	26.25	0.98
Cheek meat	4.00	67.45	2.70
Head meat	1.91	54.05	1.03
Salivary glands	1.25	10.00	0.12
Lips	1.00	11.25	0.11
Tail	2.00	56.85	1.14
Spleen	1.25	10.00	0.12
Tongue #1	3.23	58.00	1.87
Weasand meat[a]	0.43	90.00	0.48
Tripe (stomach)	8.50	11.00	0.94
Sweetbread (thymus)	0.49	75.00	0.37
Inedible			
Lung (standard trim)	5.00	9.50	0.48
Gullet	0.95	10.00	0.10
Ox gall (raw)	0.30	200.00	0.60
Pancreas glands	0.75	122.00	0.92
Inedible tallow	60.00	13.01	7.81
Meat & bone meal			
50% protein	32.00	8.64	2.76
Dried blood			
80% protein	6.50	13.56	0.88
Hide/cured			
(Colorado steer)	64.00	74.54	47.71
		Total Value	$73.80

[a] Weasand is the muscular lining which surrounds the esophagus from larnyx to first stomach.

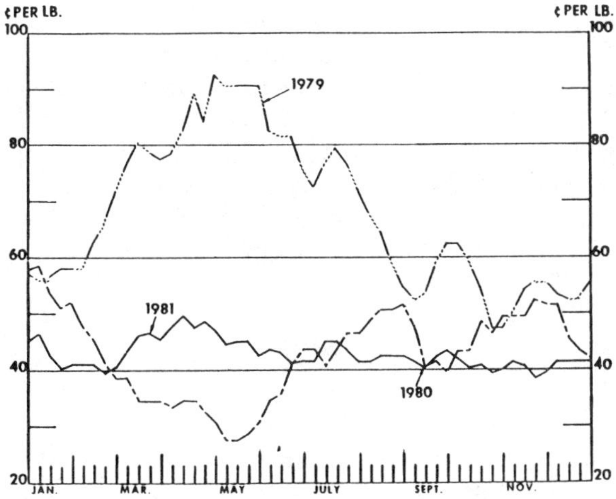

Figure 4.11. Steer hide prices 1979–1981. Courtesy of the Western Livestock Round-Up.

COSTS AND RETURNS

During the past several years, beef industry publications have carried the following headlines:

ABP (AMERICAN BEEF PACKERS) BANKRUPTCY CLOSES 7 PLANTS

SWIFT AND COMPANY CLOSES BEEF PLANTS IN DES MOINES, IOWA, AND GRAND ISLAND, NEBRASKA

MONFORT OF COLORADO CLOSES GREELEY PLANT, BUYS NEBRASKA PLANT

These headlines, plus several others, identify poor profitability for many in beef packing business. Reasons for closings are cited as high labor costs and inefficiencies in plant operations. Some packing companies claim that union labor has driven hourly labor costs so high that the costs cannot be covered. Therefore, there is no alternative but to sell the plant. Closed plants stay idle or another packing company can purchase the closed plant and negotiate a new labor contract, usually at a lower hourly rate than the previous one.

There is considerable variation in hourly wage rates paid by packing plants. A 1979 survey showed that the top nine packing companies with the highest labor costs were paying $7.85 per hour, while the lowest nine were paying $3.82 per hour. Most of the latter plants were smaller in size and were non-union. Iowa Beef Processors, considered by many as the leading beef packers, had an hourly labor cost in their nine plants of $6.57 per hour which compared to an average of $5.96 for all 69 plants which were surveyed.

The meat packing industry reports net earnings of less than one percent of total sales which is considerably lower than many other U.S. businesses.

Amid financial woes for some packing companies, others appear to be flourishing. These packing companies have modernized their plants or built new ones with emphasis on efficiency in producing boxed beef. Their management has implemented innovative changes in labor and customer relations in producing a streamlined, efficiently produced product. There is evidence that power and influence are being concentrated in two or three large, efficient packing companies. Some industry people are raising questions about dwindling competition and pricing efficiency because of the new packing giants.

Slaughter costs per head are typically in the $20 to $25 range, while the processing and boxing costs range from $40 to $50 per carcass. Packers have to pay federal graders for carcasses graded within their plants. Currently hourly grading fees are $23.20 Monday through Friday, $28.20 for weekends and $46.20 per hour for legal holidays. Graders can grade approximately 230 carcasses per hour.

Carcass Preferences

Most packers have preference for yield grade 2, Choice quality grade carcasses weighing from 600 lbs to 800 lbs. Higher yield grades give more pounds of retail cuts and the Choice grade has a price differential over Good which has been traditionally built into the marketing system. During the past several years, some packers have established some preference for "no-rolls" (carcasses not carrying a grade stamp, but they would be mostly high good according to the grade standards). These packers are showing a preference for the "no-roll," high yield grading (1 and 2) carcasses over the Choice, yield grade 3 carcasses. Retailers with their house name beef (such as Safeway Quality, King Soopers Generic, and so forth) have been the primary impetus in establishing the preference for the "no-roll," high yield grade carcasses.

The carcass weight range of 600 lbs to 800 lbs for slaughter steers and heifers has been established primarily for two reasons. First, the carcass needs to fit the prescribed number of boxes which are used in the beef trade. These boxes are constructed to properly fit the semitrailer trucks. Carcasses lighter than 600 lbs do not use the space efficiently and carcasses too far in excess of 800 lbs will not physically fit the boxes.

Second, size of carcass relates to size and weight of the cut and the portion size. The consumer's purchase of meat is significantly influenced by number of portions needed and cost per serving. Also, size in relation to thickness of cut is important which may well dictate the ribeye size and size of other muscles. A T-bone steak, cut an inch and a quarter thick, with an 18 square inch ribeye is approximately 32 ounces. A T-bone with a 12 square inch ribeye, cut to the same thickness is a 21-ounce steak. There is a discount on ribs over 20 lbs because size of the prime rib cut is too large and the cost is excessive to many consumers.

Many packers would prefer heavier carcasses within the desired weight range because slaughter and processing cost per pound is significantly less than on lighter carcasses. For example, at a $65 per carcass processing cost, a 600 lb carcass would cost approximately 11 cents per pound while an 800 lb carcass would be approximately 8 cents per pound.

Some packers, usually the smaller ones, will slaughter cattle yielding carcass weights below 500 lbs. These cattle will usually be veal calves, slaughter calves and light steers and heifers yielding carcasses of approximately 150 lbs, 350 lbs and 500 to 600 lbs respectively.

SELECTED REFERENCES

Publications

Bowling, R. A. *Packer buyer preference.* Proc. 15th Conf. on Artificial Insemination of Beef Cattle, 17 January 1981. Denver, Colorado.

Henrickson, R. L. 1983. *Status of hot processing of meat in the United States.* Okla. Agric. Expt. Sta. MP–114.

Jennings, C. B. *The changing world of the beef packing industry.* National Beef Symposium, Okla. State Univ., 12 March 1982.

Visuals

IBP Boxed Beef Story. 20 minute video tape. Librarian, IBP, Inc., P.O. Box 515, Dakota City, NE 68731.

IBP Story. 20 minute video tape. Librarian, IBP, Inc., P.O. Box 515, Dakota City, NE 68731.

5 The Feeder and the Beef Industry

The cattle feeding industry is dependent on three primary factors: (1) source of feeder cattle, (2) market outlet for fed cattle, and (3) source of feed. Since feed cost comprises most of the expense in a feedlot operation, excluding cost of feeder cattle, location of feeding operations will generally be concentrated where the major portion of feed is produced.

Fed cattle refers to those cattle which have been fed concentrates (grains). Typically, for several weeks prior to slaughter, these cattle are fed in confinement areas or feedlots where distance to feed, water, and possibly shelter is minimized.

The major cattle feeding areas in the United States are shown in Figure 5.1. These areas reflect feed producing regions primarily where cultivated roughage and grains are grown. These locations are determined by soil type, growing season and available moisture or irrigation water. The heaviest concentration of fed cattle in the United States is eastern Colorado, Nebraska (along the Platte River), the high plains of Texas, southwestern Kansas and the western half of Iowa.

A review of Figure 1.12, in Chapter 1, shows the dramatic growth of the fed cattle industry during the past few decades. Some cattle had been grain fed during the early development of the United States; however, the large commercial cattle feeding industry began to emerge after World War II. Most of the large growth during the next two decades was due to an oversupply of cheap feed grains produced in the United States. The development of the cattle feeding industry has closely followed corn production which is the major feed grain in the United States.

Availability and price of feed grains dictates the number of cattle that are grain-fed and where changes in number of cattle fed occur in various parts of the United States. Certain Great Plains states (Texas, Kansas, and Nebraska) and Corn Belt states (Iowa, Illinois, and Indiana) continue to have a surplus of feed grain production over that which is currently fed. The increase in fed cattle numbers during the past 10 years has occurred primarily in these states having surplus feed grains. This relationship

Figure 5.1. Major cattle feeding areas representing location but not volume of cattle fed. The area circled in the plains and corn belt states shows where approximately three-fourths of U.S. cattle are fed. Courtesy of USDA.

between changes in fed cattle numbers and grain supply is tempered by grain exports to other countries, weather conditions, and source of feeder cattle. Also, the efficiency of production is important because for many years (during the 1960s), cattle feeders in California demonstrated that they could bring in outside feeder cattle and feed grain and still make cattle feeding a profitable enterprise. Development of additional irrigation water in several of the plains states has greatly enhanced feed grain production which in turn has increased number of cattle fed in that area.

Types of Cattle Feeding Operations

There are two basic types of cattle feeding operations; (1) commercial feeder and (2) farmer-feeder. These two types are generally distinguished by type of ownership and size of feedlot. The farmer-feeder operation is usually owned and operated by an individual or family (Figure 5.2). The commercial feedlot may be owned by an individual or

Figure 5.2. A farmer-feedlot operation. In this feedlot the feed is fed in troughs located within the pen. Courtesy of BEEF Magazine.

partnership, but quite often a corporation is involved, especially as feedlot size increases (Figures 5.3 and 5.4). Commercial feedlots may own the cattle, feed cattle owned by someone else (called custom cattle feeding or custom feedlots), or have a combination of the two.

The two types of feeding operations are usually referred to as those over 1000 head capacity (commercial) and those under 1000 head (farmer-feeder). A comparison of these two are shown in Figure 5.5.

Some commercial cattle feeding operations have become extremely large in recent years (Table 5.1). Assuming these top 20 companies maintained an 80% capacity in their feedlots and turned over the number of cattle 2.5 times each year, they would feed approximately 5 million head of cattle on a yearly basis. This represents approximately 20% of the total number of fed cattle marketed each year in the United States.

Figure 5.3. Aerial view of a 35,000 head capacity commercial feedlot. Some of the more important component parts of the feedlot are: (1) feedmill (grain and supplement storage, feed mixing facility), (2) collection area for feedlot run-off—primarily for pollution control, (3) hay storage, (4) open pit storage of feed—usually silage or ground, high moisture grain, (5) hospital area—for isolation and treatment of sick cattle, (6) alley-way for movement of cattle from unloading and processing area, to loading area for marketing, and movement of cattle to hospital pens, and (7) feed alleys—each side has a fence-line feedbunk into which feed is augered from feed trucks. Other important areas not shown are unloading-loading areas and handling facilities for processing recently acquired feeder cattle. Courtesy of Successful Farming Magazine.

Figure 5.4. A large commercial feedlot where thousands of cattle can be easily fed in fence line bunks using feed trucks. Courtesy of BEEF Magazine.

Even though the number of cattle fed in feedlots over 1000 head capacity is increasing, there are some definite advantages for the farmer-feeder. These advantages and disadvantages are identified with the following comparisons:

1. Farmer-feeder can utilize cattle as a market for homegrown feeds.
2. Farmer-feeder can effectively utilize high roughage feeds in a backgrounding or warm-up operation.
3. Farmer-feeder can distribute available labor over several different types of operations.
4. When cattle feeding is unprofitable, the farmer-feeder can totally close out the cattle feeding and divert time and dollars into other phases of the farming operation. The large commercial feedlot has much higher overhead costs which are fixed, regardless of whether the pens are filled with cattle or are empty.

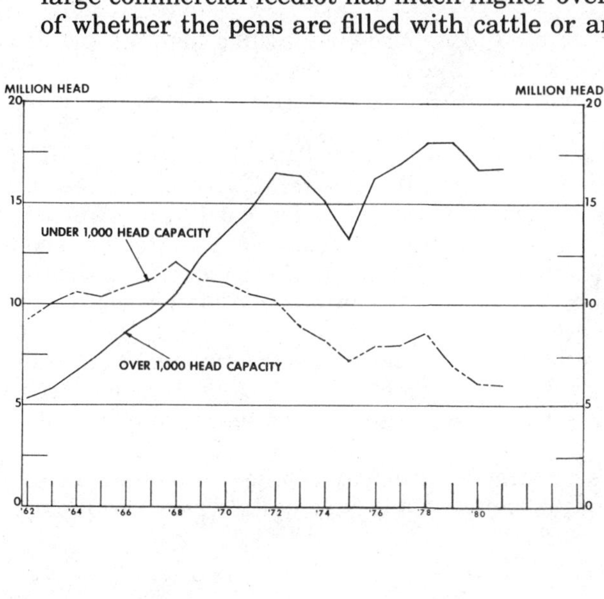

Figure 5.5. Fed cattle marketing by size of feedlot (23 states). Courtesy of Western Livestock Round-Up.

Table 5.1. Top 20 Cattle Feeding Companies in the United States

Rank	Company	Location	One Time Capacity (No. Head)
1	Cargill, Inc. (Caprock, Ind.)—5 lots	TX, KS	216,000
2	Monfort of Colorado—2 lots	CO	200,000
3	Foxley Cattle Co. (Flavorland, Ind.)—6 lots	TX, NE	195,500
4	AZL Resources (Procor, Inc.)—5 lots	TX, NE	186,500
5	Hitch Enterprises—3 lots	OK, KS	182,000
6	Barrett-Crofoot—3 lots	TX	130,000
7	Western Beef, Inc.—5 lots	TX, CA, NM	123,500
8	Friona Ind.—3 lots	TX	122,000
9	Continental Grain Co. (Allied Mills)—3 lots	CO, TX	105,000
10	J. R. Simplot, Inc.	ID, OR	100,000
11	Harris Feeding Co.	CA	100,000
12	DeKalb AgResearch, Inc.—3 lots	KS, TX, AZ	95,000
13	Stratford-Pratt—2 lots	TX	95,000
14	Cactus-Frontier—2 lots	TX, OK	93,000
15	Hi Plains Feedyard	TX	90,000
16	Miller Family—3 lots	CO, WY, TX	85,000
17	Red River Land & Cattle	AZ	85,000
18	Wilhelm Co.—2 lots	CO	82,000
19	Granada Land & Cattle Co.—2 lots	NE	82,000
20	Valley View Cattle Co.—2 lots	TX	81,000
	TOTAL		2,448,500

Source: Western Livestock Journal, Nov. 12, 1979.

5. Commercial feedlots are generally operated in a more businesslike manner, particularly in maintenance of detailed records and analysis of records.
6. Typical commercial feedlots utilize more professional expertise in nutrition, health and marketing.
7. Financing becomes a larger challenge for commercial feedlots. Approximately 50% of cattle marketed through commercial feedlots are custom-fed with larger feedlots having a higher percentage of custom-fed cattle. Custom-fed cattle are owned by other cattle feeders, speculators, farmers, ranchers, or packers and are fed on a contractual arrangement. Custom cattle feeding for others greatly reduces the operating capital requirement for the commercial feedlot and shifts some of the risk to the customer.

   One of the greatest risks for commercial feedlots, highly involved in custom feeding, is keeping the lots full. Customers will stop putting cattle into custom lots when financial losses occur over a period of time. This poses the problem as identified in Number 4.
8. There are some definite economies of size (Figure 5.6). Similar economies of size are shown for cost per pound of gain as well as total fixed costs. There are reports,

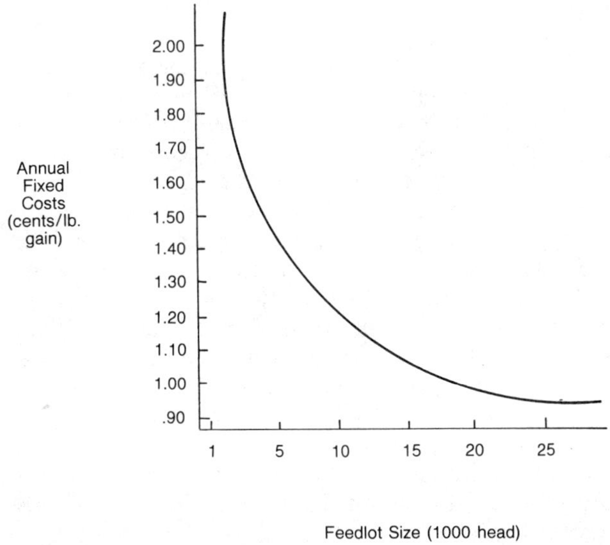

Figure 5.6. Feedlot size and total annual fixed costs (Texas Agric. Expt. Sta. Bull. 1083, 1969).

however, which show that higher costs can be incurred by larger lots (over 25,000 head) if they are not extremely well-planned, organized, and managed.

9. Farmer-feeders usually feed just one group of cattle per year which leaves the facilities vacant for several months. Commercial feedlots attempt to keep their lots full and have a cattle turnover rate of two to two and one-half times per year.

ESSENTIAL FACTORS NEEDED IN ANALYZING A FEEDLOT OPERATION

The five primary factors needed in making an assessment of a feedlot operation are shown in Figure 5.7. The major component parts of each factor are listed under each heading. Each of these factors will be discussed in further detail.

Facilities Investment	Cost of Feeder Cattle	Feed Cost Per Pound of Gain	Non-Feed Costs Per Pound of Gain	Total Dollars Received
Land Pens Equipment Feed Mill Office	Grade Weight Shrink Transportation Gain Potential	Ration Performance of the Cattle —Gain —Efficiency Length of Feeding	Death Loss Labor Taxes Interest Insurance Utilities Veterinary Repairs	Market Choice Transportation Shrink Dressing % Quality Grade Yield Grade Manual Value

Figure 5.7. Major factors utilized in analyzing a feedlot operation. Drawn by CSU Graphics.

Facilities Investment

Commercial feedlot facilities are similar throughout the United States. The general layout is open lot, dirt pens with capacities varying from 100 to 500 head. The pens are sometimes mounded in the center for cattle to have a dry resting area. The fences are pole, cable, or pipe. There are fenceline feedbunks with concrete aprons inside the pens where cattle stand while eating. A feedmill to process grains and other feeds is usually a part of the facility. Feed is distributed to the fenceline bunks with specialized feed trucks which can mix while traveling to the feed bunks. Bunker trench silos hold corn silage and other roughages. Grains may also be stored in these silos; however, they are often stored in steel bins above the ground. During the 1960s, the cost for these facilities was $30 to $60 per head. Today the approximate cost is $100 per head.

Facilities for farmer-feeder operations vary from unpaved, wood-fenced pens to paved lots with shelter (sheds) to semi- or total-confinement sheds. The latter may have manure collection pits under the cattle with feed stored in airtight facilities. Most feeds are stored in upright silos and grain bins. Feeds are typically processed on the farm and distributed to feedbunks located either inside and outside the pens with tractor-powered equipment. Current costs of these facilities are shown in Table 5.2. These costs are computed for farmer-feeders under Midwest feeding conditions. This is where the majority of farmer-feeders are located, although it appears that these costs would approximate those in other farmer-feeder areas.

Most total confinement feedlots are farmer-feeder feedlots located in the Midwest with fewer than a 1000-head capacity. The world's largest confinement feedlot, however, has a capacity of 32,000 head and was built at an estimated cost of $6.2 million. This is a commercial feedlot owned by J. R. Simplot Company located in eastern Oregon.

Cost of Feeder Cattle

Cattle feeders usually calculate what they can pay for feeder cattle based on projected feed costs and anticipated selling price of slaughter cattle (Table 5.3).

Figures 5.8 and 5.9 show breakeven prices and slaughter steer prices for Corn Belt and Great Plains feeding over the past several years. Breakeven costs would include cost of feeder cattle and cost of gain. When breakeven costs are compared to market

Table 5.2. Feedlot Facilities Costs for Farmer-Feeder

Type of Facilities	Approx. Cost Per Head
Open lot (windbreak fence on one side)	$200–300[a]
Open lot (pole shed)	$250–350
Total confinement (uninsulated, deep pit, concrete stone silo)	$350–400
Total confinement (insulated deep pit, airtight silos)	$450–500

Source: Adapted from *Feedlot Management,* 1980 Planner Issue.

[a] Higher costs are for 300 head; lower costs are for 1000 head with bunker silos.

Table 5.3. Computation of Projected Price for Feeder Cattle

Item	Price Per Cwt.	Total Dollars
Sell 1152 lb steer (1200 lbs with 4% shrink after 140 days on feed)	$60	$691.20
Cost of 500 lbs gain feed cost ($0.40/lb) + nonfeed cost ($0.14/lb)	$54	$270.00
Value of feeder steer (sale price − gain cost)		$421.20
Price feeder can pay for 700 lb steer and break even (per cwt)	$60	

price, profit and loss margins can be computed. Market price has to be higher than breakeven price before a profit can be realized. These margins have been quite erratic over the past several years. This demonstrates the tremendous risks involved in feeding cattle.

Grain prices have a tremendous effect on cost of gains for the cattle feeder. It is

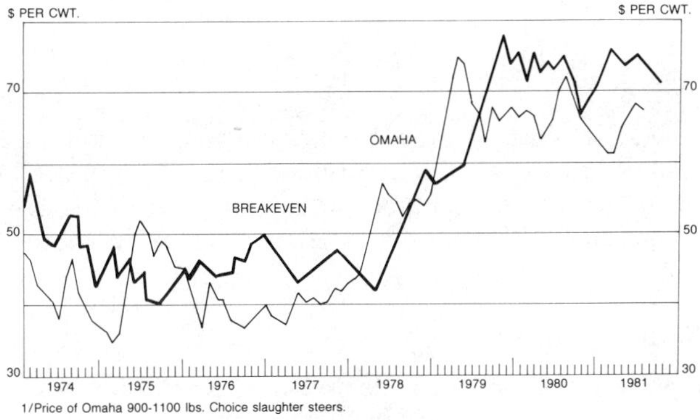

Figure 5.8. Breakeven corn belt cattle feeding, 1974–1981. Courtesy Western Livestock Round-Up.

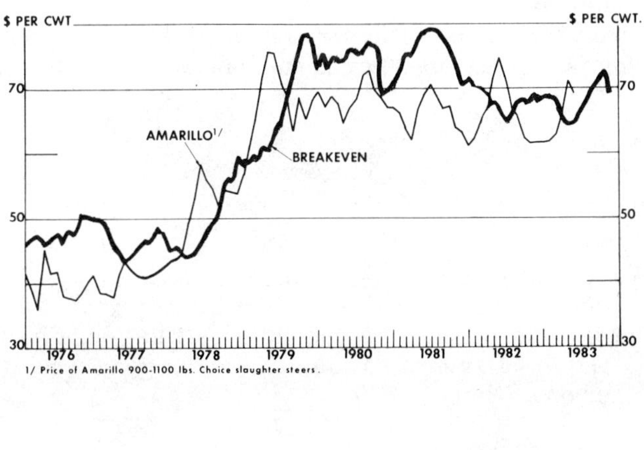

Figure 5.9. Breakeven Great Plains Custom Cattle Feeding, 1976–1983. Courtesy of Western Livestock Round-Up.

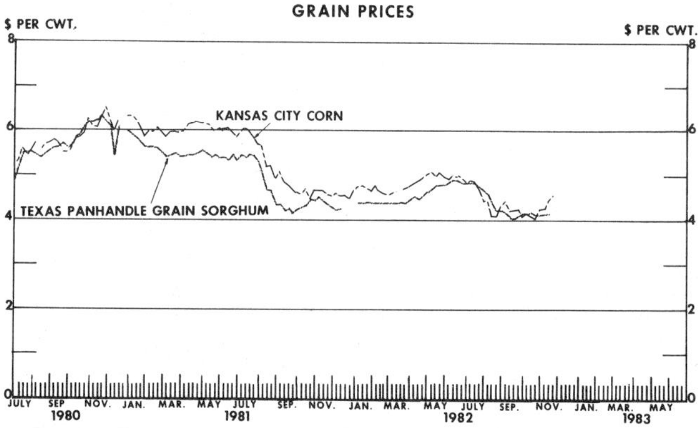

Figure 5.10. Corn and sorghum prices have wide fluctuations. Courtesy of Western Livestock Round-Up.

not uncommon to see the price of grain increase or decrease 50% from one year to the next (Figure 5.10).

Cattle feeders should have a record system that permits an analysis of costs and returns with anticipated breakeven prices. The various cost and return items are identified in Table 5.4 (Corn Belt cattle feeding) and Table 5.5 (Great Plains custom cattle feeding).

Cost of Gain

Once feeder cattle are placed in a feedlot, cost of gain becomes a major economic concern for the cattle feeder. Total cost of gain typically is divided into two parts: (1) feed cost per pound (or cwt) gain and (2) non-feed cost per pound (or cwt) gain. The latter is sometimes referred to as yardage cost. While there is some overlap in factors affecting these two areas, they will be discussed separately.

Factors Affecting Feed Cost Per Pound of Gain

Most factors affecting feed cost per pound of gain are reflected in the animals' daily rate of gain or pounds of feed required per pound of gain. There are many factors affecting rate and efficiency of gain, the value of which can be determined by a simple cost-return comparison as shown in Table 5.6.

Feed efficiency is a very economically important trait for the cattle feeder. It has been determined that a 5% improvement in feed efficiency is equivalent to any one of the following:

1. Reducing ration costs on a dry matter basis by $8 per ton.
2. Reducing purchase costs of the feeder animal by $1.75 per cwt.
3. Increasing daily gain by 0.6 pound per day.
4. Reducing interest rate on operating capital from 15% to 9.5%.

Daily gain and feed efficiency have a relatively high relationship (more rapid gains result in less feed required per pound of gain) where cattle are fed from the same in-weights to the same slaughter weights. The primary reason for this relationship is that cattle which gain more rapidly require less feed to maintain their body weight (Table 5.7) as they are fed a shorter period of time.

Table 5.4. Corn Belt Cattle Feeding

	Selected Costs at Current Rates[a]									
Purchased During Marketed During	**June Dec.**	**July Jan. 83**	**Aug. Feb.**	**Sept. Mar.**	**Oct. Apr.**	**Nov. May**	**Dec. June**	**Jan. 83 July 83**	**Feb. Aug.**	**Mar. Sept.**
	Dollars Per Head									
Expenses:										
600-lb feeder steer	393.42	391.56	407.10	398.88	380.70	383.28	374.10	391.80	404.10	415.14
Transportation to feedlot (400 miles)	5.28	5.28	5.28	5.28	5.28	5.28	5.28	5.28	5.28	5.28
Corn (45 bu)	114.30	112.50	102.60	88.65	90.00	95.40	101.25	103.95	113.85	120.60
Silage (1.7 tons)	38.69	37.52	34.24	31.20	31.59	33.30	34.27	35.34	37.21	38.49
Protein supplement (270 lb)	34.56	34.16	34.02	33.34	33.08	33.34	34.02	34.02	34.16	34.02
Hay (400 lb)	12.20	11.60	10.60	10.30	10.40	10.90	10.80	11.20	11.20	11.20
Labor (4 hours)	14.80	14.80	14.80	14.80	14.80	14.80	14.80	15.20	15.20	15.20
Management[b]	7.40	7.40	7.40	7.40	7.40	7.40	7.40	7.60	7.60	7.60
Vet medicine[c]	5.04	5.06	5.07	5.05	5.03	5.05	5.04	5.09	5.11	5.13
Interest on purchase (6 months)	34.03	33.54	34.99	34.28	29.61	29.82	29.10	28.11	28.99	29.77
Power, equip., fuel, shelter, depreciation[c]	23.52	23.61	23.63	23.56	23.48	23.56	23.52	23.74	23.85	23.91
Death loss (1% of purchase)	3.93	3.90	4.07	3.99	3.81	3.83	3.74	3.92	4.04	4.15
Transportation (100 miles)	2.31	2.31	2.31	2.31	2.31	2.31	2.31	2.31	2.31	2.31
Marketing expenses	3.35	3.35	3.35	3.35	3.35	3.35	3.35	3.35	3.35	3.35
Miscellaneous & indirect costs[c]	10.17	10.21	10.22	10.19	10.15	10.19	10.17	10.27	10.31	10.34
Total	703.00	696.80	699.68	672.58	650.99	661.81	659.15	681.18	706.56	726.49
Selling price required to cover:										
Feed and feeder costs (1,050 lb) $/cwt	56.49	55.94	56.05	53.56	51.98	52.97	52.80	54.89	57.19	59.00
Selling price required to cover all costs (1,050 lb) $/cwt	66.95	66.36	66.64	64.06	62.00	63.03	62.28	64.87	67.29	69.19
Feed costs per 100-lb gain $/cwt	44.39	43.51	40.32	36.33	36.68	38.43	40.08	41.00	43.65	45.40
Choice steers, Omaha $/cwt	58.92	59.33	61.20	64.03						
Net margin $/cwt	−8.03	−7.03	−5.44	−.03						

Table 5.4 (*Continued*)

	Selected Costs at Current Rates[a]									
Purchased During **Marketed During**	**June** **Dec.**	**July** **Jan. 83**	**Aug.** **Feb.**	**Sept.** **Mar.**	**Oct.** **Apr.**	**Nov.** **May**	**Dec.** **June**	**Jan. 83** **July 83**	**Feb.** **Aug.**	**Mar.** **Sept.**
						Dollars Per Head				
Prices:										
Feeder steer, Choice (600–700 lb) Kansas City $/cwt	65.57	65.26	67.85	66.48	63.45	63.88	62.35	65.30	67.35	69.19
Corn $/bu[d]	2.54	2.50	2.24	1.97	2.00	2.12	2.25	2.31	2.53	2.68
Hay $/ton[d]	61.00	58.00	53.00	51.50	52.00	54.50	54.00	56.00	56.00	56.00
Corn silage $/ton[e]	22.76	22.07	20.14	18.35	18.58	19.59	20.16	20.79	21.89	22.64
32–36% protein supp. $/cwt[f]	12.80	12.65	12.60	12.35	12.25	12.35	12.60	12.60	12.65	12.60
Farm labor $/hour	3.70	3.70	3.70	3.70	3.70	3.70	3.70	3.80	3.80	3.80
Interest rate, annual	17.30	17.19	17.19	17.19	15.56	15.56	15.56	14.35	14.35	14.35
Transportation rate $/cwt per 100 miles[g]	0.22	0.22	0.22	0.22	0.22	0.22	0.22	0.22	0.22	0.22
Marketing expenses $/cwt[h]	3.35	3.35	3.35	3.35	3.35	3.35	3.35	3.35	3.35	3.35
Index of prices paid by farmers (1910–14 = 100)	1,073	1,077	1,078	1,075	1,071	1,075	1,073	1,083	1,088	1,091

[a] Represents only what expenses would be if all selected items were paid for during the period indicated. The feed ration and expense items do not necessarily coincide with experience of individual for management, production level, and locality of operation.
[b] Assumes 1 hour at twice the labor rate.
[c] Adjusted monthly by the index of prices paid by farmers for commodities, services, interest, taxes, and wage rates.
[d] Average price received by farmers in Iowa and Illinois.
[e] Corn silage price derived from an equivalent price of 5 bushels corn and 330 lb hay.
[f] Average price paid by farmers in Iowa and Illinois.
[g] Converted from cents/mile for a 44,000-pound haul.
[h] Yardage plus commission fees at a Midwest terminal market.
Source: Livestock and Meat Situation, LPS–4, May, 1983.

Table 5.5. Great Plains Custom Cattle Feeding

Purchased During Marketed During	June Dec.	July Jan. 83	Aug. Feb.	Sept. Mar.	Oct. Apr.	Nov. May	Dec. June	Jan. 83 July 83	Feb. Aug.	Mar. Sept.
	Selected Costs at Current Rates[a]									
	Dollars Per Head									
Expenses:										
600 lb feeder steer	380.28	391.74	402.66	386.58	379.50	371.28	376.14	396.36	403.68	423.00
Transportation to feedlot (300 miles)	3.96	3.96	3.96	3.96	3.96	3.96	3.96	3.96	3.96	3.96
Commission	3.00	3.00	3.00	3.00	3.00	3.00	3.00	3.00	3.00	3.00
Feed:										
Milo (1,500 lb)	79.05	80.55	72.00	65.40	64.20	65.55	71.10	72.60	78.60	83.55
Corn (1,500 lb)	88.95	83.25	76.05	71.25	75.30	80.85	82.05	81.60	87.75	91.50
Cottonseed meal (400 lb)	46.00	46.00	46.00	46.00	46.00	48.00	48.00	48.00	50.00	48.00
Alfalfa hay (800 lb)	45.20	45.60	45.20	46.40	45.20	46.80	49.60	48.40	50.40	45.20
Total feed cost	259.20	255.40	239.25	229.05	230.70	241.20	250.75	250.60	266.75	268.25
Feed handling & management charge	21.00	21.00	21.00	21.00	21.00	21.00	21.00	21.00	21.00	21.00
Vet medicine	3.00	3.00	3.00	3.00	3.00	3.00	3.00	3.00	3.00	3.00
Interest on feeder & ½ feed	47.16	47.40	43.55	38.84	35.88	34.15	33.85	34.56	34.91	34.82
Death loss (1.5 percent of purchase)	5.70	5.88	6.04	5.80	5.69	5.57	5.64	5.95	6.06	6.34
Marketing[b]	f.o.b.	f.o.b.	f.o.b.	f.o.b.	f.o.b.	f.o.b.	f.o.b.	f.o.b.	f.o.b.	f.o.b.
Total	723.30	731.38	722.46	691.23	682.73	683.16	697.34	718.43	742.36	763.37
Selling price required to cover:[c]										
Feed and feeder costs (1,056 lb) $/cwt	60.56	61.28	60.79	58.30	57.78	58.00	59.36	61.27	63.49	65.46
All costs $/cwt	68.49	69.26	68.41	65.46	64.65	64.69	66.04	68.03	70.30	72.29
Selling price $/cwt[d]	61.64	61.80	62.77	65.68						
Net margin $/cwt	−6.85	−7.46	−5.64	+.22						
Cost per 100-lb gain:										
Variable costs less interest $/cwt	57.78	57.06	53.86	51.77	52.08	54.15	56.08	56.11	59.16	59.72
Feed costs $/cwt	51.84	51.08	47.85	45.81	46.14	48.24	50.15	50.12	53.35	53.65

Table 5.5. (*Continued*)

Purchased During / Marketed during	Selected Costs at Current Rates[a]									
	June / Dec.	July / Jan. 83	Aug. / Feb.	Sept. / Mar.	Oct. / Apr.	Nov. / May	Dec. / June	Jan. 83 / July 83	Feb. / Aug.	Mar. / Sept.
	Dollars Per Head									
Prices:										
Choice feeder steer 600–700 lb Amarillo $/cwt	63.38	65.29	67.11	64.43	63.25	61.88	62.69	66.06	67.28	70.50
Transportation rate $/cwt/100 miles[e]	0.22	0.22	0.22	0.22	0.22	0.22	0.22	0.22	0.22	0.22
Commission fee $/cwt	0.50	0.50	0.50	0.50	0.50	0.50	0.50	0.50	0.50	0.50
Milo $/cwt[f]	5.27	5.37	4.80	4.36	4.28	4.37	4.74	4.84	5.24	5.57
Corn $/cwt[f]	5.93	5.55	5.07	4.75	5.02	5.39	5.47	5.44	5.85	6.10
Cottonseed meal $/cwt[g]	11.50	11.50	11.50	11.50	11.50	12.00	12.00	12.00	12.50	12.00
Alfalfa hay $/ton[h]	113.00	114.00	113.00	116.00	113.00	117.00	124.00	121.00	126.00	113.00
Feed handling & management charge $/ton	10.00	10.00	10.00	10.00	10.00	10.00	10.00	10.00	10.00	10.00
Interest, annual rate	18.50	18.25	16.75	15.50	14.50	13.75	13.50	13.25	13.00	12.50

[a] Represents only what expenses would be if all selected items were paid for during the period indicated. The feed ration and expense items do not necessarily coincide with experience of individual feedlots. For individual use, adjust expenses and prices for management, production level, and locality of operation. Steers are assumed to gain 500 lbs in 180 days at 2.8 lbs per day with feed conversion of 8.4 lbs per pound gain.

[b] Most cattle sold f.o.b. at the feedlot with 4 percent shrink.

[c] Sale weight 1,056 lbs (1,100 lbs less 4 percent shrink).

[d] Choice slaughter steers, 900–1100 lbs. Texas-New Mexico direct.

[e] Converted from cents per mile for a 44,000-lb haul.

[f] Texas Panhandle elevator price plus $0.15/cwt handling and transportation to feedlots.

[g] Average prices paid by farmers in Texas.

[h] Average price received by farmers in Texas plus $30/ton handling and transportation to feedlots.

Source: Livestock and Meat Situation, LPS-4, May, 1983.

Table 5.6. Evaluating the Cost and Return Comparison of a Factor Influencing Rate and Efficiency of Gain for a 100 Day Feeding Period

Item	Control Group		Added Factor Having Positive Influence on Rate and Efficiency of Gain
Feed intake (lbs)	2200		2052
Total gain (lbs)	280		320
Feed saved (lbs)		148	
Value of feed saved @ $0.06/lb		$ 8.88	
Additional gain (lbs)		40	
Value of gain @ $0.64/lb		$25.60	
Advantage of gain and feed saved		$34.48	
Additional cost (assume added factor) Cost $0.04/head/day		−$ 4.00	
Total return per head		$30.48	

Note in Table 5.7 that daily gain approaches 3.0 lbs per day before a higher percent of the feed goes to gain than to maintenance. In this example, a difference of one pound of gain per day, within the same initial weight of cattle, accounts for $25 to $30 difference in maintenance cost. Maintenance cost, in a sense, is a necessary evil in feeding cattle. It is all cost with no return; however, it is an unavoidable cost to keep the cattle alive and functioning.

Since feed efficiency has such a significant relationship to cost of gain, a discussion of major factors affecting it is warranted. The three most important variables affecting feed efficiency are: (1) energy density of the feed, (2) amount of dry matter consumed, and (3) rate of gain. Rations high in roughage fill the rumen with bulk, thus limiting

Table 5.7. Relationship of Gain to Maintenance Requirements for Cattle Fed to 1150 Pound Market Weight

In Wt	ADG	Days to 1150 Lb	Lb Corn[a] for Main.	Value of Corn for Main.	Lb Corn for Growth after Main.	Percent of Feed for Main.
400	2.0	375	2310	$92.40	1586	59
	2.5	300	1848	73.92	1629	53
	3.0	250	1540	61.60	1670	48
600	2.5	220	1421	$56.84	1309	52
	3.0	183	1182	47.28	1339	47
	3.5	157	1014	40.56	1375	42

[a] Total feedlot ration converted to corn equivalent to simplify example.
Source: Ideal Beef Memo, Jan. 1982.

feed intake and gain. Physical distention of the rumen is one of two main factors controlling dry matter intake. Adding grain to a high-roughage ration reduces fill, thus increasing feed and energy intake and improving gain.

Feed intake with high energy rations is controlled by the products of digestion curtailing the animals' appetite (called "chemostatic" regulation) rather than rumen fill. Thus increasing the energy density in an already high energy ration usually has little effect on energy intake or gain. Simply stated, improvements in feed efficiency are reflected mainly as increases in gain in low energy rations and as reductions in feed intake in high energy rations.

There are several ways in which the cattle feeder can improve feed efficiency in a feedlot operation. Some of these are as follows:

1. Increase the energy density of the feed by feeding less roughage. The highest practical energy level for a finishing ration is in the range of 0.58 to 0.61 Mcal/lb net energy for gain. Rations containing higher energy levels increase the risk of serious problems occurring, such as acidosis. Some farmer-feeders, raising their own silage, may feed high roughage levels to optimize the amount of feed produced per acre with cost of gain.

2. Processing grain is another method of improving feed efficiency. Steam flaking, reconstituting feed, and high-moisture grain harvesting will improve feed efficiency 8% to 17% depending on the processing method used and type of grain, but cost of feed is also increased.

3. Feed additives and growth stimulants can improve feed efficiency from 3% to 10%. Several of these stimulate rate of gain but also indirectly improve feed efficiency due to the relationship between gain and feed efficiency. Table 14.9, in Chapter 14, identifies the most common implants and feed additives used in feedlots.

4. Feeding cattle to the proper composition of gain. It takes more feed to deposit a pound of fat than a pound of lean. Cattle which mature too early and cattle fed beyond their logical slaughter potential will be very inefficient and gain will be costly (Figure 5.11). Cattle that are capable of gaining rapidly and producing a yield grade 2 carcass at 1100 to 1250 lbs liveweight will most likely be highly efficient in feed utilization.

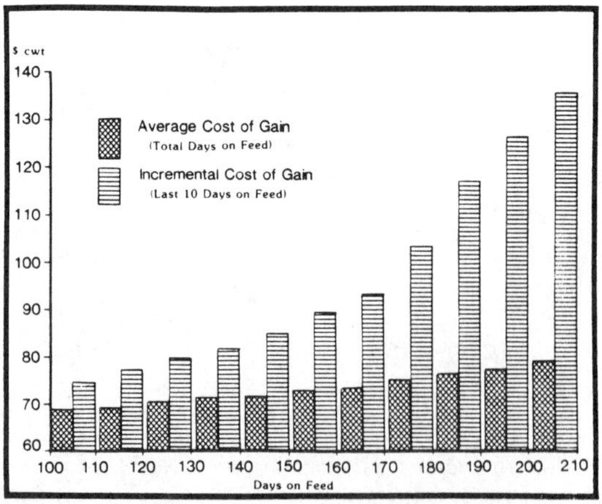

Figure 5.11. Average and incremental cost of feedlot gain with increasing days of feed. Gain costs computed using 1981 cost data. *Source:* Beef Business, Bulletin (NCA), 3 July 1981.

5. Keeping cattle free of climatic stresses such as muddy pens, and exposure to extremely cold or hot temperatures will contribute marked changes in feed efficiency. Cattle exposed to these stresses usually have slower gains through reduced feed consumption and higher maintenance requirements with resulting poor feed conversions.

6. Cattle which have genetic ability to gain rapidly will produce efficient gains. These cattle have been genetically selected for appetite which significantly increases their consumption of feed. Most cattle feeders have not tapped this genetic resource because compensatory gain has been available. Cattle with compensatory gain are typically light for their age which means they have had restricted nutrition for part or all of their life. When they are fed finishing rations in the feedlot, they compensate for the earlier restriction of feed by gaining very rapidly.

Cattle with compensatory gain are not as numerous as in past years, primarily because those producing them have found it an unprofitable venture. Also, feeders have been more competitive in the market place for these cattle, so their economic advantage has been reduced compared to several years ago.

Feeder cattle with the genetic potential to reach 1100 to 1250 lbs at 14 to 16 months of age, having desirable carcasses, are becoming more numerous. Cattle feeders who can identify and purchase these genetically superior cattle can usually have low gain costs because of excellent feed conversion. It has been estimated that cost of gain can be lowered $4.00 to $8.00 per hundred-weight by feeding cattle with high genetic gaining ability.

7. Sex differences of feedlot cattle affect feed conversion, although differences in feed efficiency vary depending on slaughter endpoint. It is generally accepted that heifers are less efficient than their counterparts, steers and bulls are more efficient than steers. These relationships appear to be valid if the different sexes are slaughtered at similar slaughter weights. If the three sexes, however, are slaughtered at similar compositional endpoints (same amount of carcass fat), differences in feed efficiency appear to be minimal.

Factors Affecting Nonfeed Costs

Of nonfeed costs per pound of gain (Figure 5.7), gaining ability, health, and interest rates are the most important. Cattle which have more rapid daily gains have less maintenance feed cost and have less overhead cost because of fewer days required to accumulate a given amount of weight gain. Table 5.7 shows that cattle gaining 2.5 lbs per day required 63 more days in the feedlot than cattle gaining 3.5 lbs per day. At a nonfeed cost of 35 cents per day, slower-gaining cattle would require approximately $22 more per head than fast-gaining steers to break even at the same slaughter weight.

Healthy cattle are less costly because they do not require labor for treating them and the cost of medication is eliminated. Also, healthy cattle will have a lower death loss and a higher rate of gain because of a consistently high feed consumption pattern.

Table 5.8 shows the importance of interest rates affecting the daily non-feed cost.

Cost per pound gain can be affected by several cents when the interest rate varies a few percentage points.

Cattle feeding is a high-risk business because of fluctuating market prices, the biological variability of the cattle being fed and the complexity of many other factors which require careful management in a cattle feeding operation.

Table 5.8. Daily Cost Per Head at Various Interest Rates (Assumes a $625 Total Cost of Animal and Feed)

Annual Interest	Daily Interest Cost Per Head
18%	$0.308
15%	$0.257
12%	$0.205

SELECTED REFERENCES

Publications

Gee, C. K., Van Arsdale R. N., and Gustafson, R. A. U.S. Fed-Beef Production Costs, 1976–77, and Industry Structure. *USDA Agric. Econ. Rpt. No. 424.* June, 1979.

Great Plains Beef Cattle Feeding Handbook. Cooperative Extension Services of CO, KS, MT, NB, NM, ND, OK, SD, TX and WY.

Madsen, A. G., Meckling, W. and Gee, C. K., 1979. *Custom Cattle Feeding in Northern Colorado.* Colo. Agric. Expt. Sta. Gen. Series 985.

McGrann, J., Ball, J., Boehlje, M. D. and Rouse, G., 1979. *Beef Cattle Feeding in Iowa 1979. Evaluation of feeding systems.* Iowa State Univ. Ext. Serv. Pamphlet, Pm-602.

Livestock and Meat Situation, LMS–243, October 1982. ESC and USDA, Wash., D.C.

Price, D. P., 1981. *Beef Production.* Dalhart, TX: Southwest Scientific.

Thompson, G. B. and O'Mary, C. C., 1983. *The Feedlot.* Philadelphia: Lea and Febiger.

Visuals

"Feedlot Production" (filmstrip and tape). Vocational Education Productions. California Polytechnic State University, San Luis Obispo, CA 93407.

6

The Stocker-Yearling Producer and the Beef Industry

Stocker or yearling operations handle cattle which are fed and managed for growth prior to going into a feedlot for finishing. These operations typically involve steers and heifers after they have been weaned. Some stocker operations utilize thin cows which can make rapid and efficient gains prior to slaughter. Replacement heifers, intended to go into the breeding herd, are typically included in the stocker category. This discussion, however, will relate primarily to steers and heifers being grown prior to their finishing phase in the feedlot. Development of replacement heifers is covered in Chapters 7 and 14.

There are several alternate stocker and yearling programs (Figure 6.1). Some operations involve the same ownership of calves from birth through the feedlot finishing phase on the same farm or ranch. Others involve the same ownership but custom feeding in the growing and finishing phases, while others involve cattle being purchased and sold several times within the interval between weaning and slaughter. There are relatively few weaned calves going through two summer grazing seasons prior to feedlot finishing because economics does not favor this system. Costs are too high for the relatively small amount of gain obtained.

The primary basis of the yearling operation is to market available forage and high roughage feeds such as grass, crop residues (corn stalks, grain stubble, beet tops and others), wheat pasture, and silage. Other reasons are to use labor more efficiently and utilize areas which are best suited for summer grazing.

A distinction is often made between wintering and backgrounding, with the primary difference being type of ration fed. A wintering program emphasizes slower and cheaper total cost of gain, with the intent that the calf go to a pasture program in the spring. Backgrounding emphasizes a faster rate of gain, with more grain and less roughage, to prepare the calf for feedlot placement.

Calves may be fed roughage on the ground (Figure 6.2) or in feed bunks (Figure 6.3). Feeding cattle in feed bunks gives more flexibility to move from a wintering to a

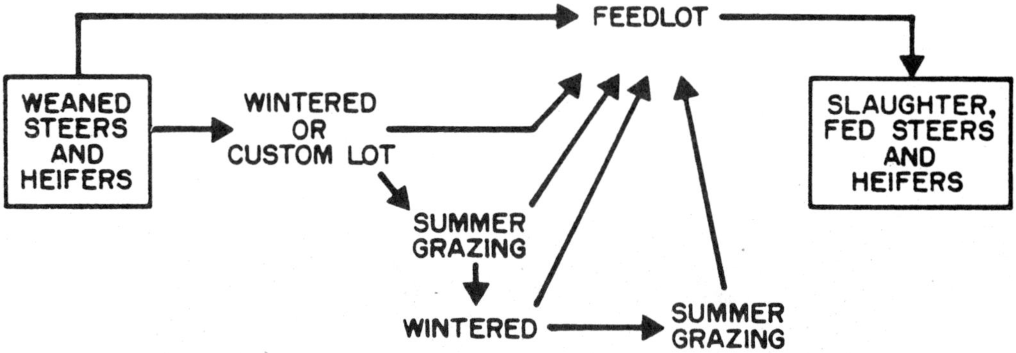

Figure 6.1. Alternative production pathways for weaned steers and heifers. Drawn by CSU Graphics.

backgrounding program since grain can be more conveniently added to the feeding program.

Some stocker or yearling cattle will not be fed harvested feeds through the wintering program. This is particularly true where climate, in combination with certain forage systems, permits green forage to be grazed on a year-round basis. In certain areas of the Southeast, winter annuals such as rye, oats, rye-grass, wheat, and barley provide a forage base for winter grazing (Figure 6.4). Also, some clovers planted in this same area in the fall will produce grazing by late January.

COMPUTING BREAKEVEN PRICES

Yearling operators, not maintaining ownership of their own home-raised calves, will assess whether they want to buy stocker cattle for their available forage, sell the forage, or rent the forage to someone else. Individuals owning backgrounding lots do not have this much flexibility in deciding between alternatives of whether or not to own cattle. Owners of specialized backgrounding lots have high overhead costs which must be covered or absorbed. During periods of poor economic returns with cattle backgrounding,

Figure 6.2. Most calves are grown on high roughage feeds during the winter. Courtesy of Elanco Products Company.

Figure 6.3. High roughage feeds can be fed in feedbunks and grain can also be fed if deemed desirable. These cattle receive alfalfa hay during the winter, then they are on pasture during the spring and summer. Courtesy of Grant Heilman.

Figure 6.4. Steers grazing a winter annual in Alabama. Courtesy of Alabama Agric. Expt. Sta.

their decisions will have to be made on how many cattle to purchase or to contract in an attempt to minimize the loss.

The first basic question to be answered is, "Can I pencil-in a profit with the best available information?" If not, the decision would most likely be one of selling or renting the forage or taking a risk on an increased selling price several months ahead. Yearling operators would consider some of the following calculations in making those decisions.

EXAMPLE 1—ESTIMATING A BREAKEVEN PRICE ON GAIN

Estimated Cost

400 lbs (Purchase Wt)
.70 (Current Price)
$280 (Dollars Paid)

Estimated Gain

780 lbs (Sale Wt)
400 lbs (Purchase Wt)
380 lbs
380 lbs ÷ 360 days = 1.05 lbs/day (Gain)

Estimated Sale Value

780 lbs (Sale Wt)
.65 (Sale Price)
$507 (Projected Sale Value)

Breakeven Price on Gain

$507 (Projected Sale Value)
280 (Dollars Paid)
$227
$227 ÷ 380 lbs = $59.74/cwt

If the cost of gain is less than $59.74 per hundred pounds, it would be a profitable venture.

EXAMPLE 2—ESTIMATING NECESSARY SALE PRICE TO BREAK EVEN, WHERE PAST COSTS OF GAIN ARE KNOWN (AVERAGE OF $.40 LB). OPERATOR EXPECTS TO PURCHASE 425 LB CALVES AND SELL 700 LB YEARLINGS A YEAR LATER

Total Cost of Gain

700 lbs (Sale Wt)
425 lbs (Purchase Wt)
275 lbs (Total Gain)
275 lbs × $0.40 = $110 (Total Cost of Gain)

Estimated Purchase Cost

425 lbs (Purchase Wt)
.68 (Price per lb)
$289 (Dollars Paid)

Necessary Selling Price

$289 (Dollars Paid)
110 (Cost of Gain)
$399 (Total Cost)
$399 ÷ 700 lbs = $57.00 per cwt

A selling price over $57 per cwt would be profit.

After making some calculations on projected cost of gain and a breakeven selling price, a producer might prepare some budgets based on different rates of gain under either a wintering or backgrounding program. This would assume that a cow–calf producer might be evaluating these alternatives rather than selling the calves in the fall or the possibility of utilizing a custom backgrounding operation. Some producers have only a forage-type operation with little flexibility of altering growth rate of the cattle grazed. These operators may want to prepare budgets to determine if they want to purchase cattle or lease their forage to another producer.

An example of the budgeting process for wintering or backgrounding operations is shown in Table 6.1.

Table 6.1. The Budgeting Process for Several Alternative Management Programs for Calves

	Wintering Rate of Gain		Backgrounding Rate of Gain	
	Low	High	Low	High
Initial weight in pounds	425	425	425	425
Average daily gain	0.9	1.25	1.5	1.75
Number of days feeding	150	150	150	150
Sale weight (after 2% shrink)	549	600	637	674
Purchase cost (425 lbs at $0.68)	$289	$289	$289	$289
Feed costs	63	77	82	90
Other variable costs	26	26	26	26
Total Costs	$378	$392	$397	$405
Breakeven sales price (total cost/market weight) × 100 =	$68.85	$65.33	$62.32	$60.09
Estimated selling price/cwt	68	67	66	65
Estimated returns/cwt	(−)$0.85	$1.67	$3.68	$4.91
Estimated returns/head	(−)$4.67	$10.02	$23.44	$33.09

Source: Western Livestock Round-Up.

In these examples, faster-gain programs show a higher return potential than slower-gain programs. Even though total costs are higher for faster-gain programs, the cost per pound of gain is less. This occurs in this situation because a higher percentage of feed is used for growth rather than maintaining body weight.

In some cases, slower-gain programs may be a useful alternative to selling calves in the fall because there is a market for feed even though it might be difficult to sell. Also, if calves are retained through the winter, they might command a higher price in the spring, or pasture gains may compensate for reduced returns of the wintering program. In this case, producers who have a wintering and summer pasture program for their own calves should do a separate budget for both winter and summer programs.

Table 6.1 is only an example of the budgeting process. Producers should prepare their own individual budgets and base cost estimates on their own values. If valid cost and return data are available from past years, this will increase accuracy of budget estimates for the current year.

Marketing

Marketing decisions will have the greatest effect on the profit potential of most yearling operations. Cattle are typically marketed twice, with a purchase and a sale involved. Purchase weight and price, along with sale weight and price, will have a major influence on total dollars returned to the operation. Price of cattle relative to shrink involved, health of cattle, and their gain potential is extremely important.

Yearling operators would prefer to purchase shrunk cattle at average prices and thin, healthy cattle which are relatively light for their age. To a cow–calf producer, these cattle would be identified as "mismanaged cattle," however, they represent potentially profitable cattle to the yearling operator. Heavy calves at weaning, possessing a potential for rapid gains, need to be utilized in the finishing phase of the feedlot rather than in yearling operations. Exceptions to this might be when feed cost and interest rates are relatively low.

Seasonal price patterns suggest fall calf prices are most often not the optimum for the marketing year. Feed prices, forage supplies, and projected calf and yearling prices are needed to determine the time of marketing. Even though spring prices (and total returns) are often higher than in the fall, potential net returns using a simple budget should be the deciding factor.

Conditions can change, especially price, from the time the budgets are constructed. These conditions, affecting cost and return, should be periodically evaluated to determine the best marketing alternatives. What worked one year may not be the best course of action the following year. Also, if large numbers of cattle are held over winter, bunched marketings of yearlings in the spring could cause the outlook to become more pessimistic. Prudent producers may want to take a hard look at forward pricing opportunities that will result in a profitable return.

Receiving Cattle—Health and Nutrition

Often cattlemen fail to recognize all costs associated with sick cattle. The average sick animal will shrink 10% to 20%. Considerable additional labor is required per sick animal. Medicine, treatment programs, and death losses are expensive. One study revealed an average three-day medication cost for a 400 lb calf was $7.08. A 5% death loss increases the cost of each calf $10 to $20 or 5% to 8%. One time through the chute is considered equivalent to a seven-day feeding period, therefore, it is cheaper to prevent disease than to treat it.

More beef cattle develop and subsequently die from respiratory diseases than from all other causes. In cattle under close confinement, respiratory diseases cause 75% of the illnesses and 64% of the deaths. Pneumonias constitute 75% of the diagnoses of respiratory diseases.

Bovine respiratory disease (BRD) is seldom the result of a single factor. BRD usually is caused by a combination of stress, virus infection, and invasion of the lungs by pathogenic bacteria such as Pasteurella and Hemophilus. Stress undermines the natural defenses built in the lining of the trachea and bronchi. Respiratory viruses (such as IBR, PI_3, BVD, and others) further damage these natural defenses. Ultimately, pathogenic bacteria find a wide-open road into the lungs where they localize, multiply, and cause the severe damage called BRD or pneumonia or shipping fever.

Besides price paid and price received for stocker or yearling cattle, health and nutrition are the most important factors affecting costs and returns. The content of Chapters 14, 19, and 20 should be carefully evaluated as it pertains to stocker or yearling operations.

SELECTED REFERENCES

Publications

Andre, P. D. Implanting—Do it Right! *BEEF,* November 1980.
Proceedings—Light Cattle Management Seminars, 1982. Colo. State Univ., Fort Collins, CO.

7

The Commercial Cow–Calf Producer and the Beef Industry

GENERAL DESCRIPTION

Commercial cow–calf operations are numerous and varied throughout the United States. The size of operation will range from fewer than ten cows located on land near an urban center to more than 10,000 cows running on thousands of acres of western range. Table 7.1 shows the number of beef cows on farms and ranches by herd size. Note that almost 50% of the operations have less than 20 cows. Those operations with less than 50 cows account for almost 85% of the farms and about 37% of the beef cows.

Increase in cow herd size does not always imply an increased efficiency of production. Several studies, however, demonstrate that there is a greater return per head as cow herds increase to 1000 head (Nelson 1976 and Gee et al. 1979).

There are types of intensive cow–calf production where cows remain in confinement year around to less intensive operations where cows are maintained on pasture throughout the entire year. Less intensive cow–calf operations are in high mountain valleys, plains, and desert areas where 30 to 100 acres are required per cow.

Most cows will calve in late winter and early spring months with the majority of calves being born in February, March and April. A few producers calve their cows in late summer and fall months primarily to reduce losses from calf scours and to complement their forage production program. Other producers may have both a spring and fall calving program to extend the use of their bulls and to use their labor and forage more efficiently. There are limited studies to evaluate spring versus fall calving in the same geographical location. A few producers continue to calve on a year around basis, however, usually a critical economic assessment will not favor this type of calving program.

Calves are usually weaned at the same time of year, their ages ranging from 5 to 10 months of age. Early weaning (1 to 3 months of age) can be managed adequately; however, it is not widely practiced because it is not economically feasible for most opera-

Table 7.1. Beef Cows and Heifers That Have Calved[a]

Farms With (Head)	No. of Farms	No. of Head (000)	Farms, Percent of Total (%)	Head, Percent of Total (%)
1–9	393,880	1,890	34.7	4.9
10–19	268,468	3,663	23.6	9.4
20–49	291,507	8,780	25.7	22.5
50–99	109,612	7,309	9.6	18.7
100–199	46,472	6,092	4.1	15.6
200–499	20,970	5,937	1.9	15.2
500–999	3,900	2,529	0.3	6.5
1000 or more	1,496	2,787	0.1	7.2
Total	1,136,305	38,987		

[a] 1981 Cattle Inventory Adjusted by 1978 Ag. Census ratios
Source: USDA and Census of Agriculture

tions. Weaned calves may go directly into the feedlot, but the majority of calves currently are grown-out on forage for several months before entering the feedlot.

Cow–calf pairs will graze thousands of acres of grasses, legumes and forbs, which are only utilized effectively by ruminants. In many wheat and other small-grain producing areas, cattle will graze green growth in late summer and fall and straw aftermath following harvesting of grain in the summer. Cows will be seen grazing untillable acres and cornstalk aftermath in the highly productive Corn Belt states.

Regardless of the type of operation, most cow–calf producers expect a cow to produce a live calf every year and do it profitably (Figure 7.1).

Figure 7.1. A live calf born and weaned is a primary goal of the commercial cow–calf producer. Courtesy of American Polled Hereford Association.

ESTABLISHING AN OPERATION

Many cow–calf operations have existed for decades, being passed down to a family member over several successive generations. During the past 10 to 20 years, land has become a major investment commodity primarily because of an expanding population with large available financial resources. Thus the price of land has exceeded its ability to be profitable under many phases of agricultural production, including the production of beef cattle (Figure 7.2).

Cost of a commercial cow–calf operation can be evaluated on a per cow basis. This would include cost of land, cow, buildings, equipment, and other minor costs. It is not uncommon at the present time to have investment cost of \$2000 to \$4000 on a per cow basis. With an interest cost of 15%, this amounts to an annual interest cost of \$300 to \$600. This cost added to the annual production cost will be approximately \$500 to \$1000 per cow per year. Even when calf prices are high, they will not cover all of these costs (e.g. \$1.00 per pound on 450 lb calves).

Total capital investments for an economical cattle operation are extremely high. Several studies show that an economical unit is approximately 300 head. At \$3000 per cow, the total investment would be \$900,000. It is evident that high total capital investment plus high annual interest costs are preventing many potential beef producers from purchasing beef operations.

An assessment of economical cow–calf operations and the data in Table 7.1 indicate that a large number of cattle operators do not depend primarily or even substantially on beef cattle for their incomes. They have other, more important sources of ag income, or they have off-farm jobs. They are not producing cattle for significant income purposes but as a means of maintaining a life style in the country, or as a means of using otherwise wasted land and crop aftermath. They are not as likely to make production and marketing decisions based on the usual economic considerations directly affecting beef cattle.

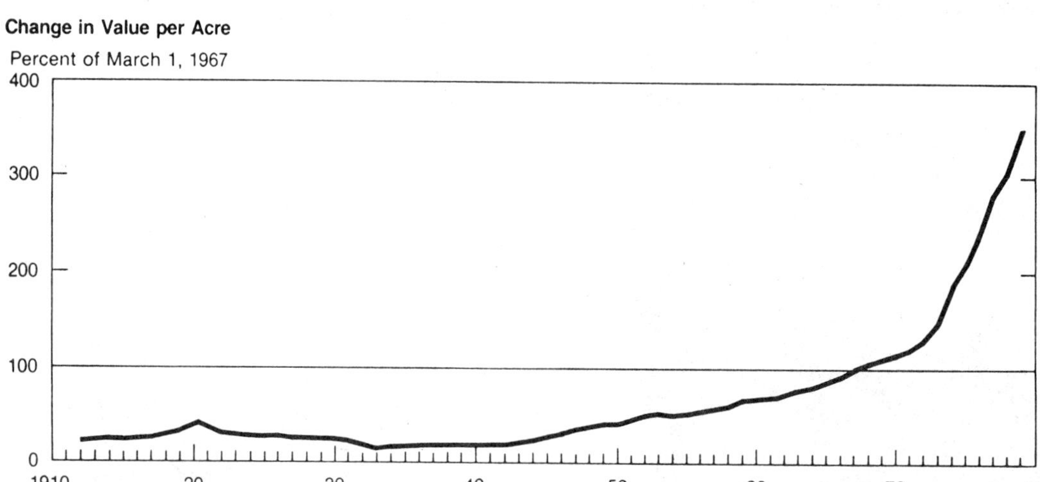

Figure 7.2. Change in farm and ranch real estate values. Courtesy of USDA.

FACTORS AFFECTING BREAKEVEN PRICES

Cow–calf producers can easily assess the profitability of their operations by utilizing three criteria: (1) percent calf crop weaned, (2) weight of calves at weaning and (3) production costs per year on a per cow basis (annual cow cost). With these three sources of information, the price needed per hundredweight of calf to cover production costs can be computed as shown in the following example:

% calf crop weaned 80%
weaning weight .... 420 lbs
annual cow cost $300

lbs of calf weaned per cow in the herd

= 420 lbs × 0.80 = 336 lbs

$$\text{breakeven price} = \frac{\$300}{420 \text{ lbs} \times 0.80} = \$89.29 \text{ per cwt}$$

Table 7.2 shows breakeven prices for various levels of calf crop percent, weaning weight, and annual cow costs.

A further examination of Table 7.2 shows the economics of different levels of management by a cow–calf producer. Obviously a producer should be attempting to improve the percent calf crop weaned, increase weaning weight per calf and at the same time, lower annual cow costs or minimize the increases. A producer having a 95% calf crop weaned, with calves weighing an average of 550 lbs, and a $250 annual cow cost can have a profitable operation if the calves will sell for over $47.85 per cwt. A producer with a 75% calf crop, 350 lb weaning weights and an annual cow cost of $350 must have over $1.30 per pound to break even. This low level of productivity must be improved if the producer is to stay in the cow–calf business.

Table 7.2. Breakeven Price (per Hundred Pounds) with Various Calf Crop Percentages, Annual Cow Costs, and Weaning Weights

Calf Crop (Weaned) (%)	Annual Cow Cost ($)	Average Calf Weight at Weaning: 350 lbs	450 lbs	550 lbs
	350	$105.26	$ 81.87	$66.99
95	300	90.23	70.18	57.42
	250	75.19	58.48	47.85
	350	117.65	91.50	74.87
85	300	100.84	78.43	64.17
	250	84.03	65.36	53.48
	350	133.33	103.70	84.85
75	300	114.28	88.89	72.73
	250	95.24	74.07	60.61

FACTORS AFFECTING POUNDS OF CALF WEANED

Pounds of calf weaned per cow in the breeding herd reflects both weaning weight and percent calf crop weaned. Table 7.3 demonstrates that both are very important in calculating pounds of calf weaned per cow exposed, and a producer should attempt to improve both traits. A further analysis of this table will show that a 1% change in percent calf crop is approximately equivalent to a 5 lb change in weaning weight.

Figure 7.3 shows the major factors a cow–calf producer needs to critically evaluate in improving pounds of calf weaned per cow exposed. Each operation will differ in levels of productivity in each area. This also implies that priorities for improvement will usually differ for different operations.

In past years, when cow–calf producers desired to produce more beef, they typically purchased additional land and cows. More emphasis needs to be placed on increasing productivity and dollar return on the productive unit—the cow. Beef producers need to carefully note the progress dairy producers have made in increasing milk production on a per cow basis while keeping total milk production rather constant (Figure 7.4).

Most of the remainder of this chapter will emphasize improving productivity and economic return on a per cow basis.

IMPROVING PERCENT CALF CROP

Calf crop percent can be calculated several ways. The most realistic calculation which reflects production efficiency is as follows:

$$\%\text{ net calf crop} = \frac{\text{no. calves weaned}}{\text{no. cows in breeding herd}} \times 100$$

Calf crop percentages for most cow herds in the United States will be between 60% and 100%. Extreme environmental conditions of drought, severe spring storms and major disease problems may cause the calf crop percent of individual herds to fall below 60%. Individual herds can document calf crop percentages over 90% for over 10 consecutive years; however, these are exceptional herds. While there is no valid data, the ex-

Table 7.3. Calculation of Pounds of Calf Weaned per Cow Exposed

Percent Calf Crop	Average Weaning of All Calves (lbs)						
	350	400	450	500	550	600	650
	Pounds of calf weaned per cow exposed (lbs)						
75	262	300	338	375	412	450	488
80	280	320	360	400	440	480	520
85	298	340	382	425	467	510	552
90	315	360	405	450	495	540	585
95	333	380	427	475	522	570	618
100	350	400	450	500	550	600	650

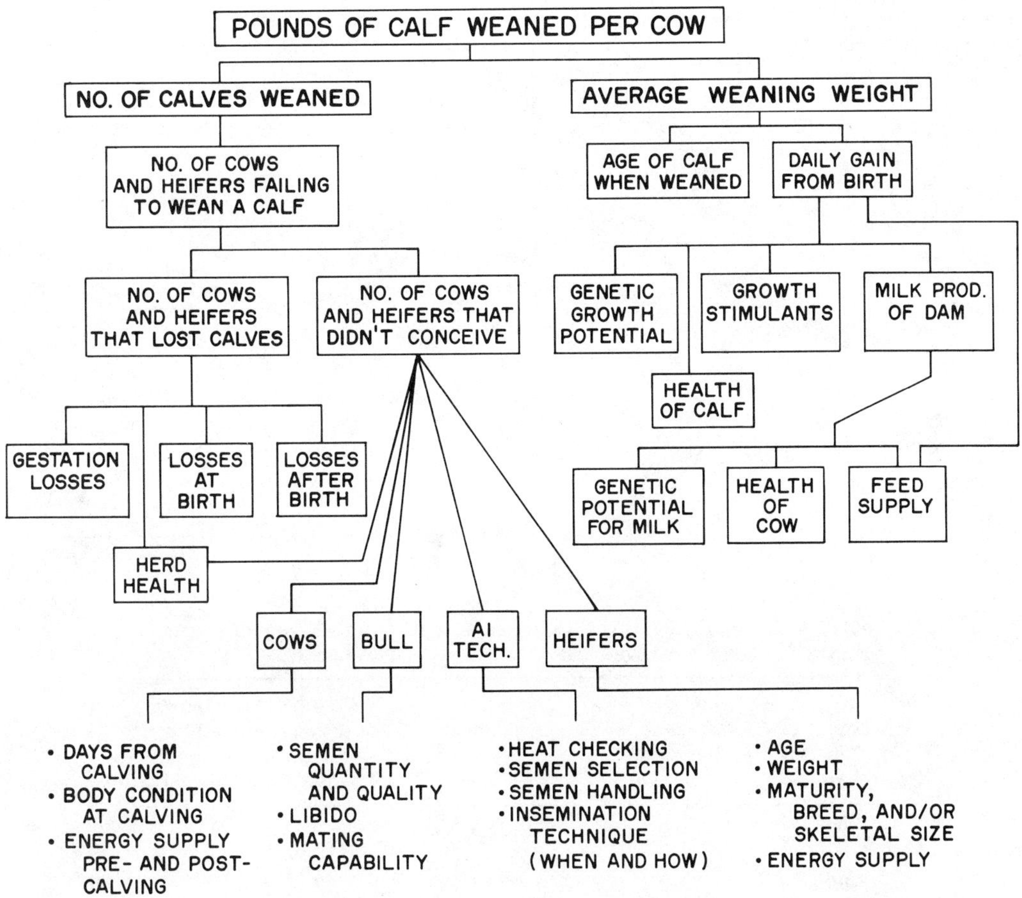

Figure 7.3. The major factors affecting pounds of calf weaned per cow bred. Drawn by CSU Graphics.

pected average percent calf crop weaned in the United States, as previously calculated, would be between 70% and 80%. The survey of 500 South Dakota cow–calf producers showed a 78% weaning rate (South Dakota Cow–Calf Day Report, 10 December 1980).

Economics of Percent Calf Crop

Reproductive performance as reflected in calf crop percent is the most economically important production trait. This trait cannot be neglected or poorly managed without having a severe economic effect on the cow–calf producer.

Table 7.4 shows how changes in percent calf crop affect the breakeven price. Note in this example that as calf crop percent improves, its effect on breakeven price is diminished.

Figure 7.4. Milk production, number of cows and milk per cow. *Source:* USDA.

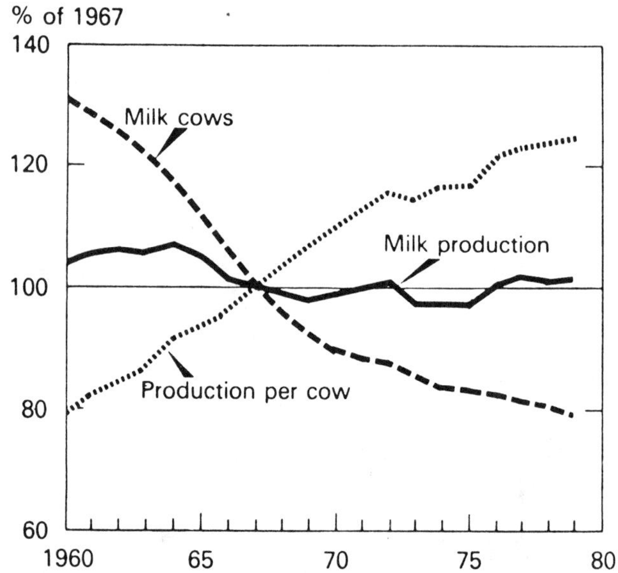

Table 7.4. Effect of Various Calf Crop Percentages on Breakeven Prices (Assumes $300 Annual Cow Cost and 420 lb Weaning Weights)

Percent Calf Crop	Lbs Calf Weaned Per Cow Exposed	Breakeven Price Per cwt	Change in Breakeven Price (Per cwt) For Each 5% Change in Calf Crop Percent
100	420	$ 71.43	
			$3.76
95	399	75.19	
			4.17
90	378	79.36	
			4.67
85	357	84.03	
			5.26
80	336	89.29	
			5.95
75	315	95.24	
			6.80
70	294	102.04	

Factors Affecting Calf Crop Percent

Table 7.5 shows the relative importance of factors affecting the net calf crop in a herd where detailed records have been kept over several years. Information from this table shows the most important factors affecting net calf crop were failure of the female to become pregnant during the breeding season and calf losses at or shortly after birth.

Table 7.5. Factors Affecting Net Calf Crop in a Disease-free Beef Herd Bred by Natural Service (14-year summary)[a]

Factor	No.	Reduction in Net Calf Crop Percent
Females not pregnant at end of breeding season	2232	17.4
Perinatal calf deaths	821	6.4
Calf deaths birth to weaning	372	2.9
Calf deaths during gestation	295	2.3
(Net calf crop weaned (%)	9107	71.0
Totals	12827	100.0

[a] Includes females 14 months to 10 years of age during breeding seasons of 45 or 60 days duration.
Source: U.S. Livestock and Range Research Station, Miles City, Montana.

These two factors accounted for 82% of the reduction in net calf crop. Therefore, management to improve calf crop percentages must reduce losses from all four loss categories, but most specifically, overcome these two major barriers.

Losses identified in Table 7.5 resulted in a research herd of cattle. These losses appear to be similar to losses occurring in producers' herds. Calf crop percent weaned in approximately 250 South Dakota cow–calf herds over a two-year period was 78%. The 22% reduction in calf crop percent resulted from a 13% loss from breeding to calving and a 9% loss from birth to weaning (South Dakota Cow–Calf Day Report, 10 December 1980).

Factors affecting calf losses will vary in individual herds. Producers should have adequate records to know where the greatest losses are occurring. A management program to improve calf crop percent can occur if a producer is knowledgeable of losses and the cause and effect relationships affecting each loss area. Some major cause and effect relationships affecting calf crop percent are explained in the next several sections.

Factors Affecting Pregnancy Rates

1. Heifers at 14 to 15 months of age need to reach minimum weights if they are to be cycling. Angus, Hereford and Angus-Hereford crossbreds (majority of available breeding females), of medium frame size, need to weigh approximately 700 lbs at this age. Heifers weighing 400 to 500 lbs at weaning would need to gain 1.0 to 1.5 lbs per day to reach the desired target weight at breeding. It may be economical to separate replacement heifers into high and low weaning weight groups and feed them different levels of feed to achieve the desired target weight at breeding (*J. Anim. Sci.* 44:165).

Weights and gains must be higher (50 to 150 lbs) for heifers from larger breeds (Table 7.6) and heifers of larger skeletal size within the same breed. Heifer calves, weaning below 375 lbs, may not be able to reach puberty in sufficient time to calve as two-year-olds. Research evidence and economics greatly favor calving heifers at two years of age unless the feed supply is very limited, especially where heifers are extremely lightweight at weaning.

Table 7.6. Breed and Weight Differences Affecting Percent of Heifers Cycling[a]

Breed[b]	Weight at Start of Breeding: 600 lbs (%)	700 lbs (%)	800 lbs (%)
Angus	70	95	100
Angus X Hereford	45	90	100
Brahman Crossbreds	15	45	85
Charolais Crossbreds	10	65	95
Chianina Crossbreds	10	50	90
Gelbvieh Crossbreds	30	85	95
Hereford	35	75	95
Limousin Crossbreds	30	85	90
Maine-Anjou Crossbreds	15	60	95
Pinzgauer Crossbreds	55	90	100
Shorthorn	75	95	100
Simmental Crossbreds	25	80	95
Tarentaise Crossbreds	40	90	100

[a] Assumes heifers are at least 13 months of age.
[b] Crossbreds are from Angus and Hereford cow base.
Source: MARC

2. Heifers which calve early in their first calving season continue to calve early throughout their lifetime performance. Heifers calving late in the calving season are more likely to be open as three-year-olds, especially in a short breeding season of 60 days or less. Heifers can be selected for early pregnancy, at pregnancy test time, where more heifers have been saved earlier as potential replacements.

3. Heifers typically have a longer postpartum interval than cows. This appears to be an age of cow effect; however, there is evidence that part of this effect is due to nutrition (Table 7.7).

Table 7.7. Effects of Gestation Feed Level on Reproduction in Heifers and Cows

Gestation Feed[a]	Dam	Post-Partum Interval (days)	In Heat by Beginning Breeding Season (%)	October Pregnancy[b] (%)
Low	Heifer	100	17	50
	Cow	58	93	83
High	Heifer	77	47	78
	Cow	60	88	81

[a] Low = 8.0 lb TDN; High = 15.0 lb TDN fed last 90 days gestation.
[b] After 45-day AI period.
Source: U.S. Livestock and Range Research Station, Miles City, Montana.

Heifers, fed together with cows, typically do not get an equal share of feed. Thus, an energy restriction delays the onset of estrus. Physical separation of two- and three-year-old cows from mature cows is a sound management practice and it will usually shorten the postpartum interval in the young cows.

A longer post-partum interval in heifers emphasizes the need to have heifers calve at the beginning of the calving season. Breeding heifers 2 to 3 weeks prior to the cow herd and selection for early pregnancy can accomplish this objective. Breeding heifers earlier than the cow herd will lengthen the total breeding and calving seasons which may create additional costs.

4. Body condition (evaluated primarily as fat over back and ribs) of cows affects length of postpartum interval. Research has shown that about half the number of cows in "thin" condition will be cycling 60 days after calving as compared to cows in "good" body condition. Cows in too high or fat body condition, however, reflect an uneconomic feed level which may negatively affect reproductive efficiency.

5. Extreme calving difficulty and delayed obstetrical assistance will extend the post-partum interval. Results from Miles City have shown that a 10 minute increase in duration of labor lengthened the interval from calving to first estrus by two days, reduced the percent of females exhibiting estrus the first 21 days of the breeding season by 7% and decreased the percent pregnant during a 45 day AI period by 6%. Pulling the calf is recommended if the cow has not had her calf after one hour of the second stage of labor. Second stage starts at the first abdominal press and ends with the birth of the calf.

6. Artificial insemination and length of breeding season can affect the number of resulting pregnant cows. A longer breeding season (120 days, under natural service) extends available bull power while a shorter breeding season (45 days) places heavy demand on available bulls. Bull to cow ratios would vary under these two extremes. Length of breeding season can affect AI technicians. Heat checking and technician efficiency can become lax with extended breeding seasons. AI technicians can tire and affect conception rates where large numbers of cows are bred in a few days with or without estrous synchronization.

The primary use of artificial insemination is to utilize genetically superior sires more effectively. Artificial insemination programs, however, will lower calf crop percent unless high levels of management are given to heat checking, semen quality and handling, and insemination technique. A producer considering implementing an AI Program, where a long breeding season is involved, may best do it in a step-wise manner of shortening the breeding season and gradually reducing the number of days of natural service. Productivity and economics need to be carefully assessed before attempting to shorten the breeding season too rapidly. For many operations, a 25-day AI plus a 25-day natural service breeding season seems to work very well.

7. Recent research work shows that approximately two-thirds of the 20% reduction in calf crop percent is due to fertilization failure and embryonic death (*MARC Beef Research Program,* Progress Report No. 1, April 1982). Part of this loss can be attributed to hormone imbalances and poor uterine environment. It is not clear how to prevent these losses through management programs. Excellent herd health, nutrition and culling programs along with other current management practices should help prevent many fertilization failures and embryonic losses.

8. The bull obviously plays an important role in affecting pregnancy rate. The bull

is expected to deposit adequate numbers of viable sperm in the reproductive tract of the female at the proper time.

Bulls should be subjected to a breeding soundness examination which evaluates physical conformation and soundness through genital palpation, scrotal circumference, and testing semen for motility and morphology. Scrotal circumference is primarily related to sperm production and semen volume. Recently, scrotal circumference, however, has been shown to be closely related (r = −0.71) to age of puberty in heifers (*CSU Expt. Sta. Gen. Series 973,* April 1978). This relationship indicates that as scrotal circumference in bulls increases, their half-sib heifers reach puberty at earlier ages. Table 7.8 shows the breed differences in scrotal size of yearling bulls. Scrotal circumference was largest in yearling bulls of breeds in which heifers reached puberty the earliest. The correlation between heifer age at puberty and yearling bull scrotal circumference was 0.98 among breed averages.

Bulls passing a breeding soundness examination may later experience fertility problems. Significant changes in scrotal temperature, caused by disease and extremely high or low ambient temperatures, can cause reduced fertility or temporary sterility.

Libido (sex drive), mating ability and mating capacity are important factors affecting calf crop percent; however, there is no simple measurement of these characteristics in individual bulls. Individual bulls can successfully impregnate 50 to 60 cows in a 60 day breeding season, however, the typical cow to bull ratio is quoted as 30 to 1. This latter ratio is best used as a base unless the mating capacity of individual bulls is known. Environmental conditions may direct an upward or downward adjustment of this 30 to 1 ratio. Dominance of individual bulls, in multiple sire breeding herds exists, and individual bulls will rank differently over several years as to the proportion of calves they sire.

Yearling bulls (14 to 18 months of age) can be used without reducing pregnancy rate if yearling bulls are properly managed. Average cow to bull ratio should be somewhat less (20 to 1) and the yearling bull should be well fed and not allowed to get too thin.

Bulls can be used naturally in a synchronization program and calf crop percentage

Table 7.8. Breed Comparisons of Testicle Size of Yearling Bulls and its Relationship to Age of Puberty in Heifers

Breed	No. Hfrs	Heifer Age at Puberty (days)	No. Bulls	Scrotal Circumference of Yearling Bulls: Average (cm)	Range (cm)
Gelbvieh	81	341	22	34.8	30.2–42.2
Brown Swiss	126	347	19	34.3	31.0–39.6
Red Poll	95	352	20	33.5	29.7–37.1
Angus	24	372	79	32.8	26.2–38.4
Simmental	157	372	28	32.8	26.2–39.1
Hereford	27	390	55	30.7	26.2–36.1
Charolais	132	398	31	30.4	25.4–37.6
Limousin	161	398	20	30.2	24.4–34.3
Average		368		32.3	

Source: MARC Beef Research Program, Progress Report No. 1, April 1982.

can be kept high. In this situation, one bull should be used on one group of 10 to 30 cows kept in a relatively small pen or drylot. It has been well demonstrated that a bull can settle a relatively large number of cows in a five day period. Giving the bull a ten day rest between cycles is recommended.

9. There are some genetic factors affecting pregnancy rate. The most noteworthy genetic factor is heterosis resulting from the use of crossbred females. Crossbred females reach puberty at an earlier age and have higher conception rates (approximately 6% to 8%).

Research has shown that as inbreeding increases in 10% intervals, number of cows pregnant will usually decrease 1% to 2%. There is, however, considerable variation between lines in response to inbreeding. Some inbred lines are high in fertility. This suggests that opportunity exists to improve fertility within a breed by selection between lines, that is, by culling the poorest lines.

Results of breed evaluation at the Meat Animal Research Center, Clay Center, Nebraska, has shown considerable variation among breed groups in both age and weight at puberty. Faster gaining breed groups, selected for milk production (e.g., Brown Swiss), reach puberty at younger ages than breed groups of similar growth rate and mature size that have not been selected for milk production. Breed groups of *Bos indicus* cattle (Sahiwal and Brahman crosses) reached puberty at an older age than any breed groups of *Bos taurus* cattle of the largest mature size and the lowest milk level. Typically the Brahman and Brahman crosses will have higher pregnancy ratios under the environmental conditions of the southeastern United States (*J. Anim. Sci.* 44:729).

10. A phosphorus deficiency, one of the most common mineral deficiencies in the United States, manifests itself by lengthening or retarding the estrous cycle; cows may fail to breed, or if they breed, they may produce weak calves (Table 7.17).

11. Heifers which are extremely heavily muscled are not as reproductively efficient as heifers more moderate in muscle expression. This relationship is well verified in double-muscled females and is observed in breeds which are noted for heavy muscling.

Factors Affecting Calf Losses at or Shortly After Birth

Calf losses at birth are typically the second most important reason for lower percent calf crops. Calving difficulty (dystocia) accounts for most calf deaths within the first 24 hours of calving (Table 7.9) and most calving difficulty occurs in two-year-old heifers.

Table 7.9. Calf Losses by Days Following Calving

Days After Calving	Percent of all Losses	Major Category	Percent Losses From Major Category
1–3	68	Dystocia	61
4–6	6	Accidents	31
7–9	5	Scours	38
10–42	12	Pneumonia	32
43–weaning	9	Missing	33

Source: U.S. Livestock and Range Research Station, Miles City, Montana.

The challenge of calving difficulty to the cow–calf producer is twofold. First, handling calving difficulties when confronted with them, and secondly, preventing or minimizing calving difficulties.

Handling Calving Difficulties

The problem in most cases is simply that the calf is too large at birth, or the birth canal (pelvic opening) of the cow is too small, or it is a combination of the two. Research at the Meat Animal Research Center (MARC) has shown a fourfold increase in calf death loss when calving difficulty is experienced. Even though it has been the most important single factor contributing to calving difficulty, calf birth weight has accounted for less than 10% of the variation observed in calving difficulty in cows 4 years old and older.

Most calving difficulty losses can be prevented with timely and correct obstetrical assistance (see Chapter 11). The majority of calves that die at birth or shortly thereafter are normal calves whose lungs have functioned for varying lengths of time.

Personal attention to calving (particularly heifers) appears to be more lax toward the end of a long calving season when calving competes with planting of crops and night-time hours. When management inputs are lowered, then greater calf losses occur. Recent interest has been expressed in the Konefal method of day-time calving. This Canadian producer claims that feeding cows at 11:00 AM to noon and again at 9:30–10:00 PM keeps the cows busy feeding which results in most calves arriving in daytime hours (Corn Belt Cow/Calf Conference, Ottumwa, Iowa, 23 February 1980). Research data verify this relationship, although not all cows respond the same way (Bellows 1981).

Preventing Calving Difficulties

Birth weight, particularly in two-year-old heifers (Figure 7.5) is the primary cause of dystocia, and birth weights have been significantly increasing in the cattle population over the past several years. This is primarily because producers have been using more bulls from larger breeds and more selection pressure for growth and frame size within all breeds.

Birth weight appears to be more highly influenced by genetics than environment within the same year. High levels of feed (15 to 18 lbs TDN) during gestation only increases birth weights 3 to 5 lbs over cows fed low levels (7 to 8 lbs TDN). There has been little difference in calving difficulty in cows and heifers on different levels of feed during gestation (Bellows 1982). There are different birth weight levels from cattle of similar genetic background raised in contrasting regions of the United States. These differences have been reported to be as large as 7 to 20 lbs (*J. Anim. Sci.* 33:923). The reason for these differences is not known, although it suggests that nutrition and temperature differences may affect birth weight. Large yearly differences in birth weights within the same herd also suggest that environment can affect birth weights.

There is evidence that level of crude protein in the gestation of the pregnant heifer can cause differences in birth weights of calves and calving difficulty. Bellows (1981) reported birth weights of 73 and 84 lbs and calving difficulties of 42% and 58% in heifers fed 86% and 145% of the NRC crude protein requirement. Recommendations

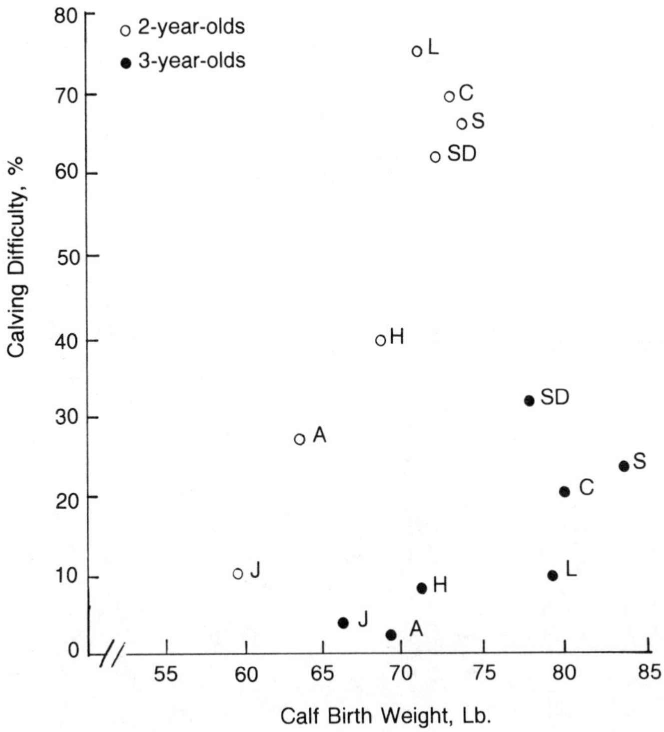

Figure 7.5. Relationship between calf birth weight and calving difficulty. (Different breeds of bulls, (J) Jersey, (A) Angus, (H) Hereford, (SD) South Devon, (L) Limousin, (C) Charolais and (S) Simmental used on Angus and Hereford cows). *Source:* MARC.

are to feed heifers a balanced ration and not feed lower levels of protein to circumvent calving problems. Producers should be aware that they can overfeed protein by not only feeding excessive amounts of protein supplement but also in feeding large amounts of a forage such as alfalfa hay.

It has been shown that selection for yearling weight of 30 lb increments has increased birth weight by almost 2.5 lbs (*J. Anim. Sci.* 39:459). This is because there is a high genetic correlation (R = 0.60) between birth weight and growth rate (yearling weight).

There are both breed of sire (Figure 7.5) and breed of dam (Figure 7.6) effects on birth weight and thus on calving difficulty. Part of the breed effect on birth weight is due to differences in gestation length (Figure 7.7).

Some producers feel that shape of the calf has a definite effect on calving difficulty. For example, longer, slimmer calves have less difficulty than shorter, thicker calves. Available research data do not support this opinion (*J. Anim. Sci.* 36:695); however, there are observations that do support shape of the calf as being important in affecting calving difficulty. Figure 7.8 verifies some of these observations where longer, slimmer Chianina and Brahman sired calves demonstrated less calving difficulty than thicker-made Charolais and Maine-Anjou sired calves. Birth weights of all four sire breeds were very similar.

Birth weight should be included in a selection program, primarily through bull selection. The level of birth weight acceptable for various producers is different. Some cow

Figure 7.6. Relationship between calf birth weight and calving difficulty in (HA) Hereford-Angus, (J) Jersey, (SD) South Devon, (L) Limousin, (S) Simmental and (C) Charolais crossbred cows. *Source:* MARC.

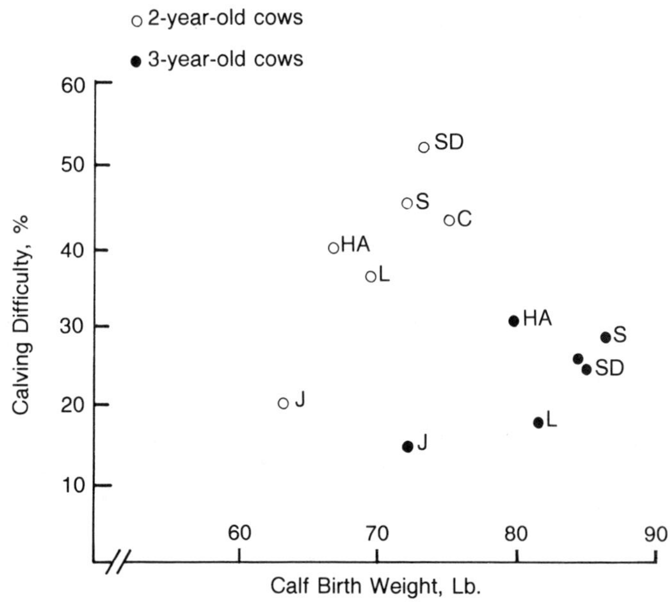

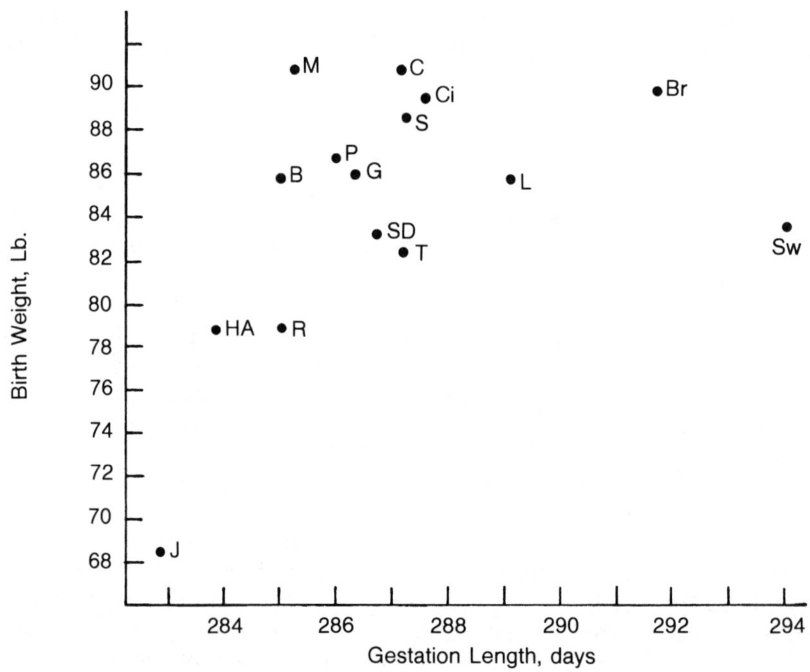

Figure 7.7. Relationship of gestation length to birth weight for various sire breeds. Same breed letter designations as Figure 7.5 with the addition of (P) Pinzgauer, (G) Gelbveih, (R) Red Poll, (M) Maine-Anjou, (Ci) Chianina, (Br) Brahman, (Sw) Sahiwal, and (T) Tarantaise. *Source:* Meat Animal Research Center.

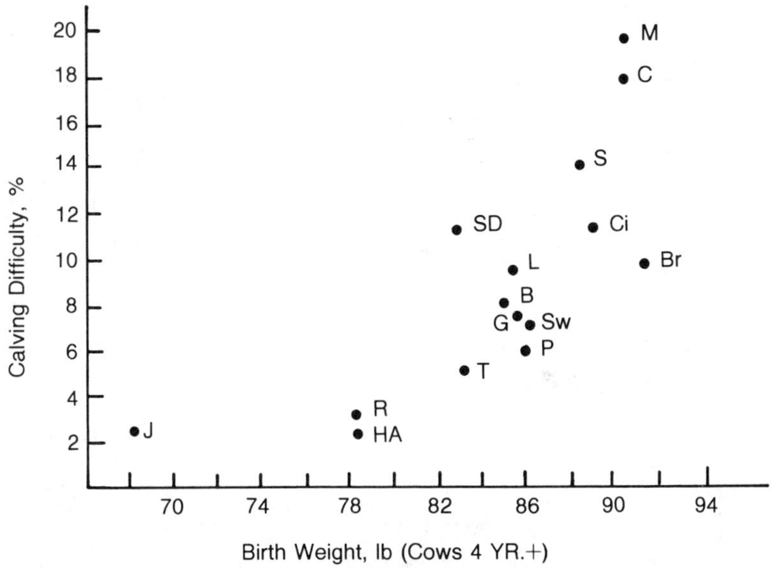

Figure 7.8. Relationship of birth weight to calving difficulty, posing question about importance of shape of the calf. Designated sire breeds are the same as those previously identified in Figures 7.6 and 7.7. *Source:* Meat Animal Research Center.

herds can easily handle 85 to 100 lb birthweights, while other herds experience marked calving difficulty with 80 lb calves. Where cows in a particular herd can easily have 90 to 100 lb calves, two-year-old heifers in the same herd would likely have difficulty with calves over 80 lbs. Therefore, more care must be used in selecting bulls for heifers than for cows.

Natural service bulls should be selected with moderate birth weights and AI bulls should have progeny test data for birth weights and calving ease. Weaning weight and yearling weight ratios can be relatively high without encountering serious birth weight problems. Most bulls with very high growth potential, however, will usually sire heavier birth weight calves. The following index has been shown to be effective in selecting for growth and against birth weight: index = yearling weight − 3.2 (birth weight). The index should be used for selection on a within herd basis.

Table 7.10 shows some selected examples of bulls whose records demonstrate varying combinations of calving ease performance with their growth records.

There is little justification to select bulls from breeds known for extremely light birth weights because a serious reduction in weaning weight and feedlot gain will result. Pounds of calf weaned per cow bred will typically be lowered which means the producer uses an extreme to solve one problem and in turn causes another problem.

Birth weight of the bull being selected should be evaluated. Some breeds known for calving ease in the past now are producing bulls with birth weights in excess of 100 lbs. These bulls may work well on cows but not on heifers in most cow herds.

An additional factor affecting birth weight is age of cow. Two-year-old heifers will

Table 7.10. Birthweight and Growth Records for Selected Angus and Simmental Bulls

Sire Breed and Bull No.	Birth wt of Calves		Calving Ease Score: Bull's own Calves		Calving Ease Score: Daughter's Calves		Calves: Weaning Wt		Calves: Yearling Wt	
	No. Prog. or Acc.	Ratio or EPD	No.	Ratio or EPD	Prog. No.	Ratio or EPD	No. Prog. or Acc.	Ratio or EPD	No. Prog. or Acc.	Ratio or EPD
	ACC	***EPD***		***EPD***		***EPD***	***ACC***	***EPD***	***ACC***	***EPD***
Angus[a]										
1	0.87	+0.1	—	—	—	—	0.86	+32.2	0.85	+77.1
2	0.87	+6.4	—	—	—	—	0.87	+28.2	0.86	+66.5
3	0.95	−2.3	—	—	—	—	0.95	+13.4	0.94	+45.0
4	0.86	−4.0	—	—	—	—	0.89	−17.5	0.89	−42.1
5	0.89	+4.8	—	—	—	—	0.91	+ 1.1	0.90	−23.8
	No.	***Ratio***	***No.***	***Ratio***	***No.***	***Ratio***	***No.***	***Ratio***	***No.***	***Ratio***
Simmental[b]										
1	2026	107	1270	113	477	102	3061	96	377	96
2	1113	104	432	108	67	91	835	102	225	102
3	263	103	27	117	NA	NA	324	107	213	109
4	4577	96	574	85	876	106	5439	103	1127	103
5	1184	99	150	99	125	117	1409	104	957	104

[a] Values obtained from Angus Sire Evaluation Report, Angus Journal, Oct. 1981.
[b] Values obtained from American Simmental Association National Sire Summary, 1980–1981. Higher ratios (above 100) for birth weight and calving ease reflect smaller birth weights and less calving difficulty.

have calves weighing on the average of 8 lbs less than calves from mature cows. This relationship needs to be taken into account when evaluating birth weight records. Bull calves will weigh about 5 lbs more than heifer calves. This is the primary reason why more bull calves will be assisted at birth.

In general, size of the pelvic opening (birth canal) ranks as the most important factor affecting calving difficulty that has been attributed to the dam. It is generally true that dams with larger pelvic openings have less trouble at calving. Within all increments in size of pelvic area, calving difficulty increases with increased birth weight. Mechanical measurements of pelvic area can be obtained to predict dystocia. The results have produced limited success. The primary reason is that individual birth weights cannot be predicted.

It is a questionable practice to include pelvic measurements in a female selection program. The end results do not apparently justify the cost (*J. Anim. Sci.* 36:695 and *Kansas State Cattleman's Day Report,* 1979). It has also been determined that cows with larger pelvic areas generally give birth to heavier calves. Pelvic area is largely a function of the size of the female, and larger females have heavier calves. It is not known for certain if selection for increased pelvic size can offset the increased birth weight. There are producers measuring pelvic size in heifers then mating the heifers with larger pelvic openings to high performing bulls typically having larger birth weights.

There is evidence that selection for skeletal size, for example, hip height, will account for most differences in pelvic area. Thus the larger the skeletal size, the larger the pelvic opening. The larger the skeletal size, the greater the mature weight which results in larger calves at birth. So the conflict between pelvic area and birth weight exists whether pelvic area is measured directly or indirectly.

It is common thinking among beef producers that adequate exercise in the beef female is needed to prevent calving difficulty. The reasoning is that muscle tone and strength is needed to properly expel the calf. Research at Miles City does not support this concept as cows kept in feedlots the last three months of gestation had essentially no difference in calving difficulty when compared to cows that exercised daily (Table 7.11). Cows that were force exercised, by walking to water, required approximately 30% more feed to maintain the same body weight compared to those cows in the feedlot. There is some information with dairy heifers, in total confinement, which indicates forced exercise might have a positive effect on reproductive performance. Exercise might benefit beef females kept in confinement facilities for extended periods of time.

Table 7.11. Effect of Exercise on Calving Difficulty

Activity	No.	Gestation Length (days)	Birth wt (lbs)	(%)	Calving Difficulty (Score)
Forced[a]	30	280	74	24	1.3
Restricted[b]	31	280	71	26	1.5

[a] Walk of two miles daily last 90 days of gestation.
[b] Held in feedlots last 90 days of gestation.

Factors Affecting Calf Losses During Gestation

Losses during gestation are usually low, averaging between 2% to 3% reduction in calf crop percentage. The embryo may fail to implant for various reasons or abortion can occur during all stages of pregnancy. Abortion may result from physical trauma or disease problems. Serious diseases such as brucellosis, leptospirosis, IBR (infectious bovine rhinotracheitis), and others can reduce the calf crop percent by few percent to more than 50%. Herd health programs (see Chapter 19) should be planned, based on herd needs and surrounding conditions. Herd health must be given serious attention or gestation losses will be much higher than the 2% to 3% which is normally expected.

Total death losses are less in crossbred than in purebred calves. This demonstrates the effect of heterosis and that inheritance is involved in some of these losses.

Factors Affecting Calf Losses One to Two Days After Birth to Weaning

Table 7.5 shows that in the Miles City herd, the calf loss was 2.9% after calving difficulties had taken their toll the first couple of days following calving. In a survey of 1922 nationwide producers taken in 1976, calves lost from birth to weaning was 5.5% (Gee, *Agric. Econ. Rpt.* No. 409). Highest losses were in the Great Plains region (9.1%) and the lowest in the southwest (3.6%). No doubt part of this regional difference is due to different weather conditions, particularly severity of spring blizzards in the Great Plains area. Categories of losses include disease, inclement weather, predators, theft, and other losses. Disease appears to cause the largest losses although there were insufficient records to accurately assess the importance of each disease category.

The primary disease problem resulting in the highest death loss of calves is infectious calf scours followed by pneumonia. Prevention and treatment of these diseases are discussed in detail in Chapter 19.

Management Program to Reduce Calf Losses

A management program was implemented in several herds of cattle in an attempt to reduce a calf death loss which had ranged from 16% to 22% over a three-year period. The calf death loss was reduced to an average of 2.8% over a three-year period by implementing management practices and calving techniques in the areas of (1) improved calving facilities, (2) improved sanitation, (3) treatment of sick calves with nutrient and electrolyte drench, (4) closer observation during calving, herd vaccination program, and (5) improved nutrition of the cow herd (Duren 1981). It becomes obvious that when producers give careful attention to calving losses, they can be reduced (Figure 7.9).

IMPROVING WEANING WEIGHTS

Calves are typically born in the spring and weaned in the fall. Most calves weaned are between 6 and 10 months of age. Some calves will be born in the fall, while in some operations, cows will calve year around.

Weaning weights will vary from under 300 lbs to more than 600 lbs with an estimated seven-month weaning weight average between 400 and 450 lbs in the United States.

Figure 7.9. Calf crop percentages improve when producers give proper attention to cows and calves during calving time. Courtesy of Norden Laboratories.

A survey of approximately 500 cow–calf producers in South Dakota reported an average weaning of 469 lbs (*South Dakota Cow–Calf Day Rpt.*, 10 December 1980). This average weight may well be higher than the U.S. average since South Dakota has had one of the more active beef cattle improvement programs for an extended period of time.

Economic Considerations

It was pointed out earlier that increasing weaning weights of calves could lower the breakeven price required to cover production costs. Table 7.12 shows that at a given calf crop percent and annual cow cost, the increments in weaning weight have a decreasing economic advantage.

There is little merit in increasing weaning weight if the cost to produce it exceeds the increased sale value. For example, a producer might increase the weight of each calf by 10 lbs. It would be an unprofitable change in productivity if the annual cow cost per cow was increased by $10.00 per cow, unless the sale price of calves was over $1.00 per pound.

Age of Calf When Weaned

Obviously, weight of the calf is a function of age where there is adequate nutrition and the calf remains healthy. Length of calving season varies greatly with different operations; however, most would fall within the 60 to 150-day range.

Figure 7.10 shows weaning weight of calves where they are born at different times during the calving season but weaned at the same time.

Table 7.12. Changes in Breakeven Prices with 50 lb Increments in Weaning Weight (Assumes an 80% Calf Crop and $300 Annual Cow Cost)

Weaning Weight (lbs)	Pounds of Calf Weaned	Breakeven Price (cwt)	Change in Breakeven Price (cwt)
350	280	$107.14	
			$13.39
400	320	93.75	
			10.42
450	360	83.33	
			8.33
500	400	75.00	
			6.82
550	440	68.18	
			5.68
600	480	62.50	

The average difference in weaning weights of calves between those born in the first and last one-third of a 60-day calving season is approximately 70 lbs. Seventy pounds at $0.70 per pound is $49. This amount assumes that the two groups of cows and calves were managed together and consumed the same amount of feed. The latter is not generally true, because later-born calves will be younger when weaned and will not have consumed as much feed. Most annual cow costs, however, will be similar for both groups of cows. Older, heavier calves have likely consumed more feed because in most cases they were older and capable of consuming more forage prior to being weaned.

Several management practices which would result in a high percentage of calves being born early in the calving season were identified in the earlier section on improving percent calf crop. These practices will increase weaning weight because the calves will be older when weaned.

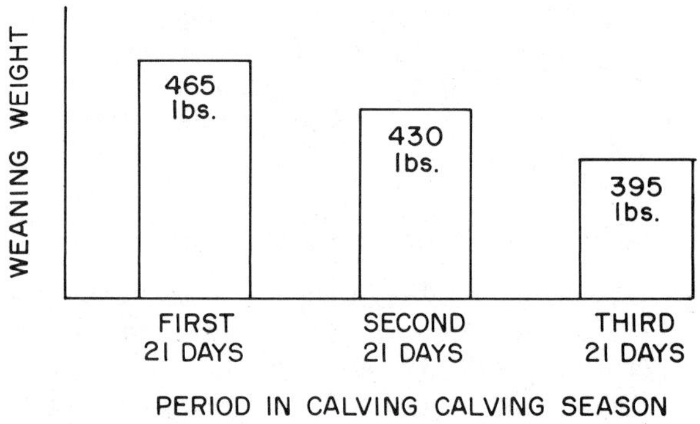

Figure 7.10. Relationship of weaning weight to calving period within the calving season. This assumes a 430 lb calf weaned at 210 days of age. The expected gain is 1.7 lb/day (430 minus 70 lb birth weight ÷ 210 days). Drawn by CSU Graphics.

Genetic Factors Influencing Calf Gains

Where there is an adequate nutrient supply, then usually the most limiting factor influencing gain is the genetic makeup of the calf. A major genetic influence on nutrient supply comes in the genetic ability of the cow to produce a heavy flow of milk. Growth potential of the calf and milk production of the cow are improved genetically through selection and crossbreeding.

Selection can change the gene combination for growth through sire, heifer, and cow selection practices. It has been well demonstrated that effective selection can result in a genetic increase of 4 to 6 lbs per year in weaning weights of calves. Most of this change will come through effective bull selection based on evidence of genetic superiority in weaning or yearling weights. Some genetic improvement can be made through cow and heifer selection but amount of progress is limited due to the number of females that must be kept.

There is a breed effect on weaning weight which is shown in Table 13.7 (Chapter 13). Generally, larger, faster-gaining breeds will have heavier weaning weights at the same age. Information in Table 13.7 shows the breed effect on pounds of calf weaned per cow bred which reflects both weaning weight and calf crop percentage. In this large research study, Brown Swiss and Brahman sires ranked highest, while Jersey, Charolais, Limousin, and South Devon were lower-ranking sire breeds, based on breeding these bulls to Angus and Hereford cows.

Crossbreeding for the average cow–calf producer can result in a 20% increase in pounds of calf weaned per cow bred (Figure 7.11). Most of this increase occurs from improved reproductive performance; however, 25% to 40% of the increase is the effect

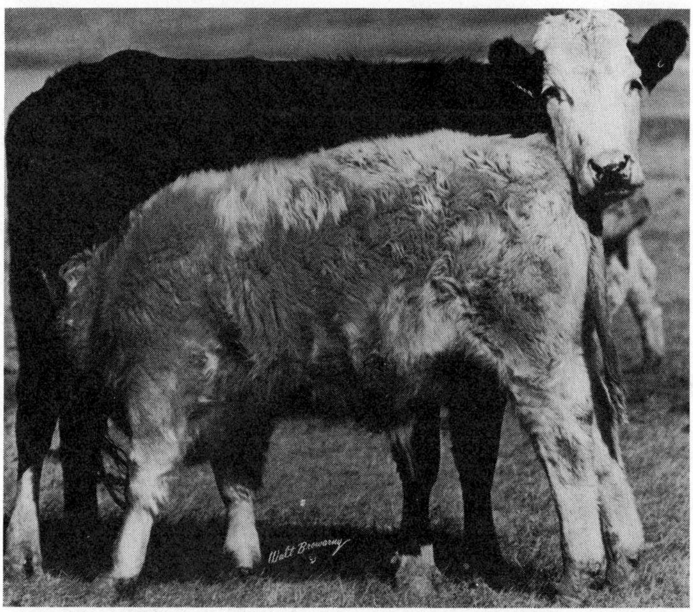

Figure 7.11. Crossbred cows, producing crossbred calves, can give the heaviest weaning weights for cow–calf producers. Courtesy of *The Charolais Banner.*

Table 7.13. Crossbreeding and Within Breed Selection

Breed and Level of Performance	Wean. Wt (lbs)	Ave. Wean. Wt (lbs)	Wean. Wt. After Crossbreeding (lbs)	Breakeven[a] Price
Low A	340			
		330	346[b]	$90.25
Low B	320			
Med. A	420			
		410	430[b]	$72.67
Med. B	400			
High A	500			
		490	514[b]	$60.82
High B	480			

[a] Assumes 80% calf crop and $250 annual cow cost.
[b] Assumes a 5% heterosis above the average of the two breeds.

of heterosis on (1) growth potential of the crossbred calf and (2) increased milk production of the crossbred cow. An economic evaluation of crossbreeding showed that a 16% to 20% increase in net income after taxes could result through effective use of crossbreeding (Univ. of Wyo, *Agric. Expt. Sta. Res. J. 54* 1971).

A cow–calf producer with an average weaning weight of 400 lbs and an 80% calf crop produces 320 lbs of calf weaned per cow bred. If this producer would implement an effective selection and crossbreeding program, the pounds of calf weaned per cow bred would be 414 lbs in six years. Assuming the annual cow cost could remain constant at $300, this would change the breakeven price from $0.94 per lb to $0.72 per lb.

Many commercial producers want to know what breeds to use in a crossbreeding program. Certain breeds do complement one another more satisfactorily, however, producers need to pursue the genetic resource in more depth. It is extremely important to understand the variation which exists within each breed (see Figures 12.7 and 13.2 in Chapters 12 and 13). In many cases the genetic superiority of cattle within breeds has more economic significance than heterosis obtained by crossing two or more breeds (Table 7.13). It should not be a question of genetic superiority versus heterosis for commercial producers. They should take advantage of both to optimize productivity and maximize profitability.

Growth Stimulants

The most common growth stimulants for calves have been ear implants of Compudose® and Ralgro®. Implants can be given to young calves when they are a few days or weeks old. Average growth response to various implants is shown in Table 14.9 (Chapter 14). Implants should not be used on replacement heifers because of possible detrimental effects on reproduction.

The cost of a 36 mg Ralgro® implant is under $1.00. The approximate 25 lb increase in weaning weight well justifies the implant and labor cost. The implant must be placed

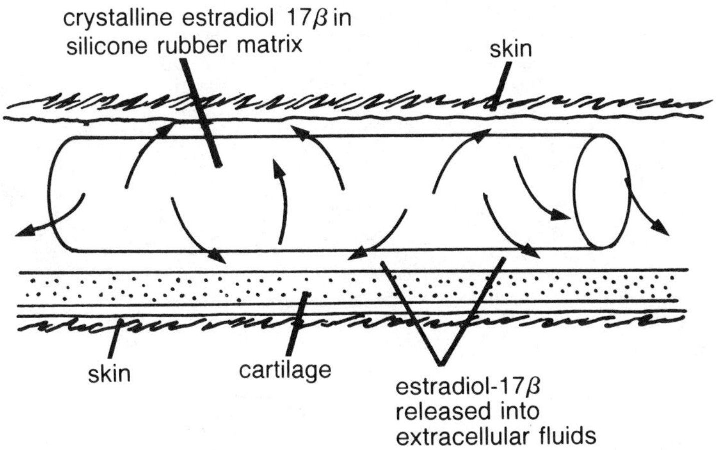

Figure 7.12. The Compudose® implant is a 1.18 inch long × 3/16 inch thick cylinder containing 24 mg. of estradiol. It is implanted under the skin on the back of the steer's ear (see Figure 14.10). Courtesy of Elanco Products Company.

at the base of the ear, close to the head, under the skin, and above the cartilage. Compudose® (Figure 7.12), an estradiol ear implant was recently made available. The growth-promoting response of this implant lasts over a 200-day period. The implant costs approximately \$2.50 and the expected response is a 5% increase in pre-weaning gain.

Proper implanting procedures are shown in Figures 14.8 through 14.11 (Chapter 14).

Feed Supply

Environment, along with the genetic potential of the calf, determines the calf's weaning weight. Environmental effects, such as health of cows and calves and feed supply are extremely important. It is estimated that approximately 60% to 70% of the weaning weight is accounted for by milk production of the cow. The remaining 30% to 40% comes from grass and other forage that the calf consumes directly. Therefore, available forage to both cows and calves will significantly affect the weights of the calves at weaning time.

Month of birth has some effect on weaning weights of calves. Highest daily gains appear to be realized when calves are born 2 to 3 months prior to green forage production. Calves born earlier, and particularly calves born later (summer), will have slower daily gains. Most cows have a peak in milk production 2 to 3 months following calving. Milk production appears to be sustained longer if cows go to lush, green pasture about the time they are peaking in milk production. Calf gains are more dependent upon milk early in the calf's life as compared to the last couple of months prior to weaning (Sims 1975, *CSU Progress Rpt.* PR75–51).

The effect of forage supply on weaning weights is vividly portrayed in Table 7.14.

Probably of greater importance than weaning weight per calf or pounds of calf weaned per cow bred is pounds of beef produced per unit of land area. This measurement reflects not only the improvement made in cattle but also range and pasture improvements accumulated over a period of several years. An excellent demonstration of this is shown in Table 7.15 where range reseeding, grazing intensity, and cropping improvements, along with improvements in productivity of cattle, are reflected in calf production at

Table 7.14. Effect of Range Forage Supply on Weaning Weight and Calf Crop

Treatment	No. Cows	Calves Born (%)	Calves Weaned (%)	Weaning Weight (lbs)
Lightly grazed (38.8 acres/cow)	80	92	90	435
Moderately grazed (30.5 acres/cow)	80	92	89	422
Heavily grazed (23.1 acres/cow)	80	77	70	372

Source: Montana Ag. Expt. Sta. Bull. 549.

the Eastern Colorado Research Center. Records show very little improvement in average weaning weights or calf crop percent over the 20 year period. Total animal units maintained on the same land area changed from 144 to 360, and saleable beef increased from 78,750 pounds in 1953 to 154,915 in 1972.

Creep feeding allows calves to have supplemental feed (usually concentrates). Feed is provided in a creep feeder, or some other physical barrier is provided which prevents cows from having access to the supplemental feed. Creep feeding can add an additional 20 to 50 lbs to the weaning weight of calves. Caution should be exercised, however, because added weight may not always be economical. A producer should compute the expected cost of gain and future production and utilization of calves before deciding to creep feed. Creep rations can be very simple where use of the cheapest concentrate is usually advisable. Addition of bulky feeds such as oats, beet pulp, and dehydrated alfalfa pellets will prevent digestive problems. Rations such as 50% corn and 50% oats, or 50% barley and 50% oats are satisfactory, with there being no advantage in feeding

Table 7.15. Calf Production at the Eastern Colorado Range Research Center for the Years 1953–1972 From Cows Three Years Old and Older

Years	No. Cows	Weaning Wts (lbs)	Percent Weaned	Pounds Weaned per Cow	Acres per Animal Unit	Lbs of Calf per Acre
1953–1957	75	430	90	387	23	17
1958–1962	81	434	95	412	21	20
1963–1967	96	431	94	405	19	22
1968–1972	118	431	93	401	15	27
20-year average		432	93	402	19.5	21

Source: Colo. State Univ. Expt. Sta. Progress Report, PR73–80

Table 7.16. Creep Feeding Considerations

Advantages	*Disadvantages*
1. Added weaning weight.	1. Cost of added weight usually higher than the return.
2. When price of calves is higher relative to feed costs.	2. Not recommended for replacement heifers—impairs future milk production. Thus weaning lighter calves and few calves compared to non-creep fed heifers (J. Anim. Sci 53:33). Therefore, producer needs to separate male and female calves.
3. Useful for calves with skeletal frame and growth potential, particularly if they will go into feedlot shortly after weaning.	3. Creep fed yearlings will weigh very little more than non-creep fed yearlings.
4. Purebred breeder can better measure growth potential (yearling weights) of bull calves. Added bloom and weight will give an advantage to the purebred breeder's merchandising program.	4. Creep feed may be converted very inefficiently. Some studies show that the feed conversion may be as high as 15 lbs of creep fed per pound of grain. This poor conversion usually occurs when calves are on high milking cows and have access to ample supplies of high-quality forage.
5. Less stress at weaning time because calves adjust to their post-weaning ration sooner. Producer may want to creep feed only 2 to 3 weeks before weaning the calves.	5. Market price of calves might be less if they are too "fleshy" or fat at sale time.
6. Most useful during periods of drought of low feed supply to maintain body condition of cows for good reproductive rate the following year.	

a complex creep ration. The creep feeder should be located near the area where cows and calves spend a considerable amount of time, such as near water or shade. Advantages and disadvantages of creep feeding are pointed out in Table 7.16.

The nursing calf usually has an adequate supply of nutrients when receiving cow's milk and an ample amount of forage. Salt and a phosphorus supplement should be provided to cows on a year-around basis. Most mature forages are low in phosphorus. Therefore, cows will generally show a positive response to phosphorus supplementation (Table 7.17).

Feed supply can be significantly affected by weather conditions, particularly rainfall. Moisture effects on weaning weights, through forage supply, is shown in Table 7.18.

Table 7.17. Effect of Phosphorus Supplementation on Cow and Calf Performance

Treatment	No. Cows	Calves Born (%)	Calves Weaned (%)	Weaning Weight (lbs)	Calf Weaned per Cow (lbs)
No phosphorus suppl.	48	64	58	425	244
Bone meal suppl.	49	84	78	494	370
Dicalcium phosphate suppl.	47	85	83	507	420
Bone meal & trace mineral suppl.	47	87	85	481	408

Source: USDA Tech. Bull. 981.

Table 7.18. Effect of Rainfall During the Grazing Season on Weaning Weights at the U.S. Range Livestock Experiment Station (Montana)

Year	No. Calves Born	Wean. Wt (lbs)	180-day Adj. Wean. Wt (lbs)	Annual Rainfall (in.)	Rainfall During Growing Season (in.)
1932	229	379	375	15.0	10.1
1934	229	301	310	5.5	3.5
1948	472	418	399	16.1	13.5
1949	379	297	289	8.8	4.0

Source: USDA Tech. Bull. 1181.

Herd Health

Any herd health problem that affects milk supply of the cow will obviously affect the weaning weight of her calf. Also some calves may not fully recover from diseases (e.g., scours, pneumonia) that occur during the first few weeks after birth. These effects will be reflected in reduced calf weights at weaning.

Serious disease problems of the calf that affect feed consumption or normal body processes will reduce weaning weights. Minor health problems of the calf (parasites, pinkeye, foot rot) can easily result in the calf weighing 5 to 25 lbs less at weaning (see Chapter 19).

ANNUAL COW COSTS

Assessing costs of production is the most neglected area in many commercial cow–calf operations. Adequate expense records must be maintained so that cost areas can be carefully analyzed. Each producer should know where major costs occur and how these costs can be lowered or held in line with the level of productivity of the cattle (Figure 7.13).

Table 7.19 shows costs and returns of average cow–calf operations typical of several areas in the United States. These costs are computed using market values of products produced, including hay and pasture at current selling prices. Similar calculations, using actual Western farm and ranch management association records, showed actual cow costs of $150 and $225 in 1970 and 1975 respectively. This demonstrates the marked rise in annual cow costs over a 10 year period of time.

Many cow–calf producers, not keeping detailed records, will feel that a $300 annual cow cost is high. It may be high if a producer is only considering out-of-pocket or cash costs on a per cow basis. In this case, producers are not giving full value to pasture, hay, and other roughages and to their labor. The combined cost of these items is usually over $100 per cow. Another way to calculate annual cow cost is on a cost basis only, not using the market value of products produced. Assuming land and cattle are paid for (no interest charges), no cost for labor (all contributed by the manager-operator), hay and pasture at cost of production, the annual cow cost could be less than that

Figure 7.13. Cow–calf producers must keep accurate income and expense records to assess annual cow costs and the profitability of their operations. Photo by Duane Dailey, University of Missouri.

shown in Table 7.19. A return in this situation would have to be considered a return to capital, labor, and management. On the other hand, there are times when the market value of home-grown hay, pasture, and other feed is less than the cost of production. An example is the analysis of the Missouri M.I.R. (Mail-in Record) program involving 110 beef cow herds and 10,000 cows. The 1980 costs and returns are shown in Table 7.20. Pasture costs were $22.36 per AUM or 338% of the $6.62 per AUM actually credited for the crop. A drought in 1980 lowered crop yields; however, pasture production costs would be expected to be at least twice the market price for rented pasture.

The Missouri data emphasize that the method used in pricing pasture, hay, and silage made a very significant difference in the breakeven price needed on 477 lb calves which actually sold for $75.68 per hundredweight (Table 7.21). Table 7.21 also shows the effect of computing interest on land investment cost ($2700 per cow) at varying rates of interest. Each additional 5% of interest adds an additional $135 per cow.

Methods of computing annual cow costs or using other cost accounting records can vary considerably. Regardless of method, the most important concerns should be: (1) valid information to critically analyze costs relative to levels of productivity so that inefficiencies can be noted and hopefully changed, and (2) information to make decisions on income tax during the year in addition to the end of the year.

Improving Annual Cow Costs

An analysis of average annual cow costs in Table 7.19 shows the following breakdown of component costs: feed (54%), interest (20%), labor (10%), and other (approximately 16%). If interest rates were computed at 15% instead of 8%, then interest would comprise 31% of the total cow cost instead of 20%. For operations with no interest costs, feed costs will comprise two-thirds to three-fourths of the total annual cow cost. Where interest

Table 7.19. Costs per Breeding Cow for Cow–Calf Operations by Regions in the United States, 1979

Item	West	South-West	Great Plains	North Central	South East	Average All Regions
	Region					
Variable Costs:	($)	($)	($)	($)	($)	($)
Supplementary feed[a]	94	46	58	70	45	58
Range and pasture	18	24	14	53	71	43
Veterinary and medicine	3	4	4	4	4	4
Marketing	2	5	4	4	4	4
Utilities and feed	9	10	10	12	12	11
Labor hired[b]	11	15	8	3	10	10
Machinery, buildings and equipment repair	7	13	12	3	14	14
Interest on operating capital (8% interest)	7	5	3	9	9	7
Overhead[c]	9	10	7	9	9	9
TOTAL VARIABLE COSTS	160	132	120	167	178	160
Ownership costs:						
Machinery, buildings and equipment[d]	39	42	36	59	47	47
Livestock[e]	59	61	65	54	46	56
Land tax	18	9	13	20	11	13
TOTAL OWNERSHIP COSTS	116	112	114	133	104	116
Operator and family labor[f]	30	29	27	40	38	34
TOTAL ALL COSTS	306	273	261	340	320	310

[a] Hay computed at cost of production, exclusive of a land charge.
[b] Range was from 1 to 4 hours per cow.
[c] Included in overhead is accounting service, telephone, road maintenance, etc.
[d] Replacement reserve, interest, taxes, and insurance.
[e] Depreciation is charged for herd bulls only. Brood cows are assumed to be raised from heifer calves born in the operation, and salvage values are recovered through sale of cull cows.
[f] Range was from 8 to 12 hours valued at hired labor wage.
Source: USDA Report to United States Senate Committee on Agriculture.

must be paid on the total investment per cow, interest will be more than 60% of the total annual cow cost.

The greatest consideration in reducing annual cow costs should be directed toward feed and interest costs since they comprise most of the total cost. Cows and replacement heifers should be fed according to their nutrient requirements. Nutrient costs of various available feed should be continually evaluated. Cow–calf operations having availability of aftermath feeds (corn stalks, grain stubble, and straw) will usually have the greatest opportunity to keep feed costs lower than other operations. Interest charges on operating

Table 7.20. Costs and Returns per Cow in Missouri (1980), Average Sale Weights of Young Cattle Less than 600 lb

	Quantity per Cow	Price per Unit	Dollars per Cow	Your Figures
INCOME PER COW				
Net production (.817 calf/cow):				
.630 calves sold × 477 lbs @ $75.68			$ 227	$_____
.123 heifers transferred to breeding herd × 764 lbs @ $54.26			51	_____
0.64 head increase in inventories × 473 lbs @ $62.44			19	_____
.817 calf (425 lbs) produced per cow			$ 297	$_____
Inventory change due to price effects, purchases, etc.			−11	_____
Miscellaneous income			0	_____
TOTAL GROSS INCOME PER COW			$ 286	$_____
COSTS PER COW:				
Feed				
Commercial feed & ingredients	0.203 tons	$172	$ 35	$_____
Grains	0.195 tons	107	21	_____
Hay & haylage	1.820 tons	38.42	70	_____
Silage and green chop	0.360 tons	18.64	7	_____
Pasture	8.09 AUM	6.49	52	_____
Total Feed			$ 185	$_____
Other Variable Costs:				
Feed storage, hauling, preparation			$ 1	$_____
Veterinary & medicine			6	_____
Other livestock services & materials			5	_____
Labor (8.84 hrs. @ $3.91 per hour)			35	_____
Machine hire			0	_____
Machinery & equipment operating costs			16	_____
Utilities			3	_____
Insurance			2	_____
Personal property & sales taxes			1	_____
Real estate repair & maintenance			6	_____
Miscellaneous overhead			1	_____
Interest on operating costs ($250 @ 8%)			20	_____
Total Variable Costs			$ 281	$_____
Fixed Costs				
Real estate interest, taxes & depreciation			$ 7	$_____
Interest on breeding herd ($474 @ 8%)			38	_____
Interest & depreciation on machinery & equipment			14	_____
Total Fixed Costs			$ 60	$_____
TOTAL COSTS			$ 341	$_____
RETURN TO MANAGEMENT			$(−55)	$_____

Source: Jacobs, V. E. and Kirtley, C. L. Farm Management Newsletter, 26 Aug., 1981.

Table 7.21. Estimated Breakeven Prices Required for Calves Sold in Missouri in 1980[a]

Costing Method Employed	Breakeven Price for 477 lb Calves (cwt)
(1) Forage at market prices	$ 87.03
(2) Forage at cost of production	137.85
(3) Same as No. (2) with interest on land omitted	98.38
(4) Forage at full cost of production with land and cattle investment charged at:	
0%	89.30
4%	116.55
8%	144.55
12%	172.55
16%	200.55

[a] 1980 was a drought year.
Source: Jacobs, V. E. and Kirtley, C. L. Farm Mgt. Newsletter, 26 August, 1981.

capital can vary several percent. Producers should evaluate several sources of credit to determine lowest interest charges and other services provided.

Labor costs can be evaluated by looking at the amount needed and if labor is being utilized in areas of greatest priority. Table 7.22 shows the labor needed for cattle operations in Colorado. Labor use per A.U. (Animal Unit) drops rapidly as herd size increases but plateaus at approximately 500 animal units. It is not uncommon to find in an operation two full-time men handling the same size cow herd that one person is handling for another operation.

Size of cow is an important factor affecting annual cow costs. Table 7.23 shows how cow weight and level of milk production affect energy requirements. These requirements, established by the NRC, have been shown in several research studies to be somewhat

Table 7.22. Amount of Annual Labor per A.U. for Ranches in Colorado

Animal Units	Hours by Regions			Aver. for State	Percent Ranches in State
	Mtn.	Mtn. Fringe	Eastern		
100 and less	43	41	37	39	17
100–200	21	21	18	19	20
200–300	15	17	10	12	14
Ave. per A.U.	16	21	16	18	—

Source: Gee, C. K. Colo. Expt. Sta. Bull. 550S.

Table 7.23. TDN Requirements for Cows of Various Weights and With Different Milk Production Levels (NRC)

Cow wt (lbs)	Annual TDN for Maintenance (lbs)	Percent of a 1000 lb Cow	Daily lbs of TDN Required for Maintenance and Milk at Various Milk Production Levels		
			8 lb	16 lb	24 lb
800	2315	86	9.4	12.0	14.7
1000	2709	100	10.5	13.1	15.8
1200	3098	114	11.5	14.2	16.8
1400	3472	128	12.6	15.2	17.9
1600	3832	142	13.6	16.3	18.9

below the actual amount of TDN required by the postpartum cow for high reproductive performance. The requirements, however, are comparable to demonstrate that more feed is required when weight or milk production increases.

There is not complete agreement as to the size of cow needed today for commercial production. Many producers feel that cow size will and should vary according to environmental conditions, particularly feed supply. There are observations that large cows which wean heavier calves will not become pregnant where the feed supply is limited. Also where cow size is increased, the amount and nutrient quality of feed is increased to heifers to reach puberty in sufficient time to calve at 2 years of age. Therefore, cow size is best determined by the amount and nutrient quality of available feed and the desired carcass weight and carcass quality of the resulting progeny. In the latter category, carcass weights of 600 to 800 lbs, yield grades 1 and 2, and a quality grade of high good to low choice are the beef industry preferences for the majority of the slaughter steers and heifers.

Generally speaking, a larger cow will wean a heavier calf; however, smaller cows can wean heavy calves if proper genetic and environmental conditions exist. Moderate cow size, with a terminal sire crossing program, seems most economically feasible considering energy requirements associated with size and the fact that medium sized cows can wean heavy calves. There are numerous observations of moderate sized cows (900 to 1100 lbs in average body condition) that can reproduce regularly and wean a 500 to 600 lb calf which has an excellent feedlot and carcass performance.

The most apparent way to achieve economic efficiency of smaller sized cows which produce calves with high feedlot and carcass performance, is through a terminal crossbreeding program. Cow lines should be selected for moderate size and high maternal performance. Sire lines should receive selection emphasis for growth and cutability. Table 7.24 gives economic support for this type of crossbreeding program.

The breed-cross of the cow has been shown to affect the amount of feed required for maintenance of the same amount of body weight. Angus-Hereford cross cows required 9%, 14%, and 16% less feed than Jersey, Charolais, and Simmental cross cows, respectively (Ferrell and Jenkins 1982).

Table 7.24. Percent Return on Investment[a] from Combinations of Mature Size for Two Breeding Systems Under a Pasture Program

Breeding System and Size of Sire	Size of Cow		
	Small	Medium	Large
Crossbreeding			
Small	15.9	15.9	15.0
Medium	16.7	16.6	15.4
Large	17.7	17.4	16.0
Straightbreeding	14.8	14.9	14.9

[a] Return on investment is defined as total income less expenses, other than interest, expressed as a percentage of total expenditures other than interest.
Source: J. Anim. Sci. 41:1238.

SELECTED REFERENCES

Publications

Beef Improvement Federation Proceedings (Genetic effects on reproduction in beef cattle), 26 April 1972.

Bellows, R. A. *Dystocia and associated factors.* Proceedings, The Range Beef Cow Symposium VI, 1979, page 33.

Bellows, R. A. *Managing for higher calf crop percentages.* Angus Journal, January 1980.

Bellows, R. A. *Some nutrition effects on birth weight and calving difficulty and relationships of feeding time and calving.* Proceedings, Fifteenth Conference on Artificial Insemination of Beef Cattle. Denver, CO., 17 January 1981.

Bellows, R. A., Short, R. E., and Richardson, G. V., 1982. *Effects of sire, age of dam and gestation feed level on dystocia and postpartum reproduction.* J. of Anim. Sci. 55:18.

Beverly, J. R. and Spitzer, J. C., 1980. *Management of replacement heifers for a high reproductive and calving rate.* Texas A and M Ext. Public. B-1213.

Cartwright, T. C. *Potentials and problems of genetic improvement of fertility of cows and bulls.* Polled Hereford World, September 1979.

Costs of producing feeder cattle in the United States. 1979. USDA Report to United States Senate Committee or Agriculture, Nutrition, and Forestry.

Cow–Calf Management Guide, 1980. Cattleman's Library. Published by the Cooperative Extension Services of CO, ID, MT, OR, UT, WA and WY. Copies available from the Univ. of ID.

Duren, E. *The Pegram Project. Integrated calf management.* Proc., The Range Beef Cow Symposium VII. December 7–9, 1981, Rapid City, SD.

Ferrell, C. L. and Jenkins, T. G., 1982. *Energy utilization by mature cows.* U.S. Meat Anim. Res. Ctr. Beef Res. Program Progress Rpt. No. 1.

Gee, C. K., 1978. *Cattle and calf losses in feeder cattle production.* USDA Agric. Econ. Rpt. No. 409.

Gee, C. K., Haider, M., and Madsen, A. G., 1979. *Economics of size for mountain and plains cow-calf ranches.* Colo. Agric. Expt. Sta. Bull. 569S.

Lasley, J. F., 1981. *Beef Cattle Production.* Englewood Cliffs: Prentice-Hall, Inc.

Martin, T. G., Lemenager, R. P., Srinivasan, G., and Alenda, R., 1981. *Creep feed as a factor influencing performance of cows and calves.* J. of Anim. Sci. 53:33.

Maurer, R. R., Echternicamp, S. E. and Pope, W. F. *Conception and embryonic development in*

the beef female. U.S. Meat Anim. Res. Ctr. Beef Res. Program Progress Rpt. No. 1, April 1982.
Minish, G. L. and Fox, D. G., 1979. *Beef Production and Management.* Reston, Va.: Reston Publishing Co., Inc.
Neuman, A. L., 1977. *Beef Cattle.* New York: John Wiley and Sons.
Nelson, T. R., 1976. *Economies of size in cow–calf production.* Great Plains Beef Cattle Handbook, GPE-4950.
O'Mary, C. C. and Dyer, I. A., 1972. *Commercial Beef Cattle Production.* Philadelphia: Lea and Febiger.
Preston, T. R. and Willis, M. B., 1970. *Intensive Beef Production.* New York: Pergamon Press.
Price, D. P., 1981. *Beef Production.* Dallas: Southwest Scientific Publishing Co.
Totusek, R. *Principles of profitable cowherd management.* Ark. Agric. Expt. Sta. Spl. Rpt. 93, April 1981.

Visuals

Cow/Calf Production (filmstrip and tape). Vocational Education Productions. Calif. Polytechnic State Univ., San Luis Obispo, CA 93407.
Calving Management (filmstrip and tape). Vocational Education Productions. Calif. Polytechnic State Univ., San Luis Obispo, CA 93407.

8
The Purebred Breeder and the Beef Industry

The purebred breeder is a specialized cow–calf producer, sometimes referred to as a seedstock producer or registered breeder. The purebred breeder's primary goal is to make genetic improvement which can be utilized by the entire beef industry. Seedstock producers are in reality the genetic engineers for the beef industry.

Purebred breeders can be categorized into three different groups as shown in Figure 8.1. Within any one of the major breeds there is likely not more than one to five percent of the herds of the major breeds which would be identified as *elite* herds. These are herds that direct genetic change in the breed and in the beef industry. The *primary* and *multiplier* herds take the genetic change initiated in *elite* herds and expand it into more numbers of breeding animals and units of semen.

Elite herds will sell bulls and semen to other purebred herds throughout the United States. *Multiplier* herds, which sell primarily to commercial cow–calf producers, market most of their breeding animals within a 100 to 150 mile radius of their farm or ranch. Knowledge of these relationships is important in choosing a breed or breeds and in developing a marketing program.

Establishing Goals for a Breeding Program

Specific goals in a breeding program are determined by identifying the traits and some specific numerical value or objective measurement for each trait. The intent is to make genetic improvement in the economically important traits. Table 8.1 gives a list of beef industry traits and their respective heritabilities and economic importance.

Breeders can select for a combination of all economically important traits or concentrate on improvement of a few traits, for example, maternal traits in one breed and growth-carcass traits in another breed. The latter has merit if commercial producers primarily choose a terminal crossbreeding system. Most purebred breeders emphasize

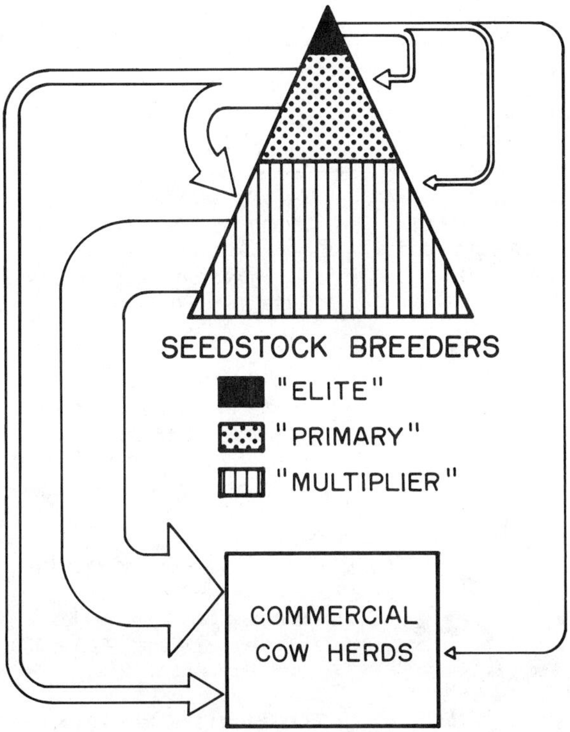

Figure 8.1. Structure of purebred herds showing the main genetic flow of bulls and semen within the categories of purebred herds and to the commercial herds. Size of arrow reflects the magnitude of numbers involved. Drawn by CSU Graphics.

traits which can be easily merchandised for the greatest financial return. Maternal traits are not easily marketed, therefore the bull traits of growth, frame size, cutability and some other visual traits are usually emphasized.

Table 8.2 shows most of the specific beef cattle traits included in breeding programs of most progressive breeders.

Performance Records

Many breeding cattle currently are evaluated and selected totally on visual appraisal. It is a well-known fact that visual appraisal alone is a very poor method of improving the economically important traits identified in Table 8.1. Use of the scales, backfat probe, and tape measure gives breeders the objective measurements (Table 8.2) needed to identify beef cattle superiority or inferiority.

The primary traits measured in most performance record programs are weaning weight and yearling weight. Weaning and yearling weight ratios, calculated within the herd are meaningful and valid only when animals being compared are given similar environmental treatment. Along with individual weight ratios, the concept of breeding values is included in most records programs. Basically, breeding values combine weight ratios for traits such as weaning and yearling weights and heritabilities for these traits. A breeding value is an accumulation of all production information on an individual,

Table 8.1. Considerations for Trait Selection in the Goal Setting Process

Broad Classification of Traits	Comparative Economic Importance	*Heritability* High > 0.4 Med. 0.2 to 0.39 Low < 0.2	Beef Industry Goal[a]
Reproduction	High	Generally low except medium for ease of calving, birth weight, first service conception and scrotal circumference.	*Calf crop weaned*—95%+ *Calving interval*—cow calves every 365 days or less. *Calving season*—90 days or less (preferably 60 days).
Growth	Average	Weaning weight is medium, others are high.	*205-day wt*—500 lbs + *Feedlot gain*—3.0 lb +/day *365-day wt*—1000 lbs (steer basis, intensive production). *Feed efficiency*—6.0 lbs feed (or less) per lb of gain (high energy).
Carcass	Low	High	*Quality grade*—Min. low choice *Yield grade*—2 *Lean yield per day*—0.80 lbs *Preferred carcass wt*—600 to 800 lbs
Conformation	Average	Medium to high	*Frame size* of 4.5 to 5.5 (sl. steer) or hip ht of 47 to 49 in. at 365 days (for desired yield grade at preferred carcass wts) *Soundness*—skeletal, udders, eyes. Putting a *productive* animal into an *attractive* package.
Genetic defects	Depends on incidence	Highly heritable	Freedom (or extremely low incidence) of *inherited abnormalities.*

[a] Goals for individual breeders may differ because of various environmental conditions and different economic pressures.

the sire, dam, and all half sibs. The breeding value is what a commercial breeder can expect from a seedstock producer's product.

As performance records are taken, the big challenge is to get them into usable form and then effectively utilize the records. Some purebred and commercial producers hand calculate their records, prepare, and keep their own record forms. Other producers submit their collected records to a state beef cattle improvement association, breed association, or other location where the records are handled by computer and returned in usable form. Performance records are most useful where breeding values are included; therefore, purebred breeders should use record systems where breeding values are calculated.

Table 8.2. Specific Trait Identification to Genetically Improve Specific Production Areas

Production Area	Traits		
	Bull	**Cow**	**Heifer**
Reproduction	Reproductive soundness exam. (scrotal circum. and semen evaluation) Sex drive (libido) Mating capacity (skeletal soundness) Progeny data on birth wt and calving ease	Pregnancy status Calving interval Skeletal soundness	Age and weight (British, 650 to 700 lbs at 15 mo, European, 700 to 800 lbs at 15 mo) Pregnancy status
Pre-weaning growth	Indiv. and progeny adj. 205-day wt ratio Maternal Breeding Value (MBV) Weaning Breeding Value (WBV)	Progeny record—Adj. 205-day wt ratio Most Probable Producing Ability (MPPA), MBV and WBV	Indiv. adj. 205-day wt ratio Maternal Breeding Value (MBV) Weaning Breeding Value (WBV)
Post-weaning growth	140-day gain ratio Indiv. and progeny adj. 365-day wt Yearling Breeding Value (YBV)	Individual and progeny ratios for daily gain or adj. 365-day wt, YBV	Indiv. adj. 365-day wt ratio Yearling Breeding Value (YBV)
Feed efficiency	Lb feed/lb gain ratio for indiv. and progeny (cost limitations) Use correlation to daily gain on weight constant basis	Some progeny records but usually of limited value	Not practical
Conformation and carcass	Hip height (frame size) Backfat probe Visual score for yield grade Progeny test data for yield grade, quality grade, lean yield/day Breeding soundness including freedom from genetic abnormalities	Limited progeny records for hip height and carcass traits. Breeding soundness including freedom from genetic abnormalities	Hip height (frame size) Visual score for yield grade Breeding soundness including freedom from genetic abnormalities

Figures 8.2, 8.3, 8.4, 8.5, 8.6, and 8.7 show several performance records forms and data from the American Hereford Association's Total Performance Record (TPR) program and the American Angus Association's Angus Herd Improvement Records (AHIR) program. Most major breed associations have similar performance records programs providing similar types of records.

T.P.R. WEANING AND YEARLING GROUP REPORT
AMERICAN HEREFORD ASSOCIATION
HEREFORD DRIVE KANSAS CITY, MISSOURI 64101

DATE APRIL 1, 1981 PAGE 1
BREEDER BEST HEREFORDS HEREFORD KANSAS FILE NO. B033016

CALF Herd Identity Number	Date Calved (MO DAY YR)	Sex	Adjusted Birth Weight	Birth Weight Ratio *	EZ	SIRE Herd Identity Number	DAM Herd Identity Number	Age	WEANING RECORD Mgt Code	Days of Age When Weighed	Weight	Weight per Day of Age	Adjusted 205 Day Weight	Weight Ratio *	Adjusted Height	FR SC	Optional 1	2	3	4
0001	012180	B	82	1107	1	6713	76007	4	1	249	710	285	621	1026	445	54		4		5
0002	012180	B	83	1096	1	7761	78087	2-	1	249	560	225	537	887	425	44		6		3
0003	012280	B	90	1020	1	6713	75051	5	1	248	610	246	530	876	437	50		4		2
0004	012280	B	86	1063	1	6713	75004	5	1	248	725	292	627	1036	445	54		5		7
0005	012380	C	84	955	1	7729	78040	2-	1	247	510	206	498	1038	422	46		4		5
0006	012480	B	103	878	2	7761	78030	2-	1	246	710	289	673	1112	441	52		5		4
0007	012480	C	80	1050	1	7509	73041	7	1	246	530	215	464	967	426	48		3		2
0008	012480	B	89	1031	1	7729	78066	2-	1	246	650	264	619	1023	427	45		7		7
0009	012580	C	80	1050	1	7509	74056	6	1	245	470	192	414	863	420	45		6		4
0010	012680	C	82	1026	1	7761	78090	2-	1	244	500	205	493	1028	430	50		3		6
0011	012880	B	98	932	2	6713	75046	5	1	242	690	285	612	1011	433	48		5		4
0012	012880	C	86	979	1	7729	78006	2-	1	242	515	213	510	1063	430	50		7		6
0013	012980	C	92	907	1	6713	77067	2+	1	241	460	191	447	932	416	43		6		8
0014	013080	C	82	1026	1	7509	73007	7	1	240	515	215	457	952	426	48		7		7
0015	013080	B	83	1096	1	7729	78029	2-	1	240	580	242	565	933	423	43		5		7
0016	013080	C	86	979	1	7509	73031	7	1	240	470	196	419	873	414	42		7		9
0017	013080	B	86	1063	1	6713	76074	4	1	240	680	283	610	1008	439	51		4		4
0018	013180	B	102	889	2	7509	74002	6	1	239	800	335	710	1173	455	59		5		6
0019	013180	B	98	932	1	7509	75058	5	1	239	780	326	692	1143	445	54		5		7
0020	013180	C	76	1097	1	7761	78017	2-	1	239	500	209	496	1034	422	46		7		5
0021	013180	B	90	1020	1	6713	73016	7	1	239	690	289	612	1011	433	48		6		4
0022	020180	B	91	1009	1	7729	78085	2-	1	238	630	265	613	1013	431	47		3		1
0023	020180	B	102	889	2	7509	73020	7	1	238	630	265	564	932	435	49		2		8
0024	020180	C	92	907	1	6713	76031	4	1	238	520	218	475	990	428	49		7		5
0025	020180	C	94	884	1	6737	74041	6	1	238	580	244	519	1082	430	50		5		4
0026	020180	B	94	976	2	6737	75078	5	1	238	710	298	632	1044	437	49		3		5
0027	020280	C	86	979	1	6713	73008	7	1	237	515	217	463	965	428	47		4		6

CALF Herd Identity Number	YEARLING RECORD Mgt Code	Days of Age When Weighed	Weight	Weight per Day of Age	Adjusted Weight 365 Days	Adjusted Weight 452 Days	Weight Ratio *	Adjusted Height	FR SC	Optional 5	6	7	8	SC	COMMENTS
0001	7	417	1220	293	1107		987	500	55			5		34	
0002	7	417	1080	259	1033		921	478	44			4		33	
0003	7	416	1140	274	1034		922	488	49			3		34	
0004	7	416	1230	296	1108		989	496	53			5		36	
0005	5	427	820	192	773		1039	462	46			6			
0006	7	414	1320	319	1253		1117	496	53			7		35	
0007	5	426	895	210	788		1059	466	48			4			
0008	7	414	1190	287	1132		1009	480	45			5		37	
0009	5	425	780	184	689		926	462	46			2			
0010	5	424	865	204	817		1098	470	50			4			
0011	7	410	1280	312	1173		1045	488	49			3		37	
0012	5	422	835	198	794		1067	468	49			4			
0013	5	421	810	192	757		1017	456	43			5			
0014	5	420	820	195	727		977	464	47			6			
0015	7	408	1095	268	1056		941	476	43			7		35	
0016	5	420	725	173	646		868	458	44			5			
0017	7	408	1175	288	1082		964	490	50			4		33	
0018	7	407	1405	345	1286		1146	506	58			4		36	
0019	7	407	1315	323	1200		1070	496	53			5		35	
0020	5	419	785	187	748		1005	462	46			7			
0021	7	407	1320	324	1212		1080	486	48			8		34	
0022	7	406	1080	266	1041		928	484	47			5		36	
0023	7	406	1160	286	1086		952	486	48			4		37	
0024	5	418	830	199	750		1008	466	48			9			
0025	5	418	845	202	754		1013	468	49			5			
0026	7	406	1245	307	1140		1016	490	40			6		35	
0027	5	417	830	199	743		999	484	49			7			

Figure 8.2. Total Performance Records (TPR) weaning and yearling report. Note particularly the columns marked (*) which show ratios for birth, weaning, and yearling weights. Courtesy of the American Hereford Association.

Hand Calculated Performance Record Systems for Commercial Producers

Commercial producers, in many areas, have available to them computerized record systems through their state's beef improvement association. Most genetic improvement will come through effective bull selection; however, a commercial producer may desire to keep individual records on cows and calves. This will permit a small added increase in genetic improvement, but more important, will allow for overall improvement in management when cows are identified as individual productive units.

Pre-weaning and post-weaning growth have considerable economic importance to the beef industry. They have direct effects on net return and these traits are also positively associated with some aspects of maternal ability, efficiency of gain, and pounds of retail

T.P.R. GROWTH BREEDING VALUE SELECTION WORK SHEET REPORT BASED ON YEARLING DATA DATE APRIL 1, 1981 PAGE 1
AMERICAN HEREFORD ASSOCIATION BREEDER BEST HEREFORDS HEREFORD KANSAS FILE NUMBER B033016

CALF Herd Identity Number	Sex	Date Calved (MTH DAY YEAR)	SIRE Herd Identity Number	DAM Herd Identity Number	Age of Dam	DATA ON FILE Adjusted Weight	Weight Ratio	Pat Half Sibs Number Head	Pat Half Sibs Average Ratio	Mat Half Sibs Number Head	Mat Half Sibs Average Ratio	Progeny Number Head	Progeny Average Ratio	Breeding Value Ratio	SELECTION DECISIONS
BULL CALVES															
0001	B	012180	6713	76007	4	1107	987	51	988	2	1058			995	
0002	B	012180	7761	78087	2-	1033	921	21	1021					964	
0003	B	012280	6713	75051	5	1034	922	51	990	3	1048			964	

Figure 8.3. TPR sheet showing Yearling Breeding Values computed for bull calves. Similar forms exist for heifer calves, cows, and herd sires. Courtesy of the American Hereford Association.

T.P.R. WEANING BREEDING VALUE SELECTION WORK SHEET
AMERICAN HEREFORD ASSOCIATION TYPE REPORT BULL CALVES BREEDER BEST HEREFORDS HEREFORD KANSAS DATE SEPT 1, 1980 PAGE 1 FILE NUMBER B033016

Animal					Maternal Breeding Value Data: Dam			Maternal Grand Sire			Paternal Grand Sire			Sire			Growth Breeding Value Data: Individual		Paternal Half Sibs		Progeny		Growth Breeding Value Ratio	Maternal		Selection Decisions
Herd Identity Number	Sex	Date Calved Mo	Day	Yr	Number of Daughters	Number of Calves	Average Ratio	Number of Daughters	Number of Calves	Average Ratio	Number of Daughters	Number of Calves	Average Ratio	Number of Daughters	Number of Calves	Average Ratio	Adjusted Weight	Weight Ratio	Number Head	Average Ratio	Number Head	Average Ratio		Breeding Value Ratio	Percent Accuracy	
0001	B	01	21	80		3	1058	73	112	1032	2	2	897	3	3	939	621	1026	56	949			988	996	57	
0002	B	01	21	80		1	887	22	22	1050	21	31	1008				537	887	26	1032			975	1004	55	
0003	B	01	22	80		4	1005	14	42	1023	2	2	897	3	3	939	530	876	56	952			939	980	57	

T.P.R. WEANING BREEDING VALUE SELECTION WORK SHEET
AMERICAN HEREFORD ASSOCIATION TYPE REPORT HEIFER CALVES BREEDER BEST HEREFORDS HEREFORD KANSAS DATE SEPT 1, 1980 PAGE 4 FILE NUMBER B033016

Animal					Maternal Breeding Value Data: Dam			Maternal Grand Sire			Paternal Grand Sire			Sire			Growth Breeding Value Data: Individual		Paternal Half Sibs		Progeny		Growth Breeding Value Ratio	Maternal		Selection Decisions
Herd Identity Number	Sex	Date Calved Mo	Day	Yr	Number of Daughters	Number of Calves	Average Ratio	Number of Daughters	Number of Calves	Average Ratio	Number of Daughters	Number of Calves	Average Ratio	Number of Daughters	Number of Calves	Average Ratio	Adjusted Weight	Weight Ratio	Number Head	Average Ratio	Number Head	Average Ratio		Breeding Value Ratio	Percent Accuracy	
0005	C	01	23	80		1	1038	12	12	1043	21	26	979				498	1038	31	1015			1020	1007	54	
0007	C	01	24	80		6	950	150	350	1057	19	41	1035	22	22	1050	464	967	78	1010			982	1042	64	
0009	C	01	25	80		5	1025	150	350	1057	19	41	1035	22	22	1050	414	863	78	1010			976	1062	64	

T.P.R. WEANING BREEDING VALUE SELECTION WORK SHEET
AMERICAN HEREFORD ASSOCIATION TYPE REPORT MOTHER COWS BREEDER BEST HEREFORDS HEREFORD KANSAS DATE SEPT 1, 1980 PAGE 7 FILE NUMBER B033016

Animal					Maternal Breeding Value Data: Individual			Maternal Grand Sire			Paternal Grand Sire			Sire			Growth Breeding Value Data: Individual		Paternal Half Sibs		Progeny		Growth Breeding Value Ratio	Maternal		Selection Decisions
Herd Identity Number	Sex	Date Calved Mo	Day	Yr	Number of Daughters	Number of Calves	Average Ratio	Number of Daughters	Number of Calves	Average Ratio	Number of Daughters	Number of Calves	Average Ratio	Number of Daughters	Number of Calves	Average Ratio	Adjusted Weight	Weight Ratio	Number Head	Average Ratio	Number Head	Average Ratio		Breeding Value Ratio	Percent Accuracy	
73041	C	02	08	73		6	950	11	42	978	232	759	1008	150	350	1057					6	950	960	996	87	
74056	C	02	11	74		5	1025	50	195	985	232	759	1008	150	350	1057	487	961	54	1066	5	1025	1012	1036	86	
75004	C	01	20	75		4	1058	11	42	978	190	584	1021	6	18	1020	456	975	17	1041	4	1058	1021	1032	84	

T.P.R. WEANING BREEDING VALUE SELECTION WORK SHEET
AMERICAN HEREFORD ASSOCIATION TYPE REPORT HERD SIRES BREEDER BEST HEREFORDS HEREFORD KANSAS DATE SEPT 1, 1980 PAGE 9 FILE NUMBER B033016

Animal					Maternal Breeding Value Data: Individual			Maternal Grand Sire			Paternal Grand Sire			Sire			Growth Breeding Value Data: Individual		Paternal Half Sibs		Progeny		Growth Breeding Value Ratio	Maternal		Selection Decisions
Herd Identity Number	Sex	Date Calved Mo	Day	Yr	Number of Daughters	Number of Calves	Average Ratio	Number of Daughters	Number of Calves	Average Ratio	Number of Daughters	Number of Calves	Average Ratio	Number of Daughters	Number of Calves	Average Ratio	Adjusted Weight	Weight Ratio	Number Head	Average Ratio	Number Head	Average Ratio		Breeding Value Ratio	Percent Accuracy	
6713	B	03	29	76	3	3	939	40	60	1040	111	259	996	2	2	897	712	1076	85	999	57	951	984	967	65	
6737	B	02	16	76	4	4	1022	150	350	1057	97	328	986	70	116	1025	743	1188	51	1000	58	1032	1064	1025	70	
6758	B	05	07	76				27	121	1031	3	6	1018				676	1056	48	1015	8	968	969	1016	60	

Figure 8.4. TPR sheets showing Maternal Breeding Values (MBV) and Weaning Breeding Values (WBV) computed for bull calves, heifers, cows, and herd sires. Note the data which are used to compute MBV and WBV (identified as maternal breed value ratio and growth breeding value ratio respectively). Courtesy of the American hereford Association.

product. Calf crop percentage is also very important economically. More improvement in this trait, however, can be made through environmental changes (see Chapter 7).

Pre-weaning growth is best measured and recorded as the adjusted 205-day weight. Calves should be weighed as near to 205 days as possible with 160 to 250 days being the recommended acceptable range. Calves outside this range will give information that cannot be meaningfully compared to the other records without risk of considerable error. Weaning weights are age adjusted by the following formula:

$$\text{Adjusted 205 day wt (lbs)} = \frac{\text{actual wt} - \text{birth wt}}{\text{age in days}} \times 205 + \text{birth wt}$$

It is not necessarily recommended that commercial producers weigh calves at birth. The added information would not likely offset the time and expense of taking the weights. A commercial producer producing crossbred calves could use a standard birth weight of 75 or 80 lbs in the formula.

Weaning weights need to be adjusted for age of dam so that records can be more

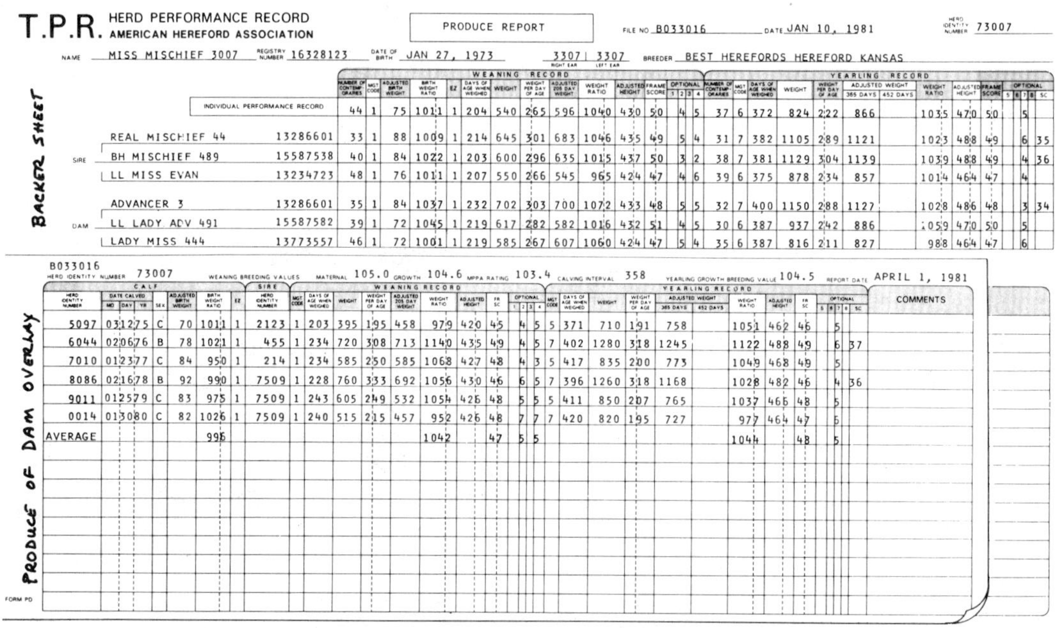

T.P.R. HERD PERFORMANCE RECORD — AMERICAN HEREFORD ASSOCIATION

PRODUCE REPORT — FILE NO B033016 — DATE JAN 10, 1981 — HERD IDENTITY NUMBER 73007

NAME MISS MISCHIEF 3007 — REGISTRY NUMBER 16328123 — DATE OF BIRTH JAN 27, 1973 — 3307 | 3307 (RIGHT EAR / LEFT EAR) — BREEDER BEST HEREFORDS HEREFORD KANSAS

BACKER SHEET

	Registry number	Weaning: Number of contemporaries	MGT code	Adjusted birth weight	Birth weight ratio	EZ	Days of age when weighed	Weight	Weight per day of age	Adjusted 205 day weight	Weight ratio	Adjusted height	Frame score	Optional 1 2 3 4	Yearling: Number of contemporaries	MGT code	Days of age when weighed	Weight	Weight per day of age	Adjusted weight 365 days	Adjusted weight 452 days	Weight ratio	Adjusted height	Frame score	Optional 5 6 7 8	SC
INDIVIDUAL PERFORMANCE RECORD		44	1	75	1011	1	204	540	265	596	1040	430	50	4 5	37	6	372	824	222	866		1035	470	50	5	
REAL MISCHIEF 44	13286601	33	1	88	1009	1	214	645	301	683	1046	435	49	5 4	31	7	382	1105	289	1121		1023	488	49	6	35
SIRE BH MISCHIEF 489	15587538	40	1	84	1022	1	203	600	296	635	1015	437	50	3 2	38	7	381	1129	304	1139		1039	488	49	4	36
LL MISS EVAN	13234723	48	1	76	1011	1	207	550	266	545	965	424	47	4 6	39	6	375	878	234	857		1014	464	47	4	
ADVANCER 3	13286601	35	1	84	1037	1	232	702	303	700	1072	433	48	5 5	32	7	400	1150	288	1127		1028	486	48	3	34
DAM LL LADY ADV 491	15587582	39	1	72	1045	1	219	617	282	582	1016	432	51	4 5	30	6	387	937	242	886		1059	470	50	5	
LADY MISS 444	13773557	46	1	72	1001	1	219	585	267	607	1060	424	47	5 4	35	6	387	816	211	827		988	464	47	6	

PRODUCE OF DAM OVERLAY

B033016 — HERD IDENTITY NUMBER 73007 — WEANING BREEDING VALUES MATERNAL 105.0 GROWTH 104.6 — MPPA RATING 103.4 — CALVING INTERVAL 358 — YEARLING GROWTH BREEDING VALUE 104.5 — REPORT DATE APRIL 1, 1981

Calf: Herd identity number	Date calved MO DAY YR	Sex	Adjusted birth weight	Birth weight ratio	EZ	Sire: Herd identity number	Weaning: MGT code	Days of age when weighed	Weight	Weight per day of age	Adjusted 205 day weight	Weight ratio	Adjusted height	FR SC	Optional 1 2 3 4	Yearling: MGT code	Days of age when weighed	Weight	Weight per day of age	Adjusted weight 365 days	Adjusted weight 452 days	Weight ratio	Adjusted height	FR SC	Optional 5 6 7 8	SC	Comments
5097	031275	C	70	1011	1	2123	1	203	395	195	458	979	420	45	4 5	5	371	710	191	758		1051	462	46	5		
6044	020676	B	78	1021	1	455	1	234	720	308	713	1140	435	49	4 5	7	402	1280	318	1245		1122	488	49	6	37	
7010	012377	C	84	950	1	214	1	234	585	250	585	1068	427	48	4 3	5	417	835	200	773		1049	468	49	5		
8086	021678	B	92	990	1	7509	1	228	760	333	692	1056	430	46	6 5	7	396	1260	318	1168		1028	482	46	4	36	
9011	012579	C	83	975	1	7509	1	243	605	249	532	1054	426	48	5 5	5	411	850	207	765		1037	466	48	5		
0014	013080	C	82	1026	1	7509	1	240	515	215	457	952	426	48	7 7	7	420	820	195	727		977	464	47	6		
AVERAGE				996								1042		47	5 5							1044		48	5		

FORM PD

Figure 8.5. Cow record forms. The backer sheet issued once gives the individual performance of the cow and her sire, dam, and four grandparents. A produce of dam overlay is issued each time a new weaning or yearling record is generated on a calf. Courtesy of the American Hereford Association.

comparable. The appropriate number of pounds (Table 8.3) are added to the age adjusted 205-day weight.

A weaning weight ratio is then computed by dividing each calf's adjusted 205-day weaning weight by the average of its sex group. This record is expressed as a percentage of its sex group.

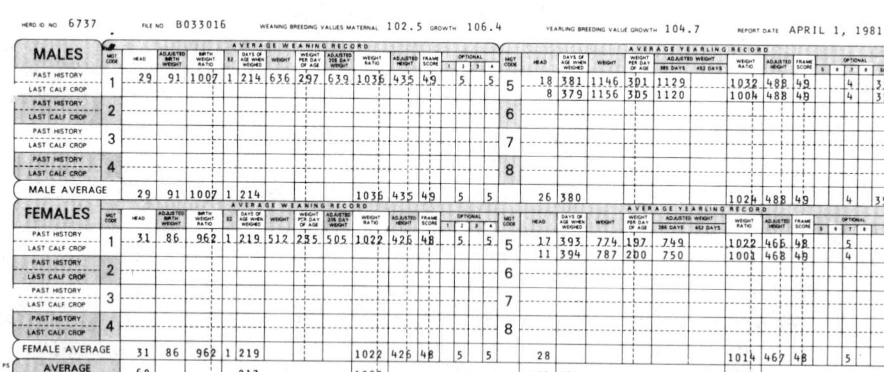

HERD ID NO 6737 — FILE NO B033016 — WEANING BREEDING VALUES MATERNAL 102.5 GROWTH 106.4 — YEARLING BREEDING VALUE GROWTH 104.7 — REPORT DATE APRIL 1, 1981

	Weaning: MGT code	Head	Adjusted birth weight	Birth weight ratio	EZ	Days of age when weighed	Weight	Weight per day of age	Adjusted 205 day weight	Weight ratio	Adjusted height	Frame score	Optional 1 2 3 4	Yearling: MGT code	Head	Days of age when weighed	Weight	Weight per day of age	Adjusted weight 365 days	Adjusted weight 452 days	Weight ratio	Adjusted height	Frame score	Optional 5 6 7 8	SC
MALES																									
PAST HISTORY	1	29	91	1007	1	214	636	297	639	1036	435	49	5 5	5	18	381	1146	301	1129		1032	488	49	4	35
LAST CALF CROP	1													5	8	379	1156	305	1120		1004	488	49	4	35
PAST HISTORY	2													6											
LAST CALF CROP	2													6											
PAST HISTORY	3													7											
LAST CALF CROP	3													7											
PAST HISTORY	4													8											
LAST CALF CROP	4													8											
MALE AVERAGE		29	91	1007	1	214				1036	435	49	5 5		26	380					1024	488	49	4	35
FEMALES																									
PAST HISTORY	1	31	86	962	1	219	512	235	505	1022	426	48	5 5	5	17	393	774	197	749		1022	466	48	5	
LAST CALF CROP	1													5	11	394	787	200	750		1001	468	49	4	
PAST HISTORY	2													6											
LAST CALF CROP	2													6											
PAST HISTORY	3													7											
LAST CALF CROP	3													7											
PAST HISTORY	4													8											
LAST CALF CROP	4													8											
FEMALE AVERAGE		31	86	962	1	219				1022	426	48	5 5		28						1014	467	48	5	
AVERAGE		60				217				1029					54	387					1018				

FORM PS

Figure 8.6. Form used to summarize progeny data on herd sires and to update breeding values (shown at the top of the form). Courtesy of the American Hereford Association.

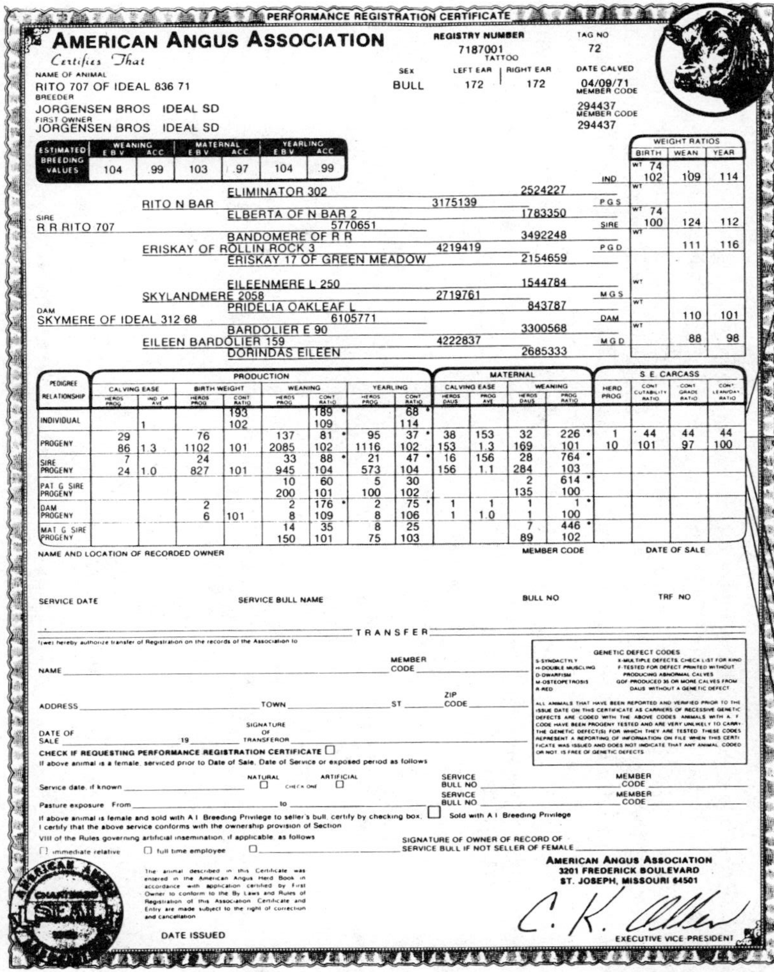

PERFORMANCE REGISTRATION CERTIFICATE

AMERICAN ANGUS ASSOCIATION

Certifies That

NAME OF ANIMAL	REGISTRY NUMBER	TAG NO
RITO 707 OF IDEAL 836 71	7187001	72

BREEDER: JORGENSEN BROS IDEAL SD

FIRST OWNER: JORGENSEN BROS IDEAL SD

SEX: BULL — TATTOO: LEFT EAR 172 | RIGHT EAR 172 — DATE CALVED 04/09/71 — MEMBER CODE 294437 — MEMBER CODE 294437

ESTIMATED BREEDING VALUES	WEANING E B V	ACC	MATERNAL E B V	ACC	YEARLING E B V	ACC
	104	.99	103	.97	104	.99

SIRE: R R RITO 707 (5770651)

- RITO N BAR 3175139
  - ELIMINATOR 302 2524227
  - ELBERTA OF N BAR 2 1783350
- ERISKAY OF ROLLIN ROCK 3 4219419
  - BANDOMERE OF R R 3492248
  - ERISKAY 17 OF GREEN MEADOW 2154659

DAM: SKYMERE OF IDEAL 312 68 (6105771)

- SKYLANDMERE 2058 2719761
  - EILEENMERE L 250 1544784
  - PRIDELIA OAKLEAF L 843787
- EILEEN BARDOLIER 159 4222837
  - BARDOLIER E 90 3300568
  - DORINDAS EILEEN 2685333

WEIGHT RATIOS	BIRTH	WEAN	YEAR
IND	WT 74 / 102	109	114
P G S			
SIRE	WT 74 / 100	124	112
P G D		111	116
M G S			
DAM		110	101
M G D		88	98

PEDIGREE RELATIONSHIP	PRODUCTION: CALVING EASE HERDS PROG	IND OR AVE	BIRTH WEIGHT HERDS PROG	CONT RATIO	WEANING HERDS PROG	CONT RATIO	YEARLING HERDS PROG	CONT RATIO	MATERNAL: CALVING EASE HERDS DAUS	PROG AVE	WEANING HERDS DAUS	PROG RATIO	S E CARCASS: HERD PROG	CONT CUTABILITY RATIO	CONT GRADE RATIO	CONT LEAN/DAY RATIO
INDIVIDUAL		1		193 / 102		189 • / 109		68 • / 114								
PROGENY	29 / 86	1.3	76 / 1102	101	137 / 2085	81 • / 102	95 / 1116	37 • / 102	38 / 153	153 / 1.3	32 / 169	226 • / 101	1 / 10	44 / 101	44 / 97	44 / 100
SIRE PROGENY	7 / 24	1.0	24 / 827	101	33 / 945	88 • / 104	21 / 573	47 • / 104	16 / 156	156 / 1.1	28 / 284	764 • / 103				
PAT G SIRE PROGENY					10 / 200	60 / 101	5 / 100	30 / 102			2 / 135	614 • / 100				
DAM PROGENY			2 / 6	101	2 / 8	176 • / 109	2 / 8	75 • / 106	1 / 1	1 / 1.0	1 / 1	1 • / 100				
MAT G SIRE PROGENY					14 / 150	35 / 101	8 / 75	25 / 103			7 / 89	446 • / 102				

NAME AND LOCATION OF RECORDED OWNER — MEMBER CODE — DATE OF SALE

SERVICE DATE — SERVICE BULL NAME — BULL NO — TRF NO

TRANSFER

NAME — MEMBER CODE

ADDRESS — TOWN — ST — ZIP CODE

DATE OF SALE 19 — SIGNATURE OF TRANSFEROR

CHECK IF REQUESTING PERFORMANCE REGISTRATION CERTIFICATE ☐

GENETIC DEFECT CODES

AMERICAN ANGUS ASSOCIATION
3201 FREDERICK BOULEVARD
ST. JOSEPH, MISSOURI 64501

C. K. Allen
EXECUTIVE VICE PRESIDENT

DATE ISSUED

PERFORMANCE REGISTRATION CERTIFICATE

The Performance Registration Certificate is a pedigree that includes all available performance information from Angus Herd Improvement Records. Included is individual performance and progeny performance of the individual, parents and grandparents. Information on the certificate is accumulated from all herds involved on A.H.I.R.

Weight ratios for birth weight, weaning weight and yearling weight are reported in the pedigree section for the individual, sire, dam, paternal and maternal grandparents. The actual birth weight from A.H.I.R. records is printed above the birth weight ratio.

Individual

Individual performance is listed for calving ease, birth weight, weaning weight and yearling weight. The number of contemporaries is listed for each trait. As an example:

193 There were 193 calves in the contemporary group measured for birth weight and
102 the individual birth weight ratio was 102. Also printed is individual information for weaning and yearling.

Progeny

Performance of offspring of the individual. First is calving ease information from **first-calf heifers only.** Listed is the number of herds that contributed to the data and the total number of progeny such as: 29 herds
86 progeny

Next is listed the average calving ease score from first-calf heifers, which is 1.3 for this bull. A score of 1.0 would mean no calving difficulty.

Birth weight information includes the number of herds, the number of progeny and the average birth weight ratio of calves. For weaning information, the number of herds that contribute to the performance information is listed and the total number of progeny followed by the average contemporary group size and the average ratio of all calves weighed. For example, the average contemporary group size for weaning weight is **81 head** and the **average weaning ratio is 102.** Information for the yearling section is listed in the same manner.

The **maternal** section reports the production information of daughters. In the maternal section, calving ease is reported from first-calf heifers. Next the number of herds and the total number of daughters in production are recorded. And last the total number of offspring of the daughters is reported and their average weaning ratio. For example, 169 daughters representing 32 herds contributed to the maternal weaning data. The daughters have produced 226 offspring with an average weaning ratio of 101.

Sire Evaluation carcass data is the only carcass information reported on the Performance Certificate. Listed are the number of herds that contributed to the carcass data and the number of progeny. Next the number of contemporaries and the ratios for cutability, USDA carcass quality grade and lean per day of age are listed. The ratios are based on within-herd information.

Sire Progeny

Performance of all offspring produced by the sire, including commercial progeny.

Paternal Grandsire Progeny

Performance of the offspring of the paternal grandsire (sire's sire).

Dam Progeny

Performance of the dam's offspring. She has produced six calves in two herds with an average birth weight ratio of 101; eight calves measured through weaning with an average ratio of 109 and eight calves measured through yearling with an average ratio of 106. She has one daughter in production.

Maternal Grandsire Progeny

Performance of offspring of the maternal grandsire.

The Estimated Breeding Value for weaning, maternal and yearling is printed in the upper left hand corner of the certificate. The breeding value combines in one figure all the individual performance measures of an animal and the average performance of relative groups. It is a mathematical method of measuring genetic potential based on individual performance and productivity of related animals such as the sire, dam and half-sibs. Estimated Breeding Values become more meaningful as generations of cattle and more performance information have been recorded on A.H.I.R. Also included is an accuracy figure that reflects how much reliance should be placed on each estimated breeding value. An accuracy figure close to 1.0 is preferred.

Figure 8.7. Angus breeders have the option of obtaining a regular registration certificate (presents primarily the names and registration number of ancestors) or a performance registration certificate as shown above. Courtesy of the American Angus Association.

Table 8.3. Age of Dam Adjustment Factors

	Additive Factors	
Age of Cow	**Male Calves**	**Female Calves**
2 yr olds (21-33 mo)	60 lbs	54 lbs
3 yr olds (34-46 mo)	40 lbs	36 lbs
4 yr olds (47-59 mo)	20 lbs	18 lbs
5-10 yr olds	0 lbs	0 lbs
11 yr olds	20 lbs	18 lbs

Birth and weaning data can then be recorded on individual cow record cards or sheets (Figures 8.8 and 8.9) if records are hand calculated. Computerized records will usually provide cow summaries. This allows a producer to evaluate cows on maternal ability which would include calving interval. Such records are used in culling cows and selecting replacement heifers.

An MPPA (Most Probable Producing Ability) record can be calculated on each cow. If a commercial producer culled cows within age groups, the MPPA could easily be ignored. MPPA is used to compare cows which do not have the same number of calves in their records (see Table 8.4). In the example, cow A has the lowest lifetime average. This, however, is for only a single calf for which environmental conditions or the calf's genetic potential for growth might have been below the average of what the cow would normally produce. One or more calves from cows B or C could also have had a record of 85 or less. All three cows are probably low producers, but use of MPPA enables more accurate culling and, in this example, indicates that cows B and C are slightly lower-producing cows than A.

MPPA for weaning weight is computed by the following formula:

$$\text{MPPA} = \overline{\text{H}} + \frac{\text{NR}}{1 + (\text{N} - 1)\text{R}}(\overline{\text{C}} - \overline{\text{H}})$$

where $\overline{\text{H}} = 100$, the herd average weaning weight ratio,
N = the number of calves included in the cow's average,
R = 0.4, the repeatability factor for weaning weight ratio,
and $\overline{\text{C}}$ = average for the weaning weight ratio for all calves the cow has produced.

MPPA of cow D in the example (Table 8.4) above is computed as follows:

$$\text{MPPA (cow D)} = 100 + \frac{3 \times 0.4}{1 + (3 - 1)\ 0.4}(110 - 100) = 106.7$$

After obtaining a post-weaning gain over a 140 to 160 day feeding period, an adjusted yearling weight can be calculated as follows:

$$\text{Adjusted 365 day wt} = 160 \times \frac{\text{actual final wt} - \text{actual weaning wt}}{\text{number of days between wts}} + \text{adj. 205 day wt}$$

Tag. No. ______________
Tattoo R ________ L ________
Birth Date ________ Birth Wt. ________
Adj. Wn. Wt. ______________
Wn. Ratio ________ Wean. Frame Score ____
Adj. Yr. Wt. ______________
Yrl. Ratio ________ Yrl. Frame Score ____

Date MBV __________
MPPA __________
WBV __________
YBV __________
CI __________

Calf Record

Birth Date	Tag No.	Sex	Sire	BW / CE	Date Wnd.	Adj. Wng. Wt.	WWR	Gain / Ratio	Adj. Yrl. Wt.	Yrl. Wt. R	Fr. Size		CWT. / SC	C.Gr. / BF	YG	Disposition
											Wng.	Yrl.				

Comments ______________________________________

Date	Wt.	H.Ht.		

Figure 8.8. Individual Cow Record Form applicable primarily to purebred breeders. Courtesy of CSU.

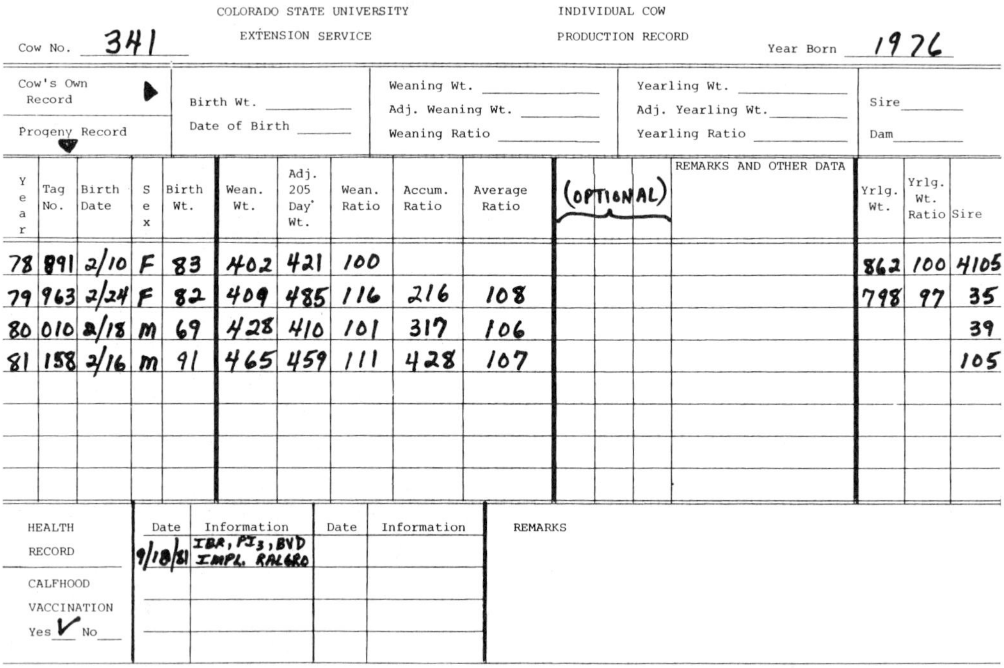

COLORADO STATE UNIVERSITY EXTENSION SERVICE — INDIVIDUAL COW PRODUCTION RECORD

Cow No. 341 Year Born 1976

Cow's Own Record ▶ Birth Wt. ____ Date of Birth ____ Weaning Wt. ____ Adj. Weaning Wt. ____ Weaning Ratio ____ Yearling Wt. ____ Adj. Yearling Wt. ____ Yearling Ratio ____ Sire ____ Dam ____

Progeny Record ▼

Year	Tag No.	Birth Date	Sex	Birth Wt.	Wean. Wt.	Adj. 205 Day Wt.	Wean. Ratio	Accum. Ratio	Average Ratio	(OPTIONAL)			REMARKS AND OTHER DATA	Yrlg. Wt.	Yrlg. Wt. Ratio	Sire
78	891	2/10	F	83	402	421	100							862	100	4105
79	963	2/24	F	82	409	485	116	216	108					798	97	35
80	010	2/18	M	69	428	410	101	317	106							39
81	158	2/16	M	91	465	459	111	428	107							105

HEALTH RECORD	Date	Information	Date	Information	REMARKS
	9/18/81	IBR, PI3, BVD IMPL. RALGRO			

CALFHOOD VACCINATION Yes ✓ No ___

Figure 8.9. Individual cow record form for utilization primarily by commercial producers. (This is intended to be a 5″ x 8″ card). Courtesy of CSU.

Yearling weight ratios are computed separately for each sex group by dividing the individual's record by the sex group average. All animals in a sex group should have been given a similar environmental treatment if weights and ratios are to be meaningful and comparable.

Breeding Values

Reference has been made several times previously to breeding values. Breeding values most readily available to breeders at the present time are maternal, weaning, and yearling breeding values.

Table 8.4. MPPA Records on 6 Cows

Cow	No. Calves	Ave. Wean. Wt. Ratio	MPPA
A	1	85	94.0
B	2	88	93.2
C	4	90	92.7
D	3	110	106.7
E	4	112	108.8
F	1	115	106.0

Maternal Breeding Values (MBV) measure milk production potential. MBV ratios give information which estimate the milking ability that animals can pass to their offspring. For example, MBV on a young bull estimates in a comparative manner (100 is average) how his daughters will milk. This record is usually accompanied by an accuracy computation (how much confidence you can have in the ratio) of only 0.50 to 0.60. When the bull's daughters produce calves of their own, the accuracy figure will go to 0.90 and above.

Weaning Breeding Values (WBV) measure primarily the growth aspect of weaning weights. High WBV ratios give estimates that these individuals can pass on to their offspring genetic superiority for weaning growth.

Yearling Breeding Values (YBV) measure growth at a year of age. The YBV ratios estimate breeding potential for post-weaning growth. Accuracy values for both WBV and YBV are important to know how much confidence to have in the breeding value.

Other breeding values can be calculated and future performance record programs may include breeding value ratios for such traits as birth weight, hip height, and scrotal circumference.

SIRE SELECTION

Importance of the Bull

Effective bull selection should be focused on two primary areas: (1) contribution to producing live calves and (2) making genetic improvement in economically important traits. This section will give attention to number (2) since number (1) was discussed earlier in Chapter 7.

Even though a bull contributes 50% of the genetic material to each calf, the magnitude of the bull's contribution is greater because of the increased number of offspring sired. The genetic importance of successive bulls used in a herd is shown in Figure 8.10.

This information shows that the three bulls selected have contributed nearly 90% of the genetic material to the calves currently in the herd. Results from the U.S. Meat Animal Research Center, as shown in Table 8.5, clearly demonstrate that 80% to 90% of genetic improvement in a herd will come from bull selection.

Emphasis on bull selection does not diminish the importance of good beef females because genetically superior bulls typically do not have poor dams. Most genetic superiority or inferiority of cows, however, will depend on the bulls previously used in the

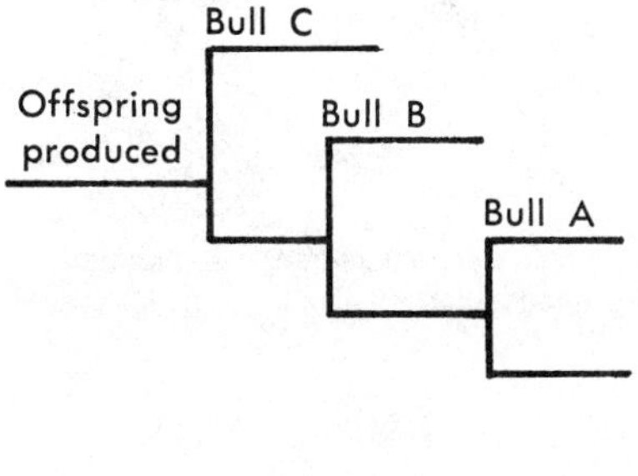

Bull	% genes contributed to offspring produced
A	12½
B	25
C	50

Figure 8.10. Genetic contribution of three bulls to the offspring currently produced in the herd. Drawn by CSU Graphics.

Table 8.5. The Genetic Contribution of Bull Selection in Cattle Selected for Weaning Weight or Yearling Weight

	Selection Differential (Selection Practiced)	
	***(lb)* Weaning Weight**	***(lb)* Yearling Weight**
Bulls	78	140
Heifers	19	18
Total	97	158
% from bull	78/97 = 80	140/158 = 89

Figure 8.11. Purebred breeders must produce genetically superior bulls which will complement well-planned crossbreeding programs. Courtesy of the American Brahman Breeders Association.

herd. High genetic input coming from effective sire selection dictates where priorities should be placed (Figure 8.11).

Traits to be Emphasized

Most commercial beef producers will emphasize adjusted yearling weights in their genetic improvement programs. This will allow them to increase the pounds they sell and also provide the feeder with cattle that have improved feedlot gains. Adjusted 205-day weight and post-weaning gain can be also considered as these records have been obtained to calculate the adjusted 365-day yearling weight. While emphasis on growth traits is important, producers should recognize that there is antagonism between growth and some maternal traits.

Other records which are emphasized, particularly on yearling bulls, would be Maternal Breeding Values, scrotal circumference, backfat probe, and hip height.

Seedstock producers will put different emphasis on various traits depending on their breed and the genetic base they have. For example, the average Angus or Hereford breeder may set priorities on yearling weight and carcass composition. Simmental breeders with high levels of growth and yield grade in their cattle may put major emphasis on maternal traits, for example, birth weight, calving ease, and milk production.

Locating Genetically Superior Bulls

Commercial and seedstock producers should identify breeders who have honest, comparative records on their cattle. Most performance-minded seedstock producers will have maternal breeding values, weaning weights, yearling weight, and birth weights on their bulls. More seedstock producers are obtaining feedlot and carcass data on their own bulls or are using bulls that have these test data behind them.

Seedstock breeders should provide meaningful performance data on their bulls and commercial producers should request this information. Excellent performance records can be obtained and made available from the farm or ranch. Ratios should be evaluated, and they are meaningful if the bulls have been treated alike in their environment.

The best approach in selecting outstanding bulls is to evaluate the records first and select several bulls that are superior in performance. From that group, bulls can be selected that meet the buyer's visual appraisal for skeletal size, freedom from predisposition to fat, scrotal size, and other soundness traits.

Bull testing stations provide a common environment where bulls from several sources can be compared. Comparisons are valid for rate of gain, skeletal size, fat probe, and scrotal circumference but not valid for weaning weight ratios and maternal traits such as milk production. Weight per day of age comparisons have some validity; however, caution should be exercised. Examples of records available from bull testing stations are shown in Tables 8.6 and 8.7. Note in Table 8.6 the CSU Experiment Station bulls have individual feed efficiency and some inbreeding coefficients given. The cooperator bulls (Four Corners BCIA Test) have pen feed efficiencies along with other performance information.

Table 8.8 shows the higher levels of productivity from using high performance rather than low performance bulls. This applied evaluation shows that the genetic theory has practical application.

Table 8.6. Selected Bull Test Data on Bulls (CSU and Cooperator) Sold in the 1981 Sale at the San Juan Basin Research Center, Hesperus, Colorado.

AUCTION—MONDAY, 27 APRIL 1981

12:30 P.M.

Performance of Experiment Station Bulls

Final test results on individually fed station-bred bulls for the 140-day performance test. The adjusted weaning weights are adjusted to 205 days of age, 6–9 year age of dam and zero inbreeding. The dam's Most Probable Producing Ability was calculated from all calves weaned by each dam. Ratios are based on the bull's record relative to the average of 100. Average daily gain was 2.88 for bulls fed from electronic self-feeders. Adjusted yearling weight is equal to adjusted weaning weight plus the average daily gain from weaning to final weight then multiplied by 160.

Test No.	Tattoo	1980 Birthdate	Percent Inbred or LC with	Dam's MPPA	Calf Birth Wt	Actual Weaning Wt	Adjusted Weaning Wt	Weaning Wt Ratio	On-Test Wt Nov. 18–19	Ave. 140-Day Wt, Apr. 7–8	140-Day Daily Gain	Daily Gain Ratio	Feed/lb Gain Ratio	Adjusted Yearling Wt	Yearling Wt Ratio	Sale Price
HEREFORD																
	Sire: Ouray 8120															
2–8	0212	4–6	Tar	98	91	451	524	105	553	1014	3.29	114	101	1027	108	$1550
4–10	0074	3–9	14	102	84	501	497	100	633	1102	3.35	116	118	1034	109	2150
6–2	0122	3–19	7	95	75	390	395	79	519	932	2.95	102	104	879	92	850
6–7	0236	4–15	13	96	89	415	490	98	521	929	2.91	101	104	949	100	850
6–8	0144	3–21	18	100	86	450	502	101	566	1011	3.18	110	125	1003	105	
6–15	0170	3–27	13	112	105	548	617	124	665	1071	2.90	101	107	1084	114	
			Sire Average		88	459	504	101	576	1010	3.10	107	109	996	104	
	Sire: Prospector 7042															
5–15	0004	2–7	12	110	84	577	502	101	723	1192	3.35	116	106	1052	111	$3200
6–5	0020	2–20	Elec	98	92	484	475	95	620	1016	2.83	98	98	951	100	
			Sire Average		88	530	488	98	672	1104	3.09	107	102	1001	105	

Table 8.6 (*Continued*)

COOPERATORS

Final test results on bulls in pen-groups for the 140-day test. Ratios are based on the bull's record relative to the average of 100. Ratios for average daily gain were calculated within breed groups. Average daily gains by breed groups were: Angus, 3.34; Red Angus 3.22; Polled Hereford, 2.87; Hereford, 3.12; Simmental, 3.40; Limousin, 3.43; Charolais, 3.62; Salers, 3.52. Weaning weight and yearling weight are adjusted for age (205-day and 365-day age) and for age of dam. Yearling weights were calculated as adjusted weaning weight plus average daily gain from weaning to final weight × 160. These adjusted weights and the adjusted weaning weight ratio are useful for within herd comparisons. Critical bull comparisons should be made on daily gain on test because of preweaning differences between ranches. It will be decided prior to the sale which bulls will sell.

Test No.	Tattoo	Breed	1980 Birthdate	Sire	Sire of Dam	Actual Weaning Wt	Adjusted Weaning Wt	Adjusted Weaning Wt Ratio	On-Test Wt Nov. 18–19	Ave. 140-Day Wt Apr. 7-8	140-Day Daily Gain	Daily Gain Ratio	Feed/lb Gain	Adjusted Yearling Wt	Sale Price
	Craig, J.; Durango, CO														
1–1	JCLR12M	Lim	2–26	Gendarme	Challenger CFM7	725	710	102	713	1250	3.84	112		1172	$2750
1–2	JCLR5M	Lim	4–2	Eros	Email Cim 55	620	690	98	689	1203	3.67	107		1203	2600
	Neal, G. A.; Dolores, CO														
3–1	213M	Sim	4–23	Barvarian	Toni	590	755	118	639	1173	3.81	112		1249	1600
3–2	202M	Lim	3–6	El Toro	Decor	680	672	105	682	1190	3.63	106		1104	2150
						Pen Average			661	1181	3.72	109	5.00	1176	

Table 8.7. Performance and Sale Prices of a Few of the Angus and Simmental Bulls at the 1981 Northeast Colorado Bull Test Association.

Sale Lot	Cooperator	Birth Date	Birth Wt	Adj 205 Wt	Adj 365 Wt	SC	Back-fat	Adj Hip Ht	ADG(Ratio)	WDA(Ratio)	Index	Sire	Sire of Dam (Sale Order April 11, 1981)	Sale Price
	Angus													
	Average of 89 Angus Bulls				1043	33.0	.22	48.4	3.19(100)	2.83(100)	100	Average of 89 Angus Bulls		
11–3	Goeglein Angus	4–06		628	1259	35.0	.28	50.4	3.76(118)	3.45(122)	120	Emulous 2515T	Commodore of Wye	$4500
27–4	Wardell Angus Ranch	2–11			1209	36.5	.34	50.3	3.83(120)	3.23(114)	117	Shoshone Viking GD60	Rito 72	$1700
26–1	Dave & Ruth VanKem-pen	3–20	74	592	1133	36.0	.20	48.8	3.72(117)	3.08(109)	113	Jones Emulous 232	O Emulous 22	$1500
11–2	Goeglein Angus	3–29		609	1158	36.0	.18	48.3	3.66(115)	3.15(111)	113	Emulous 2515T	Spur TN Emulous 20	$1600
27–1	Wardell Angus Ranch	1–22			1184	37.0	.20	49.8	3.62(114)	3.21(113)	113	Shoshone Viking GD60	Marshall Pride	$2000
21–1	Keith Russell	4–07	74	621	1164	34.0	.28	50.4	3.52(110)	3.20(113)	112	Shoshone Titan FD60	Mon Reposa Lancer	$1800
	Simmental													
	Average of 44 Simmental Bulls				1135	34.2	.09	51.6	3.48(100)	3.08(100)	100	Average of 44 Simmental Bulls		Sale Price
50–1	W.I. or Helen Kingsley 7/8 Simmental	2–06	112	810	1264	38.5	.06	52.9	4.15(119)	3.57(116)	118	Wolf Creeks Parisien	Beat #216	$1800
42–2	Allen & Arick Davis 3/4 Simmental	4–15	90		1345	38.0	.12	53.4	3.73(107)	3.70(120)	114	Signal	Toni	$2950
58–1	Seward Cattle Co. 3/4 Simmental	4–13	95		1338	36.5	.16	52.5	3.71(107)	3.68(119)	113	Oil	Toni	$1600
43–2	Rollie & Ward Deering 3/4 Simmental	4–11	95		1245	37.0	.06	54.1	3.92(113)	3.42(111)	112	Polled Power	Toni	$2600
44–3	Charles Frush 3/4 Simmental	3–22	121	577	1185	39.5	.04	50.5	4.06(117)	3.24(105)	111	Cezon	Canadian Salvator	$1150

Table 8.8. Progeny Performance of High and Low Performing Angus and Hereford Bulls Selected for Weaning Weight and Post-Weaning Gain (Bulls Used for Four Years)

	Bulls Used		
	High Perform.	Low Perform.	Diff. (Lbs)
No. of bulls selected	8	8	
Aver. birth wt (lbs)	70	59	11
Aver. 250-day wt (lbs)	554	486	68
Aver. daily test gain (lbs/dam)	2.47	1.89	0.58
No. calves produced	137	125	
Aver. birth wt (lbs)	62	59	3
Aver. 250-day wt (lbs)	424	405	19
Aver. gain on pasture (lbs)	60	56	4
Aver. gain in feedlot (lbs)	286	269	17
Aver. final wt (lbs)	830	793	37
Aver. carcass wt (lbs)	484	468	16

Source: Auburn Agric. Expt. Sta. Bull. 418.

National Sire Evaluation

A brief description of National Sire Evaluation is given in Chapter 13. The use of a Reference Sire allows many bulls to be tested in several different herds and environments. Most leading breeds have National Sire Evaluation programs, and summaries are available in breed journals or breed association offices (Figure 8.12). Each year there are an

Figure 8.12. Sire Summaries are available for many of the more numerous breeds of beef cattle. Courtesy of CSU Photo Lab.

HOW TO READ THE REPORT

Accuracy *(ACC) is the reliability that can be placed on the* ***expected progeny difference*** *(EPD). An accuracy of close to 1.0 indicates higher reliability.*

Expected progeny difference *(EPD) is the estimate of how future progeny of each sire are expected to perform in each of the traits listed. EPD is expressed in pounds, either plus or minus.*

Carcass ***MG*** *reflects the number of* ***management groups,*** *and* ***P*** *indicates the number of progeny.*

Carcass cutability *(EPD) is expressed in percent.*

USDA carcass quality grade is calculated to one-third of a grade.

Retail yield *reflects the actual difference in pounds among the bulls evaluated. It is determined by multiplying the actual EPD by 485, the average age in days of all cattle slaughtered.*

Maternal breeding value *is taken from the Angus performance pedigree. This estimated breeding value is a prediction of how the daughters of a bull will milk based on weaning weight information. Here again, the accuracy figure is the amount of reliability that can be placed on the breeding value.*

Birth Weight		Weaning Weight		Yearling Weight		Carcass	Carcass Cutability	USDA Carcass Quality Grade	Retail Yield	Maternal Breeding Value	
ACC	EPD	ACC	EPD	ACC	EPD	MG—P	EPD (%)	EPD (⅓ Grade)	EPD (Lb.)	ACC	EBV
72	−1.9	.82	+7.0	.82	+18.9	1−3	+.19	+.04	−19.4	.92	101

Bull with highest MBV with accuracy over .90.

Southolm Bar Lad 50D Registration #7809655
Birth Date: 5-3-72 **Sire:** Southolm Bar Lad 88Z
Breeder: Frank Slezina, Coaldale, Alta, CN
Owner(s): Burtons Highland Meadows, Sisters, OR
Great Plains Western Corp., Staunton, VA
Frank Slezina, Coaldale, Alta, CN

Birth Weight	Weaning Weight	Yearling Weight	Carcass	Carcass Cutability	USDA Grade	Retail Yield	Maternal Breeding Value
ACC EPD	ACC EPD	ACC EPD	MG—P	EPD	EPD	EPD	ACC EBV
.43 + 1.3	.90 + 6.6	.89 + 25.2					.97 110

Bull with highest EPD for both weaning and yearling weights.

Rito Excel 809 Registration #8781383
Birth Date: 3-18-76 **Sire:** PBC 707 1M E0046
Breeder: Pioneer Beef Cattle Co., Johnston, IA
Owner(s): Byron G. Beukema, Newton, IA

Birth Weight	Weaning Weight	Yearling Weight	Carcass	Carcass Cutability	USDA Grade	Retail Yield	Maternal Breeding Value
ACC EPD	ACC EPD	ACC EPD	MG—P	EPD	EPD	EPD	ACC EBV
.87 + .1	.86 + 32.2	.85 + 77.1	1 − 16	+.23	−.22	+17.0	.74 102

Bull with one of the most desirable EPDs for birth weight.

Sum Up Bloage D29 Registration #4451292
Birth Date: 5-12-64 **Sire:** Bloage of Craigie
Breeder: David R. Miller, Smithville, MO
Owner(s): 3 6 Angus Ranch, Burwell, NB
Sitz Angus Farm, Harrison, MT

Birth Weight	Weaning Weight	Yearling Weight	Carcass	Carcass Cutability	USDA Grade	Retail Yield	Maternal Breeding Value
ACC EPD	ACC EPD	ACC EPD	MG—P	EPD	EPD	EPD	ACC EBV
.76 − 4.1	.82 + 4.9	.80 + 1.8					.94 101

Two bulls with the best combination of MBV and EPD for yearling weight.

Jetliner 707 of Conanga Registration #7795918
Birth Date: 4-3-73 **Sire:** Rito 707 of Ideal 836 71
Breeder: Grant Ranch, Mott, ND
Owner(s): Knop Angus Farm, Ida Grove, IA
Jorgensen Ranches, Ideal, SD

Birth Weight	Weaning Weight	Yearling Weight	Carcass	Carcass Cutability	USDA Grade	Retail Yield	Maternal Breeding Value
ACC EPD	ACC EPD	ACC EPD	MG—P	EPD	EPD	EPD	ACC EBV
.95 + 3.3	.95 + 25.3	.95 + 55.1	25 − 170	+.26	−.27	+13.1	.97 104

Schearbrook Shoshone Registration #7011424
Birth Date: 2-7-71 **Sire:** Beaufort of Wye
Breeder: Henry Leonhardt, Jr., & Sons, Cowley, WY
Owner(s): Henry Leonhardt, Jr., & Sons, Cowley, WY
Schearbrook Angus, Clayton, OH

Birth Weight	Weaning Weight	Yearling Weight	Carcass	Carcass Cutability	USDA Grade	Retail Yield	Maternal Breeding Value
ACC EPD	ACC EPD	ACC EPD	MG—P	EPD	EPD	EPD	ACC EBV
.94 + 5.6	.95 + 25.2	.94 + 56.0					.98 109

Bull with a unique combination of moderate birth weight and high yearling weight.

Nichols Benchmark J32 Registration #9107078
Birth Date: 4-10-77 **Sire:** Bold Bell Boy G90
Breeder: Merrill Nichols & Sons, Bridgewater, IA
Owner(s): Bell & Sons, Osceola, IA
Nichols Farms, Bridgewater, IA

Birth Weight	Weaning Weight	Yearling Weight	Carcass	Carcass Cutability	USDA Grade	Retail Yield	Maternal Breeding Value
ACC EPD	ACC EPD	ACC EPD	MG—P	EPD	EPD	EPD	ACC EBV
.90 + .8	.89 + 21.1	.89 + 64.8	2 − 6	−.14	−.05	−2.4	.58 101

Figure 8.13. Selected information from the 1981 Angus Sire Evaluation Report which includes data on 673 bulls. Courtesy of American Angus Association.

increased number of progeny evaluated, and as more numbers accumulate, the sire comparisons become more meaningful.

Figure 8.13 shows some sire evaluation data on several bulls listed in the 1981 Angus Sire Evaluation Report. The top part of the figure gives an explanation of the data in the report. Additional future progeny on the bulls in the report may change in the EPDs if the accuracy figure is around 0.50 to 0.80. Future data on bulls with present accuracy figures of 0.90 and higher, however, will likely make very little change.

Each bull listed in the Angus Sire Evaluation Report can be compared to every other bull listed. The data shown take into account only differences expressed in each herd in which bulls were used and do not deal with absolute weights or ratios. For example, bull 1 has a weaning EPD of +10.0 lb (Acc. of 0.97), and bull 2 has a weaning EPD of −10.0 lb (Acc. of 0.97). If these two bulls were used in the same herd, each on an equal group of cows, bull 1's calves would be expected to weigh 20 lbs more at weaning than the calves from bull 2.

Figure 8.14 shows selected information taken from the 1980–1981 Simmental National Sire Summary on approximately 1000 bulls, representing the summary of over 400,000 progeny records. In this summary, a ratio is presented for each trait, and the accuracy is computed as an expected deviation from the ratio.

Purebred breeders should consider using bulls that have been tested under the National Sire Evaluation Program or sons of these bulls where the sons have high individual trait ratios. Commercial producers can in turn use high performing bulls purchased

Here is an explanation of the performance information that appears in the Sire Summary.

All the Data that prints for the bulls in the Sire Summary is located in this section. The bulls are arranged in alphabetical order by their official name.

* = Reference Sire

(1) NAME OF BULL (2) BULL'S SIRE / DAM'S SIRE	(3) Horned Polled Status	(4) Registration Number	Bull's Birthdate and Country of Origin	(5)	(6) CALVING EASE FIRST CALF	CALVING EASE 2ND CALF & OLDER	BIRTH WEIGHT	CALVING EASE INDEX	WEANING WEIGHT	WEANING GAIN	YEARLING WEIGHT	YEARLING GAIN	DAUGHTER'S 1ST CALF EASE	DAUGHTER'S 1ST CALF BIRTH WEIGHT	DAUGHTER'S 1ST CALF WEAN WEIGHT	USDA QUALITY GRADE	CARCASS—YIELD GRADE	RETAIL CUTS PER DAY AGE
Firn *	H	8		RATIO	98.64	97.05	98.14	97.93	99.03	98.47	98.63	100.57	98.11	99.86	99.49	102.65	105.08	101.18
Dursli			3-12-67	ACCURACY	8.59	1.59	.41	2.22	.42	.69	1.75	3.55	2.89	.75	.63	4.96	5.64	2.30
			SWITZERLAND	PROGENY	72	1296	2904	3508	4449	1974	289	289	718	851	1339	84	83	83
Fisch 20F	H	240467		RATIO		100.91	97.28	97.92	101.11									
Bar 11 Ueli * (H)		5339	3-2-74	ACCURACY		* 14.81	5.30	19.01	4.80									
Tino			CANADA	PROGENY		23	28	37	30									
Florian *	H	11		RATIO	98.19	98.95	101.63	100.27	98.23	97.99	98.57	98.27	80.985	97.186	100.16	99.31	98.40	100.84
Ami			3-2-68	ACCURACY	2.23	.76	.24	.75	.25	.40	1.01	2.04	1.73	.51	.45	2.87	3.26	1.33
			SWITZERLAND	PROGENY	1554	* 7147	9126	12670	13647	6514	973	973	2159	1857	2772	249	253	246

1. Bull's Official Name. **Remember many bulls have nicknames but will appear only under their official registered name including prefixes if they were used.**

   A Bull listed on a shaded line indicates he is a Reference Sire. A star(*) by a bull's name also indicates he is a Reference Sire.

2. Names of his Sire and his Dam's Sire, **along with a Reference Sire designation and their horned-polled status.**
3. Bull's Horned-Polled Status.
4. Bull's ASA Registration Number, **and ASA registration numbers (if any) of his sire and dam's sire.**
5. Bull's Date of Birth and Country of Origin.
6. These are all 14 Traits **that are summarized in the Sire Summary.**
7. Ratio. **A trait ratio compares the performance of a bull's progeny to progeny of all other sires in the breed for this trait. A ratio of 100 would mean his progeny are exactly average for the breed. Ratios over 100 indicate that his progeny are above average. Obviously, half of all Simmental sires will be above 100 and half below 100 in each trait; this does not mean that bulls with ratios of less than 100 are not worth using.**
8. Accuracy. **Random changes (+ or -) in a trait ratio should be expected as additional progeny are reported. The majority of trait ratios should change less than the accuracy value. When a bull has had 300 progeny reported for any trait, his performance in that trait is rather firmly established and it's not likely to change much in the future.**
9. Progeny. **Number of progeny sired by this bull that were compared to a Reference Sire in this trait.**

Figure 8.14. Data from the Simmental National Sire Summary showing how the sire information is presented. Courtesy of American Simmental Association.

from seedstock producers who have directly or indirectly used the National Sire Evaluation Program or a similar testing program. These different methods of sire selection will give assurance to a higher predictability of selection for genetic superiority (Figure 8.15).

Figure 8.15. This bull, pictured as a yearling, has excellent individual, progeny test, and National Sire Evaluation records. Courtesy of CSU.

Individual:		
	Adj. 205 Day Wt. Ratio	116
	Adj. 365 Day Wt. Ratio	116
	160 Day Gain Ratio	116
	Yearling Frame Size	6.5
	Backfat Probe at 1015 lbs	0.25 in.
	Yearling Scrotal Circumference	37.2 cm

	Sire	No. Calves	Test Gain	Weight Per Day of Age	Yield Grade	Quality Grade	Lbs Lean Per Day
Progeny Test:	Pictured	12	3.25	2.61	2.7	Ch–	.82
	Ref. Sire	13	3.00	2.51	2.6	Ch–	.79
	Pen Ave.	313	2.88	2.42	2.9	Ch–	.75

National Sire Evaluation Record:

Distribution		*Yearling*		*Weaning*		*Birth*		*Maternal*		
Herds	Progeny	EPD	ACC	EPD	ACC	EPD	ACC	DGT	MBV	ACC
4	142	+92.1	.82	+44.2	.84	+0.3	.82		104.6	.63

Current Breeding Values:

Maternal	*Weaning*	*Yearling*
105	106	109

Bulls Versus Steers

Seedstock and commercial producers should be fully aware of how sex differences affect productivity and performance of the economically important traits. Sexual maturity as reflected in bulls, steers, and heifers has a marked effect on growth and carcass characteristics. Most beef industry goals are reflected as a steer endpoint since most cattle going to slaughter are steers. Table 8.9 compares the expected productivity differences between steers and bulls.

The data in Table 8.9 are provided to demonstrate that selected bulls will need to be more extreme than what is expected in their counterpart steers. Producers that select bulls at the same level of productivity anticipated in the steers will generally be disappointed as to siring ability of the bulls.

The growth data in Table 8.9 are compared at the same age, while carcass data are compared at the same slaughter weight.

What is a Bull Worth?

This question is typically posed by producers evaluating bulls with high performance records. The base price for a bull is his salvage or slaughter value. Any value of the bull, above slaughter price, is based on the bull's ability to sire live calves and the genetic superiority of his calves.

Using the example previously shown in Table 8.8, the value of high-performing versus lower performing bulls can be compared (Table 8.10).

Some commercial producers use the guideline that a new bull should cost them the value of three weaning steers or two yearling steers. From the above example, it is obvious that value of commercial bulls can be \$2000 to \$3000 above slaughter value.

Using an average price of \$2000 for commercial bulls, a seedstock producer (multiplier) could well justify a price up to \$50,000 for a bull. This is assuming that half of the \$2000 is for expenses and only 50 bulls would be sold over a four-year period averaging \$2000 (50 bulls @ \$1000 = \$50,000).

Table 8.9. Comparative Productivity of Bulls and Steers[a]

Trait	Bulls	Steers
205-day wt (lbs)	475	450
Feedlot gain (lbs/day)	3.0	2.75
365-day wt (lbs)	955	890
Yearling hip ht (in.)	47	45
Feed/lb gain	6.8	7.8
Backfat (in.)	0.3	0.6
Ribeye area (sq in.)	13.0	11.0
Cutability (%)	51.0	48.0
Quality grade	Gd+	Ch–

[a] The growth data in Table 8.9 is compared at same age, while carcass data is compared at same slaughter weight. Data used have been adapted from several steer and bull comparisons where half-brothers were evaluated.

Table 8.10. Value Differences of High and Low-Performing Bulls Used 4 Years

Item	Bulls	
	High Perform.	*Low Perform.*
Salvage value after 4 years (1800 lb @ $0.60)	$1080	$1080
Weaning advantage (19 lbs @ $0.70 × 140 calves)	$1862	—
Feedlot advantage (37 lbs @ $0.65 × 140 head)	$3367	—

Elite seedstock producers can well justify prices in the several hundred thousand dollar category, especially if they are selling semen and AI certificates. Some of these breeders are selling two units of semen and an AI certificate for $150 or higher. Some bulls have generated over a million dollars with semen and certificate sales.

Using Bulls in Crossbreeding Programs

A genetically superior bull will make genetic improvement for one or more traits in a herd, but not necessarily improve these same traits in all herds. Each herd has a different level of productivity, and part of this productivity is determined by genetic superiority or inferiority in the herd.

Commercial producers can use genetic improvement most effectively in their herds by using a combination of sire selection and crossbreeding. There are several crossbreeding systems; however, the one which appears to maximize net return is shown in Figure 8.16. This is true if a producer keeps cow size small to medium in the Breed A to Breed B cross and uses larger, terminal cross bulls of Breed C. Approximately 60% of the cowherd would need to be used to raise the smaller A to B females and all the 40% of terminal cross steers and heifers would be sold. Table 8.11 shows the bull selection criteria for each cross to be made.

Producers using a 2-breed or 3-breed rotation crossbreeding system (see Chapter 13) should emphasize the following traits in a bull selection program: (1) keep frame size of progeny in 4.0 to 6.0 range, (2) emphasize maternal traits, particularly MBV in all bulls and calving ease for first-calf heifers, and (3) selection for heavy weaning weights and yearling weights (ratios and breeding values). It should be recognized, however, that these crossbreeding systems will not be as high in net return as the one shown in Figure 8.16. In these crossbreeding programs, as a producer selects for heavier weaning and yearling weights in the calves, cow size will increase proportionally. Thus, increased maintenance costs of heavier cows typically offset the economic advantage of heavier calves. Cow size and calf weights, therefore, are usually determined by amount of feed resources (Figure 8.17).

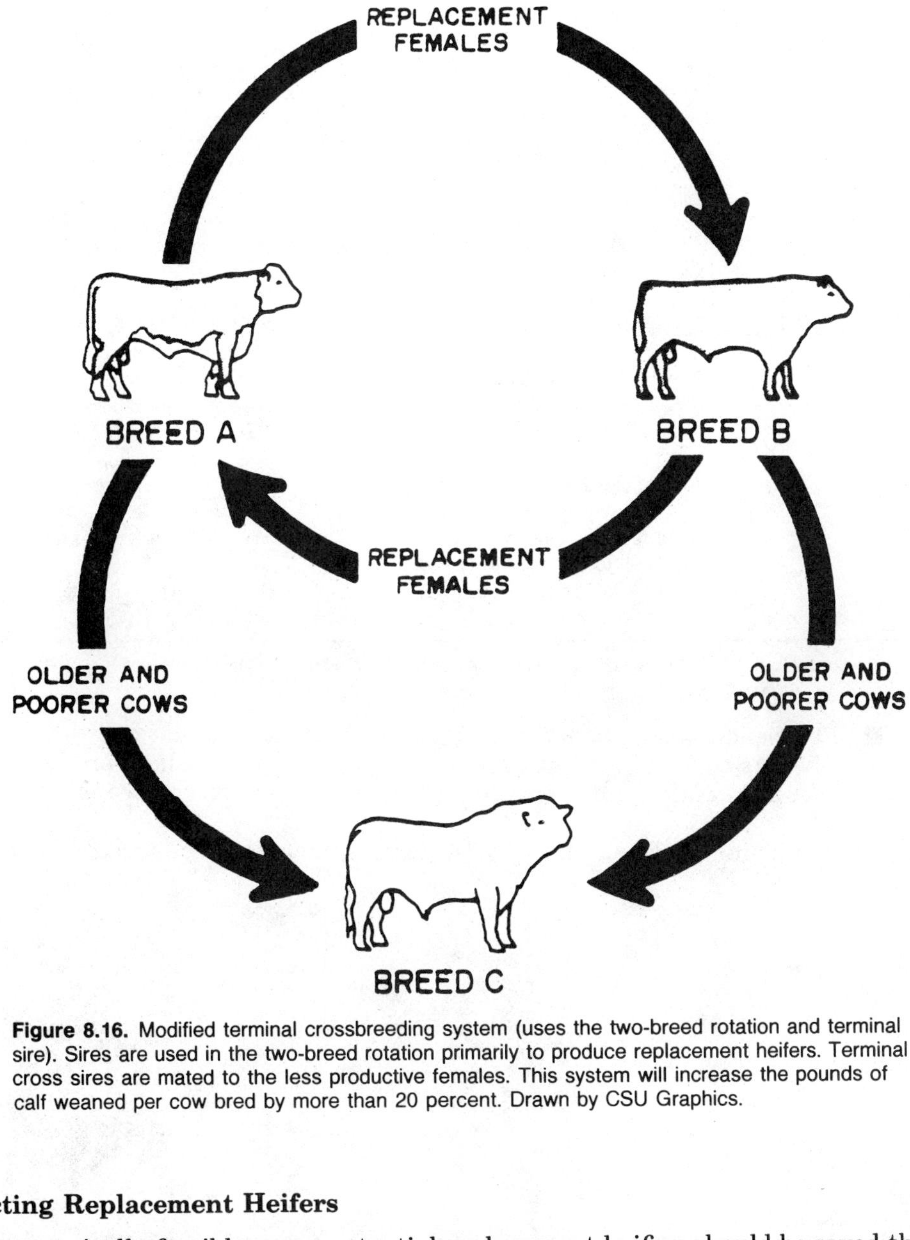

Figure 8.16. Modified terminal crossbreeding system (uses the two-breed rotation and terminal sire). Sires are used in the two-breed rotation primarily to produce replacement heifers. Terminal cross sires are mated to the less productive females. This system will increase the pounds of calf weaned per cow bred by more than 20 percent. Drawn by CSU Graphics.

Selecting Replacement Heifers

If economically feasible, more potential replacement heifers should be saved through breeding before a final culling is made. This favors a higher calf crop percentage by culling for pregnancy and particularly early pregnancy. Culling for pregnancy is based more on how it affects economic return rather than making known genetic change in

Table 8.11. Bull Selection Criteria for the Modified Terminal Crossbreeding Program where Replacement Heifers are Raised

Type of Cross	Level of Productivity Desired	Traits Emphasized in Bull Selection[a]
A × B Cows	Small to medium size (900–1100 lbs mature wt).	Frame size 3.5 to 4.5.
	Heavy milk flow (consistent with feed supply).	MBV of 101 or higher (0.90 or higher accuracy if used AI).
	Ease of calving.	WBV of 101 or higher min.—wean ratio of 103 min.
	Steers YG2, Qual. Grade Gd+ to CH− at min. 600 lb carcass wt.	Birth wt under 80 lbs (for heifers) Scrotal circum. 34 to 40 cm.
Breed C Bulls to use on A × B Cows	Heavy calves at weaning	Frame size 6.0 to 7.5 (steer basis)
	Rapid feedlot gains, YG1 to 2, Gd+ to Ch− quality grade at 650 to 750 lb carcass wts.	WBV 103 Min.—wean ratio 105 min.
		YBV 103 min.—yrl. ratio 105 min. Birth wt 85 to 100 lbs.

[a] All breeding values (MBV, WBV, YBV) should be higher than those given if artificial insemination is used.

reproductive efficiency. Some producers, particularly seedstock producers, will save more replacement heifers beyond breeding and pregnancy evaluation. These producers practice a more complete culling after evaluating the weaning performance of the heifer's first calf.

Replacement heifer selection for genetic improvement of economically important

Figure 8.17. This Jersey-Angus crossbred cow weighs 650 lbs as a 3-year-old. The Charolais-sired calf weighed 480 lbs at 210 days and 680 lbs at 10 months. Terminal crossing programs may have advantages not yet recognized by many commercial beef producers. Courtesy of Texas A and M University.

Table 8.12. Replacement Heifer Selection

	Stage of Production			
	Wean.	**Yrl. or Breeding**	**Preg. Check**	**After 1st Calf**
Trait				
Actual wean. wt	***[a]			
Adj. 205-day wt, MBV, and WBV	**		*	
Sire and dam records (MPPA, MBV, WBV, YBV)	**	*	*	
Hip ht and visual (frame, fat and soundness)	**	*	*	
Actual yrl. wt		***		
Adj. 365-day wt and YBV		**	*	
Pregnancy status			***	
Weaning performance of first calf				***

[a] Asterisk identifies trait to consider, while number of asterisks give recommended emphasis (greater number = more emphasis)

traits is not a simple task. Most of the problem is centered in evaluating future maternal performance. Some heifers, having high weaning weights and weaning weight ratios, do not perform as anticipated or as ranked by their record. Fat accumulation, reflecting itself in weight, appears to be detrimental to future performance in maternal traits.

Heifer selection can occur in part or totally at different stages in the early productive life of the heifer. Most of these stages occur at the same time of marketing or when performance records are taken. These stages are weaning, yearling, breeding, pregnancy checking, and after weaning of first calf. Table 8.12 shows the recommended emphasis to place on traits at various stages of the heifer's productive life. The number of heifers culled at any one stage is dependent on marketing alternatives and feed resources. These two factors will have the greatest effect on economic return which must be balanced with objectives of improving calf crop percentage and making genetic change in several traits.

Culling Cows

Unless a producer is involved in a herd reduction program, relatively few cows can be culled to enhance genetic progress in the herd. Most cows basically cull themselves because of death loss and age. Producers can practice some culling on performance records if there is a very low death loss and if cows over 9–10 years of age can sustain high levels of productivity for several more years.

Cows can be culled on pregnancy status and calving interval. The basis for this culling appears to be more for economic justification rather than practicing selection to make genetic progress. There is evidence that open cows can rebreed and produce calves regularly during the remainder of their productive life (*KSU Cattlemen's Progress Rpt.* 230). The economic decision to cull open heifers and cows should be based more on

Table 8.13. Cow Records on High and Low Producing Cows in an Angus Herd

	Breeding Values of Cow			Performance Records of Calves		
Cow No.	MBV	WBV	YBV	No. Calves	Wean. Wt Ratio	Yrl. Wt Ratio
3591	109	106	107	4	109	108
1644	95	97	95	3	93	91

Table 8.14. Cow Records on High and Low Producing Cows in a Hereford Herd

	Records on Cows					Individual Records on Own Calves	
Cow No.	MBV	WBV	YBV	MPPA	Calving Interval (Days)	Weaning Wt Ratio	Yearling Wt Ratio
3201	108	107	108	109	351	113	105
						119	118
						109	110
						119	116
3215	101	99	100	97	382	98	102
						91	NA
						96	97
						78	82

replacement cost. It appears more obvious that the commercial producer should cull the open females; however, the decision is not as obvious or easy for a seedstock producer to cull open females with high breeding values. This decision is most controversial in the minds of many producers. The decision should, however, be made on known biological relationships and not on opinions.

Culling cows to assist in making genetic change in a herd is relatively simple if the cows performed together under similar environmental conditions. One of the biggest problems in comparing cow records is that the cows likely have produced calves from different bulls. Relatively large genetic differences in bulls used in the herd would make the cows' records less comparable as when all cows were bred to the same bull. This usually will cause no serious problem where producers might be culling the bottom 10% to 20% of their cows on performance records. Tables 8.13 and 8.14 show typical available records that can be used in culling cows. The lower breeding values and lower weight ratios of calves rather easily identify lower producing cows in the herd.

SELECTED REFERENCES

Publications

Cartwright, T. C., 1979. *Size as a component of beef production efficiency: Cow–calf production.* J. of Anim. Sci. 48:974.

Cundiff, L. V. and Gregory, K. E., 1977. *Beef Cattle Breeding.* USDA Agric. Inform. Bull. 286.

Guidelines for Uniform Beef Improvement Program. Beef Improvement Federation Recommendation, 1981. USDA Ext. Serv. Program Aid 1020.

Lasley, J. F., 1978. *Genetics of Livestock Improvement.* Englewood Cliffs, NJ: Prentice-Hall, Inc.

Performance Record Pamphlets and National Sire Evaluation Summaries (available from most beef breed associations), 1983.

Smith, G. M., 1979. *Size as a component of beef production efficiency: Feedlot production and integrated efficiency.* J. of Anim. Sci. 48:966.

Warwick, E. J. and Legates, J. E., 1979. *Breeding and Improvement of Farm Animals.* New York: McGraw Hill.

Visuals

"Purebred Operations" (filmstrip and tape). Vocational Education Productions. California Polytechnic State University, San Luis Obispo, CA 93407.

9

Integrated Beef Cattle Operations

In Chapter 1, the beef industry was visually portrayed and described as several loosely interlocking segments which take the beef animal and its resulting products through several stages of production, processing, and marketing. These segments, which depend on one another for product flow, have different economic parameters, and thus in a sense, operate somewhat independently from one another.

Vertical integration refers to complete ownership or management control of two or more of these segments or stages of production and marketing. While there has been a considerable amount of vertical integration in many industries in the American economy and in certain areas of agriculture (e.g., poultry industry), there has been relatively little vertical integration in the beef industry. No doubt this is due to strong feelings of independence of cattle producers and lack of incentives because of the wide geographical dispersement of the industry. Also, the management and resource base is quite different for each segment of the industry.

Purebred and Cow–Calf

Although in a minority, there are several cow–calf operations which contain both purebred (seedstock) and commercial herds. Resources and management are similar for both types of cattle operations which favors this type of integration. One of the primary advantages is that it allows producers to control genetic input into their commercial herds while reducing variability of environmental differences between purebred and commercial herds. Also, bulls produced in the seedstock herd can be developed under similar environmental conditions as the commercial herd. This is believed by some producers to improve adaptability and performance of breeding cattle.

Cow–Calf/Yearling (Growing)/and Feedlot

Little precise information exists which describes the amount of integration existing between cow–calf and feedlot segments. Figure 9.1 gives results from one report which

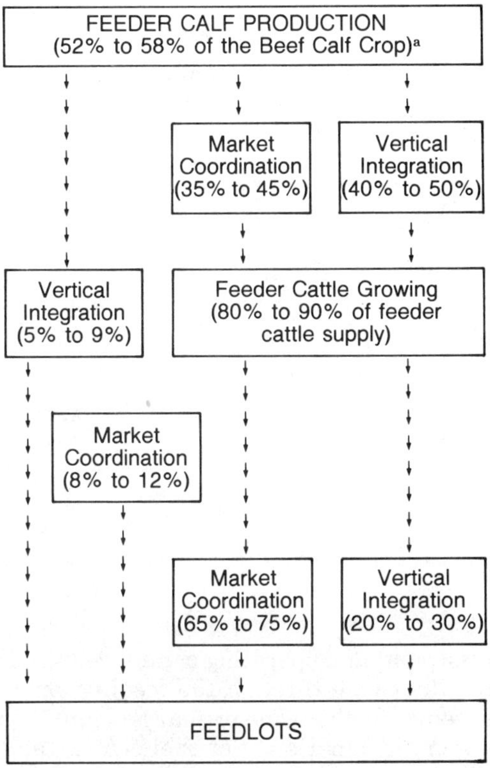

[a] Feeder calf production equals the best calf crop less death loss (approx. 8%), replacement of breeding stock (approx. 20%), veal slaughter (approx. 5%), and slaughter of cattle not finished on grain (approx. 12%).

Figure 9.1. Vertical integration and market coordination of feeder cattle to the feedlots. *Source:* Henderson and Schwart (1977).

estimates that 40% to 50% of the feeder calves produced are integrated with yearling or growing operations. Also, approximately 20% to 30% of cattle placed on feed are integrated with yearling or growing segments. Fewer than 10% of feeder calves produced are integrated through all three stages of production.

Combining of cow–calf and yearling operations is one of the most common types of integration in the beef industry. In fact, some may argue that they are to be considered a single production segment rather than two. Land, feed, and management resources are quite similar for both types of operations. Over a period of time, a single ranch may produce cows and calves, shift to only a yearling operation, or have a combination of the two. This shifting appears to be directed by an economic assessment as to which would be most profitable. Profitability is highly related to the proportion of calves to yearlings over the years and the annual cost of maintaining a cow.

Cow–Calf (Yearling) and Feedlot

Custom cattle feeding has permitted vertical integration of the cow–calf or yearling producer and the feedlot. Rarely is there one owner of the cow–calf/yearling operation

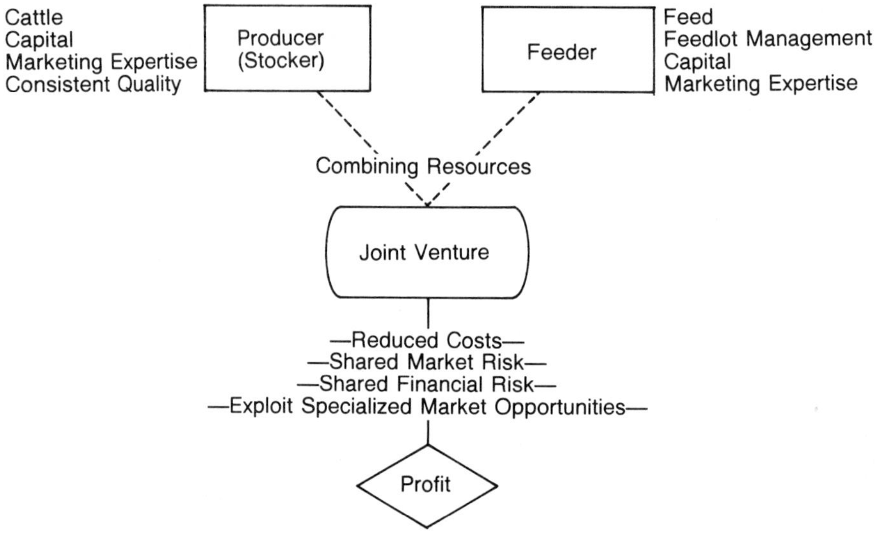

Figure 9.2. An example of a joint venture where producers in two segments combine their resources to enhance their profits. *Source:* Beef Business Bulletin (NCA), 26 August 1983.

and the feedlot because of the large amount of investment and operating capital required. Custom feedlots do allow cattle producers to own the cattle through the feeding phase while paying for feed, labor, and cost of using the feedlot facility. The primary advantages to cow–calf producers allow them to utilize the genetic superiority of their cattle, and it also spreads the high investment and operating cost (on a per cow basis) over more pounds of calf produced. Joint ventures exist between some cow–calf producers and feedlot operations (Figure 9.2).

Custom feeding gives cow–calf producers some revenue stability because fed cattle prices appear to fluctuate less from month to month than do feeder cattle prices. Although cow–calf producers retain ownership of their calves through the feedlot, they do assume risks of health problems and market price changes. It also changes the pattern of cash flow which may not be agreeable with lenders of borrowed money, and income tax problems may be encountered the first year or two.

Some feedlot operations desire to maintain ownership of growing cattle to (1) assure a more consistent flow of feeder cattle into the feedlot, (2) utilize high roughage feeds which may be part of a farming operation, and (3) background cattle for health and feeding adaptation prior to the finishing phase in the feedlot.

Feedlot and Packer

Some packers have found it desirable to integrate into the feeding segment primarily to stabilize the slaughter cattle supply for their operation. The primary reasons for packers owning their own slaughter cattle are: (1) to have a reliable source of slaughter cattle which permits an efficient use of their labor supply, (2) to have a ready supply of cattle of desired weights and grades, (3) to meet unexpected orders, and (4) to have a supply of cattle when other slaughter cattle are not readily available.

An example of this type of integration, on a large scale basis, is Monfort of Colorado who feed approximately 400,000 head of cattle each year in two feedlots near Greeley, Colorado. Monfort has an annual cattle slaughter of approximately 1.5 million head in company-owned packing plants located in Greeley, Colorado and Grand Island, Nebraska. Monfort also is integrated into retailing of beef.

There has been concern, primarily by cattle feeders, that if packers also feed cattle, packers could adversely affect market prices through this type of arrangement. This concern became magnified when custom feedlots were rapidly expanding during the 1960s and early 1970s. As a result of these concerns, the Packers and Stockyards Administration (USDA) implemented a regulation that prohibited packers from owning, financing, or participating in the management of a custom feedlot. This regulation prohibited packer control over custom fed cattle—cattle owned by someone other than the packer. Packers are not prohibited from hiring the services of custom feedlots as long as the packers own the cattle being fed.

Packers may own feedlots as long as they feed only their own cattle. The number of cattle owned and fed by packers has ranged from 5% to 10% of the total fed cattle supply since 1955 with no consistent trend, either up or down. It appears there is a lack of economic incentive to expect this number to increase significantly in the near future.

Feedlots may own packing plants as long as they feed only their own cattle. There are some of these integrated operations in existence, although there are no valid data on numbers of operations and cattle involved. Feedlot owned packing plants are rather few in number, primarily because economic incentives do not favor them, especially the amount of capital needed by feeders for both feedlot and packing plant construction and operation.

Packer and Retailer

Integration of the packer and retailer segment does exist, however, the magnitude of this type of integration is not known. It does not involve a high percentage of cattle slaughtered.

Total Integration

There are very few examples in the beef industry of total integration from the purebred segment to direct sale of retail product to consumers. Periodically, beef publications will identify some cow–calf producers or feeders who market beef directly to consumers. These integrated operations represent a very small proportion of the beef marketed in the United States. Most of these operations only exist a few years, and then one or more of the segments of the total operation are phased out. Owners of these totally integrated operations state that it is extremely difficult to understand and manage the different and somewhat unique phases of the total integrated operation. Some examples of totally integrated operations are:

1. Schmitt–Burnet, Inc., Near Great Falls, MT
   –30,000 acre ranch.
   –500 head crossbred cows and crossbred bulls.
   –fed in Schmitt–Burnet Feedlot

–slaughter in company slaughter house in Stanford, MT.
–company makes door to door deliveries of frozen retail beef to 1200 families in Montana area.

2. Cedar Breaks Beef Co., Fort Pierre, SD.
   –cow–calf, feedlot, packing operation.
   –marketing primarily frozen beef from 20 carcasses per week to customers in the general area.
3. Commanche Land and Cattle Co., Commanche, TX.
   –raise approximately 500 calves.
   –operate 5,000 head feedlot.
   –market their home-raised calves plus 50% of the cattle in the feedlot through 5 retail stores (Commanche Meats).
4. Triple C Meats, near Devine, TX.
   –11,000 head custom feedlot.
   –slaughter, processing and retail meat market near feedlot.
   –sells 250 head per week through retail outlet.
5. Stuart Anderson's Black Angus Restaurants.
   –has 50 restaurants in 10 states, using four and one-half million pounds of beef per year.
   –owns 2500 acre ranch in Yakima Valley, WA.
   –has purebred Angus cattle.
   –feed 1000 head Angus crossbreds per year.

The beef industry, for the most part, remains a rather volatile industry, coordinated primarily by the open market. This is true despite the marked increase in size of feedlots, packing plants, concentration of cow–calf operations, feedlots, and packing plants in the midsection of the United States. A variety of contractual and integrated operations exist, but clear advantages which exist over time are confined to special cases.

Because of variability in profits and prices, there are incentives to tie one production segment to another agricultural or business operation to reduce risk. Risk management in many cases encourages diversification.

Integration of two or more segments can increase efficiency in some cases, but often these advantages are small when compared to other management and financial considerations. Both vertical and horizontal integration in the beef industry appear to be increasing slowly over time; however, it appears that the open market will continue to coordinate beef industry segments in the next several years ahead.

SELECTED REFERENCES

Publications

A feedyard supermarket. Feedlot Management, January 1983.

Farris, D. E. and Couvillion, W. C., 1975. *Vertical coordination of beef in the south.* Southern Coop. Series Bull. 192.

Gee, C. K. and Pursley, J. K., 1972. *The economics of retained ownership of calves on eastern Colorado cattle ranches.* CSU Expt. Sta. Bul. 551S.

Henderson, D. R. and Schwart, R. B., Jr., 1977. *Vertical coordination between beef cow–calf enterprises and feedlots in the United States.* Research Report of Beef Subsector Task for Vertical Integration. NC–117.

Petritz, D. C., Erickson, S. P. and Armstrong, J. H., 1982. *The cattle and beef industry in the United States: Buying, selling, pricing.* Purdue Coop. Ext. Serv. CES Paper 93.
Producer Breeds the Calves He Wants. Farm Journal, April 1982.
Progressive conception-to-consumer marketing. Livestock, July 1981.
Rancher-Restauranteur Has Unique Niche. The Record Stockman, 7 September 1982.
Ross, J. *Joint ventures—avoiding misadventures.* Beef Business Bulletin (NCA), 26 August 1983.
Williams, E. and Farris, D. E., 1975. *Economics of vertical coordination in beef cattle systems.* Texas Agric. Expt. Sta. Bull. MP–1205C.

10
Beef Cattle Management Decisions

Effective management of beef cattle operations implies that available resources are used to maximize continuing net profit while conserving and improving the resources. The fixed resources are land, labor, and capital, and management involving the renewable biological resources of animals and plants. Effective management involves a manager who knows how to make timely decisions based on a careful assessment of management alternatives. Modern technology is providing useful tools to make more rapid and accurate management decisions.

Many of the chapters in this book discuss biological relationships of beef cattle which determine their efficiency of production. It is important to identify other resources, which combined with the efficiency of animal production will determine the profitability of an operation.

The Manager

The manager may be an owner–operator with minimum additional labor or the manager of a more complex organizational structure involving several other individuals (Figure 10.1). An effective manager, whether involved in a one-person operation or a complex organizational structure, needs to (1) be profit oriented, (2) keep abreast of current knowledge related to the operation, (3) know how to use time effectively, (4) attend to physical, emotional, and financial needs of those employed in the operation, (5) incorporate incentive programs to motivate employees to perform at their full capacity each day, (6) have honest business dealings, (7) communicate responsibilities effectively to all employees and make employees feel they are part of the operation, (8) know what needs to be done and at what time, (9) be a self starter, (10) set priorities and allocate resources accordingly, (11) remove or alleviate high risks, (12) manage oneself so that others can see and follow a good example, and (13) identify objectives with both short- and long-range goals to achieve those objectives.

It is imperative that beef producers conduct each operation as a business and not just as a way of life. Current economic pressures associated with keener competition, force more producers to manage their operations as business enterprises.

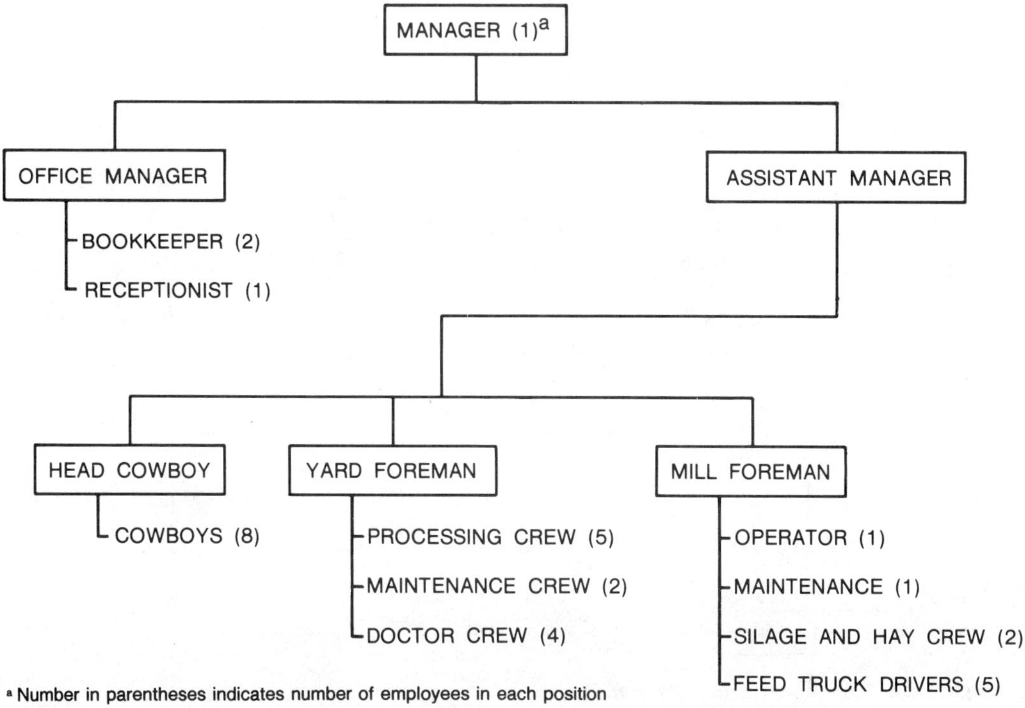

[a] Number in parentheses indicates number of employees in each position

Figure 10.1. Organizational structure of a large commercial cattle feeding operation. Drawn by CSU Graphics.

The manager is the individual responsible for planning and decision-making. The simplified management process is to plan, act, evaluate, then repeat the process. This process is given in more detailed steps in Figure 10.2. It is intended that this process not go in circles but that meaningful progress can be made with time (Figure 10.3).

Managers have the responsibility of developing an effective management plan. The plans should be written so that managers can crystallize their ideas more concisely. A written plan permits a more effective identification of goals, objectives, plan of action, and periodic evaluation.

Goals are important because they give direction to a management program. Written goals, along with objective evaluations, allow producers to measure progress of improvement and to give priority to needed management changes.

Realistic goals can partially be determined by asking and answering the following questions:

1. What do I want my cattle to do for me?
2. What are my cattle presently doing for me? (Current level of productivity and profitability.)

The answers to these questions should move from opinions to numbers and objective statements which are written down. There is generally a gap between the two questions, which allows producers to determine a course of action in moving the present level of productivity and profitability toward the desired goals.

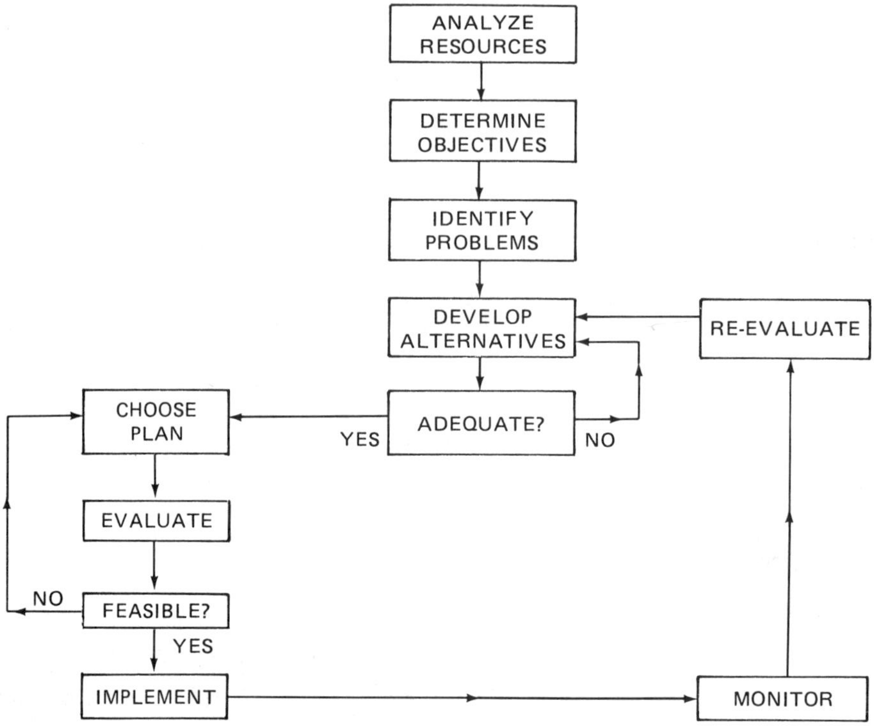

Figure 10.2. Major component parts of the planning process. Drawn by CSU Graphics.

Meaningful goals should also be consistent with goals of the beef industry (Table 8.1, Chapter 8) and the restraints imposed by the variability of individual operations.

Managers understand that it is the end results which really count. To achieve the desired end results, priority must be given to the use of time. Every manager has the same resource—a fixed amount of time. How that time is used is the difference between good managers and poor managers.

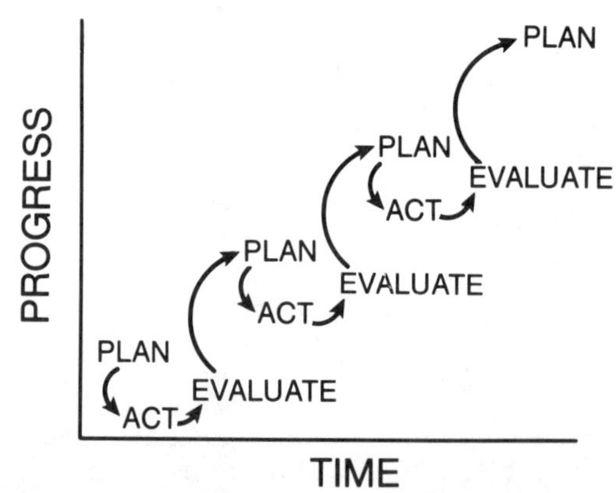

Figure 10.3 Progress is intended to be accomplished in the management decisions process. Drawn by CSU Graphics.

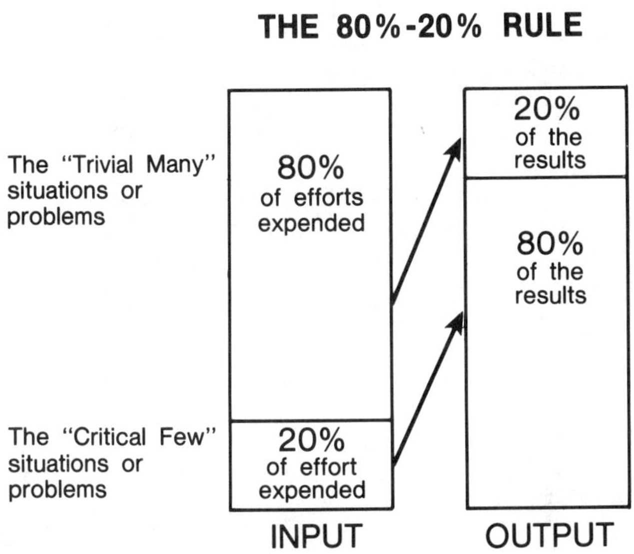

Figure 10.4. The 80%–20% rule exists in each beef operation. Courtesy of NCA.

All beef cattle operations have the "trivial many" and the "critical few" situations or problems. It is important that the manager be able to determine the difference between these two categories and how each affects the end results. An understanding of the 80% to 20% rule is important to the success of a manager (Figure 10.4).

Some managers construct management calendars which are unique to their operations. These calendars would list important activities which must be accomplished during a specific day of the month. A management calendar used by a selected cow–calf producer is shown in Table 10.1. This producer can review the major activities which must be planned and accomplished for each month, then a day calendar is planned which will accomplish these and additional activities.

IDENTIFICATION OF CATTLE

To manage cattle properly, they need to be identified. Identification may be only for ownership purposes where all cattle may have the same brand or mark. Cattle can be identified with their own unique individual identification for purposes of breed registration, recording of performance records, health treatment, or other records needed to make effective management decisions.

Ownership Identification

Hot iron branding, which has been used for centuries (see Chapter 24, years 2500 BC, 1519 AD, and 1529 AD), is the most common method of showing and proving ownership. Brands appear on most parts of the body; however, the hip, rib, or shoulder are most common. Letters, figures, or characters are used for brands with the most simple brands being the easiest to read. Complicated designs often make unreadable, blurred brands.

Table 10.1. Management calendar for a selected cow–calf operation. Operation has both spring and fall calving herds.[a]

Major Activities	Jan	Feb	Mar	Apr	May	Jun	Jul	Aug	Sep	Oct	Nov	Dec
Calving		ss	ssss	ss				ff	ffff	f		
Breeding preparation (semen, equipment, bulls)			s						f			
Vaccinate cows (Leptospirosis & Vibriosis)												
Vaccinate heifers (Clostridium C, IBR, Lepto, BVD)				s						f		
Sychronize heifers				s						f		
Sychronize cows				s						f		
Breed heifers					sss	ssss					ffff	ff
Breed cows					ss	ssss	ss				fff	fff
Brand, castrate, dehorn calves				s						f		
Vaccinate calves (Blackleg, Malignant edema)				s						f		
Pregnancy test		f							s			
Fertilize pastures				x								
Control Pinkeye						xxxx	xxxx	xxxx				
Fly control tags						x						
Wean Calves			f							s		
Process weaning weights			f							s		
Vaccinate at Weaning (IBR, PI_3, Leptospirosis)			f							s		
Retag cows		f	f						s	s		
Grub control (cows & bulls)									x			
Grub control (calves)									s			
Vaccinate heifers for brucellosis				f							s	
Cull cows (sell or feed)				f						s		
Vaccinate cows (Clostridium C)	s						f					
Select replacement heifers (wean, yrl. and Preg test time)		f	s	f					sf	s		
Increase pre-calving level of feed	s						f					
Keep facilities and equipment repaired	x	x	x	x	x	x	x	x	x	x	x	x
Income tax preparation and filing	x											
Income tax evaluation and adjustment									x			
Cash flow up-date and evaluation	x	x	x	x	x	x	x	x	x	x	x	x
Check salt and mineral supply	x	x	x	x	x	x	x	x	x	x	x	x
Evaluation target weights for replacement heifers		s				f		f				s
Evaluate management plan	x	x	x	x	x	x	x	x	x	x	x	x
Spend time planning, both short and long term	x	x	x	x	x	x	x	x	x	x	x	x
Check water supply	xxxx	xxxx	xxxx	xxxx	xxxx	xxxx	xxxx	xxxx	xxxx	xxxx	xxxx	xxxx
Attend seminar, workshop or educational conference	x	x									x	x
Plan next years budget and cash flow statement											x	

[a] s = spring calving heard, f = fall calving herd, x = applies to both herds. Placement of letter within the month shows the week designation. Type of activities and timing of activities may be different for other operations

Brands are legally registered in certain states which have brand laws. The brand must be located on a specified part of the body and it is read from top to bottom and from left to right. Description of brands includes such words as "lazy," "flying," "walking," "rocking," "box," "quarter-circle," and many others (Figure 10.5).

Earmarks, wattles, and dewlaps, made with a knife, are used by producers to denote ownership. Sometimes they are used in conjunction with brands. Many times they are easier to see than brands at a distance or when the cattle are facing a person.

Earmarks may be recorded or registered with brands. Some of the more common earmark descriptions are underslope, overslope, crop, split, swallow fork, steeple fork, bit (under or upper), half crop (under or upper), shoestring, or jingle bob (Figure 10.6).

Dewlaps are cut on the loose skin just above and in front of the brisket (Figure 10.7). If neither end or middle is cut, it becomes a "jug handle." If the skin is cut in the middle, it becomes a "double dewlap."

Wattles are cut pieces of skin about an inch wide and 3 to 6 inches long. The upper part of the skin is cut entirely free from the animal leaving the lower end intact. Common places for wattling are the nose, jaw, chin, throat, neck, shoulder, and thigh (Figure 10.7).

Individual Identification

Each head of cattle is a productive unit and should, if economically feasible, be individually identified. In recent years more cows and calves have been individually identified for management purposes. Individual identification, however, is not a common practice for stocker-yearling and feedlot operators. Large commercial feedlots identify groups of cattle according to pen numbers and give individual identification to cattle pulled into hospital pens.

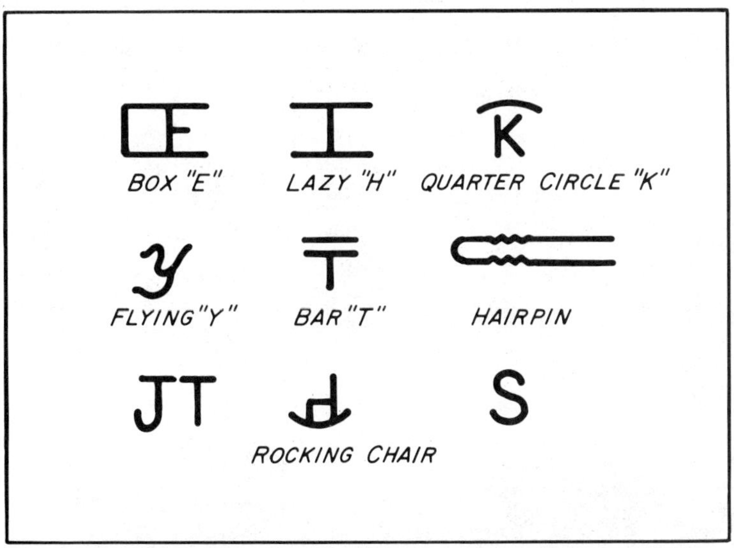

Figure 10.5. Selected hot iron brands which show different shapes and descriptive terminology. Several of the brands belonged to the author's great-grandfathers. Drawn by Dennis Giddings.

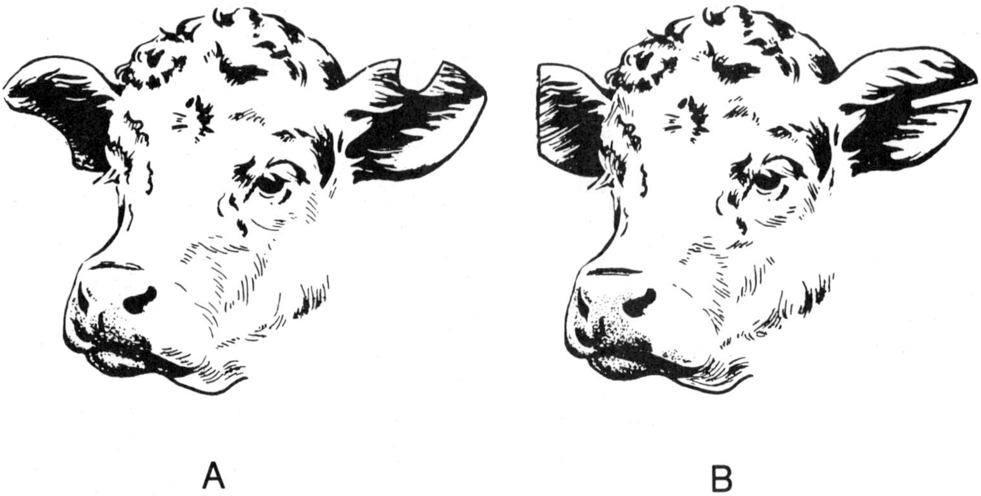

Figure 10.6. Some common ear marks. **(A)** Underslope in right ear; over-bit in left ear. **(B)** Crop in right ear; split in left ear. Courtesy of Division of Agric. Sci., Univ. of Calif.

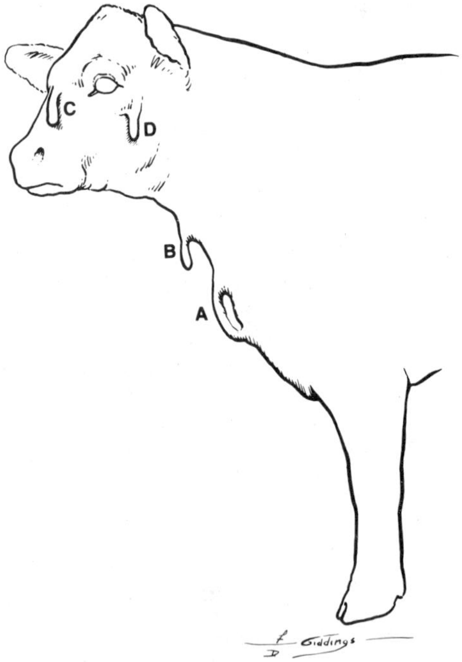

Figure 10.7. (A) Jug handle dewlap. **(B)** Throat or bell wattle. **(C)** Nose wattle. **(D)** Jaw wattle. Drawn by Dennis Giddings.

Horn brands can be used to burn into the horns. This procedure must be periodically repeated and the numbers can only be read on docile cattle.

Neck chain can be used with an individual number tag connected to the chain. This method of identification is not widely used as chains have to be adjusted for weight changes in the cattle. Also, cattle can choke when the chain catches on various objects.

Tail bobbing (cutting off part of the switch) is occasionally used to identify cattle which may not be pregnant, have weaned poor calves, need to have a health treatment, or some other management practice.

Cattle nose prints have been shown to be as individually unique as finger prints in humans. Certain livestock shows are nose printing steers to keep the identity of steer consistent with show rules.

Number brands can be applied as either hot iron brands or freeze brands. Hot iron brands are usually effective, although the time needed to apply several numbers and the need to clip long hair to make numbers readable are definite disadvantages.

Freeze brands are applied to an area where the hair has been clipped and the brands are cooled in dry ice and alcohol or liquid nitrogen. The hair, where the brand was applied, comes in white after a few weeks. There are advantages over fire branding in that the brands do not need hair clipped, it is a painless process and the value of the hide is not reduced. The disadvantages are only a 50% to 75% legibility with best results on black cattle and very poor results on white cattle.

Ear tattoos are commonly used in purebred herds as in many cases a tattoo is required for registration purposes. While the animal has to be restrained to read the tattoo, permanent identification (especially if the same tattoo is placed in both ears) is a definite plus.

Ear tags are common methods for individual cattle identification, and their use has significantly increased during recent years. Small metal tags, having the same identification numbers, can be effectively used with the larger, more easily read plastic ear tags. This prevents loss of identification in case one of the tags is lost. Ear tags in combination with ear tattoos (all three having the same number) are also an effective method to prevent loss of individual identification (Figure 10.8).

Numbers and letters on ear tags can be provided by the company, or producers can write their own with available marking pens. A numbering system should be used which will contribute to the efficiency of the management system. Some producers use the last digit of the year as the first number for the ear tag. For example, the first calf born in 1984 would be given tag number 401 (4 for 1984 and 01 for being the first calf). Calves could be numbered up through 4999 quite conveniently with this system. A problem can be encountered. After 10 years, some females may have duplicate numbers. This is particularly challenging to a computer which cannot distinguish between a 1984 cow numbered 401 and a 1994 cow numbered 401.

Some producers use letters to represent the years to prevent the 10 year duplication of numbers, e.g., M = 1982, N = 1983, O = 1984 . . . W = 1992, X = 1993 and Y = 1994. The difficulty with this system is that unless all producers are on the same letter-year system, it is hard for them to understand the ages of the cattle.

There is no single fool-proof system of individual identification for beef cattle. Two separate methods of identification are required; otherwise some cattle will lose their identity. Some of the more common dual identification systems are: (1) plastic ear tag

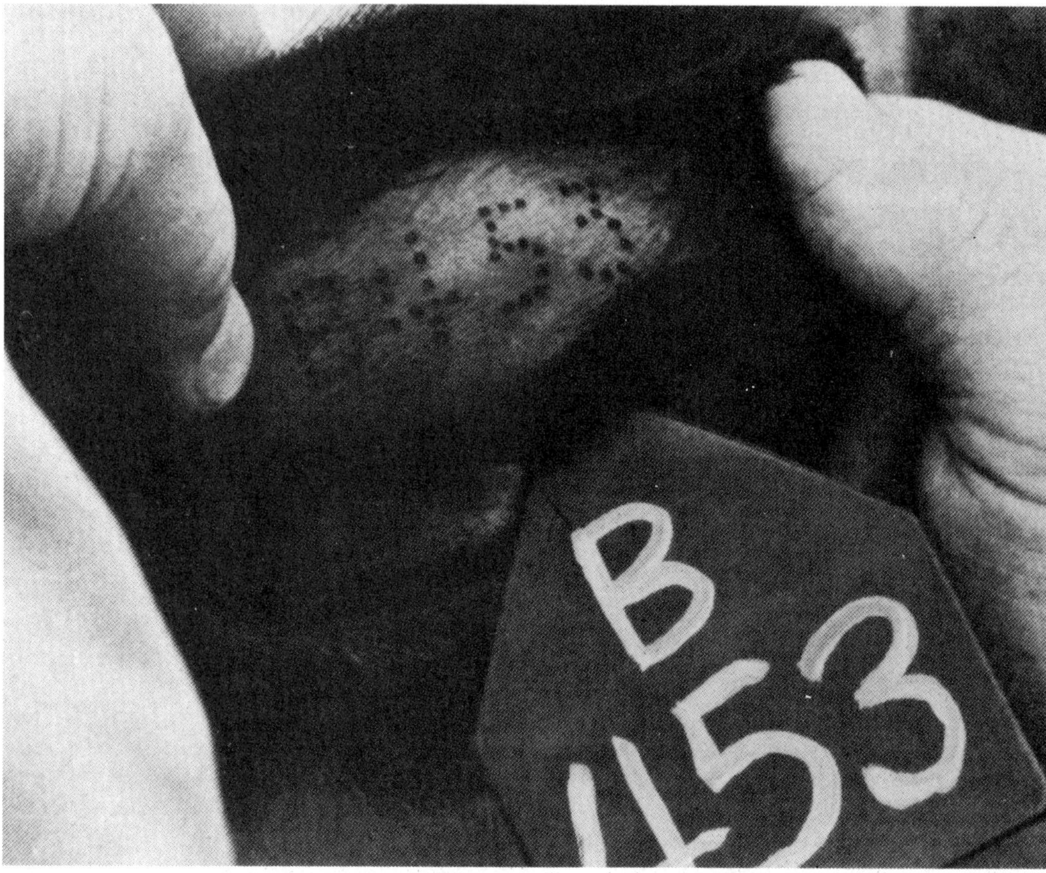

Figure 10.8. Double identification in the calf's ear using a tattoo and plastic ear tag. Note the same identification is used with both methods. Courtesy of the American Angus Association.

and metal ear tag, (2) plastic ear tag and tattoo in both ears, (3) plastic ear tag and fire brand number, and (4) fire brand number and metal ear tag. There are numerous alternatives; therefore producers need to select the combination which works best for them. The primary criteria for selecting the best system should be (1) economics (initial cost of materials and labor and cost of replacement), (2) visible at a distance, (3) numbers should stay readable during the lifetime of the animal, (4) small amount of loss, (5) minimal duplication of numbers every 10 years, and (6) an identification system which complements the record system on the cattle.

Electronic identification systems have been developed and may become a reality in the space and computer age. For electronic identification, each animal is implanted with a small capsule, placed under the skin behind the shoulder. Scientists say they could develop a unit which could be implanted with a gun—much like a growth implant gun.

The capsule contains circuits known as a transponder which will respond to signals from an outside unit some 5 to 30 feet away.

Some advantages of the electronic identification systems are: (1) data on the individual animal could be fed directly into the computer, (2) temperature can be monitored without restraining the animal, (3) improve speed and accuracy of marketing transactions, (4) deters theft, (5) more effective disease control, (6) complements performance and other record systems, and (7) could easily become a national identification system.

Disadvantages are cost ($1 to $3 per animal) and it is a very intense system being applied to rather extensive beef systems. There is also a fear that a national identification system would take away the independence of producer because of the "big brother is watching you" concept.

Financial Management

"Cattlemen must become businessmen specializing in cattle, rather than cattlemen functioning only occasionally as businessmen" is a statement continually repeated by beef industry leaders. This statement summarizes the need for beef producers to understand financial management.

Costs, returns, and profitability of a beef operation can only be critically assessed with a meaningful set of records. A financial record system should include a financial statement, a yearly budget of anticipated income and expenses, adequate income and expense records to complete the IRS tax forms (primary form shown in Figure 10.9), and a cash flow statement which shows the income and expenses by months. The latter would reflect the adequacy of operating money, and this could save interest by borrowing and repaying at certain critical times of the year. Cash flow statements can also identify growth limiting factors in the operation and determine whether or not a particular investment decision should be accepted or rejected.

Profits should be determined by evaluating all cash costs of the operation, all inventory increases or decreases, and the value of opportunity costs. Opportunity costs represent the returns given up if debt-free resources (such as owned land, livestock, and equipment) were used in their next best level of employment; for example, the value of the pastureland if it was leased, all capital represented by equipment and livestock invested in a certificate of deposit (or similar investment), and all family labor and management utilized in other employment. A price can be established for living on a farm or ranch and a goal for increasing profits can be established.

Credit and money management become most crucial during periods of inflation, high interest rates, and relatively low cattle prices. Prudent use of credit can enable a cattle operation to grow more rapidly than it could through the use of reinvested earnings and savings, so long as borrowed funds return more over time than they cost. Thus, farmers and ranchers have to look to credit as a financial tool and learn to use it effectively.

The net cost of credit may be less than the interest rate depending on taxable income bracket (Table 10.2).

SCHEDULE F (Form 1040)

Department of the Treasury Internal Revenue Service

Farm Income and Expenses

▶ Attach to Form 1040, Form 1041, or Form 1065.
▶ See Instructions for Schedule F (Form 1040).

OMB No. 1545-0074

1981 16

Name of proprietor(s): JAMES A. BROWN — Social security number: 579 : 28 : 6685

Farm name and address ▶ JAMES A. BROWN, RR #1, Box 25, HOMETOWN YOUR STATE 02115 — Employer identification number: 7 1 9 3 6 7 9 7 4

Part I Farm Income—Cash Method

Do not include sales of livestock held for draft, breeding, sport, or dairy purposes; report these sales on Form 4797.

Sales of Livestock and Other Items You Bought for Resale

a. Description	b. Amount	c. Cost or other basis
1 Livestock ▶ HEIFERS BOUGHT FOR RESALE	6,900	4,000
2 Other items ▶		
3 Totals	6,900	4,000
4 Profit or (loss), subtract line 3, column c, from line 3, column b ▶		2,900

Sales of Livestock and Produce You Raised and Other Farm Income

Kind		Amount
5 Cattle and calves		3,316
6 Sheep		
7 Swine		
8 Poultry		
9 Dairy products		91,454
10 Eggs		
11 Wool		
12 Cotton		
13 Tobacco		
14 Vegetables		
15 Soybeans		
16 Corn		
17 Other grains		
18 Hay and straw		2,266
19 Fruits and nuts		
20 Machine work		1,258
21 a Patronage dividends	272	
b Less: Nonincome items	22	
c Net patronage dividends		250
22 Per-unit retains		
23 Nonpatronage distributions from exempt cooperatives		
24 Agricultural program payments: a Cash		80
b Materials and services		324
25 Commodity credit loans under election (or forfeited)		550
26 Federal gasoline tax credit		100
27 State gasoline tax refund		
28 Crop insurance proceeds		624
29 Other (specify) ▶ FAIR PRIZES		700
30 Add amounts in column for lines 5 through 29		100,922
31 **Gross profits*** (add lines 4 and 30) ▶		103,822

Part II Farm Deductions—Cash and Accrual Method **F**

Do not include personal or living expenses (such as taxes, insurance, repairs, etc., on your home), which do not produce farm income. Reduce the amount of your farm deductions by any reimbursement before entering the deduction below.

Items		Amount
32 a Labor hired		4,438
b Jobs credit		
c WIN credit		
d Total credits		
e Balance (subtract line 32d from line 32a)		4,438
33 Repairs, maintenance		3,236
34 Interest		4,124
35 Rent of farm, pasture		1,200
36 Feed purchased		24,182
37 Seeds, plants purchased		2,286
38 Fertilizers, lime, chemicals		4,498
39 Machine hire		1,200
40 Supplies purchased		3,204
41 Breeding fees		1,078
42 Veterinary fees, medicine		2,020
43 Gasoline, fuel, oil		3,070
44 Storage, warehousing		
45 Taxes		3,050
46 Insurance		2,106
47 Utilities		2,050
48 Freight, trucking		2,056
49 Conservation expenses		1,580
50 Land clearing expenses		
51 Pension and profit-sharing plans		
52 Employee benefit programs other than line 51		
53 Other (specify) ▶		
ADVERTISING		468
FINANCIAL RECORDS		166
FARM ORG. DUES		350
DEATH LOSS-HEIFER BOUGHT FOR RESALE		450
54 Add lines 32e through 53		66,812
55 Depreciation (from Form 4562)		8,483
56 **Total deductions** (add lines 54 and 55) ▶		75,295

57 Net farm profit or (loss) (subtract line 56 from line 31). If a profit, enter on Form 1040, line 18, and on Schedule SE, Part I, line 1a. If a loss, go on to line 58. (Fiduciaries and partnerships, see the Instructions.) 57 — 28,527

58 If you have a loss, do you have amounts for which you are not "at risk" in this farm (see instructions)? ☐ Yes ☐ No
If you checked "No," enter the loss on Form 1040, line 18, and on Schedule SE, Part I, line 1a.

*Use amount on line 31 for optional method of computing net earnings from self-employment. (See Schedule SE, Part I, line 3.)

For Paperwork Reduction Act Notice, see Form 1040 Instructions.

Figure 10.9. Internal Revenue Service form showing income and expense items for which documented records must be kept.

Table 10.2. Net Cost of Borrowed Dollars

Income Tax Bracket Percent	Annual Rate of Interest Loan, %				
	6	8	10	12	14
	Net Rate of Interest After Tax Savings, Percent				
19	4.86	6.48	8.10	9.72	11.34
25	4.50	6.00	7.50	9.00	10.50
32	4.08	5.44	6.80	8.16	9.52
39	3.66	4.88	6.10	7.32	8.54
48	3.12	4.16	5.20	6.24	7.28

Income Tax Considerations

In addition to keeping accurate records for income tax pusposes, there are some other management decisions that are important in this area. A few of these are only briefly mentioned for awareness. The details and utilization of these should be more critically evaluated, especially with a tax accountant.

1. Capital gains can be taken on cattle held for breeding purposes. For example, of the $500 slaughter value of a breeding cow, only 40% or $200 is taxable. Management decisions are needed to determine number of replacement heifers saved, age of breeding animals in the herd, and culling decisions based on the level of productivity of each individual breeding animal.
2. Pay spouse a salary. This can save social security taxes, enhance retirement and estate planning, and provide eligibility for child care tax credit. It may have a disadvantage of reducing social security benefits.
3. Investment credit. A 10% deduction can be used on amount paid for new or remodeled livestock confinement buildings, other single purpose livestock structures, fences, paved roads, water wells, and drainage tiles.
4. Depreciation can be taken on breeding cattle, barns, sheds, grain storage bins, corrals, fences, irrigation wells and pumps, silos, farm lighting systems, and several others. For example, a bull purchased for $1500 could be depreciated $300 per year over the expected productive lifetime of 5 years. His slaughter value would have to be claimed as income when sold.
5. Children, working on a family farm, can be paid up to $3,300 per year without their having to pay any federal income taxes.
6. Mileage costs or actual expenses of operating and maintaining the farm share of the automobile or pickup can be deducted.

Estate and Gift Tax Planning

Many cattle operations have a large amount of debt-free capital invested in land, livestock, buildings, and equipment. Adequate knowledge and proper planning are necessary for farmers and ranchers to pass viable economic units to their heirs.

In 1981, the Economic Recovery Tax Act made major revisions to the previous Tax Code. When the new estate tax amendments are fully phased in and implemented, a husband and wife who own farm or ranch property other than in joint tenancy, will be able to transfer $2.7 million of this property (fair market value) to their children free of federal estate taxes. Also under the new law, each parent will be permitted to transfer $10,000 to each child and spouse, for a total of $40,000 per couple, or a total of $160,000 of property annually, without incurring any federal gift tax.

Beef producers should review their tax plans—including estate, gift and income—in light of the amendments contained in the 1981 tax bill, and consult with professionals in estate planning.

Risk Management

Beef production is considered a high-risk business. There are certain aspects of producing beef cattle, however, which need more attention in managing the risks involved. For example, weather and fluctuating market prices are two areas over which producers have little control.

Weather can reduce calf crop percent by causing high death losses of young calves when they are exposed to severe spring blizzards. These risks can be reduced by providing additional shelter and by avoiding a large concentration of calves when chances of severe storms are highest.

Drought can significantly affect feed supply causing a necessary reduction in herd size (Figure 10.10). Producers in drought-prone areas manage drought risk by keeping herd size below the numbers which could be grazed in the best forage years. Also, some producers will maintain a year's supply of stored forage to compensate for drought years which occur on occasion.

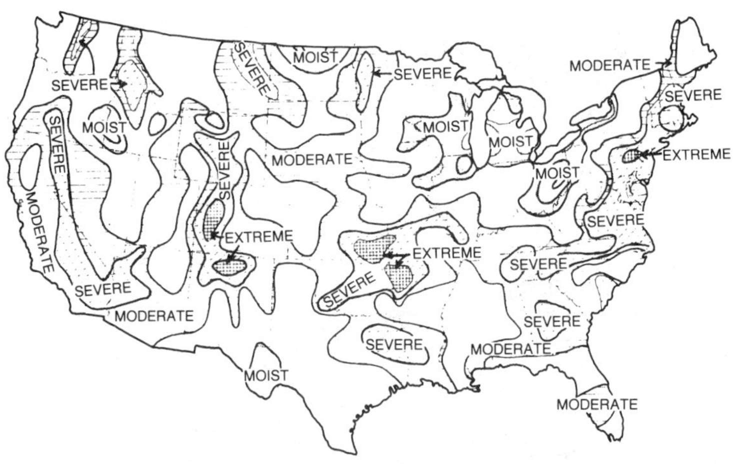

Figure 10.10. Droughts result in large annual fluctuation in feed supply. Drought occurs with greater frequency in certain areas of the United States and has to be managed accordingly. Courtesy of Western Livestock Round-Up.

Drought can reduce feed supply at critical times which can cause reduction in next year's calf crop. This is particularly true if cows are thin at calving time or feed supply is low from calving through rebreeding. This risk can be managed by providing supplemental feed to cows or supplemental feed to calves. Calves can be creep fed during periods of low forage supply (Figure 10.11) which allow the cows to maintain a higher body condition and thus have good reproduction. Calves can also be weaned early to help assure a higher calf crop next year.

Market prices have a history of wide fluctuations on a yearly, monthly or even a weekly basis (Figure 10.12). Producers can study price trends, keep current on market prices, forward-contract their cattle, and hedge their cattle, using the future market (see Chapter 18) to manage risks of price changes.

Computers and Programmable Calculators

Cattle operators are faced with a voluminous amount of information needed to keep financial and production records, inventories, and to critically assess management alternatives. Excellent managers, with conventional records systems, have an inate ability to "use their heads" to make good management decisions with a high degree of accuracy.

Figure 10.11. Creep feeding, when drought has reduced the feed supply, can help reduce the risk of a low percent calf crop the following year. Calves will use the supplemental feed, allowing the cows to maintain a satisfactory condition prior to and during breeding. Note the calf in the creep area consuming feed while the larger cow is unable to enter through the small opening. Courtesy of Powder River Livestock Equipment.

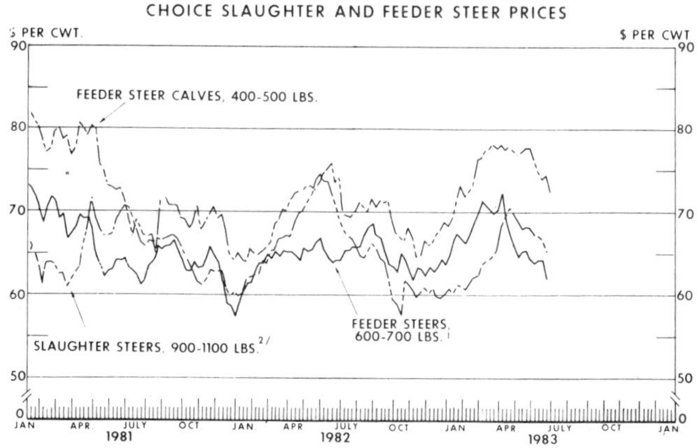

Figure 10.12. Fluctuations in cattle prices. Courtesy of the Western Livestock Round-Up.

Evidently they have a unique mind, the most complex computer, which functions exceptionally well in assimilating, storing, and recalling useful information.

Small computers (microcomputers) and programmable calculators are frequently used by farmers and ranchers as management tools. Producers must recognize what computers and calculators are and what they can and cannot do at the present time and in the

Figure 10.13. A programmable calculator with attached printer. Courtesy of Iowa State University.

future. The technology of computers is changing rapidly; therefore, producers must be ready to accept the frustration of change if they become involved with computers.

Every computer, regardless of size and shape, is nothing more than a fast adding machine and an electronic filing cabinet. In comparison, the programmable calculator (Figure 10.13) is a fast adding machine with a relatively small filing cabinet. For example, programs for these calculators have been developed at Iowa State University to assist cattle feeders in analyzing their rations. In other words, the cattle feeder: (1) enters the type and weight of cattle being fed, and (2) the amount and price of each feed included in the ration. The calculator then calculates: (1) what the cattle should gain, (2) the feed per pound of gain, and (3) total cost per pound of gain. These same calculations could be done longhand with a pencil; however, it would take 20 minutes instead of 2 minutes with the programmable calculator.

Programs for the programmable calculator are stored on small strips of magnetic tape. Each tape serves as a storage mechanism for a program, much like a tape recording.

A microcomputer system consists primarily of a system of hardware and software. The hardware for microcomputers (Figure 10.14) consists of a terminal (a typewriter keyboard and an optional TV screen), a printer, data storage devices (which are the filing cabinets), main memory, and Central Processing Unit (CPU) which is the adding machine. The data storage devices can be either magnetic tape (similar to cassette tapes), floppy discs (look like 45 rpm records), or hard discs (large storage capacity in one unit). The devices that read and write the information to and from these discs are called *disc drives* or *drives.*

Figure 10.14. Microcomputers are becoming more commonly used by cattle producers. Courtesy of *BEEF.*

Software are programs that make the hardware function. Each program typically consists of thousands of minutely detailed instructions that tell the computer exactly what data to manipulate mathematically, print, store, or display.

Good software programs is the most limiting factor in microcomputer utilization by cattle producers. Also, software developed for one microcomputer will not function on a different brand of machine. Computer utilization is expected to increase very rapidly as more compatible programs are developed. The four ways producers can obtain software are: (1) purchase a complete commercial package, (2) hire the programming done, (3) learn to program the computer themselves, or (4) obtain it from their Land Grant University.

The three major areas where microcomputers have the largest potential for cattle producers are: (1) business accounting, (2) herd performance, and (3) financial management. A computer to handle these three functions would need a relatively large memory system and would cost approximately $4000 to $5000. Some producers have access to computer terminals which have a telephone hookup to a large mainframe computer. An example of such a system is AGNET (Agricultural Computer Network) which functions through joint efforts of the Agricultural Extension Services of Wyoming, Nebraska, South Dakota, North Dakota, and Washington. An example of one of approximately 250 AGNET programs is COWCOST which evaluates costs and returns for beef cow–calf enterprises.

Computers do not simplify a cattle business or cut overhead costs. They can make a business more complex and more costly. If wisely acquired and properly utilized, a computer can increase profits but usually not by reducing expenses. The computer will not solve the problem of a poorly organized operation. Well-planned and efficient office management practices must exist and function smoothly before the addition of a computer will do anything but aggravate organizational problems of the business. The computer cannot assist the producer in managing the business effectively unless the computer receives all required data in precisely the prescribed manner. Unless the computer can provide information needed by producers at appropriate times and in an easily understood form, it should not be part of a management program regardless of price.

It is recommended that six steps be followed before a decision is reached by a producer to purchase a computer. They are: (1) what are the most important management decisions and what information is needed to make these decisions effectively? (2) what programs are available that can meet the management information needs? (3) what are the hardware needs to utilize the needed software? (4) contact local computer hardware dealers and evaluate various alternatives and service, (5) estimate the cost/benefit ratio of the proposed management information system, and (6) make the final decision after evaluating experiences of other producers who are using computers in their operations. Some producers are cautiously initiating computer-supported management by starting with a programmable calculator with the intent of expanding to a microcomputer in the near future.

Management Systems

Management systems analysis provides a method of systematically organizing information needed to make more valid management decisions. It permits many variables to be more critically assessed and analyzed as to their contribution to the total or desired end point. Individuals who have been educated to think more broadly in the framework of management systems can make reasonably valid management decisions without using data processing equipment. A pencil, hand calculator, and a well-trained mind are the primary components needed for competent management decisions. Without question, the use of computers in synthesizing voluminous amounts of information, enhances the utilization of the management system concept.

Resources are different for each operation which means there is not a fixed recipe for successful cattle production applicable to each operation. The uniqueness of one operation to another operation is identified by variables such as different levels of forage production, varying marketing alternatives, energy costs, different biological types of cattle, environmental differences such as weather, feed nutrients at various costs, different levels of competence of labor and management, and many others too numerous to mention. All of these variables and the interactions pose challenges to the producer needing to combine these variables into sound management decisions for a specific operation.

It has been a common practice to increase or maximize production of cattle by applying known biological relationships. Cattle production has typically been maximized without careful consideration of cost-benefit ratios and how increased productivity relates to land, feed and management resources. Recognition is being given to the need of optimization rather than maximization of cattle productivity. Thus there is an increased interest in management systems, which attempts to optimize production with net profit being the primary goal and possibly the only maximum involved. It should not be surprising that some of the basic biological truths will not be useful or applicable because an economic analysis will not permit them to be included in sound management decisions. For example, it is a well-known biological relationship that calves born earlier in the calving season will have heavier weaning weights. Because of this relationship, some ranchers continue to move the calving season to a period earlier in the year. A management evaluation of a specific ranch demonstrated that by changing the calving season to match forage availability, profits were increased significantly. Changing calving season alone was estimated to increase the ranch's carrying capacity by 30% to 40%, in terms of animal units. In addition, the annual cow cost was reduced 18% per cow because of reduction in winter feed requirements.

During the past several years, there has been increased activity on the part of research workers in the application of management systems to cattle management. After a system is conceptualized, it is usually described by a set of mathematical equations and called a model. Models are constructed to simulate real life situations, and attempts are made to validate them on that basis. Obviously the models are no better than the data used to construct the models and output data depends on the validity of data which is put into the model.

Computer utilization in management systems, applicable to the farmer and rancher, is quite limited primarily because of the lack of software. There are very few programs being used which combine the production data along with an economic assessment. There is increased discussion and activity in making more management system software available to producers. There are several projections which indicate that these programs will be commonly used in the next few years. Producers have to make management decisions now. It is evident that to be competitive, cattle operations will have to be operated more like a business. This means useful and accurate financial and production records that can be effectively utilized in making management decisions. It appears that the computer will continually become more important as a useful tool to improve the effectiveness of the management decision process. Until more usable computer programs become available, more producers will use the traditional tools (pencil, calculator, and a keen mind) more effectively than in the past.

REFERENCES

Publications

Albaugh, R., 1975. *How to identify livestock.* Calif. Coop. Ext. Leaflet 2297.

Bartlett, E. T. and Clawson, W. J., 1978. *Profit, meat production or efficient use of energy in ranching.* J. of Anim. Sci. 46:812.

Cartwright, T. C., 1979. *Size as a component of beef production efficiency: Cow–calf production.* J. of Anim. Sci. 48:974.

Cook, C. W., Bartlett, E. T. and Evans, G. R., 1974. *A systems approach to range beef production.* Colo. State Univ. Range Sci. Dept., Sci. Series No. 15.

Examples of benefits to farmers and ranchers resulting from new tax legislation, 1981. NCA Beef Business Bul., 4 September.

Fitzhugh, H. A., 1978. *Bioeconomic analysis of ruminant production systems.* J. of Anim. Sci. 46:797.

Harl, N., 1980. *Farm Estate and Business Planning.* 6th Edition. Skokie, Ill.: Century Communications, Inc.

Hughes, H., 1981. *The computer explosion.* Proceedings, The Range Beef Cow, A Symposium on Production, VII. Rapid City, SD., Dec.

Hughes, H., 1981. *Six steps to take in making a decision to buy a computer.* BEEF, Dec.

Joandet, G. E. and Cartwright, T. C., 1975. *Modeling beef production systems.* J. of Anim. Sci. 41:1238.

Klinefelter, D. A. and Hottel, B., 1981. *Farmers' and ranchers' guide to borrowing money.* Texas Agric. Expt. Sta. MP–1494.

Luft, L. D. *Sources of credit and the cost of credit.* Great Plains Beef Cattle Handbook, GPE–4351 and 4352.

Maddux, J., 1981. *The man in management.* Proceedings, The Range Beef Cow, A Symposium on Production, VII. Rapid City, SD., Dec.

McDowell, R. E. and Hernandez-Urdaneta, A., 1975. *Intensive Systems for Beef Production in the Tropics.* J. of Anim. Sci. 41:1228.

Miller, W. C., Brinks, J. S. and Sutherland, T. M., 1978. *Computer assisted management decisions for beef production systems.* International J. of Agric. Systems 3:147.

Notter, D. R., Sanders, J. O., Dickerson, G. E., Smith, G. R. and Cartwright, T. C., 1979. Simulated efficiency of beef production for a midwestern cow–calf feedlot management system. III. Crossbreeding systems. J. of Anim. Sci. 49:92.

Nelson, T., 1981. *Software—Is it available?* Proceedings, The Range Beef Cow, A Symposium on Production, VII. Rapid City, SD., Dec.

Ott, G. *Planning for profit with partial budgeting.* Great Plains Beef Cattle Handbook, GPE–4551.

Proceedings, Natl. Beef Profit Conference, Natl. Cattlemen's Assoc. Denver, Co. 28–29 June 1982.

Ross, J. W. *Cash flow management—strategy for the 80's.* Beef Business Bulletin (NCA), 12 August 1983.

Smith, G. M., 1979. *Size as a component of beef production efficiency: Feedlot production and integrated efficiency.* J. of Anim. Sci. 48:966.

Smith, G. M. and Harrison, V. L., 1978. *The future of livestock systems analysis.* J. of Anim. Sci. 46:807.

Smith, N. E. and Larue, E. L., 1974. *Interface of animal production and economic systems.* Fed. Proc. 33:198.

Tattooing. How to make it work. Amer. Angus Assoc.

Trede, L. D., Boehlje, M. D. and Geasler, M. R., 1977. *Systems approach to management for the cattle feeder.* J. of Anim. Sci. 45:1213.

What new tax legislation means to farmers and ranchers, 1981. NCA Beef Business Bul., August 14.

Whitson, R. E. and Kay, R. D., 1978. *Beef cattle forage systems analysis under variable prices and forage conditions.* J. of Anim. Sci. 46:823.

Visuals

Beef Cattle Identification. Vocational Educ. Productions, Calif. Polytechnic State Univ., San Luis Obispo, CA 93407.

11

Beef Cattle Reproduction

Reproductive efficiency in cattle, as measured by number of calves born or weaned per 100 breeding females on a yearly basis, is considered the most economically important trait in cattle production. It is essential to understand the reproductive process in the creation of new life as it is the focal point of overall animal productivity. Producers who manage animals for high reproductive rates must understand the production of viable sex cells, estrous cycles, mating, pregnancy, and birth.

Producers can manage their herds for high reproductive rates. Management decisions that cause males and females, selected for breeding, to reach puberty at early ages, to have high conception rates, and to have minimum difficulty at parturition are most critical. Keeping animals healthy, providing adequate nutrition, selection of genetically superior animals, and producer attention at parturition are some of the more critical management inputs to minimize reproductive loss. Live offspring born to each breeding female each year is the key endpoint to successful cattle reproduction.

FEMALE ORGANS OF REPRODUCTION AND THEIR FUNCTION

Figure 11.1 shows the reproductive organs and their location in the cow's body.

The reproduction organs consist of a pair of ovaries, which are suspended by ligaments just back of the kidneys, a pair of open-ended tubes, the oviducts (also called the Fallopian or ovarian tubes), which lead directly into the uterus (womb). The uterus itself has two horns, or branches, that merge together at the lower part into one structure, so that the lower opening, or exit, from the uterus is a canal (Figure 11.2). This canal called the cervix has a folded surface surrounded by muscles. The cervix opens into the vagina, a relatively large canal or passageway that leads posteriorly to the external parts, which are the vulva and its clitoris. The urinary bladder empties into the vagina through the urethral opening.

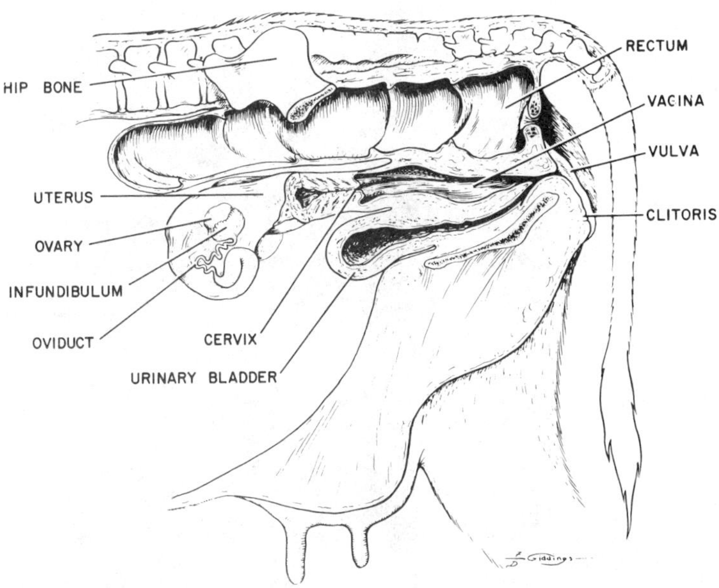

Figure 11.1. Reproductive organs of the cow. Drawn by Dennis Giddings.

Ovaries

The ovaries produce eggs (female sex cells, also called ova) and hormones (see Figure 11.3). Each egg, or ovum, is generated in a recently formed follicle within the ovary (Figure 11.4).

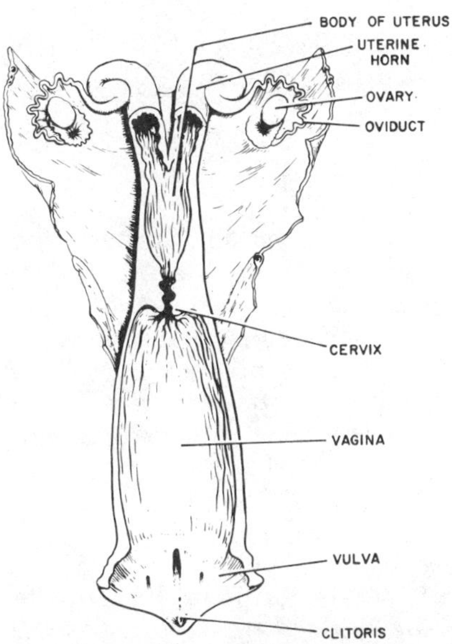

Figure 11.2. A dorsal view of the reproductive organs of the cow. Drawn by Dennis Giddings.

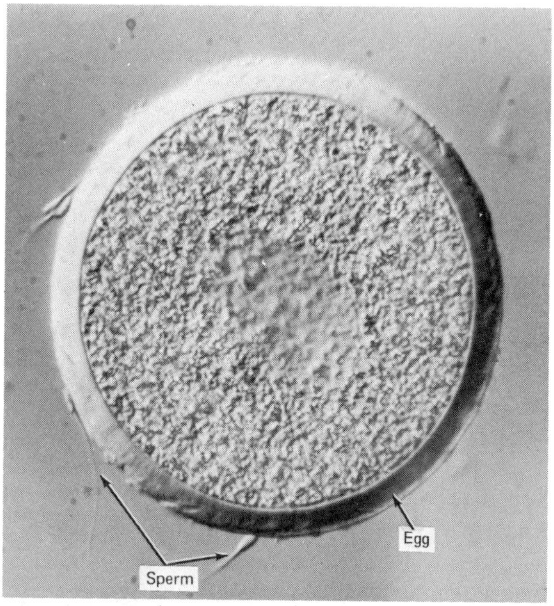

Figure 11.3. Bull sperm and cow egg, each magnified several hundred times. The sperm are located at the bottom and on the left-hand side of the egg. Note the comparative sizes. Each are single cells and each contains one-half the chromosome number typical of the other body cells. Courtesy of the Animal Reproduction Lab., Colo. State Univ.

Some tiny follicles develop and ultimately attain maximum size, about 10 mm in diameter, after having migrated from deep in the ovary to the surface of the ovary. These mature (Graafian) follicles rupture, thus freeing the ovum (ovulation). Many of the tiny follicles grow to various stages, cease growth, deteriorate, and are absorbed (Figure 11.5).

After the egg escapes from the mature follicle, the follicle develops into a "yellow body," or corpus luteum, which becomes a vitally important structure if pregnancy occurs.

Oviducts

The oviducts receive ova immediately after the ova leave the ovaries, through the open end of the oviduct which is called the funnel or infundibulum. Ova are tiny, 200 μm or less in diameter (200 μm = 1/5 mm), which is approximately the size of a dot made by a sharp pencil. Ovarian tubes also receive sperm after the female is inseminated (naturally or artificially) and therefore are sites where ova and sperm meet and where fertilization takes place. For ova to travel through the oviduct, approximately 3 to 5 days are required. From the oviduct, newly developing embryos pass to the uterus and soon attach to it.

Uterus

The uterus has a relatively short body and longer uterine horns. It is through the uterus that sperm must pass enroute to upper ends of the oviducts, there to meet the

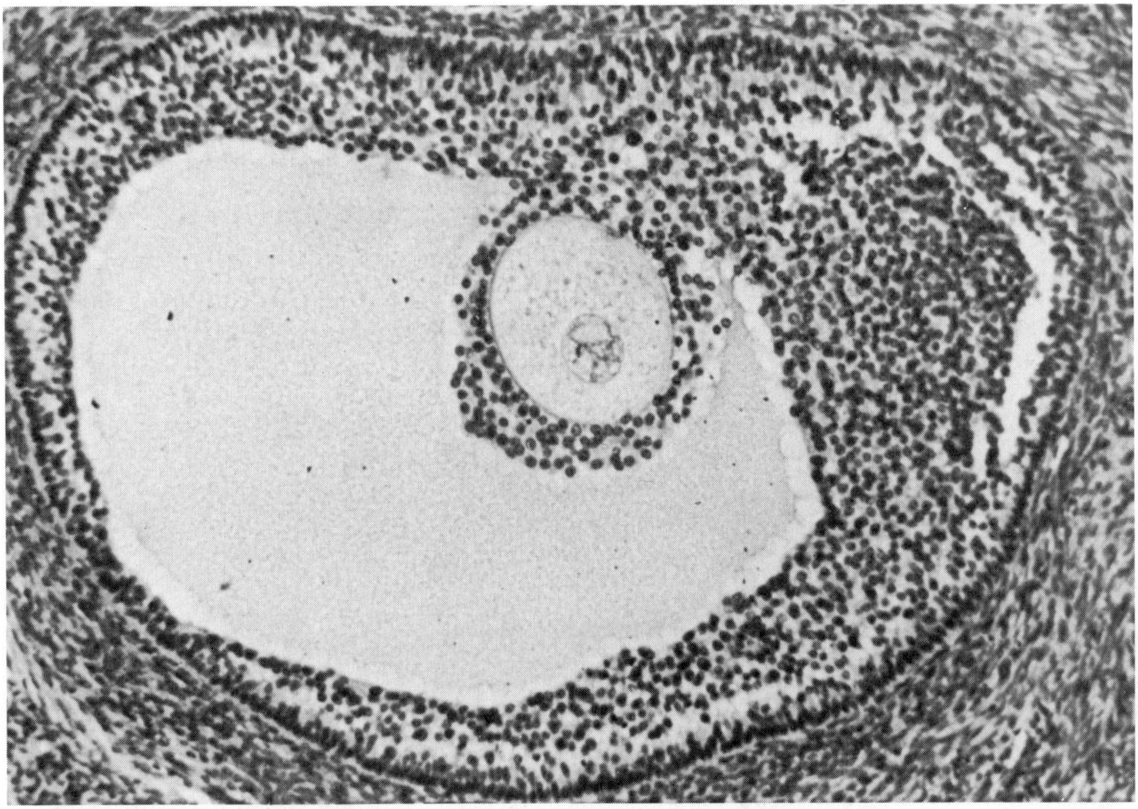

Figure 11.4. The large structure, outlined with a circle of dark cells is a follicle located on a cow's ovary. Magnified several hundred times. The smaller circle, near the center, is the egg. The large light gray area is the fluid which fills the follicle. When the follicle ruptures, the fluid will wash the egg into the oviduct (ovulation). Courtesy of the Animal Reproduction Lab., Colo. State Univ.

ova. Each embryo develops into a fetus and remains in the uterus until parturition (birth).

The lower outlet of the uterus is the cervix, an organ that constitutes a formidable gateway from the uterus to the vagina, and from the vagina to the uterus. Like the rest of the reproductive tract, the cervix is lined with mucosal cells. These cells undergo change as the animal goes from one period of estrus to another, and they also change in pregnancy. The cervical passage changes from one that is tightly closed or sealed in pregnancy, or nearly closed when the animal is not in estrus (heat), to a relatively open, very moist canal at the height of estrus.

Vagina

The vagina serves as the organ of copulation at mating and as the birth canal at parturition. Its mucosal surface changes during the estrous cycle from very moist when

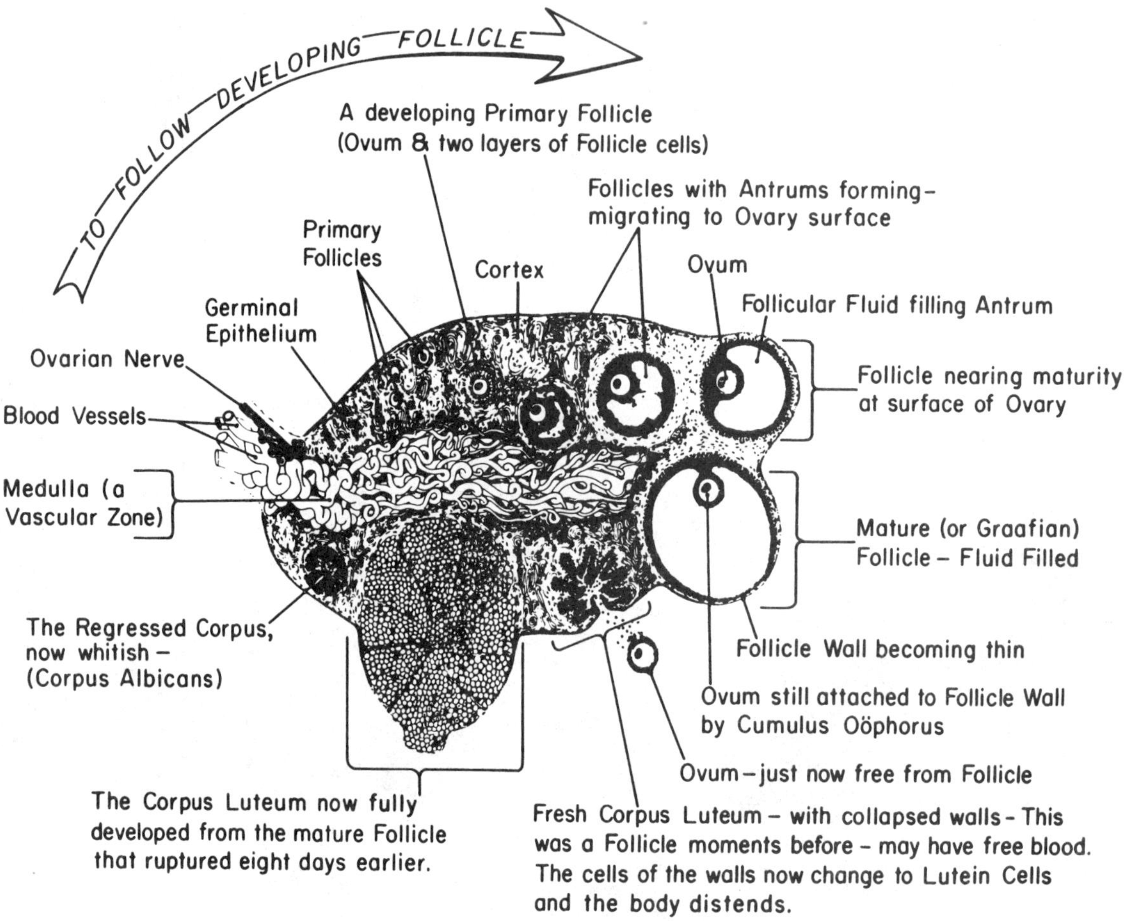

Figure 11.5. A cross section of the bovine ovary showing how a follicle develops to full size and then ruptures, thus allowing the egg to escape (ovulation). The follicle then becomes a "yellow body" (corpus luteum) which is actually orange colored in cattle. The corpus luteum degenerates in time and disappears. Many follicles cease development, stop growing, and disappear without ever reaching the mature stage. From Bone, *J. F. Animal Anatomy and Physiology.* 4th ed. Corvallis: Oregon State University Book Stores. Copyright © 1975.

the animal is ready for mating to almost dry, even sticky, between periods of heat. The tract from the urinary bladder joins the posterior ventral vagina; from this juncture to the exterior vulva, the vagina serves the double role of a passage for the reproductive and urinary systems.

Clitoris

A highly sensitive organ, the clitoris, is located ventrally and at the lower tip of the vagina. The clitoris is the homologue of the penis in the male (that is, it came from the same embryonic source as the penis). There is some research that indicates that clitoral stimulation or massage, following artificial insemination, will increase conception, but it has not been well verified.

MALE ORGANS OF REPRODUCTION AND THEIR FUNCTION

Figure 11.6 shows the reproductive organs of the bull. They consist, in part, of two testicles, which are bean-shaped organs held in the scrotum. Male sex cells (called sperm or spermatozoa) are formed in tiny seminiferous tubules of the testicles. Sperm from each testicle then pass through very small tubes into an epididymis, which is a tube that is held in a covering applied closely to the exterior of the testicle. Each epididymal tube leads to a larger tube, the vas deferens (also called the deferent duct or ductus deferens). The two vas deferens converge from the left and right sides of the body to connect with the urethral canal at its upper end, very near to where the urinary bladder opens into the urethra. The urethra is a large canal that leads through the penis to the outside of the body. The penis has a triple role: (1) that of a passage for urine, (2) that of a passage for the products originating in other organs of reproduction, and (3) that of an organ of copulation.

The seminal vesicles and prostate gland are found at the base of the urinary bladder. The left and right parts of the seminal vesicles, which lie against the urinary bladder, consist of glandular tissue that supplies a secretion that moves through the exit tube

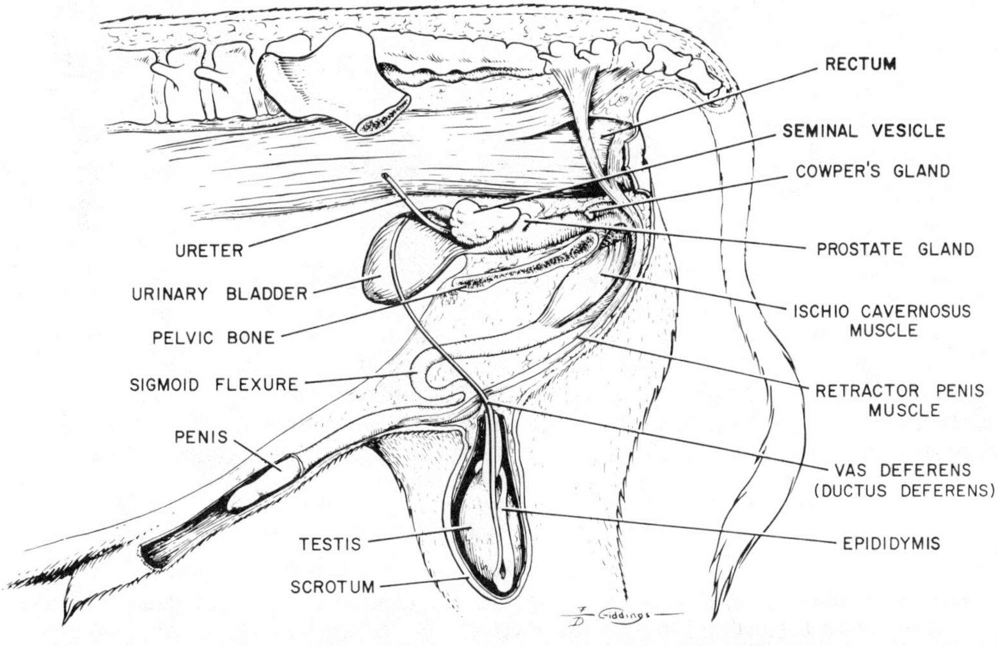

Figure 11.6. Reproductive organs of the bull. Drawn by Dennis Giddings.

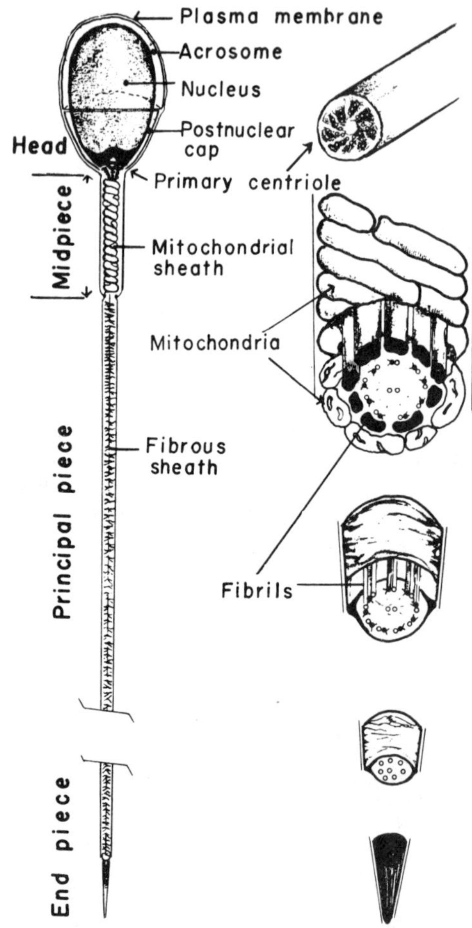

Figure 11.7. A diagrammatic sketch of structure of cattle sperm. Courtesy of Dr. Arthur S. H. Wu, Oregon State University.

of each seminal vesicle into the urethra. The prostate gland is composed of a group of some 12 or more glandular tubes, each of which empties into the urethra, also near the opening from the urinary bladder and dorsal to (above) the prostate. Another gland, bulbourethral (Cowper's) gland, which also empties its secretion into the urethral canal, is approximately 40 cm posterior to (behind) the prostate. Details of spermatozoa structure of the bull are shown in Figures 11.7, 11.8, and 11.9.

Testicles

Testicles produce (1) sperm cells that fertilize eggs, or ova, of the female, and (2) a hormone called testosterone that conditions the male so that his appearance and behavior are masculine. Thus, if both testicles are removed (as is done in castration of males),

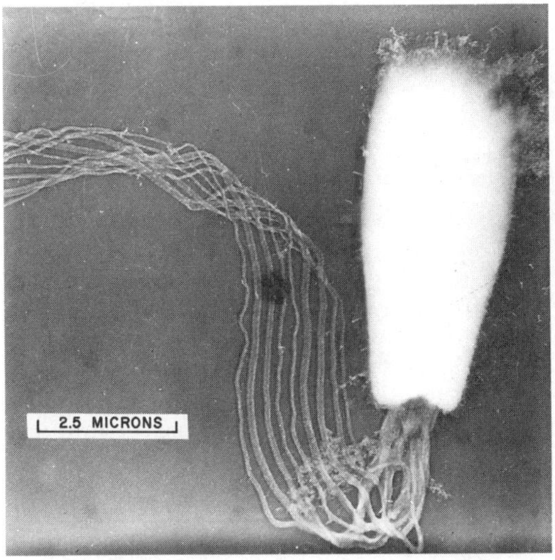

Figure 11.8. A spermatozoan (sperm cell) from a bull (magnified several thousand times) viewed through the electron microscope after treatment with 0.15 N NaOH at 25 °C for 16 hours. Note how the covering membrane of the neck region of the sperm has been removed, exposing the nine fibrils of the axial filament. Three filaments are larger than the rest. From Wu, A. S. H., and Mckenzie, F. F. Microstructure of Spermatozoa after Denudation as Revealed by the Electron Microscope. *J. of Anim. Sci.* 14:1151, 1955.

the individual loses his sperm factory and his masculine appearance. To illustrate, the effect of castration of the bull calf is to make him a steer, that is, a sterile individual lacking prominent horn development without the crest, or powerful neck, of the bull, lacking to a degree the bull's propensity to grow to his full potential and having a weaker voice than if he had been left intact. If the calf is castrated while he is immature, such organs of reproduction as the vas deferens, seminal vesicles, prostate, and bulbourethral glands all but cease further development. If castration is done in the mature animal, the remaining genital organs tend to shrink in size and function.

Occasionally during fetal development, one or both of the testicles fail to descend from the body cavity into the scrotum. The animal, known as a cryptorchid, is sterile

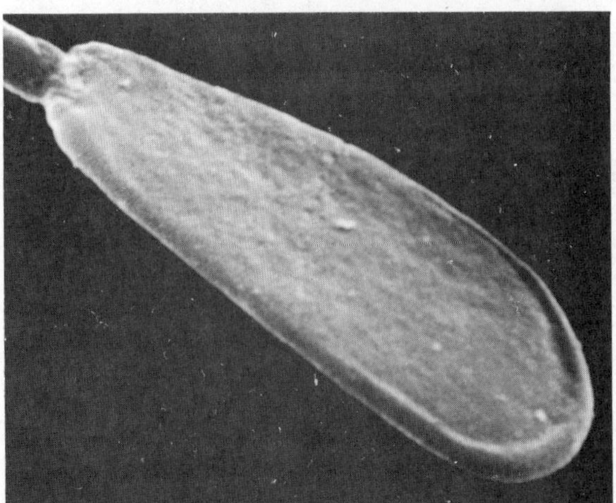

Figure 11.9. Bull sperm (magnified several thousand times) viewed with the scanning electron microscope showing the depth of the sperm head covered by the raised acrosome. Courtesy of Dr. Arthur S. H. Wu, Oregon State University.

if both testicles are retained but usually fertile if one testicle descends. Masculine characteristics develop because testosterone is produced.

Within each testicle, sperm cells are generated in the seminiferous tubules and testosterone is apparently produced in the cells between the tubules, which are called interstitial cells or cells of Leydig (Figure 11.10).

Epididymis

The epididymis affords opportunity for sperm cells, which enter it from the testicle, to mature. In passing through this very long tube (30 to 35 m), the sperm acquire more capacity to fertilize ova. Sperm taken from the part of the epididymis nearest the testicle are not likely to be able to fertilize ova, whereas those taken from areas farther along this long, winding tube increasingly show the capacity to fertilize.

In the sexually mature male animal, sperm reside in the epididymis in vast numbers. Sperm confined to the epididymis for long periods of time, degenerate and are absorbed in the part of the epididymis farthest from the testicle.

Scrotum

The scrotum is a two-lobed sac that contains and protects the two testicles—but it does more. It is the scrotum that regulates temperature of the testicles, maintaining them at a temperature lower than body temperature—3 °F to 7 °F lower. When environmental temperature is low, the tunica dartos muscle of the scrotum contracts, drawing

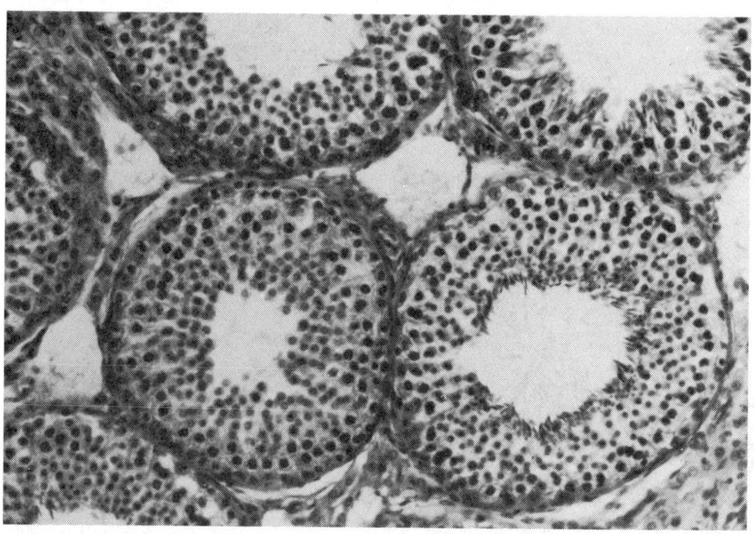

Figure 11.10. A cross-section through the seminiferous tubules of the testis of the bull. The tubule in the lower right-hand corner demonstrates the more advanced stages of spermatogenesis as the spermatids are formed near the lumen (opening) of the tubule. Courtesy of the Animal Reproduction Lab., Colo. State Univ.

the testicles toward the body and its warmth; when the environmental temperature is high, this muscle relaxes, permitting the testicles to drop away from the body and its warmth. This heat-regulating mechanism of the scrotum begins at about the time of puberty when the testicular hormone starts to function.

When the environment is so hot that the testicles cannot cool sufficiently, the formation of sperm is impeded and a temporary condition of lowered fertility is produced.

Vas Deferens

The vas deferens are essentially transportation tubes that carry sperm-containing fluid from each epididymis to the urethra. The vas deferens join the urethra near its origin as the urethra leaves the urinary bladder. In the mature bull, the vas deferens is about 3 mm in diameter except in its upper part where it widens into a reservoir, or ampulla, about 10 to 17 cm long and 1 cm wide. The ampulla of the vas deferens is profusely supplied with nerves from the pelvic plexus of the sympathetic nervous system.

Under the excitement of anticipated mating, the secretion loaded with spermatozoa from each epididymis is propelled into each vas deferens and accumulates in the ampulla of the deferent duct. This brief accumulation of semen in the ampulla is an essential part of sexual arousal. The sperm reside briefly in the ampulla until the moment of ejaculation, when the contents of each ampulla are pressed out into the urethra, and then through the urethra and the penis enroute to their deposition in the female tract.

Urethra

The urethra is a large, muscular canal extending from the urinary bladder; the bladder empties into the upper and inner end of the urethra. The urethra runs posteriorly through the pelvic girdle and curves downward and forward through the full length of the penis. Very near the junction of the bladder and urethra, the tubes from the seminal vesicles and tubes from the prostate gland join this large canal. The bulbourethral gland joins the urethra at the posterior floor of the pelvis. The urethra is lined with many tiny glands whose watery secretion is clear and high in mucoproteins.

Seminal Vesicles

The seminal vesicles are sizable organs which lie over the neck of the bladder and on either side of the pelvic urethra and which open into the urethra near the place where the vas deferens opens into the urethra. In the bull, seminal vesicles are about 10 cm long and 2.5 cm in diameter.

To the semen (the fluid that contains sperm), the seminal vesicles contribute ascorbic acid, citric acid, inorganic phosphorus, acid-soluble phosphorus and the bulk of seminal fructose.

Prostate Gland

The prostate gland also lies near the neck of the bladder and surrounds the urethra. In the bull, the main part of the prostate is about 37 mm across (left to right) and 12 mm in diameter. Additional prostate tissue is scattered among the muscles of the pelvic urethra. The prostate empties into the uretha through a small opening in the muscular wall of the urethra. It supplies antagglutin and minerals to semen plasma.

Bulbourethral Glands

The bulbourethral glands (sometimes referred to as Cowper's glands) are located on either side of the pelvic urethra, just posterior to the urethra-penis where the urethra-penis dips downward in its curve. The bulbourethral glands are covered by fibrous tissue. In the bull, they are about 25 by 12 mm in size. The secretion from the bulbourethral glands is thick and viscous, very slippery, or lubricating, and whitish in color.

The seminal vesicles, prostate, and bulbourethral glands are known as the accessory sex glands. Their primary functions are to add volume and nutrition to the sperm-rich fluid coming from the epididymis. The bull has a semen volume of 3 to 10 ml per ejaculate with 4 to 18 billion sperm in each ejaculate.

Penis

The penis is the organ of copulation. It also provides a passageway for the escape of urine. It is a muscular organ characterized especially by its spongy, erectile tissue that fills with blood under considerable pressure during periods of sexual arousal, making the penis rigid and erect.

The penis of the bull is about 1 m in length and 3 cm in diameter, tapering to the free end and it is S-shaped when relaxed. This S-curve or Sigmoid flexure, is eliminated when the penis is erect. The S-curve is restored after copulation when the relaxing penis is drawn back into its sheath by a pair of retractor muscles. The free end of the penis is termed the glans penis. Only a small portion of the penis of the bull extends beyond its sheath during erection. The full extension awaits the thrust after entry into the vagina has been made.

WHAT MAKES THE TESTICLES AND OVARIES FUNCTION?

Testicles

The testicles produce hormones under stimuli coming to them from the anterior pituitary body (AP) situated at the base of the brain. The AP elaborates two hormones important to male performance, the luteinizing hormone (LH) and the follicle-stimulating hormone (FSH). These two hormones are known as gonadotropic hormones because they stimulate the gonads (ovary and testicle). LH produces its effect on interstitial tissue of the testicle (probably the interstitial cells), causing this tissue to produce male hormone, testosterone. FSH stimulate cells in the seminiferous tubules to develop into functional spermatozoa. There is some evidence that FSH also has an influence on testosterone production.

There are genetic and environmental influences on the neurophysiological mechanism, the hypothalamus. The hypothalamus is the floor and part of the wall of the third ventricle of the brain and secretes releasing factors through portal vesicles that affect the anterior pituitary and its production of FSH and LH.

Ovaries

Ovarian hormones in sexually mature females owe the rhythmicity of their production to hormones that originate in the anterior pituitary body and the interplay between gonad-stimulating hormones produced there and ovarian hormones whose level and potency vary as the estrous cycle progresses. FSH (the same hormone as FSH in the male) circulates through the bloodstream and affects responsive follicle cells of the ovary, which respond by secreting estrogens or estrus-producing hormones, estradiol and estrone. When the amount of estradiol and estrone in the blood reaches a sufficient level, the pituitary is caused to reduce its production of FSH. With this drop in FSH production, production of estradiol and estrone subsides. When estrus-producing hormones have been lost from the body, and their depressing effect on the anterior pituitary has been spent, the anterior pituitary again steps up its production of follicle-stimulating hormone and the cycle is repeated.

Luteal cells, successors of the follicle cells, secrete a hormone called progesterone, which plays a role in the female reproductive cycle. Progesterone helps cause the occurrence and recurrence of the desire to mate, that is, the estrous cycle. Synchronized with this cycle is the important and essential phenomenon of ovulation. Ovulation occurs in the cow approximately 30 hours after the beginning of estrus.

Ovulation is controlled by hormones. The follicle of the ovary grows, matures, fills with fluid, and then softens some few hours before rupture. The follicle ruptures due to the interaction of hormones rather than bursting as a result of pressure inside.

FSH from the anterior pituitary accounts for the increase in size of the ovarian follicle and for the increase in the amount of estradiol and estrone, which are products of follicle cells. Another part of the anterior pituitary elaborates LH, which alters follicle cells and granulosa cells of the ovary, changing them into luteal cells. These luteal cells are in turn stimulated by LH from the anterior pituitary to produce progesterone.

When a certain delicate balance between high amounts of estrus-producing hormones and a minimal amount of progesterone is attained, the membranes over part of the large follicle thin out, separate, and break, thus allowing the follicular fluid to escape and carry the ovum into the open end of the oviduct. Thus, ovulation is accomplished. After the ovum escapes from the follicle, the hole through which it escaped is sealed over, and the cells that formed the ruptured follicle are luteinized by LH. In the course of 7 to 10 days, what was formerly an egg-containing follicle develops into a corpus luteum, a luteal body of about the same size and shape as a mature follicle. The egg, having escaped and migrated down the oviduct, is of course absent from the corpus luteum. The luteal cells of the corpus luteum produce a quantity of progesterone sufficient to depress the part of the anterior pituitary that produces FSH until the luteal cells reach maximum development (in nonpregnant animals), cease their development, and (in 3 weeks) lose their potency and disappear. Luteal cells lose their function because of prostaglandin which is released from the uterus.

If pregnancy occurs, the corpus luteum continues to flourish, persists in its progesterone production, and prevents further estrus-producing cycles. Thus, no more heat occurs until after pregnancy has terminated. The process of follicle development, ovulation, and corpus luteum development and regression is shown in Figure 11.5. The cow will show estrus approximately 16 hours (range of 10 to 27) and the average length of the

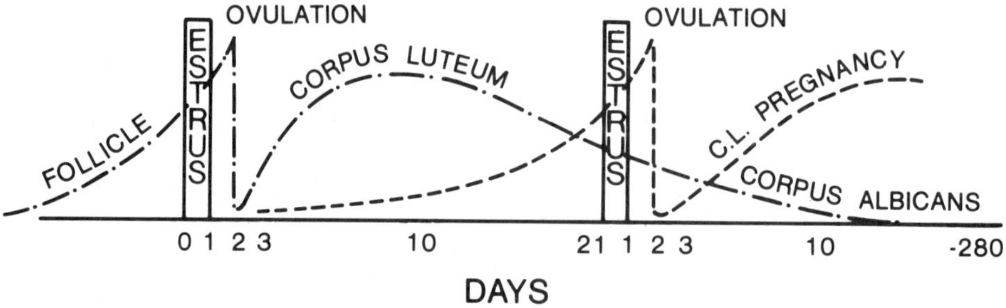

Figure 11.11. The major ovarian changes associated with the estrous cycle where pregnancy does and does not occur. Drawn by CSU Graphics.

estrous cycle is 21 days (range of 19 to 23). The major changes in ovarian structure and function, in relation to the estrous cycle, is shown in Figure 11.11.

The changing length of day is a factor that influences the estrous cycle, the onset of pregnancy and the seasonal fluctuations in male fertility. Changing day length acts both directly and indirectly on the animal—it acts directly on the hypothalmus by influencing the secretion of hormones and indirectly by affecting plant growth, thus altering the amount of quality of nutrition available. Increasing length of day is associated with increased reproductive activity in both males and females.

PREGNANCY

When sperm and egg unite (fertilization), conception occurs and pregnancy (gestaton) is initiated. The fertilized egg initiates a series of cell divisions (Figure 11.12). The new organism migrates through the oviduct to the uterus in 2 to 5 days. By then it has developed to the 16-or 32-cell stage. The chorionic and amniotic membranes develop around this new embryo and attach it to the uterus. The embryo (later called the fetus) gets nutrition and discharges wastes through these membranes. This period of attachment (thirtieth to thirty-fifth day of pregnancy) is critical and occasionally embryonic mortality will occur. Unless the environment is sufficiently favorable, the embryo dies. It is vital that management practices protect the female early in pregnancy by providing feed of sufficient quality and by minimizing stress. The female should enter breeding season in a thrifty (vigorous), gaining condition and maintain this condition throughout the first weeks of pregnancy.

The embryonic stage in the life of an individual is defined as that period in which the body parts differentiate to the extent that the essential organs are formed. At approximately 45 days, the embryo becomes designated as a fetus. The fetal period, which lasts until birth, is mainly a time of growth. The duration of pregnancy in cattle is approximately 285 days with a range of 275 to 295 days. The gestation table (Table 11.1) permits an easy calculation of projected calving dates when breeding dates are known.

PREGNANCY DETERMINATION

In many cow–calf operations, it is economically feasible to determine pregnancy in heifers and cows prior to their anticipated time of calving. This allows non-pregnant

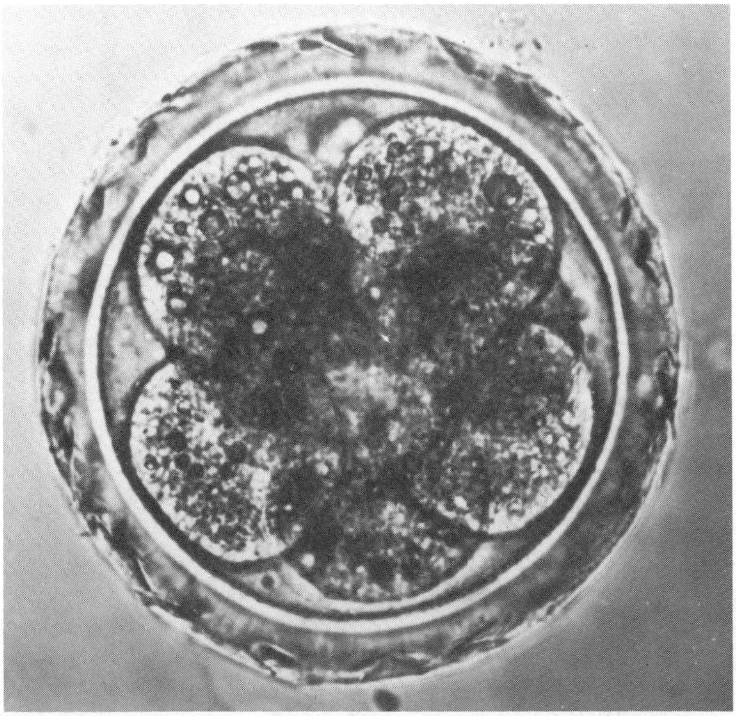

Figure 11.12. A bovine embryo in the 6-cell stage of development (magnified several hundred times). Courtesy of the Animal Reproduction Lab., Colo. State Univ.

females to be evaluated for culling since they will not produce a calf to offset the production costs.

Pregnancy in cows is most commonly determined through palpation where the person's arm is inserted into the cow's rectum and the reproductive tract is palpated for pregnancy indications.

The individual doing the palpating wears a protective covering (rubber or plastic sleeve) over the arm and hand. A lubricant such as liquid soap is used on the sleeve for ease of entry into the cow's rectum.

Palpation usually takes only a few seconds if the cattle are properly restrained and the palpator is experienced. The restraining chute should have a front wall or gate and a bar at the rear, just above the hocks. The latter prevents the animal from kicking the palpator. An entrance gate in the chute at the rear of the animal to allow entrance and exit for the palpator is desirable. This gate which swings across the chute, also provides a front gate for cows coming into the chute behind the palpator. Facilities should provide for the safety of the palpator and quiet, ease of handling of the animals.

The palpator must have a thorough knowledge of the female reproductive system and the changes which occur during pregnancy. Palpators should be experienced in palpation and be palpating cows on a regular basis. Technique and practice are important

Table 11.1. Gestation Table for Cows Based on 284 Day Gestation Length

	Find Data of Service in Upper Line. Figure Below Indicates Date Due to Calve.																															
January	1	2	3	4	5	6	7	8	9	10	11	12	13	14	15	16	17	18	19	20	21	22	23	24	25	26	27	28	29	30	31	
October	12	13	14	15	16	17	18	19	20	21	22	23	24	25	26	27	28	29	30	31	1	2	3	4	5	6	7	8	9	10	11	November
February	1	2	3	4	5	6	7	8	9	10	11	12	13	14	15	16	17	18	19	20	21	22	23	24	25	26	27	28				
November	12	13	14	15	16	17	18	19	20	21	22	23	24	25	26	27	28	29	30	1	2	3	4	5	6	7	8	9				December
March	1	2	3	4	5	6	7	8	9	10	11	12	13	14	15	16	17	18	19	20	21	22	23	24	25	26	27	28	29	30	31	
December	10	11	12	13	14	15	16	17	18	19	20	21	22	23	24	25	26	27	28	29	30	31	1	2	3	4	5	6	7	8	9	January
April	1	2	3	4	5	6	7	8	9	10	11	12	13	14	15	16	17	18	19	20	21	22	23	24	25	26	27	28	29	30		
January	10	11	12	13	14	15	16	17	18	19	20	21	22	23	24	25	26	27	28	29	30	31	1	2	3	4	5	6	7	8		February
May	1	2	3	4	5	6	7	8	9	10	11	12	13	14	15	16	17	18	19	20	21	22	23	24	25	26	27	28	29	30	31	
February	9	10	11	12	13	14	15	16	17	18	19	20	21	22	23	24	25	26	27	28	1	2	3	4	5	6	7	8	9	10	11	March
June	1	2	3	4	5	6	7	8	9	10	11	12	13	14	15	16	17	18	19	20	21	22	23	24	25	26	27	28	29	30		
March	12	13	14	15	16	17	18	19	20	21	22	23	24	25	26	27	28	29	30	31	1	2	3	4	5	6	7	8	9	10		April
July	1	2	3	4	5	6	7	8	9	10	11	12	13	14	15	16	17	18	19	20	21	22	23	24	25	26	27	28	29	30	31	
April	11	12	13	14	15	16	17	18	19	20	21	22	23	24	25	26	27	28	29	30	1	2	3	4	5	6	7	8	9	10	11	May
August	1	2	3	4	5	6	7	8	9	10	11	12	13	14	15	16	17	18	19	20	21	22	23	24	25	26	27	28	29	30	31	
May	12	13	14	15	16	17	18	19	20	21	22	23	24	25	26	27	28	29	30	31	1	2	3	4	5	6	7	8	9	10	11	June
September	1	2	3	4	5	6	7	8	9	10	11	12	13	14	15	16	17	18	19	20	21	22	23	24	25	26	27	28	29	30		
June	12	13	14	15	16	17	18	19	20	21	22	23	24	25	26	27	28	29	30	1	2	3	4	5	6	7	8	9	10	11		July
October	1	2	3	4	5	6	7	8	9	10	11	12	13	14	15	16	17	18	19	20	21	22	23	24	25	26	27	28	29	30	31	
July	12	13	14	15	16	17	18	19	20	21	22	23	24	25	26	27	28	29	30	31	1	2	3	4	5	6	7	8	9	10	11	August
November	1	2	3	4	5	6	7	8	9	10	11	12	13	14	15	16	17	18	19	20	21	22	23	24	25	26	27	28	29	30		
August	12	13	14	15	16	17	18	19	20	21	22	23	24	25	26	27	28	29	30	31	1	2	3	4	5	6	7	8	9	10		September
December	1	2	3	4	5	6	7	8	9	10	11	12	13	14	15	16	17	18	19	20	21	22	23	24	25	26	27	28	29	30	31	
September	11	12	13	14	15	16	17	18	19	20	21	22	23	24	25	26	27	28	29	30	1	2	3	4	5	6	7	8	9	10	11	October

in the accurate diagnosis of pregnancy. Individuals palpating only a few cows once a year are likely to have a large number of errors.

A palpator with skill and practice can determine pregnancy as early as 30 days after breeding. Palpation at this early stage, however, should be accompanied by good breeding records. With these records, the palpator knows the approximate breeding date of the animal. Most palpators prefer breeding to end approximately 45 days prior to palpation.

Table 11.2 gives the fetal size and some of the other identifying characteristics used

Table 11.2. Fetal Size and Characteristics Used in Determining Pregnancy

	Fetal Size		
Days of Gestation	**Weight**	**Length Inches**	**Identifying Characteristics**
30	1/100 oz	2/5	One uterine horn slightly enlarged and thin; embryonic vesicle size of large marble. Uterus in approximate position of nonpregnant uterus. Fetal membranes of 30 to 90 day pregnancy may be slipped between fingers.
45	1/8–1/4 oz	1–1 1/4	Uterine horn somewhat enlarged, thinner walled, and prominent. Embryonic vesicle size of hen's egg.
60	1/4–1/2 oz	2 1/2	Uterine horn 2 1/2″ to 3 1/2″ in diameter; fluid filled and pulled over pelvic brim into body cavity. Fetus size of mouse.
90	3–6 oz	5–6	Both uterine horns swollen (4″ to 5″ in diameter) and pulled deeply into body cavity (difficult to palpate). Fetus is size of rat. Uterine artery 1/8″ to 3/16″ in diameter. Cotyledons 3/4″ to 1″ across.
120	1–2 lb	10–12	Similar to 90-day but fetus more easily palpated. Fetus is size of small cat with head the size of a lemon. Uterine artery 1/4″ in diameter. Cotyledons more noticeable and 1 1/2 inches in length. Horns are 5″ to 7″ in diameter.
150	4–6 lb	12–16	Difficult to palpate fetus. Uterine horns are deep in body cavity with fetus size of large cat—horns 6″ to 8″ in diameter. Uterine artery 1/4″ to 3/8″ in diameter. Cotyledons 2″ to 2 1/2″ in diameter.
180	10–16 lb	20–24	Horns with fetus still out of reach. Fetus size of small dog. Uterine artery 3/8″ to 1/2″ in diameter. Cotyledons more enlarged. From sixth month until calving a movement of fetus may be elicited by grasping the feet, legs, or nose.
210	20–30 lb	24–32	From 7 months until parturition fetus may be felt. Age is largely determined by increase in fetal size. The uterine artery continues to increase in size—210 days, 1/2″ in diameter; 240 days, 1/2″ to 5/8″ in diameter; 270 days, 1/2″ to 3/4″ in diameter.
240	40–60 lb	28–36	
270	60–100 lb	28–38	

Source: Sorenson and Beverly, 1979.

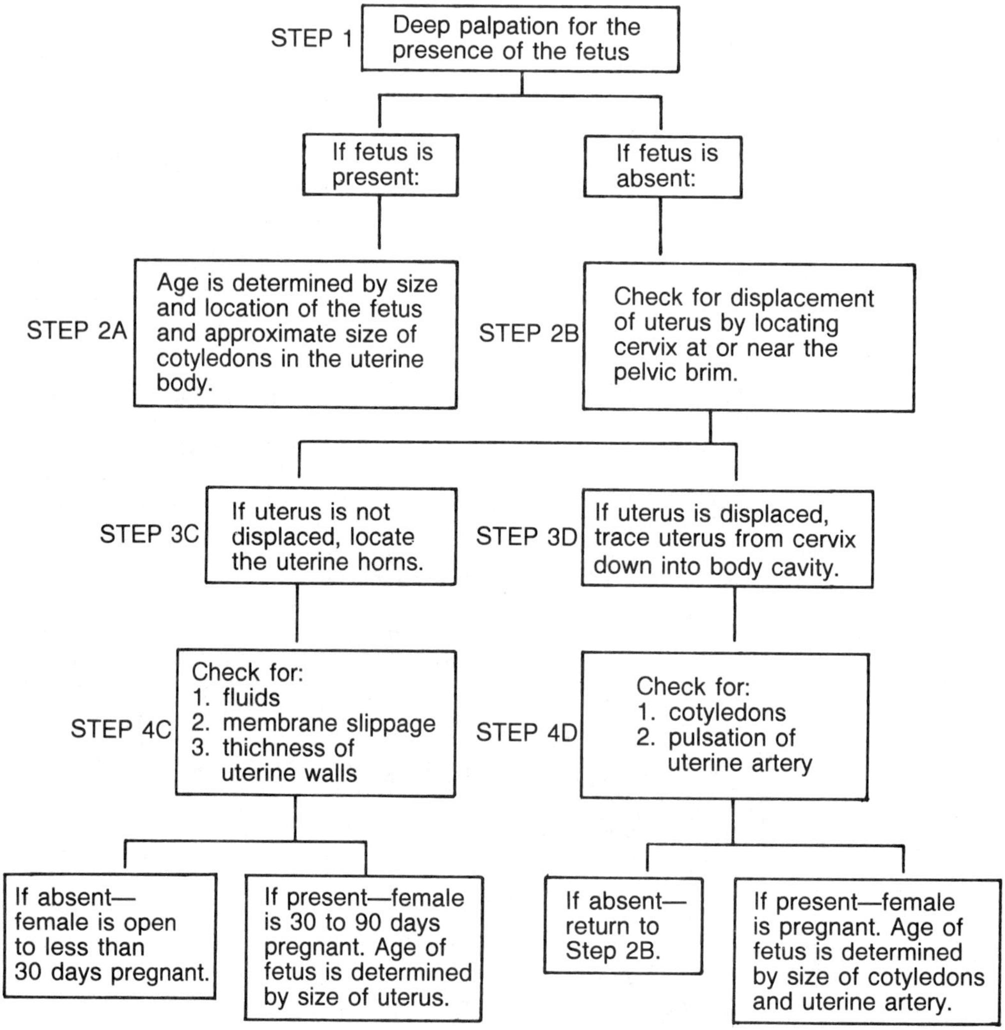

Figure 11.13. A systematic approach to pregnancy determination in the cow. *Source:* Sorenson and Beverly, 1979.

in pregnancy diagnosis. Figure 11.13 lists a stepwise procedure which is used by many palpators to determine the pregnancy status of beef females.

PARTURITION

Parturition (birth) marks the termination of pregnancy. Cows about to give birth will generally exhibit certain behavior patterns and physical characteristics. They may have a red, swollen vulva, show expulsion of the water sac (water bag), become separated from the herd, turn head toward flank, and have a full, tight udder.

The membranes that form around the embryo in early pregnancy and attach to the uterus are collectively called the placenta. This organ of pregnancy develops the capacity to produce hormones, namely, estrogens and progesterone. A balance between estrogens and progesterones is attained in which the estrogens predominate in quantity. The uterine muscles become sensitive to the hormone oxytocin, which is produced in the hypothalamus and released by the posterior pituitary. Under the stimulus of oxytocin, the weak, rhythmic contractions of the uterus that prevail through most of pregnancy become pronounced and cause labor pains, and the parturition process is initiated. Parturition is a synchronized process. The cervix, until now tightly closed, relaxes. The relaxation of the cervix, along with the pressure generated by the uterine muscles on the contents of the uterus, permits the passage of the mature fetus into the vagina and on to the exterior. Another hormone, relaxin, is thought to aid in parturition. Relaxin, which originates in the corpus luteum or in the placenta, helps to relax cartilage and ligaments in the pelvic region to help parturition occur.

At the beginning of parturition, the offspring typically assumes a position that will offer the least resistance as it passes into the pelvic area and through the birth canal. The normal presentation of the fetus is where the front feet are extended with the head between them (Figure 11.14). Occasionally calves may present themselves in a number of abnormal positions (Figures 11.14 and 11.15). In many of these situations, assistance needs to be given at parturition, otherwise, the offspring may die or be born dead, and in some instances, the mother may die as well. An abnormally small pelvis opening or an abnormally large fetus can cause some mild to severe parturition problems. Some examples of providing assistance to the cow or heifer are given. The procedure involving the most simple case of calving difficulty is shown in Figure 11.16, while procedure in some of the more difficult cases are shown in Figures 11.14 and 11.15.

Assisting the Delivery

The following steps are presented to assist relatively easy dystocia problems. The term *frontward* indicates how to proceed if the calf is positioned in the birth canal with front feet and head first. The term *backward* is used to indicate the procedure to follow if the calf is presented with the rear legs first. Those steps that do not indicate calf position apply to both presentations.

All backward presentations should be considered emergencies because the umbilical cord is cramped between the fetus and pelvis early in delivery. This means that blood and nutrient flow to the fetus is slowed, and the fetus will die or sustain brain damage unless delivery is rapid once it has entered the birth canal.

1. Make a loop in the obstetrical chain by passing the chain through the oblong ring at the end of the calving chain (Figure 11.16a).
2. Slip this loop over the gloved and well-lubricated hand. This will allow for easy application and maneuverability in the birth canal or uterus (Figure 11.16b).
3. Attach the loop of the chain to one leg of the calf and slide it up on the cannon bone 2 to 3 inches above the dewclaws. It may be necessary to maintain a slight tension on the chain so that it does not slip off the leg (Figure 11.16c).
4. Halfhitch the chain between the dewclaw and hoofhead. The halfhitch can be

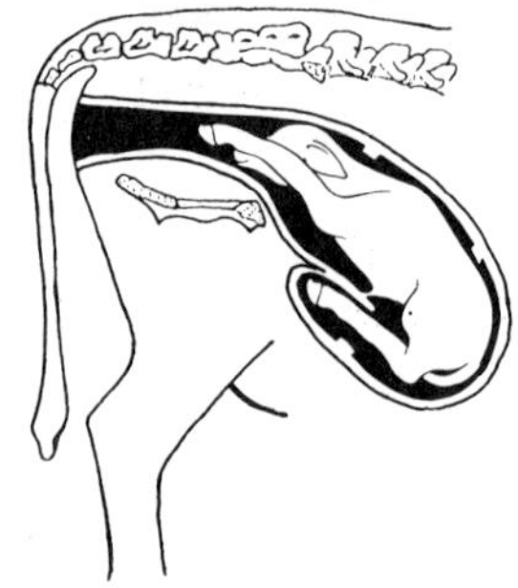

Normal Position

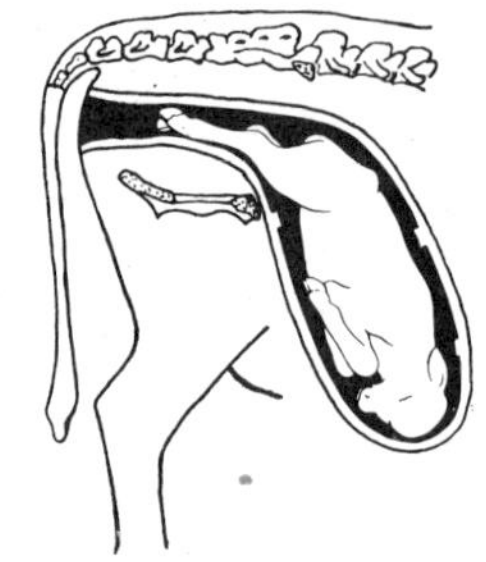

Posterior presentation *of the calf. Delivery may often proceed without complications. Assistance may be important if labor is prolonged. Death of the calf can occur due to rupture of the naval cord and subsequent suffocation.*

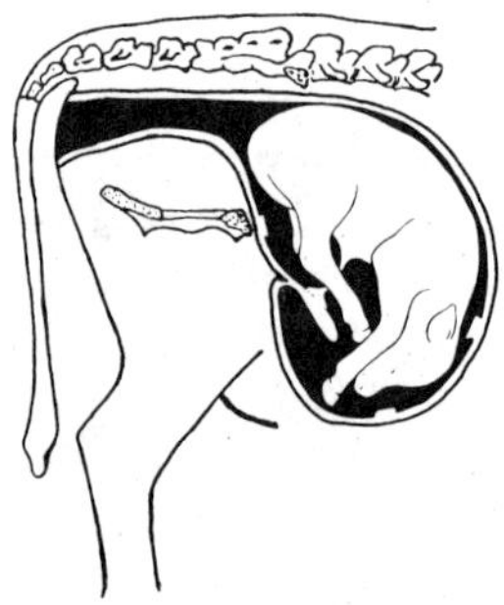

Posterior presentation *with rear legs extended under the calf's body (breech presentation). May be corrected by pushing the calf forward and grasping the legs one by one. As each leg is drawn into the birth canal, keep the hock pointed toward the cow's flank and the hoof to the midline.*

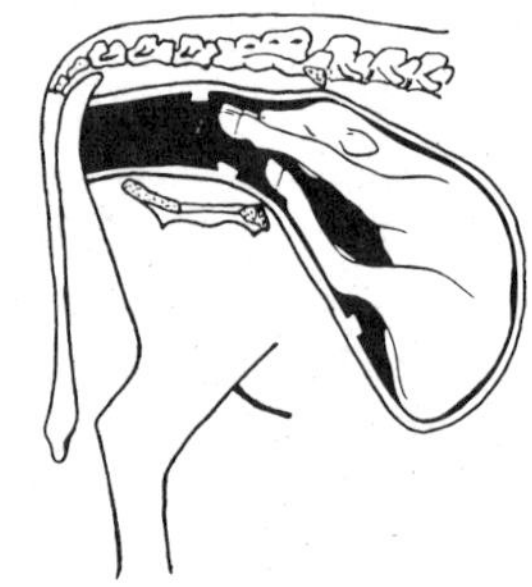

Anterior presentation *with the rear legs extended beneath the body (dogsitting* posture*). A very serious type of malpresentation. If allowed to progress into advanced labor, fetal death may result. Early professional attention may be required.*

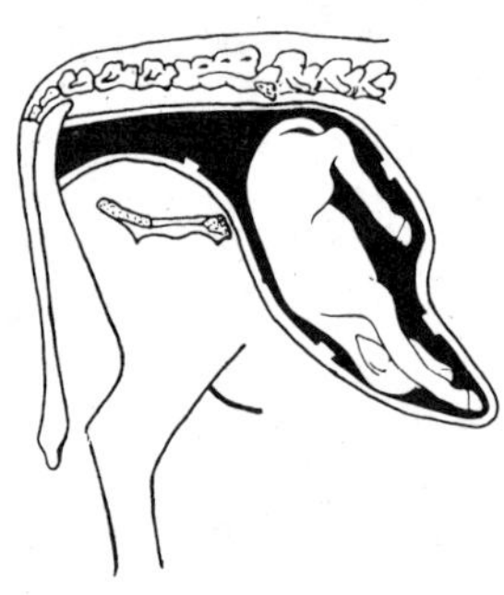

Posterior presentation *with fetus in an upside down position. This situation can be caused by twisting of the uterus or rotation of the calf. Delivery must never be attempted in this position, and professional assistance often is required.*

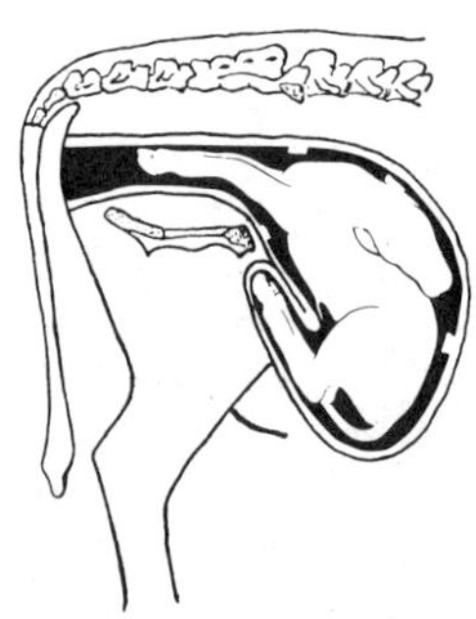

Anterior presentation *with the head and neck turned back over the body. Secure the legs with chains. Push the calf back into the body. This often brings the head into normal position.*

Figure 11.14. Normal and abnormal presentations of the calf at parturition, with guidelines in assisting with the difficult deliveries. Courtesy of Texas A and M University.

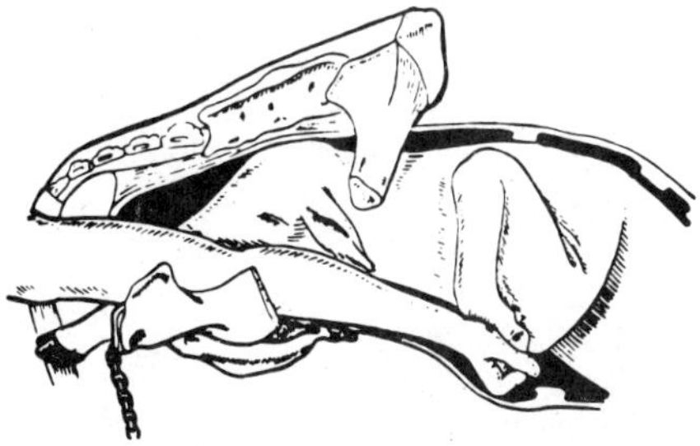

Correction of a simple leg flexion in an anterior presentation. The calf is first pushed forward and the retained foot is grasped in the cupped hand. The foot is carried outwards and then forward in an arc over the pelvic rim. More difficult cases may require that a snare be attached to the retained fetlock to help extend the leg.

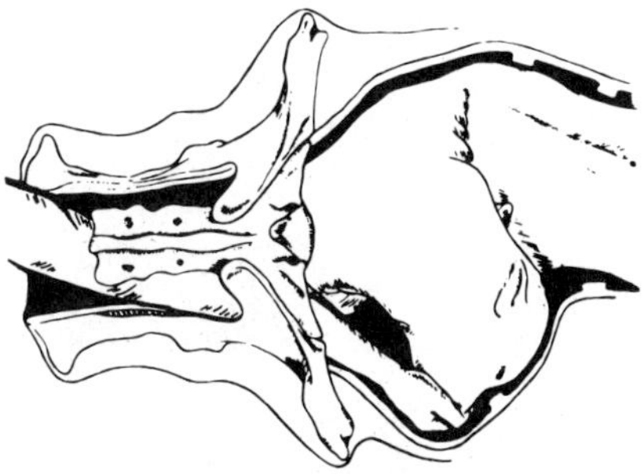

Correction of head and neck deviation in anterior presentation. Correcting difficult cases may require pulling the head and neck around with the hand. The calf is first pushed forward and the hand quickly moved to grasp the calf's muzzle. The head is pulled around and in line with the birth canal. In some cases it may be necessary to apply a snare to the lower jaw for additional traction. Excessive pressure should be avoided, as the jaw is not strong and will fracture easily. A safer and more common practice is looping the obsterical chain around the poll, under the ears and through the mouth in a "war bridle" manner. The attachment allows greater traction but will permit the calf's mouth to gape. Care should be taken to avoid cutting the birth canal with the calf's sharp incisor teeth.

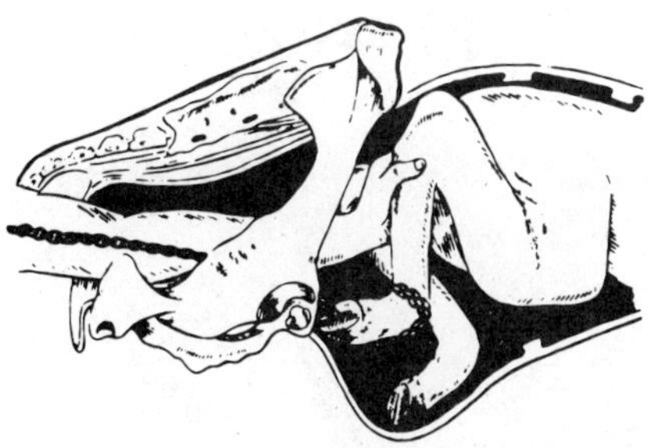

Correction of the hock flex in posterior presentations. The calf is first pushed forward. The hand grasps and cups the calf's foot, then draws it back as the hock is flexed. The foot in the cupped hand is lifted over the pelvic rim and into the vagina. An alternate method in more difficult cases is to place a snare around the pastern, attached at the front of the leg. The snare is then pulled between the digits of the foot so that when traction is applied the fetlock and pastern are flexed. The calf is pushed forward and the foot is guided over the pelvic rim as an assistant pulls the snare.

Figure 11.15. Abnormal presentations of the calf at parturition, with guidelines in assisting the difficult deliveries. Courtesy of Texas A and M University.

made on your hand outside the cow and then applied to the leg. The hitch in a hard pull helps to distribute the stress imposed on the bones over the two locations instead of one, reducing the possibility of bone fracture (Figure 11.16d).

Do not apply a single loop of the calving chain around both legs at the same time. The danger of breaking one or both legs is very high, especially if the birth is difficult and a hard pull is used.

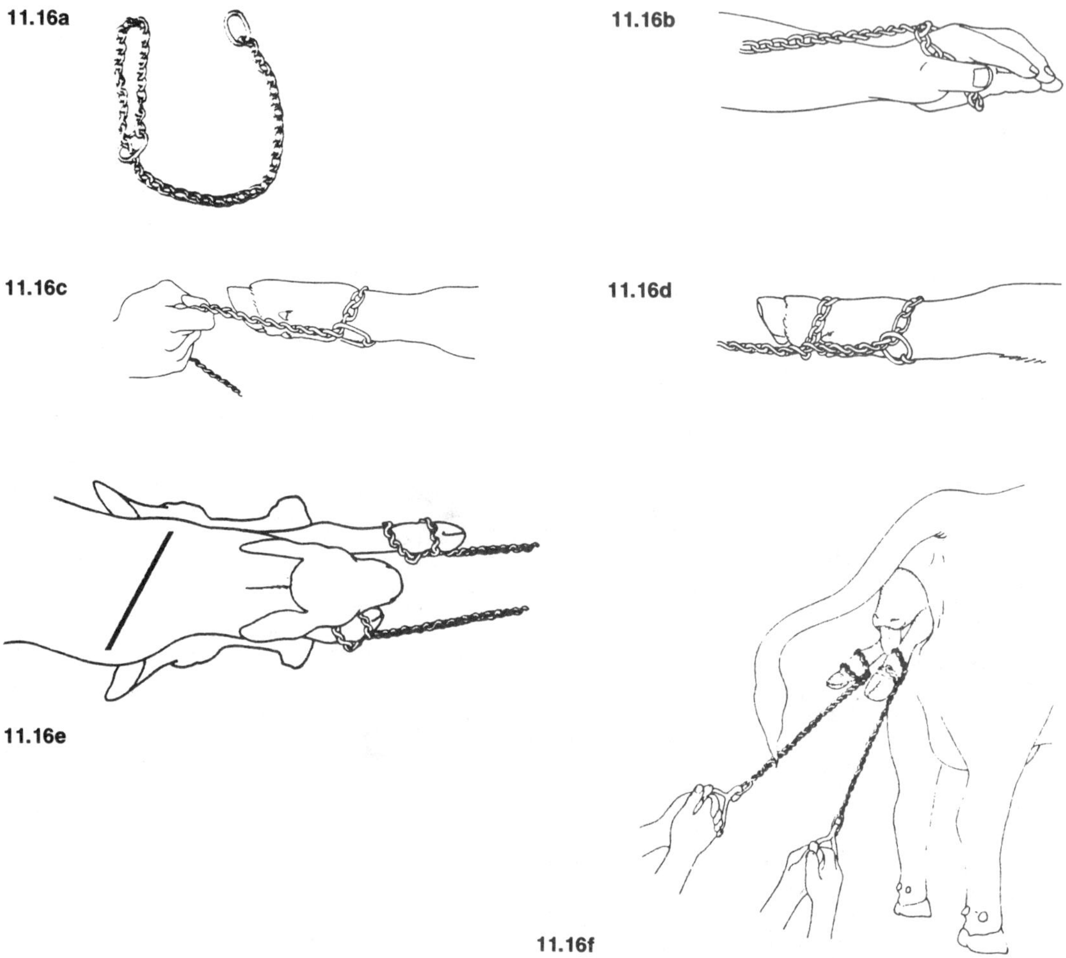

11.16a 11.16b 11.16c 11.16d 11.16e 11.16f

Figure 11.16a–f. Diagrams of providing simple assistance to a beef female experiencing difficulty calving. Content of this figure was adapted from Battaglia, R. A. and Mayrose, V. B., 1981. *Handbook of Livestock Management Techniques.* Minneapolis: Burgess Publishing Co.

5. Repeat Steps 1 to 4 for the second leg.

6. Before applying the calving chain handles to the calving chains, make sure that the chains will pull from the bottom of the leg (dewclaw side). This will insure that the legs will be pulled straight and not at an angle or with a bend.

7. Attach two calving chain handles to the chains and pull gently, making sure that the loop and half hitch of the chain have not slipped from the desired position on the calf's leg.

8. Some calves can be delivered by pulling both legs evenly; however, it is best to pull alternately on one leg and then the other a few inches at a time. When the legs

are "walked out" in this manner, the shoulders or hips are allowed to pass through the pelvic girdle one at a time (Figure 11.16e).

Calf Forward. If the shoulders should happen to become lodged in the pelvis, apply traction to the calf's head. This will reduce the compaction of the head against the sacrum (top) of the birth canal and reduce the dimensions of the shoulder and chest region. The chain can be applied by making a loop as in Step 1 and applying it to the nose and muzzle of the calf between the muzzle and the eyes. Care should be taken when pulling on the nose so that you do not break the nose or jaw.

Calf Backward. In a breech delivery, the extraction of the fetus is against the normal direction of hair growth, and the birth canal should be liberally lubricated with an obstetrical or mild household soap dissolved in warm water. It may be necessary to rotate the fetus about one-eighth of a turn to take full advantage of the greatest diameter of the cow's pelvis. If delivery proves extremely difficult, a Caesarean section is probably necessary and should not be delayed.

9. Once the calf's legs are exposed, the calf should be pulled downward (toward the cow's hocks) at a 45 degree angle (Figure 11.16f).

As the calf's head and shoulders come through the birth canal, the chance of uterine or cervical lacerations is greatest. This damage may lead to infection and future reproductive problems. Because pressure dilates the cervix and birth canal, traction that is applied gradually can usually prevent damage. If assistance is given to a cow too early, the slow application of traction would not interfere with normal cervical dilation and would again minimize the potential of damaging the birth canal.

10. One or two individuals using chains with manual strength should be able to pull a calf. If the birth is extremely difficult, however, it may be necessary to use a mechanical calf puller. When this situation arises, it is best to seek experienced help. If a calf puller is used incorrectly, it may cause permanent damage to both cow and calf. It has been demonstrated that the leverage exerted by calf pullers can pull over a ton. Indiscriminate use of calf pullers can do serious physical damage to both cow and calf.

REPRODUCTIVE SOUNDNESS IN THE BULL

To help eliminate reproductive losses due to male infertility, bulls should be evaluated for breeding soundness 30 to 60 days prior to the breeding season. Breeding Soundness Examination (BSE) consists of three basic areas: (1) a physical examination of the bull with emphasis on the reproductive system, (2) measurement of scrotal circumference, and (3) evaluation of at least one semen sample. The BSE should be done by an experienced veterinarian or a qualified reproductive physiologist.

Physical Examination

A physical examination should include observation of all conditions that might interfere with the bull's ability to locate cows in heat and breed them. A review of previous disease problems and any recent stressful conditions is useful. For example, high temperatures with infections and extremely low temperature causing frostbite of the testicles can cause structural defects in sperm cells.

A visual appraisal for unthriftiness, body condition, and structural soundness is usually made. Bulls should be in a good condition, neither thin nor too fat. Feet, legs, and eyes are among the most important structural traits which should be sound.

The reproductive system is examined rectally by palpation. Size, shape and consistency of the prostate, seminal vesicles, and ampullae are noted, and the internal inguinal rings are examined for size.

The external genitalia (penis and prepuce) are examined for any structural abnormalities or adhesions when the semen sample is collected.

Scrotal Size Circumference

Scrotal circumference is measured in centimeters (cm) by using the following simple procedure. The testicles are palpated gently but firmly down into the bottom of the scrotum. Scrotal tape is placed up over the testicles-scrotum and tightened loosely around the scrotal neck close to the body wall. The tape is then slid slowly and carefully downward. The sliding loop will enlarge as the size of the testicles-scrotum increases until the largest circumference is attained (Figure 11.17).

No additional tension should be placed on the free end of the tape. Once the tape has reached the largest circumference, it will drop off and the size can be read directly from the tape.

Table 11.3 provides the guide for determining acceptable scrotal circumference for bulls of different ages as recommended by the Society for Theriogenology. Figure 11.18 shows the relationship of increased scrotal size to improved semen characteristics. Note a threshold effect where there appears to be little advantage of having scrotal circumferences above 36 to 38 cm in yearling bulls.

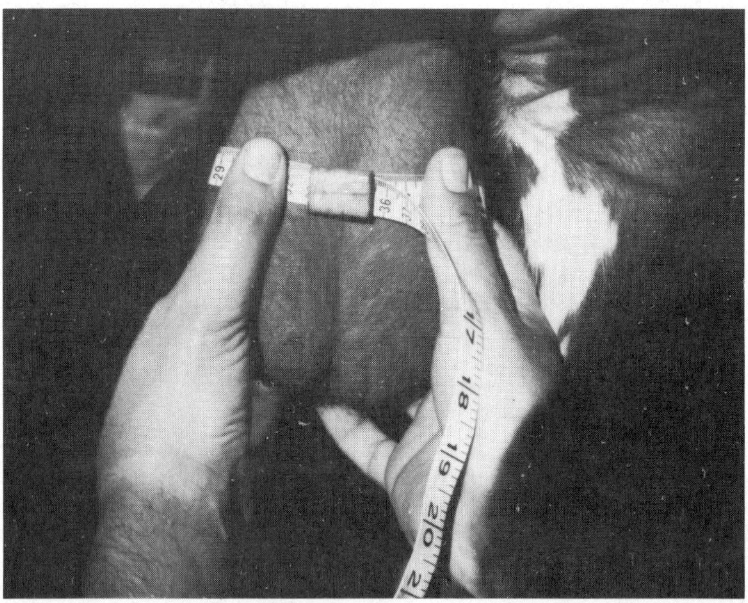

Figure 11.17. Measuring scrotal circumference.

Table 11.3. Scoring Criteria for Scrotal Circumference, Semen Morphology and Motility

Scoring Criteria	Ver. Good	Good	Fair	Poor
Scrotal Circum. by Age 12–14 mos.	>34 cm	30–34	<30	<30
15–20 mos.	>36 cm	31–36	<31	<31
21–30 mos.	>38 cm	32–38	<32	<32
over 31 mos.	>39 cm	34–39	<34	<34
SCORE FOR SCROTAL CIRCUMFERENCE:	40	24	10	10
Semen Morphology				
Primary Abnormalities	<10	10–19	20–29	>29
Total Abnormalities	<25	26–39	40–59	>59
SCORE FOR MORPHOLOGY	40	24	10	3
Gross Motility	**Rapid Swirling**	**Slower Swirling**	**Generalized Oscillation**	**Sporadic Oscillation**
Individual	Rapid Linear	Moderate Linear	Slow Linear to Erratic	Very Slow Erratic
SCORE FOR MOTILITY:	20	12	10	3

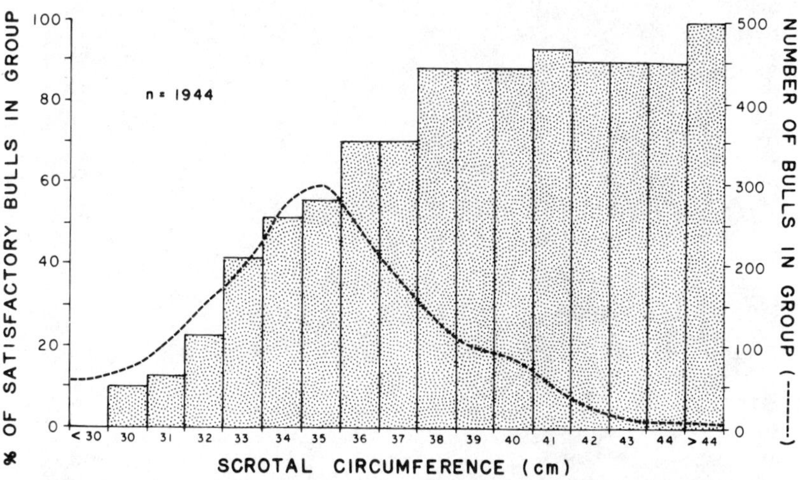

Figure 11.18. Relationship between scrotal circumference and semen characteristics in yearling bulls (Coulter, G. H. 1982 NAAB Conference on Beef AI and Embryo Transfer).

Semen Evaluation

A semen sample is collected usually through the use of an electroejaculator. Temperature control of the semen sample is necessary to properly evaluate motility. Also, the sample must be kept clean and free of water, urine, or chemical disinfectants. The semen sample is evaluated for volume, motility, and morphology.

Motility can be estimated by observing the mass movement of sperm under the microscope. Sperm cell motility is more critically evaluated by observing the movement of individual sperm cells.

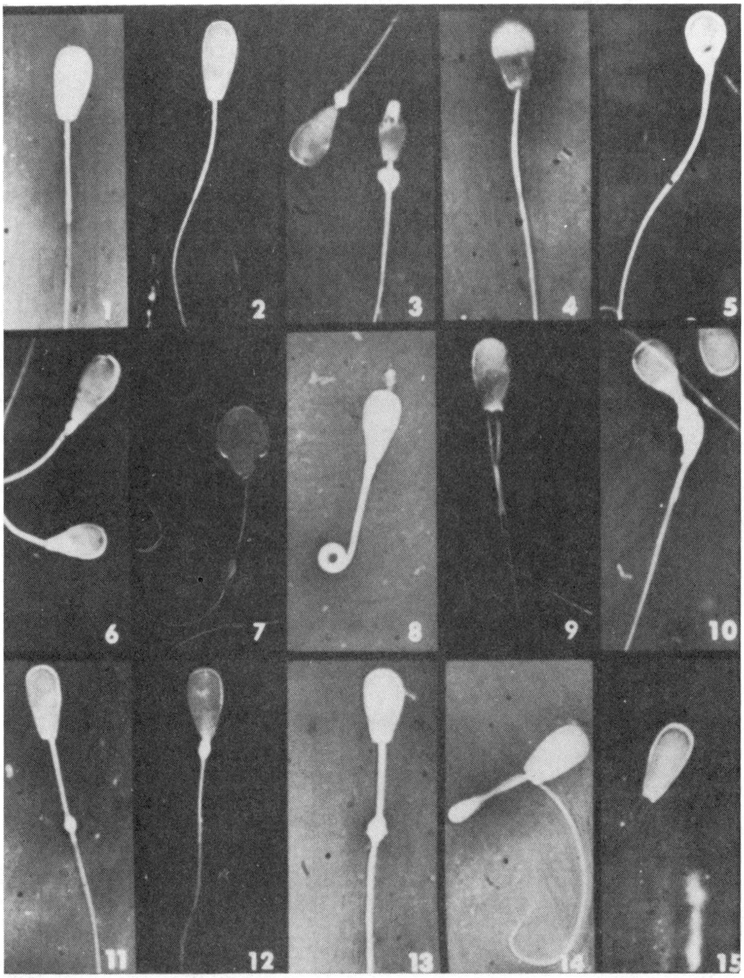

Figure 11.19. Some of the more common primary and secondary abnormalities of bull sperm. 1 & 2—Normal bovine sperm; 3—head defects; 4—large head; 5—round head; 6—pear-shaped head; 7—head defects and abnormal midpiece; 8—coiled tail; 9—double midpiece; 10—swollen midpiece; 11 & 13—distal droplet (droplet at midtail); 12—proximal droplet (droplet near head); 14—bent tail; and 15—separated head. Courtesy of Successful Farming Magazine.

Morphology or sperm cell structure is evaluated by placing a small amount of semen and stain on a slide, gently mixing them, and spreading a thin film on a glass slide. A special microscope is used to evaluate the sperm cells and at least one hundred cells are counted and classified as normal or as one of several abnormal types. Sperm cell abnormalities fall into two classifications: primary abnormalities, which are generally the more severe morphologic defects, and secondary abnormalities. Some of these abnormalities are shown in Figure 11.19.

Although many other factors are related to high conception rates, the number of abnormal sperm cells can be an additive factor in a bull's infertility if they become too numerous. There is evidence that sperm cell abnormalities may begin to affect fertility when 25% or more of the cells are abnormal.

Table 11.3 gives the scoring criteria and scores for scrotal circumference, semen morphology, and semen motility as recommended by the Society for Theriogenology (American Veterinary Society for the Study of Breeding Soundness). A breeding soundness examination form can be completed which would classify the bull into one of three categories: (1) satisfactory potential breeder, (2) questionable potential breeder, or (3) unsatisfactory breeder.

When looking at scrotal circumference and semen characteristics of young bulls near one year of age, it should be kept in mind that bulls of this age will generally improve in scrotal size and semen quality with further maturity. Scrotal circumference at 12 months of age is approximately two-thirds of the potential size at maturity (Figure 11-20). Development of the epididymis and other sperm-carrying ducts improve in most bulls up to 15 months of age. Due to immaturity, many young bulls receive "questionable"

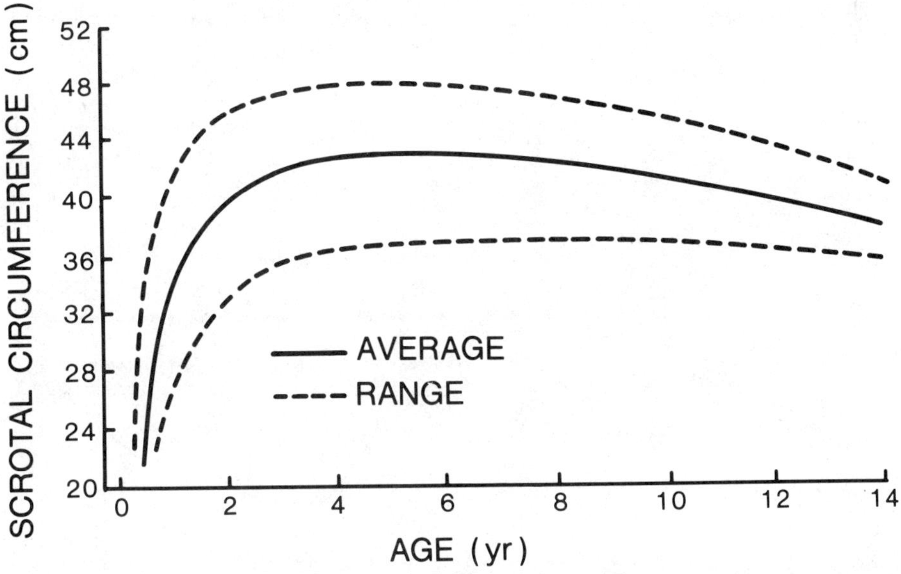

Figure 11.20. Distribution of scrotal size, by age, in 1275 Angus bulls. Note the range in size above and below the average. Adapted from Coulter, G. H., 1982. NAAB Conference on Beef AI and Embryo Transfer.

and "unsatisfactory" scores on the BSE at 12 to 14 months of age but will pass at a later time—usually a few weeks.

ARTIFICIAL INSEMINATION

Artificial insemination (AI) is a process where semen is deposited in the female reproductive tract by artificial techniques rather than by natural mating. AI was first successfully accomplished in cattle in the early 1900s.

The primary advantage of AI is the extensive use of outstanding sires to maximize genetic improvement. For example, a bull may sire 30 to 50 calves naturally per year over a productive lifetime of 3 to 8 years. In an artificial insemination program, a bull can produce 200 to 400 units of semen per ejaculate, with four ejaculates typically collected per week. If the semen is frozen and stored for later use, hundreds of thousands of calves can be produced by a single sire (one calf per 1.5 units of semen), and many of these offspring can be produced long after the sire is dead. AI also can be used to control reproductive diseases, and sires can be used that have been injured or are dangerous when used naturally.

Semen Collection and Processing

There are several different methods of collecting semen: the most common method is the artificial vagina (Figure 11.21) which is constructed to be similar to an actual vagina. The semen is collected by having the bull mount an estrus female or training him to mount another animal or object (Figure 11.22). When the male mounts, his penis is directed into the artificial vagina by the person collecting the semen and the semen accumulates in the collection tube (Figure 11.23).

Semen can also be collected by using an electroejaculator, in which a probe is inserted into the rectum and an electrical stimulation causes ejaculation. This is used most

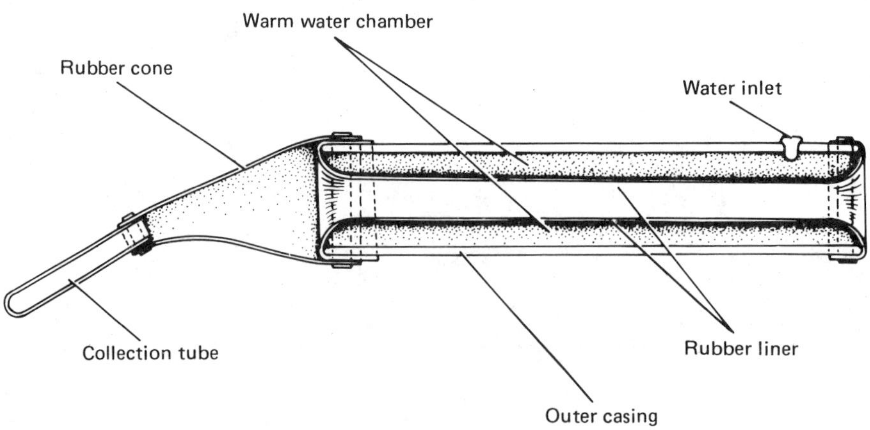

Figure 11.21. An artificial vagina used in collecting semen from the bull. Drawn by Dennis Giddings.

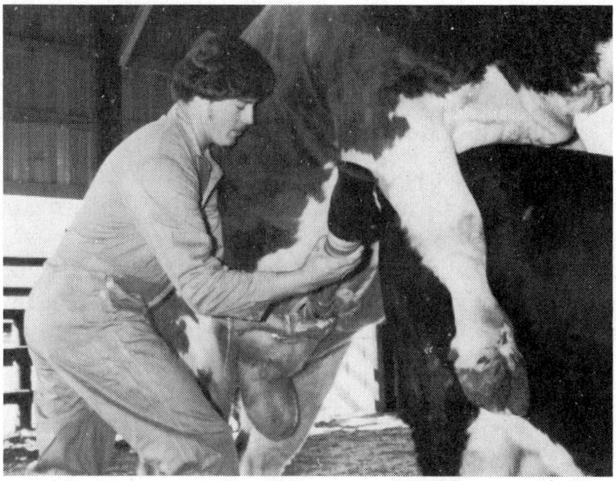

Figure 11.22. Collecting semen from a bull using the artificial vagina. Courtesy of the Animal Reproduction Laboratory, Colorado State University.

commonly in bulls that are not easily trained to use the artificial vagina and from which semen is collected infrequently.

If semen is collected too frequently, the number of sperm per ejaculate decreases. Semen from the bull is typically collected twice a day for two days a week. After the semen is collected, it is evaluated for volume, sperm concentration, motility of sperm,

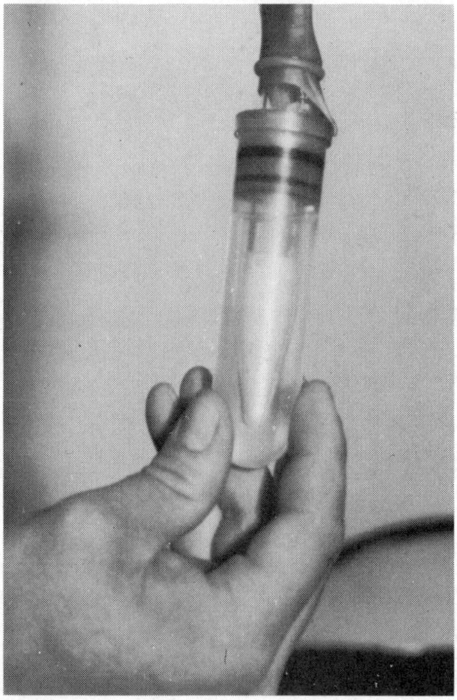

Figure 11.23. Approximately 4 ml. of bull semen in the collection tube attached to the artificial vagina. Note that the collection tube is immersed in water to control the temperature. The livability of the sperm is decreased when they are subjected to sudden temperature changes. Courtesy of the Animal Reproduction Lab., Colo. State Univ.

and sperm abnormalities (Figure 11.24). The semen is usually mixed with an extender which dilutes the ejaculate to a greater volume. This greater volume allows a single ejaculate to be processed into several units of semen, where one unit of semen is used for each female inseminated. This extender is usually composed of nutrients such as milk and egg yolk, a citrate buffer, antibiotics, and glycerol. The amount of extender used is based on the projected number of viable sperm available in each unit of extended semen. For example, each unit of semen for insemination in cattle should contain 10 million motile, normal spermatozoa.

Some semen is used fresh; however, it can only be stored this way for a day or two. Most semen is frozen in liquid nitrogen and stored in ampules (glass vials), plastic straws (Figure 11.25), or pellets. The semen can be stored in this manner for an indefinite period of time and still retain its fertilization capacity. Most cattle are inseminated with frozen semen rather than fresh semen.

Insemination of the Female

Prior to insemination, the frozen semen is thawed. Semen that is thawed should not be refrozen and used again because conception rates will be reduced.

High conception rates using AI depend on the females cycling and ovulating, detecting estrus, using semen that has been properly collected, extended and frozen, thawing and handling the semen satisfactorily at the time of insemination, insemination techniques, and avoiding extremes in stress and excitement to the animal being inseminated.

Figure 11.24. Normal bull semen. Courtesy of the Animal Reproduction Laboratory, Colorado State University.

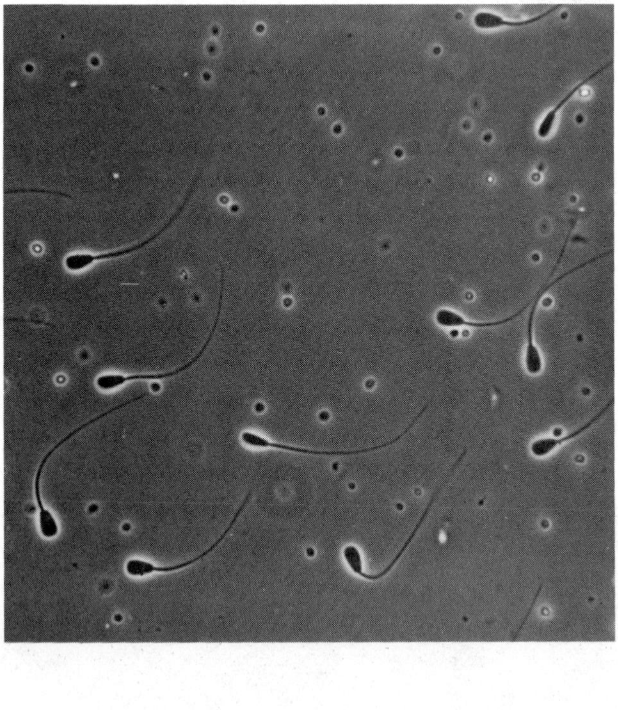

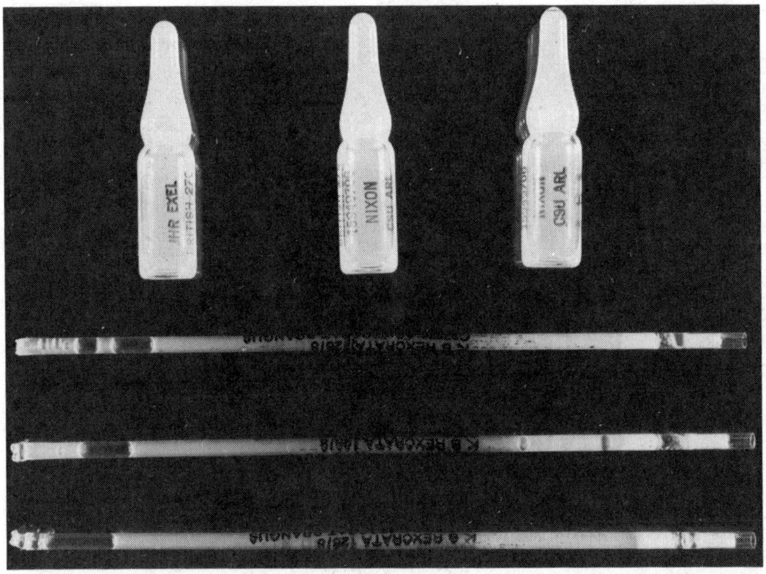

Figure 11.25. Semen is typically stored and frozen in ampules or straws. The top three in the figure are ampules and the bottom three are straws. The animal's name, registration number, location, and date of collection are printed on the ampules and straws.

Detecting Estrus

Estrus must be detected accurately because it signals the time of ovulation and determines the proper timing of insemination. The best indication of estrus is the condition called *standing heat* in which the female stands still when mounted by a male or another female.

Cows are typically checked for estrus twice daily, in the morning and evening. They are usually observed for 30 minutes to detect standing heat. Other observable signs are restlessness, roughed up tail head, being followed by a group of bull calves, attempting to mount other cows, pink swollen vulva, and a clear mucous discharge from the vagina. Some producers use sterilized bulls or hormone-treated cows as *heat checkers* in the herd. These animals are sometimes equipped with a head harness that greases or paint marks the cow's back when she is mounted (Figure 11.26).

Proper Timing of Insemination

The length of estrus and ovulation time is quite variable in cows and heifers. This variability poses difficulty in determining the best time for insemination. An additional challenge is that sperm are short-lived when put into the female reproductive tract. Also, estrus is sometimes expressed without ovulation occurring and occasionally the reverse of this occurs.

Time of insemination should be as close to ovulation time as possible. Cows found in estrus in the morning are usually inseminated the evening of the same day, and cows in heat in the evening are inseminated the following morning. Insemination, therefore, should occur toward the end or after estrus has been expressed in the cow (Figure 11.27).

Cows are penned and inseminated in a chute that restrains the animal. The most

Figure 11.26. Bull equipped with chin ball marker. Courtesy of Select Sires.

common insemination technique in cattle involves the inseminator having one arm in the rectum to manipulate the insemination tube through the cervix (Figure 11.28). The insemination tube is passed just through the cervix and the semen is deposited into the body of the uterus.

The number of cows inseminated each year is not well documented. It is estimated that in the United States approximately 3% of the beef cows (a total of approximately 1 million) are inseminated each year. This compares to approximately 60% of the dairy cows (approximately 7 million) that are inseminated each year.

The amount of semen, for the various breeds, which is sold domestically, exported, and custom frozen is shown in Table 11.4. It is not known how much beef semen is used to breed dairy cows.

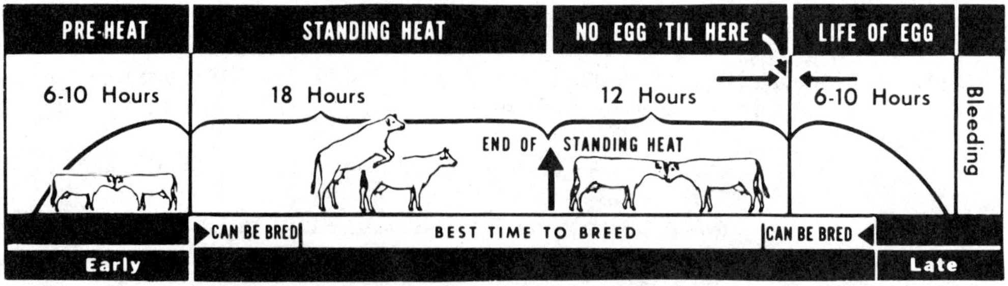

Figure 11.27. Identifying the preferred time for AI breeding. Courtesy of American Breeders Service.

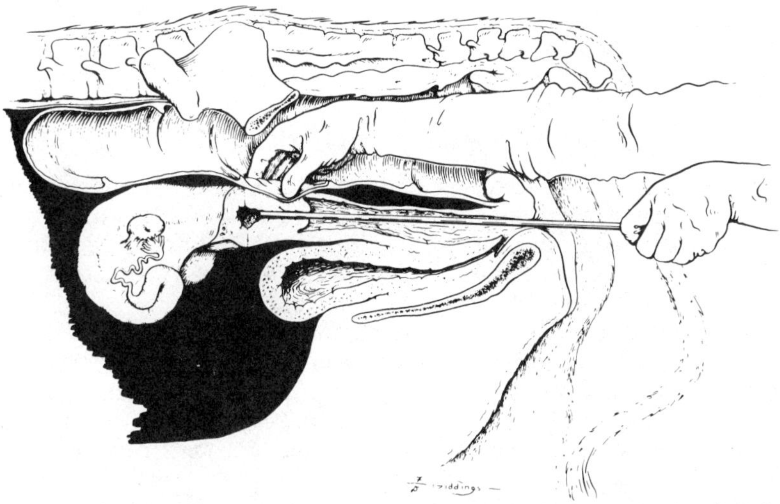

Figure 11.28. Artificial insemination of the cow. Note that the insemination tube has been manipulated through the cervix. The inseminator's forefinger is used to determine when the insemination rod has entered into the uterus. Drawn by Dennis Giddings.

Table 11.4. Beef Semen, by Breeds, Sold Domestically, Exported, and Custom Frozen, 1979

	Units of Semen (thousands)		
Breed	**Domestic Sales**	**Export Sale**	**Custom Frozen**
Angus	430.2	25.9	220.3
Simmental	225.7	43.9	172.1
Polled Hereford	172.0	52.9	120.9
Hereford	43.7	7.3	108.2
Charolais	34.0	9.6	35.0
Brahman	28.2	52.6	
Limousin	27.7		51.4
Red Angus	23.7	11.3	23.7
Maine Anjou	22.0		
Chianina	19.1		74.1
Santa Gertrudis		16.1	
Brangus			54.0
Beefmaster			35.6
Other	60.0		230.1
Total	1,086.0		1,125.0

Source: Natl. Assoc. of Animal Breeders.

Semen is available commercially from several artificial insemination (AI) companies (see Beef AI Organizations, Chapter 23).

ESTROUS SYNCHRONIZATION

Estrous synchronization is controlling or manipulating the estrous cycle so that females in a herd express estrus at approximately the same time. Estrous synchronization is a useful part of an AI program because checking heat and breeding animals, particularly under range conditions, is time consuming and expensive. Also, estrous synchronization is a tool in making embryo transfer programs successful.

Considerable research has been done on estrous synchronization over the past few decades, where several different products have been evaluated. Only recently, however, have the products been available commercially because they have not previously been cleared by the Food and Drug Administration. Also, some of the earlier research work demonstrated that conception rates were lowered when estrous synchronization products were used.

In 1979, a prostaglandin was cleared for use in cattle. Prostaglandins are naturally occurring fatty acids that appear to have important functions in several of the body systems. The prostaglandin that has a marked effect on the reproductive system is prostaglandin F_2 alpha (PGF_2).

Earlier in the chapter, it was pointed out that the corpus luteum (CL) controls the estrous cycle in the cow by secreting the hormone progesterone. Progesterone prevents the cow from expressing heat and ovulation. The prostaglandin destroys the CL, thus eliminating the source of progesterone. About three days after the injection of prostaglandin, the cow will be in heat. For prostaglandin to be effective, the cow must have a functional CL. It is ineffective in heifers that have not reached puberty or in non-cycling mature cows. Also, prostaglandin is ineffective if the CL is immature or has already started to regress. Prostaglandin is, then, only effective in heifers and cows that are in days 5 to 18 of their estrous cycle. Because of this relationship, prostaglandin is given on either a one- or two-injection system. These two systems are described using the prostaglandin, Lutalyse®.

One-injection System

This system requires one prostaglandin injection and an 11-day AI breeding season. The first 5 days are a conventional AI program of heat detection and insemination. After 5 days, a calculation can be made to determine what percentage of females in the herd have been in heat. If a smaller-than-expected percentage are cycling, the cost of the prostaglandin may not justify the anticipated benefit. If the decision is to proceed with the injection, the remaining animals not previously bred are injected on day 6. Then the AI breeding season continues 5 more days for a total of 11 days.

Field trial results in herds where a high percent of the females are cycling show a 50% to 60% pregnancy rate at the end of 11 days of AI breeding for those receiving prostaglandin versus a 30% to 40% pregnancy rate for those cows not receiving it. The amount of prostaglandin required per calf that is produced by AI, will be between 1.2 and 2.0 units.

Two-injection System

In this system all cows are injected with prostaglandin at two different times. Counting the first injection as day 1, the second injection is administered on day 11, 12, or 13. All cows capable of responding to the drug should be in heat during the first five days after the second injection of prostaglandin. Heat detection and insemination can occur each of these five days or a fixed-time insemination can be performed 76 to 80 hours after the second injection.

Field trial results in herds in which a high percent of the females are cycling show a 35% to 55% pregnancy rate at the end of the fifth day or 76 to 80 hour one-time insemination versus 10% to 12% in the cows not receiving prostaglandin. The two-injection system will require 4 to 7 units of prostaglandin per AI calf.

Other Estrous Synchronization Compounds

Syncro-Mate-B® (SMB) is another compound which is currently being used to synchronize estrus in cattle. It is a treatment consisting of subcutaneous ear implant and an intramuscular injection. The implant contains a synthetic form of progesterone called Norgestomet®. The injection contains an estrogen compound, estradiol valerate, plus Norgestomet®.

SMB does cause some non-estrous cycling females, both heifers and lactating cows, to start having estrous cycles. There are little data, however, to suggest any higher pregnancy (particularly early pregnancy) rates from using SMB over other estrous synchronization products.

Estrumate® is a prostaglandin analog (chemical variation of the same compound) which is currently marketed as an estrous synchronization product. The primary advantage of an analog is a lower dosage level, otherwise, the results should be similar to $PGF_{2\alpha}$.

Practical Guidelines

The following guidelines are given by Van Dervoort 1982 who has combined estrous synchronization and AI on thousands of cows:

1. Know how estrous synchronization works on the cow.
2. Use proper method of injection. Intramuscular injection with 18 gauge, 1¼ to 1½ inch needle. Shorter needles and fat cows will reduce success.
3. Recommended dosage is 5cc. Heavy cows should have dosage increased, for example, 1200 lb (6cc) and 1400 lb (7cc).
4. Avoid as many stress factors as possible, for example, moving just prior to and after injection, adverse weather conditions, removal of calves after injection, and other similar stresses.
5. When force breeding, consider the "peak of heat activity" as the guide to when to breed.
6. Heat check for three weeks prior to breeding if possible.
7. As many as 25% of the herd can show heat in the two or three days following a forced breeding program at the so-called 80 hour point. These cows must be bred or rebred as the case may be.

8. Be aware of shorter periods of standing heat.
9. Give serious consideration to breeding the repeats.

Prostaglandin is not a wonder drug. It will work only in well-managed herds where a high percentage of the females are cycling. Biologically, it has been well demonstrated that prostaglandin can synchronize estrus. Producers, however, must weigh the cost against the economic benefit. Caution should be exercised in administrating prostaglandin to pregnant cows as it may cause abortion. Estrous synchronization in cattle may not be advisable in areas where inadequate protection can be given to young calves during severe blizzards. Also, herd health programs must be excellent to prevent high losses from calf scours and other diseases that become more serious where large numbers of newborn calves are grouped together.

Economic Assessment

A computer simulation model was developed to evaluate the following estrous synchronization systems using PFG2: Whether or not to palpate to identify cycling cows; use a one injection, 5-day prebreeding, or two injection system; breed by detection, 80-hour appointment, or a combination of the two (Whitman 1981). The results were: (1) For most cycling rates, 5-day prebreeding systems resulted in higher total pregnancy rates and lower costs per pregnancy cow, (2) single injection systems were usually lower in cost and lower in cost per pregnancy than double injection systems, (3) palpation was desirable in the majority of systems except when 90% or more of the herd were cycling, and (4) breeding detection resulted in a lower cost per pregnancy than breeding by appointment or a combination of detection and appointment except when all cows were cycling or when cycling cows were identified by palpation prior to the breeding season.

EMBRYO TRANSFER

Embryo transfer is sometimes referred to as ova transplant or embryo transplant. In this procedure, an embryo in its early stage of development is removed from its own mother's (the donor) reproductive tract and transferred to another female's (the recipient) reproductive tract.

The first successful embryo transfer was accomplished in rabbits in 1890 and in cattle in 1951. In recent years, commercial embryo transfer companies have been established in the United States and several foreign countries. It is estimated that approximately 20,000 pregnancies from embryo transfers in cattle occurred in North America in 1980. This compares to 20 embryo transfer pregnancies in 1972.

Superovulation is the production of a greater-than-normal number of eggs. Females that are donors of eggs for embryo transfer are injected with fertility drugs which usually cause several follicles to mature and ovulate. Injections are started in the mid part of the estrous cycle, around 9 to 14 days after estrus. Follicle stimulating hormone (FSH) is injected twice per day, 5 mg per dose for 4 days. In addition, on the fourth day of FSH injections, a prostaglandin (e.g., Lutalyse) is injected, which causes the death of the corpus luteum, triggering the next estrus, usually 2 days after the Lutalyse. Approximately 12 to 24 hours after first standing estrus, the donor is bred artificially.

Usually frozen-thawed semen is used and a double dose of high quality semen is recommended. Recipients are usually synchronized with a prostaglandin. The recipient cow must be in estrus within one day of the donor's estrus. Recipients should be selected for calving ease, high milk production, and an excellent health status.

The average donor produces approximately 7 transferable embryos per superovulation treatment; however, a range of 0 to 33 embryos can be expected (Figure 11.29). This procedure gives embryo transfer its greatest advantage, that is, increasing the number of offspring from a superior female. A potential donor cow will usually produce more embryos if she is from 3 to 10 years of age, has calved regularly each year of her productive life, usually conceives in 2 services or less, has exhibited regular estrous cycles, comes from fertile blood lines, did not retain her placenta, and has no history of calving difficulties.

The bull, to be bred to the donor cow, should have a history of excellent semen production and a high conception rate where his semen has been used in an AI program.

The key to the justification of embryo transfer is the identification of genetically superior cows and bulls. Procedures for identifying genetically superior cattle is given in detail in Chapters 8 and 13. Much of the guess work can be eliminated by making repeat matings where genetically superior calves have previously been produced. Embryo transfer is usually confined to seedstock herds where genetically superior females can be more easily identified and high costs can be justified. Embryo transfer calves have an estimated average breakeven price of $2000 to $3000 at a year of age. Seedstock producers must evaluate the marketability of the embryo transfer calves. Some genetically superior embryo transfer calves would not sell for a sufficient amount to cover costs. Some embryo transfer calves which are not genetically superior can be merchandised for high prices because of a demand that has been previously created.

Donor selection, which combines expected high levels of financial return, genetic superiority, and high reproductive potential, indicates that the number of donors worthy of an embryo transfer program are relatively few. This "relatively few" number could be justified into several thousand head which is less than one percent of the total U.S. cowherd.

Figure 11.29. A Limousin donor cow with her 11 embryo transfer calves, the result of one superovulation. Courtesy of Dr. Peter Elsden, Colorado State University.

A non-surgical embryo collection procedure occurs where a flexible rubber tube (Foley Catheter) with three passageways is passed through the cervix and into the uterus (Figure 11.30). A rubber balloon, which is built into the anterior end of the tube, is then inflated to about half the size of a golf ball so that it expands to fill the uterine lumen and prevent fluid from escaping around the edges. There are two holes in the tube anterior to the balloon which lead into separate passageways, one for fluid entering the uterus and the other for fluid draining from the uterus. A balanced salt solution (antibiotics and heat-treated serum are added) is placed in a container and held about 3 feet above the cow. The container is connected to the Foley Catheter by an inflow tube. A second tube is connected to the other passageway of the catheter to drain off the medium which has washed (flushed) the ova and embryos out of the uterus. This solution is then collected in tall cylinders holding about 2 pints of fluid. Unfertilized ova and embryos drop to the bottom, then the bulk of the medium is siphoned off, and the last cupful of fluid is searched under the microscope.

Until recently, embryo transfer was done surgically, but now nonsurgical techniques can be used. Embryos usually are transferred with an artificial insemination straw gun shortly after being collected. The recipient females need to be in the same stage (within 24 hours) of the estrous cycle for successful transfers to occur. Large numbers of females must be kept for this purpose or estrous synchronization of a smaller number

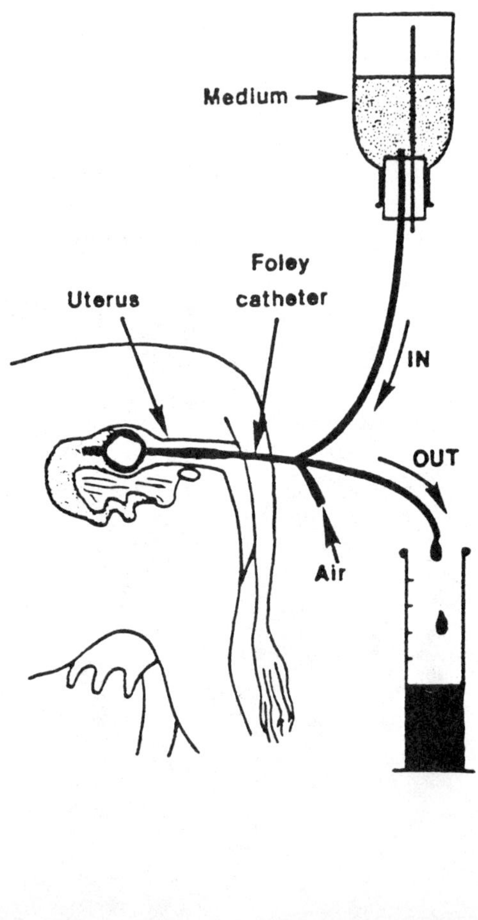

Figure 11.30. Nonsurgical collection of embryos. Courtesy of Colorado State University.

of females is necessary. Embryos can now be frozen in liquid nitrogen and remain dormant for months or years. Conception rates are lower when frozen embryos are used compared to fresh embryos, but research is narrowing the difference in conception rates. In cattle, approximately 50% to 60% of the frozen embryos will be normal after thawing. Thirty to 40% of the normal embryos will result in confirmed pregnancies at 60 to 90 days. This compares to a pregnancy rate of 55% to 65% from fresh embryos transferred the same day of collection.

There is no decline in embryo recovery rate the first three times a donor is superovulated, however, additional superovulation treatments result in a reduced number of embryos. Most donors can be superovulated three to four times a year. There appears to be little effect of season, breed, or lactating versus dry donors on the success of embryo transfer. From a survey of CSU clients (Elsden et al. 1980), 2% of the recipients aborted between 3 and 9 months of gestation, 4% of the embryo transfer calves died at birth and 4% died between birth and weaning, for a total of 10% loss between 3 months gestation and weaning.

Recent advances in reproduction technology are the world's first test tube calf (fertilization occurred outside the cow's body) born in 1981 and the first identical twin calves born in the United States resulting from embryo splitting. Splitting one embryo to produce two identical offspring is one of the initial steps in the potential cloning of beef cattle and other intriguing aspects of genetic manipulation (Figures 11.31 and 11.32). Table 11.5 shows pregnancy results from embryo splitting at Colorado State University.

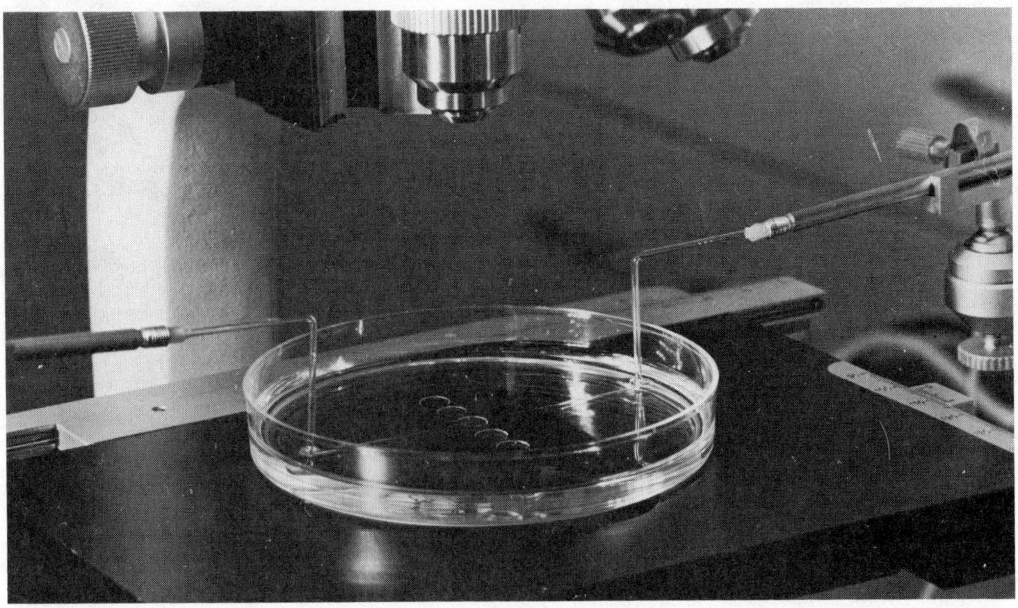

Figure 11.31. Close-up of the microtools used for splitting embryos and the microdrops in the center of the dish which contain the embryos. Courtesy of Williams, et al., 1983 NAAB Conference on Beef AI and Embryo Transfer.

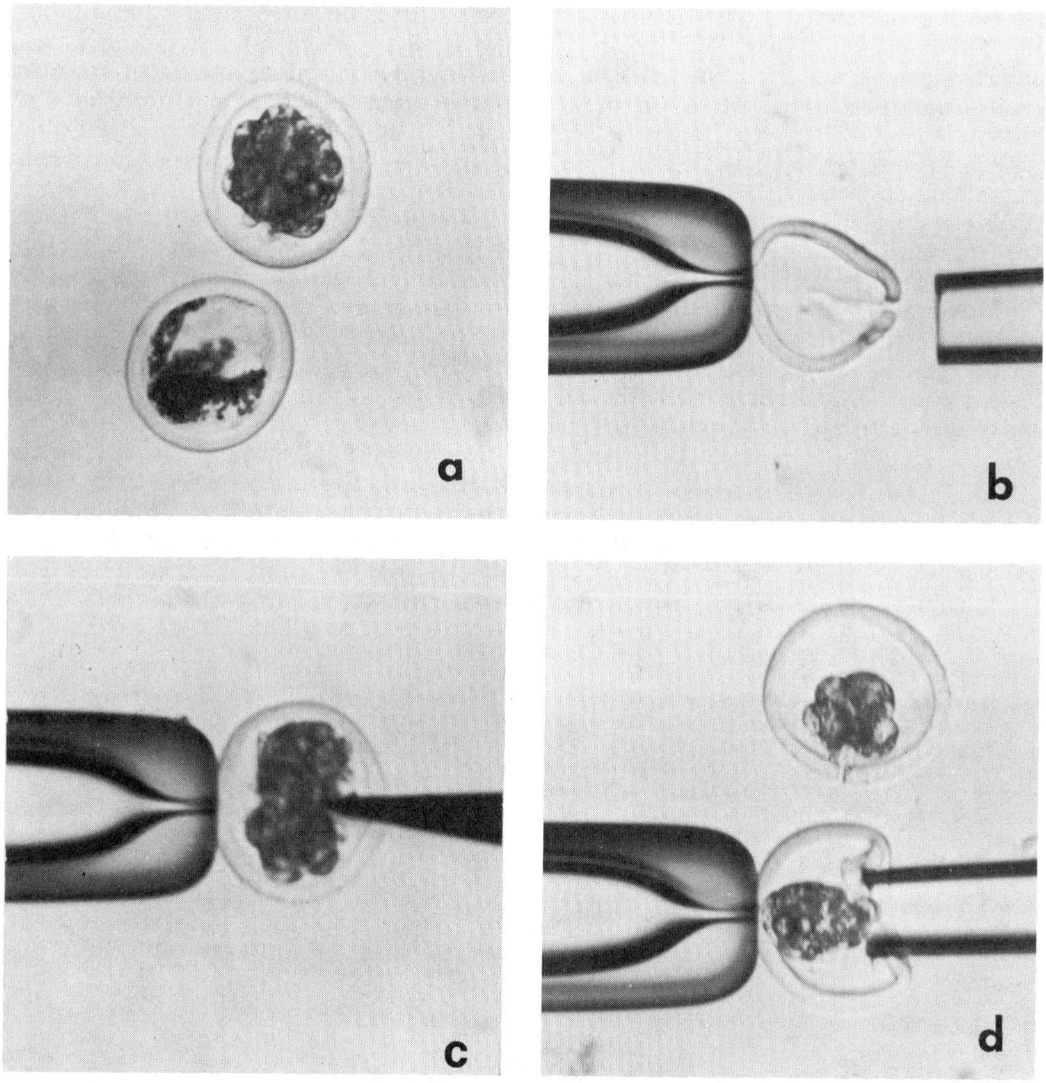

Figure 11.32. The process of embryo splitting. a. A morula (40 to 70 cells of the fertilized egg) and an unfertilized egg; b. the unfertilized egg with the cellular contents sucked out of it; c. the morula is divided into two groups of cells with the microsurgical blade; d. one-half of the cells are left in the morula while the other one-half of the cells are placed inside the other unfertilized egg. The result is two genetically identical embryos ready for transfer. Courtesy of Williams, et al., 1983 NAAB Conference on Beef AI and Embryo Transfer.

Table 11.5. Non-surgical Transfer of Bovine Demi-embryos

No. of whole embryos bisected	72
No. of demi-embryos transferred	144
No. of pregnancies/no. of demi-embryos transferred	72/144 (50%)
No. of pregnancies/no. of embryos split	72/72 (100%)
No. of embryos split resulting in 2 pregnancies	18 (25%)
No. of embryos split resulting in 1 pregnancy	36 (50%)
No. of embryos split resulting in 0 pregnancies	18(25%)

Source: Colorado State University.

Perhaps in the future an embryo of a desired mating and sex can be selected from an inventory of frozen embryos. This embryo (or embryos, if twins were desired) would then be thawed and transferred nonsurgically in the cow when she comes into heat. It is possible that some of these embryos, through micromanipulation and microsurgery, could contain all the genetic material from the same cow or the same bull, or it's possible that a heifer could receive the embryo and raise a calf which is her identical twin. This is the way reproduction technology is presently developing in the industry.

SELECTED REFERENCES

Publications

Beverly, J. R., 1979. *Recognizing and Handling Calving Problems.* MP-1203, Texas A and M Univ.

Brackett, B. G., Seidel, G. E., Jr., and Seidel, S. M., 1981. *New Technologies in Animal Breeding.* New York: Academic Press.

Brownson, R., 1983. *Breeding Soundness Examination of Beef Bulls.* Great Plains Beef Cattle Handbook, GPE-8252.

Cole, H. H. and Cupps, P. T., 1977. *Reproduction in Domestic Animals.* New York: Academic Press.

Elsden, R. P. and Seidel, G. E., Jr., 1982. *Embryo Transfer Procedures for Cattle.* Colo. Agric. Expt. Sta. General Series 1011.

Elsden, R. P., Nelson, L. D., Case, L. G., and Seidel, G. E., Jr., 1980. *Sex ratio and mortality of calves from embryo transfer.* Theriogenology 13:95 (abstr.).

Elsden, R. P., Baker, R. D., Seidel, G. E., Jr., Schneider, H. J., and Radakovich, S. Methodology of embryo transfer; Current applications of embryo transfer; Costs and success rates of embryo transfer with beef cattle; Freezing embryos and new technology; Use of embryo transfer in managing a beef herd. Proc. Ann. Conf. on Artificial Insemination and Embryo Transfer in Beef Cattle. Denver, CO., 16 January, 1982.

Hafez, E. S. E., ed., 1980. *Reproduction in Farm Animals.* Philadelphia: Lea and Febiger.

Herman, H. A., 1981. *Improving Cattle by the Millions, NAAB and the Development and Worldwide Application of Artificial Insemination.* Columbia: Univ. of Missouri Press.

Salisbury, G. W., VanDemark, N. L., and Odge, J. R., 1978. *Physiology of Reproduction and Artificial Insemination of Cattle.* 2nd Ed. San Francisco: W. H. Freeman and Co.

Seidel, G. E., Jr., 1981. *Superovulation and embryo transfer in cattle.* Science, 211:351.

Sorenson, A. M., Jr., 1979. Animal Reproduction. *Principles and Practices.* New York: McGraw-Hill Book Co.

Sorenson, A. M., Jr., and Beverly, J. R., 1979. *Determining pregnancy in cattle.* Texas Agric. Ext. Serv., B-1077.

Synchronizing estrus in beef cattle, 1980. Montana Agric. Expt. Sta. Bull. 730.

Van Dervoort, P. *10,000 synchronized cows later.* Proc. Ann. Conf. on Artificial Insemination and Embryo Transfer in Beef Cattle. Denver, CO., 16 January, 1982.

Whitman, R. W., 1981. *Evaluating beef estrous synchronization systems.* Mont. Agric. Expt. Sta. Res. Rpt. No. 180.

Williams, T. J., Elsden, R. P., and Seidel, G. E., Jr., 1983. *Bisecting Bovine Embryos: Methods, Applications and Success Rates.* Proc. Annual Beef Artificial and Embryo Transfer Conference (NAAB). Denver, CO.

Visuals

Artificial Insemination of Cattle (slides and tape). Select Sires Inc. 11740 U.S. 42, Plain City, OH 43064.

Artificial Insemination of Beef and Dairy Cattle (filmstrip and tape). Vocational Education Productions. Calif. Polytechnic State Univ. San Luis Obispo, CA 93407.

Embryo Transfer of Beef and Dairy Cattle (filmstrip and tape). Vocational Education Productions. Calif. Polytechnic State Univ. San Luis Obispo, CA 93407.

Frozen Semen Handling (slides and tape). Select Sires Inc., 11740 U.S. 42, Plain City, OH 43064.

Pre-Natal Development of the Calf (filmstrip and tape). Vocational Education Productions. Calif. Polytechnic State Univ. San Luis Obispo, CA 93407.

Reproductive Anatomy and Physiology of Cattle (slides and tape). Select Sires Inc., 11740 U.S. 42, Plain City, OH 43064.

12
Beef Cattle Genetics

Beef cattle type and performance are determined by heredity (genetic makeup) and the environment to which cattle are exposed. A knowledge of genetic principles allows cattle producers to be creative in changing the type and performance of cattle to meet man's needs and desires.

Cells, Chromosomes, and Genes

Bodies of cattle are composed of millions of tiny units called cells. While these cells can vary in size and shape, depending on type of tissue involved, they are basically the same because they have similar component parts.

One of these component parts, the nucleus (Figure 12.1), is very important because it directs the function of the cell through the genetic material it contains. Within the nucleus are rod-shaped, thread-like bodies called chromosomes. Chromosomes occur in pairs with each species having a consistent number, cattle having 30 pair (Figure 12.2). Each chromosome has the intrinsic ability to duplicate itself. When a cell divides for growth or repair purposes, the 30 pairs (60 chromosomes) duplicate themselves. Therefore in the division process, each of the resulting two cells has the same number of chromosomes (Figure 12.3).

The testicle of the bull and ovary of the cow have a different type of cell division in production of the sperm and egg. In this situation, chromosomes pair-up in the center of the cell. One chromosome of each pair randomly goes into each of the two new cells when division occurs (Figure 12.4). Thus the sperm and the egg each contain 30 chromosomes or one-half the number in other cells.

Fertilization, the union of the sperm and egg, restores the typical chromosome number (Figure 12.5).

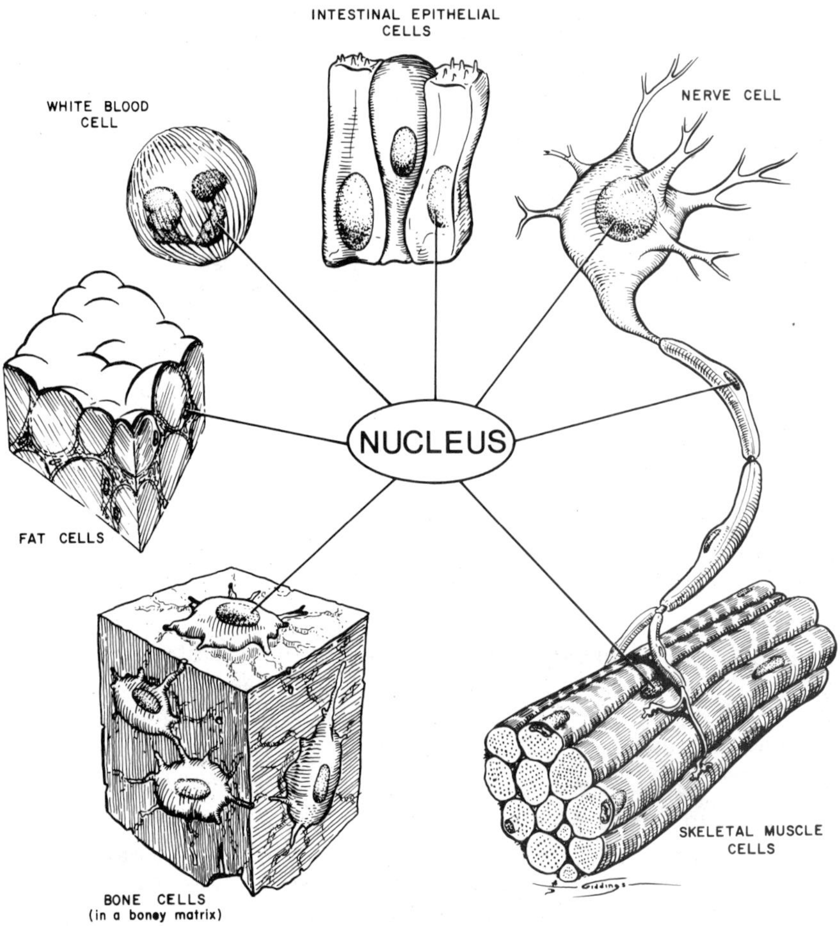

Figure 12.1. Identification of the nucleus in cells of several selected tissues. Drawn by Dennis Giddings.

Determination of Sex

The genetic sex of the calf is determined at the time the sperm and egg unite at fertilization. There is one pair of the 30 chromosomes known as sex chromosomes. They are designated as the X and Y chromosomes where the cow has two X chromosomes and the bull has an X and a Y (Figure 12.2). The sex of the calf depends upon whether or not the sperm fertilizing the egg has an X or a Y chromosome. Since the cow transmits only an X chromosome to each of her offspring through the ovum, an egg fertilized by an X-bearing sperm will produce an XX calf which is a heifer. An egg fertilized by the Y-bearing sperm produces a XY calf which is a bull. The male parent, then, determines the sex of the calf.

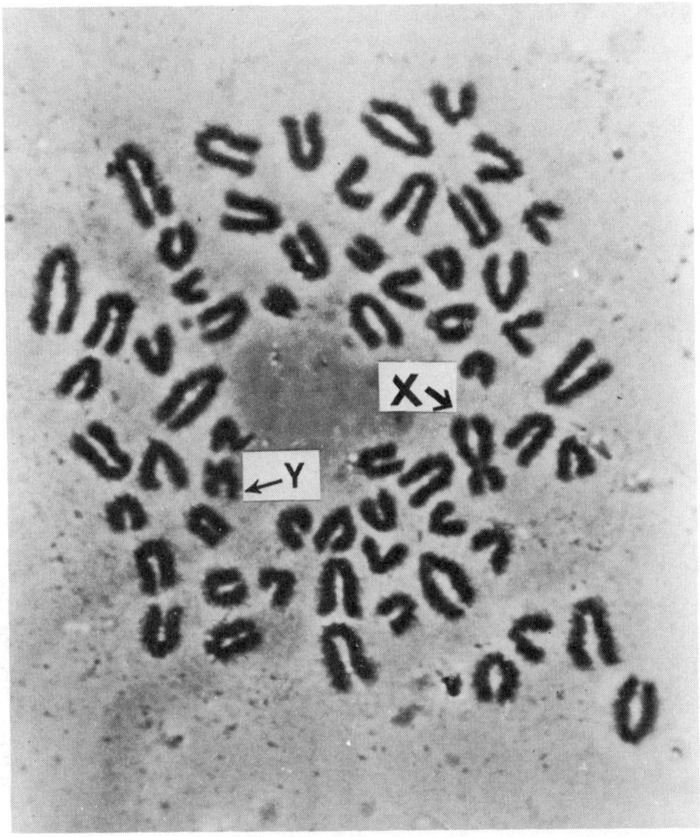

Figure 12.2. The 30 pairs of chromosomes of a bull. Magnified several hundred times. Note the X and Y chromosomes. Courtesy of Nat M. Kieffer, Texas A&M University.

Since the bull can transmit either an X or a Y chromosome to each calf, the probability of him transmitting any one is 50%. The sex ratio of all the calves would be expected to be 50% bulls and 50% heifers. There is evidence, however, that shows about 105 bull calves to every 100 heifer calves at birth and even a higher ratio of bulls at conception. The exact reason for these differences are not known.

The bull possesses a smaller amount of genetic material than the cow because the Y chromosome is only about 40% as long as the X. In the XX female, about 6% of the total genetic material is carried on the sex chromosomes. In the XY male, only about 4%. Traits carried on the sex chromosomes are called sex-linked traits. Little is known about sex-linked traits in cattle, although several of these traits are known in humans (hemophilia, muscular dystrophy, and red-green color blindness).

Genes

Genes are specific units of inheritance which are located on or as a part of the chromosome. This gene location is called a locus which is Latin for *a place.* Since chromosomes

Figure 12.3 Typical cell division (mitosis) using 2 pair of chromosomes.

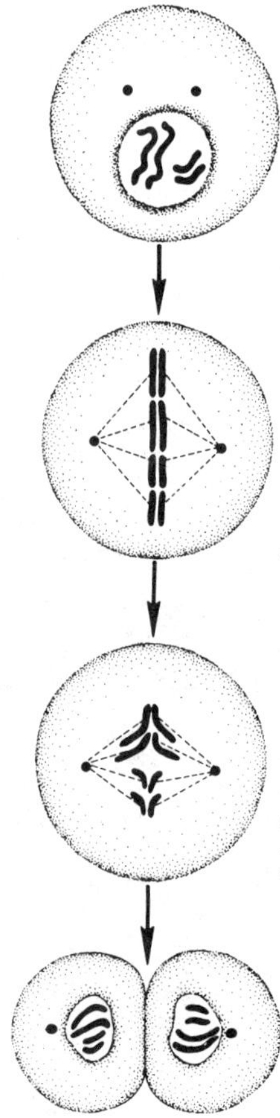

occur in pairs, so do genes. Pairs of genes occupying the same loci (position) on the chromosomes (Figure 12.6) are called alleles.

These genes control metabolic function and body development which the cattle producer measures as reproductive ability, body weight, carcass composition, and several other traits. The exact number of pairs of genes existing in cattle is not known; however, there appears to be hundreds, if not thousands, of pairs. Most beef cattle traits (for example, growth) are controlled by many pairs of genes located on several pair of chromosomes.

Inheritance With One Pair of Genes

An example of a trait controlled by a single pair of genes is black and red color. This is a simple example to show how basic genetic principles work.

Example: The letter B represents the gene for black color. The letter b represents the gene for red color. All possible combinations of these two genes can exist.

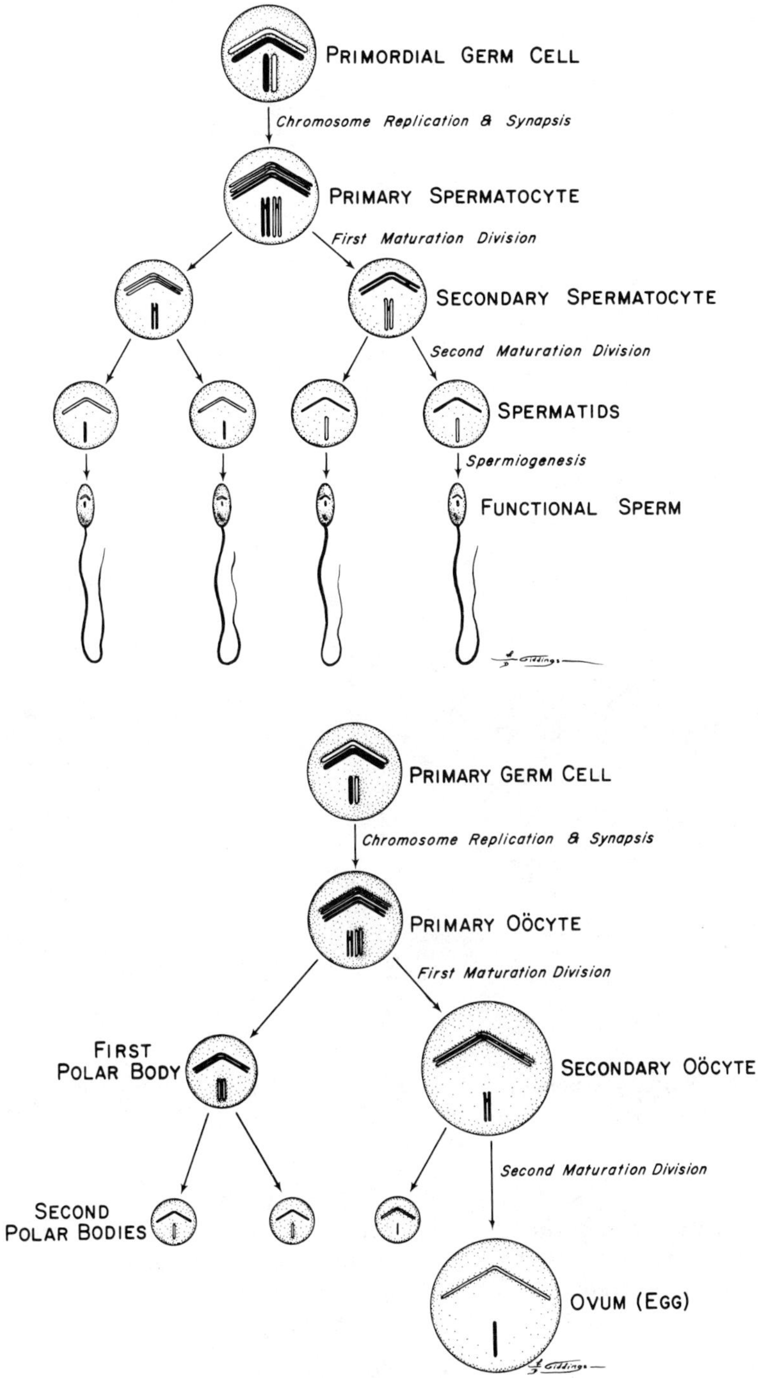

Figure 12.4. Meiosis or reduction cell division occurring in the testicle and ovary (example with 2 pair of chromosomes). Drawn by Dennis Giddings

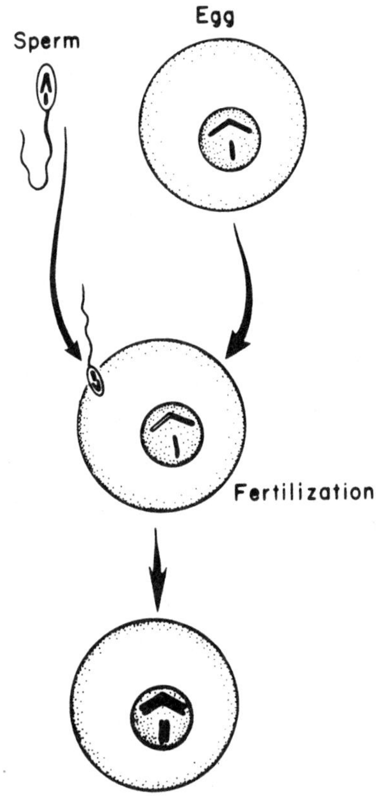

Figure 12.5. Combining of chromosomes through fertilization (2 pairs of genes used for simplification of example). Drawn by Dennis Giddings.

Number of different animals

B | | **B** This animal is black in color. Genetically it is homozygous black, which means that identical genes occupy the same locus on each chromosome.

or

B | | ***b*** This animal is black in color. Genetically it is heterozygous black, which means that different genes occupy the same locus on each chromosome. In this case, the black gene (B) is dominant to the recessive red gene (b). Dominance means that this gene will express itself while it masks the expression of the recessive gene.

or

b | | ***b*** This animal is red in color. Genetically it is homozygous red.

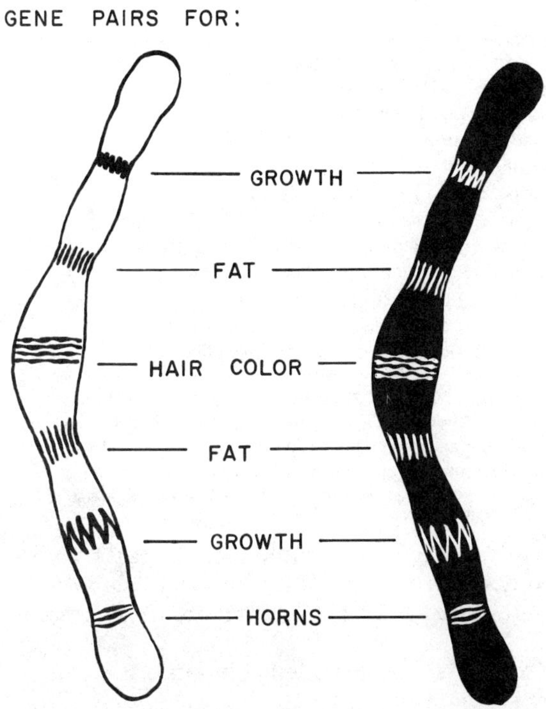

Figure 12.6. A simplified example showing a pair of chromosomes containing several pairs of genes. Drawn by Dennis Giddings.

Various matings of the three types of animals for black and red color will produce different phenotypes and genotypes (Table 12.1). The phenotype is what a producer sees or measures in the animal, while the genotype is the actual gene composition of the animal.

It should be evident, from this table, that cattle with the same phenotype can differ

Table 12.1. Different Phenotypes and Genotypes with Black and Red Gene Inheritance Resulting from Mating the Following Bulls and Cows

Bull		×	Cow(s)		=	Calves	
Genotype	**Phenotype**		**Genotype**	**Phenotype**		**Genotypes**	**Phenotypes**
BB	black	×	BB	black	=	all BB	black
BB	black	×	Bb	black	=	½ BB ½ Bb	black black
BB	black	×	bb	red	=	all Bb	black
Bb	black	×	Bb	black	=	¼ BB ½ Bb ¼ bb	black black red
Bb	black	×	bb	red	=	½ Bb ½ bb	black red
bb	red	×	bb	red	=	all bb	red

Table 12.2. Traits Controlled or Largely Influenced by One Pair of Genes

Trait	Type of Gene Action
black, red color	black (B) dominant to red (b)
polled, horned condition[a]	polled (P) dominant to horned (p)
color in Shorthorns	red (R) gene has no dominance over white (r) gene
snorter dwarf, normal size	normal size (D) dominant to dwarf (d)
pigmentation, albino	normal pigmentation (A) dominant to albino (a)
hydrocephalus, normal	normal (H) dominant to hydrocephalus (h)
osteopetrosis (marble bone disease)	normal (O) dominant to osteopetrosis (o)
syndactyly (mulefoot)	normal (S) dominant to mulefoot (s)
arthrogryposis (palate-pastern syndrome)	normal (A) dominant to palate-pastern (a)
double muscling[b]	normal (D) dominant to double muscling (d)

[a] Scurs (small, horny growths) have other genes controlling their inheritance. Also, an additional gene affects the inheritance of horns in zebu-type cattle (see the 1976 American Simmental Assoc. pamphlet, "Polled Simmentals. What are they Worth to You" for an excellent discussion of scurs and zebu horn inheritance).

[b] Recessive inheritance shown in this table is typical of the British breeds. In other breeds (for example, Piedmontese), the double museling gene appears to be dominant. Other pairs of genes also modify the expression of double museling.

in their genotype and thus their breeding performance. For example, Bb x Bb (all parents are black) gives a 3:1 phenotypic ratio (3 black, 1 red) and three different genotypes in the calves (BB—homozygous black, Bb—heterozygous black, and bb—homozygous red).

Not all genes have dominant or recessive effects. An excellent example is color in Shorthorns which is controlled primarily by a single pair of genes.

Example: R = red gene   RR = red
r = white gene   Rr = roan
rr = white

Red is not dominant to white in this case. The white gene will express itself. There are, however, other genes involved that determine the amount of red and white in roan animals (Rr).

Table 12.2 identifies several common cattle characteristics which appear to be controlled by a single pair of genes.

INHERITANCE WITH TWO PAIR OF GENES

An example using two pairs of genes permits a producer to sense more clearly what is occurring in the cattle breeding.

Example:

B = black gene   ***P*** = polled gene
b = red gene   ***p*** = horned gene

Assume bull is

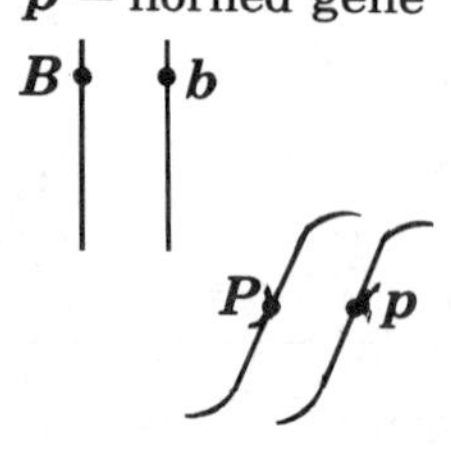

and cows are

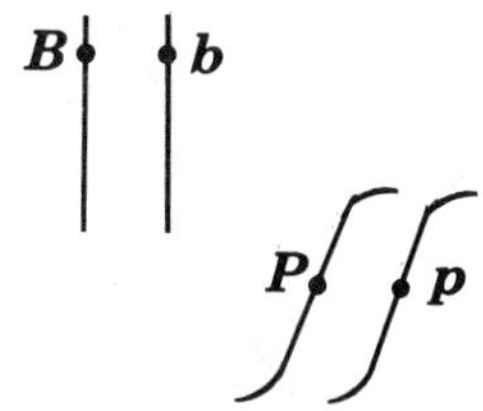

Since one chromosome of each pair goes into the formation of sperm and eggs, there are four genetically different kinds of sperm and four genetically different kinds of eggs that can be formed. When fertilization occurs, each of the genetically different kinds of sperm has an equal chance of uniting with each of the genetically different kinds of eggs. With these various matings, different genotypes and phenotypes can result in the offspring (Table 12.3).

Table 12.3. Genotypes and Phenotypes with Two Heterozygous Gene Pairs

		Eggs			
		BP	**Bp**	**bP**	**bp**
Sperm	**BP**	BBPP[a] black, polled[b]	BBPp black, polled	BbPP black, polled	BbPp black, polled
	Bp	BBPp black, polled	BBpp black, horned	BbPp black, polled	Bbpp black, horned
	bP	BbPP black, polled	BbPp black, polled	bbPP red, polled	bbPp red, polled
	bp	BbPp black, polled	Bbpp black, horned	bbPp red, polled	bbpp red, horned

[a] Genotype. Number of different genotypes: 1 (BBPP), 2 (BBPp), 2 (BbPP), 4 (BbPp), 1 (BBpp), 1 (bbPP), 2 (Bbpp), 2 (bbPp), 1 (bbpp)

[b] Phenotype. Number of different phenotypes:
- 9 black, polled
- 3 black, horned
- 3 red, polled
- 1 red, horned

A different type of gene action will give the same genotypic ratio, but the phenotypic ratio will be different. An example would be color in Shorthorns with polled, horned condition.

Example: RrPp (roan, polled bull) x RrPp (roan, polled cows)

phenotypic ratio of the calves:
- 3 red, polled
- 6 roan, polled
- 3 white, polled
- 1 red, horned
- 2 roan, horned
- 1 white, horned

geonotypic ratio of the calves: same as previous example with black, red color and polled, horned condition (1:2:2:4:1:1:2:2:1)

Where both pairs of genes show no dominance, the phenotypic and genotypic ratios would be the same (1:2:2:4:1:1:2:2:1).

Continuous Variation and Many Pairs of Genes

Most economically important traits in beef cattle such as growth rate and carcass composition are controlled by hundreds of pairs of genes; therefore, it is necessary to expand one's thinking beyond inheritance involving one and two pairs of genes. Consider even a simplified example of 20 pairs of heterozygous genes (one gene pair on each pair of 20 chromosomes) affecting yearling weight. The estimated number of genetically different gametes (sperm or eggs) and genetic combinations are shown in Table 12.4. Remember that for one pair of heterozygous genes, there are three different genetic combinations, and for two pairs of heterozygous genes, there are nine different genetic combinations.

Most beef cattle would likely have some gene pairs heterozygous and some homozygous, depending on the mating system being utilized. Table 12.5 shows the number of gametes and genotypes where eight pairs of genes are either heterozygous or homozygous and each gene pair located on a different pair of chromosomes.

Earlier a statement was made that many economically important traits in beef cattle show continuous variation primarily because they are controlled by many pairs of genes. As these many genes express themselves, and the environment also influences these traits, producers usually observe and measure large differences in the performance of animals for any one trait. For example, if a large number of calves were weighed at weaning (approximately 205 days of age) in a single herd, there would be considerable variation in the calves' weights. Distribution of weaning weights of the calves would be similar to examples shown in Figure 13.1 (Chapter 13) and Figure 12.7. The bell-shaped curve distribution demonstrates that most of the calves are near the average for all calves, with relatively few calves having extremely high or low weaning weights when compared at the same age.

Table 12.4. Number of Gametes and Genetic Combinations with Varying Numbers of Heterozygous Gene Pairs

No. Pairs of Heterozygous Genes	No. of Genetically Different Sperm or Eggs	No of Different Genetic Combinations (Genotypes)
1	2	3
2	4	9
n	2^n	3^n
20	$2^n = 2^{20} =$ approx. 1 mil.	$3^n = 3^{20} =$ approx. 3.5 bil.

Table 12.5. Number of Gametes and Genetic Combinations with Eight Pair of Genes with Varying Amounts of Heterozygosity and Homozygosity

Bull Cow	Aa aa	Bb Bb	Cc CC	Dd Dd	Ee Ee	FF FF	GG gg	Hh Hh	Total
No. different sperm for bull[a]	2	2	2	2	2	1	1	2	64
No. different eggs for cow[b]	1	2	1	2	2	1	1	2	16
No. different genetic combinations possible in calves[c]	2	3	2	3	3	1	1	3	324

[a] Total = 2 × 2 × 2 × 2 × 2 × 1 × 1 × 2. . . = 64
[b] Total = 1 × 2 × 1 × 2 × 2 × 1 × 1 × 2. . . = 16
[c] Total = 2 × 3 × 2 × 3 × 3 × 1 × 1 × 3. . . = 324

Figure 12.7 shows the use of the statistical measurement of standard deviation (SD) which describes the variation or differences in a herd where the average weaning weight was 440 lbs and the calculated standard deviation was 40 lbs. Using herd average and standard deviation, the variation in weaning weight is shown in Figure 12.7 and can be described as follows:

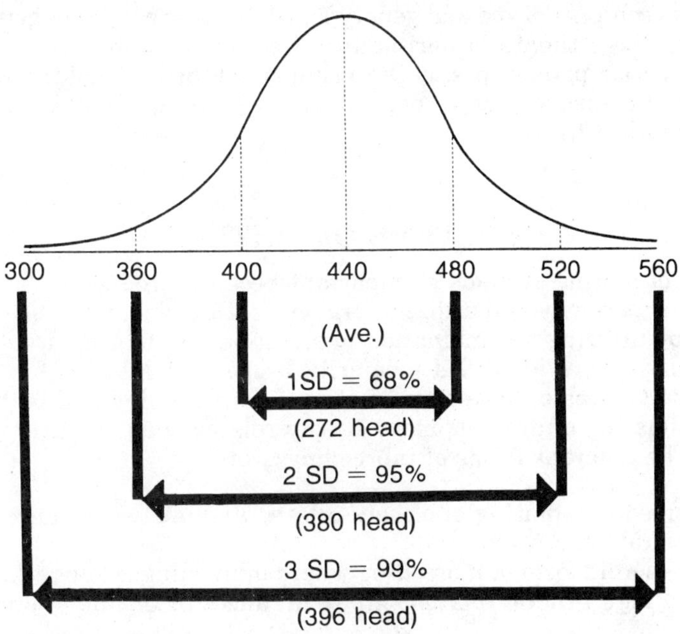

Figure 12.7. A normal, bell-shaped curve for weaning weight showing the number of calves in the area under the curve. (400 calves in the herd.) Drawn by CSU Graphics.

440 lbs ± 1 SD(40 lbs) = 400 to 480 lbs
(68% of the calves are in this range)

440 lbs ± 2 SD(80 lbs) = 360 to 520 lbs
(95% of the calves are in this range)

440 lbs ± 3 SD(120 lbs) = 300 to 560 lbs
(99% of the calves are in this range)

One percent of the calves (4 calves in a herd of 400) would be on either side of the 300 to 560 lb range. Most likely 2 calves would be below 300 lbs and 2 calves would weigh more than 560 lbs.

Weaning weight is a phenotype since the expression of this characteristic is determined by the genotype (genes received from the sire and the dam) and the environment to which the calf is exposed. The genetic part of the expression of weaning weight is obviously not simply inherited. There are many pairs of genes involved and, at the present time, the individual pairs of genes cannot be identified similar to traits controlled by one to two pair of genes.

Consider the following formula:

$$\text{phenotype} = \text{genotype} \pm \text{environment}$$

The phenotype will more closely predict the genotype if producers expose their animals to a similar environment, however, the latter must be within economic reason. This resemblance between phenotype and genotype, where many pairs of genes are involved, is predicted with the estimate of heritability. Use of heritability, along with selecting animals with superior phenotypes, is the primary method of making genetic improvement in traits controlled by many pairs of genes. The application of this method is discussed in detail in Chapter 13.

SYSTEMS OF MATING

A knowledge of mating systems is important because producers can utilize them to preserve genetic superiority and utilize hybrid vigor. Genetic improvement is maximized in many herds by utilizing a combination of selection and mating systems.

Systems of mating are identified primarily by genetic relationship of the animals being mated. The two major systems of mating are inbreeding and outbreeding.

Inbreeding is mating of animals more closely related than the average of the breed or population. The different forms of inbreeding are:

1. intensive inbreeding—mating of closely related animals whose ancestors have been inbred for several generations.
2. linebreeding—a mild form of inbreeding where inbreeding is kept relatively low while maintaining a high genetic relationship to an ancestor or line of ancestors.

Outbreeding is mating of animals less closely related than the average of the population. The different forms of outbreeding are:

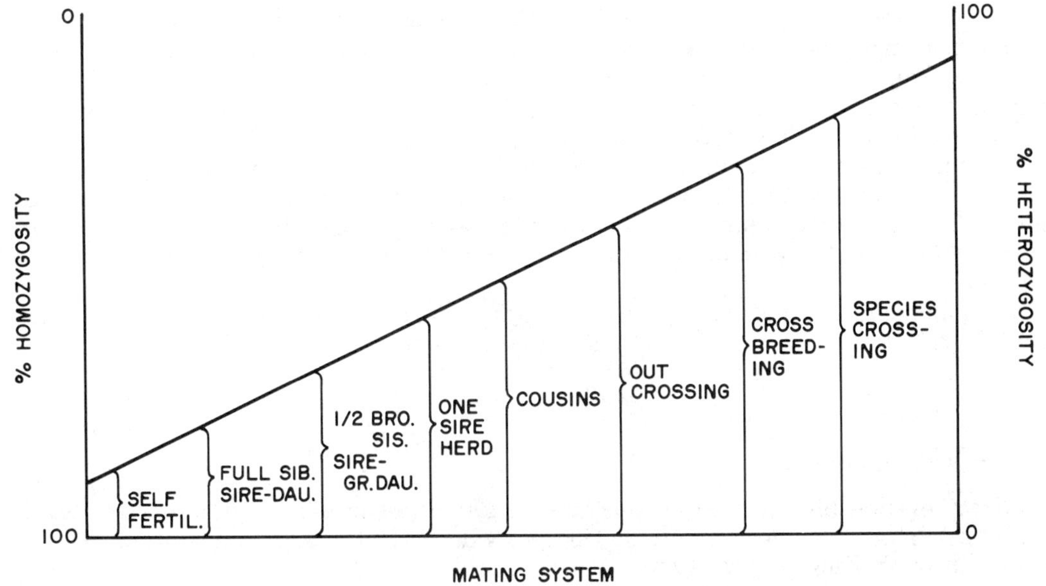

Figure 12.8. Relationship of mating system to the amount of heterozygosity or homozygosity. Self fertilization is currently not on available mating system in cattle. Drawn by CSU Graphics.

1. outcrossing—mating of unrelated animals within the same breed.
2. grading up—mating of purebed sires to nondescript or native females and their female offspring generation after generation.
3. linecrossing—crossing of rather distinct lines (may or may not be inbred) of the same breed.
4. crossbreeding—mating of animals of different established breeds.
5. species cross—crossing of animals of different species (e.g., cattle to bison).

Since mating systems are based on the relationship of animals being mated, it is important to understand more detail about genetic relationship. Proper pedigree evaluation also involves understanding relationship. Relationship is best described in knowing whether the genes in an animal or animals exist primarily in a heterozygous or homozygous condition. Figure 12.8 shows the mating systems and their relationship to a homozygous or heterozygous condition.

COMPUTING AND DEFINING RELATIONSHIP

Relationship is based on the fraction of ½. One-half is used because of the sample halving of the genetic material as it is passed from parent to offspring.

The formula for computing the relationship between two animals (x and y) is:

$$R_{XY} = \frac{\Sigma[(\tfrac{1}{2})^{n+n^1}(1+F_a)]}{\sqrt{(1+F_X)(1+F_Y)}}$$

In calculating most relationships, only the numerator is used unless both animals in question are inbred. In the latter case, the denominator is used because the two animals likely would be homozygous for different pairs of genes. This would decrease genetic similarity between the two animals. The following examples will use only the numerator in computing the relationship.

Animals are more closely related than the average of the population when they have common ancestors within the first five generations of their pedigree. Common ancestors beyond five generations contribute very little increased genetic similarity when compared to the genetic relationship of two animals chosen at random from the same population (e.g. breed).

The following examples identify several different kinds of genetic relationship, showing how relationship is calculated and giving an interpretation to the answer.

Direct Relationship

Direct relationship exists where one animal is an ancestor of the other. In the following example, Prince is a descendant of George, where George has transmitted genes to Prince through King (Figure 12.9).

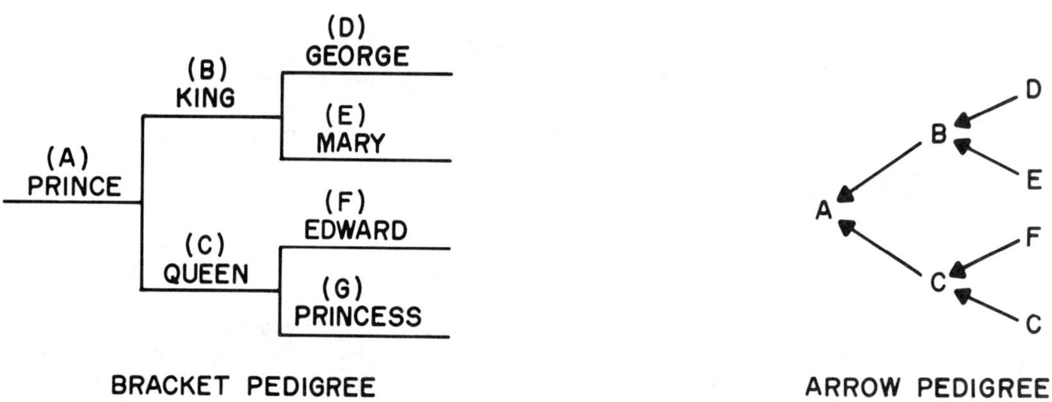

Figure 12.9. Drawn by CSU Graphics.

Pedigrees are typically shown in bracket form. An arrow pedigree can be constructed by assigning a different letter to each animal in the pedigree. If the same animal appears more than once in the pedigree, the same letter should be assigned. An arrow represents a sample half (½) of genes passed from one animal to another. The point of the arrow gives the direction of genetic flow.

Example of calculating relationship or $\mathbf{R}_{AD}$ from Figure 12.9:

Part of bracket pedigree needed

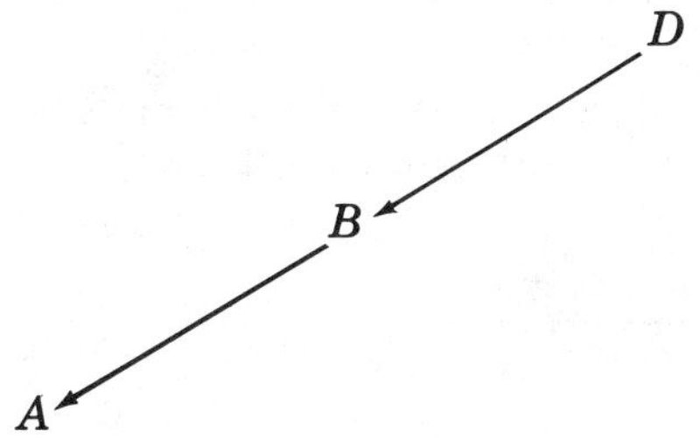

Problem

$R_{AD} = \Sigma(\frac{1}{2})^{n+n^1}(1 + Fa)$
R_{AD} = asks what is relationship between *A* and *D*.

From the formula, the relationship is computed where:

ancestor = ancestor responsible for the relationship of *A* and *D*?
Answer is *D*.

n = number of generations (arrows) from the ancestor to one of the animals for which relationship is being computed?
Answer is 2 (2 arrows from *D* to *A*).

n^1 = number of generations (arrows) from the ancestor to the other animal for which relationship is being computed?
Answer is 0 (0 arrows from *D* to *D*. The ancestor is the "other" animal).

$(1 + Fa)$ = is the ancestor inbred or in this case is *D* inbred?
Answer is no, with the information given.

Solving: R_{AD} =

ancestor	n	n^1	$(1 + F_a)$
D	2	0	—

$$= (\tfrac{1}{2})^{n+n^1}$$
$$= (\tfrac{1}{2})^{2+0} = 25\%$$
$$R_{AD} = 25\%$$

The answer means: Animals *A* and *D* have *25% more* genes the same (identical) as compared to two other animals chosen at random from the population. Relationship then measures the increased genetic similarity between the two animals.

Collateral Relationship

Collateral relationship exists when two animals have a common ancestor in their pedigrees.

Example No. 1

This is not direct relationship because *A* is not an ancestor or a descendant of *M*. They are, however, related because of an ancestor common to both.
Solving R_{AM} from Figure 12.10:

ancestor	n	n^1	$(1+F_a)$
F	2	2	—

$$= (\tfrac{1}{2})^{n+n1}$$
$$= (\tfrac{1}{2})^4$$
$$R_{AM} = 6.25\%$$

F gives a sample half of genes to *C* which in turn gives a sample half of those genes to *A*. *F* gives another sample half of genes to 0 which in turn gives a sample half of those genes to *A*. The best estimate is that *A* and *M* have 6% more of their genes identical because of common ancestor *F*. *F* is not inbred so the $(1 + F_a)$ is left blank in the formula.

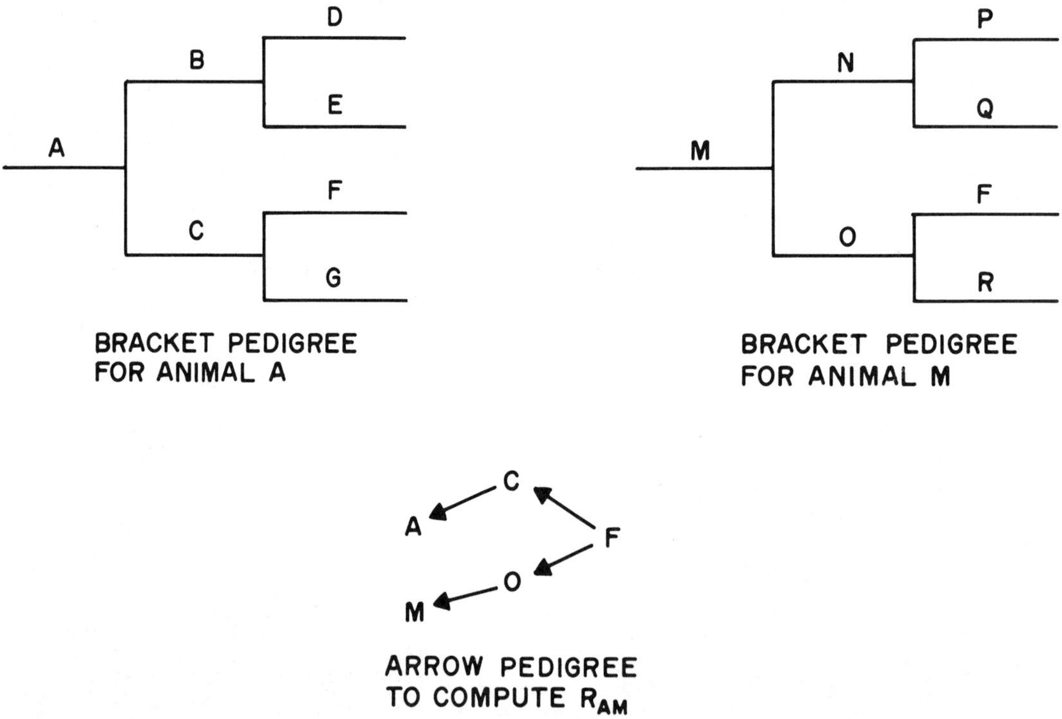

Figure 12.10. Drawn by CSU Graphics.

Example No. 2

Calculating R_{XY} from Figure 12.11 (X and Y are half sibs):

ancestor	n	n^1	$(1+F_a)$	
A	1	1	—	$=(\frac{1}{2})^2 = 25\%$

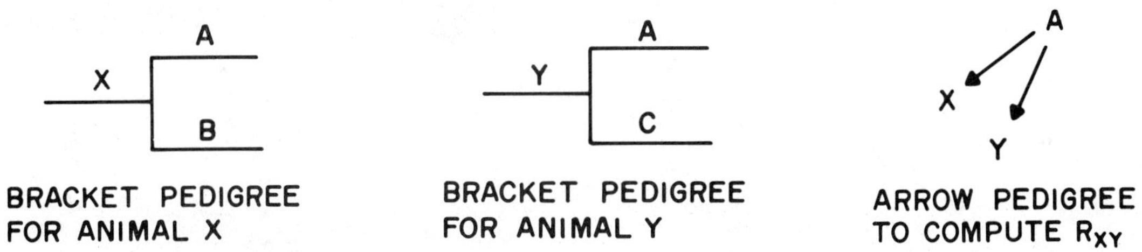

Figure 12.11. Drawn by CSU Graphics.

Example No. 3

Calculate R_{XY} (X and Y are full sibs)

Solving R_{XY} from Figure 12.12:

ancestor	n	n^1	$(1+F_a)$	
A	1	1	—	$=(\frac{1}{2})^2 = 25\%$
B	1	1	—	$=(\frac{1}{2})^2 = 25\%$
				$R_{XY} = 50\%$

The sign in the relationship formula (Σ) implies that all separate pathways influencing the relationship will be summed or totaled.

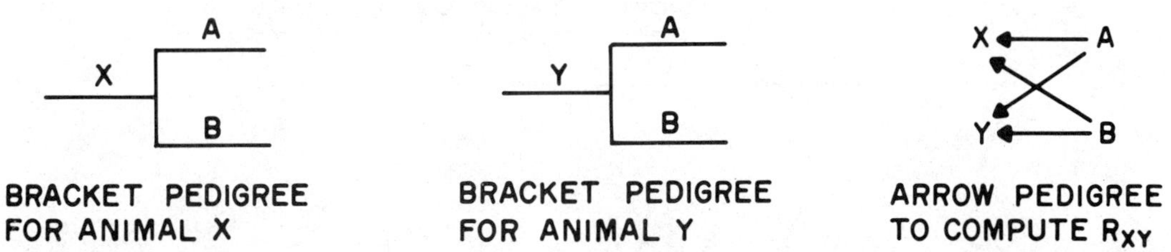

Figure 12.12. Drawn by CSU Graphics

Both Direct and Collateral Relationship

Computation of R_{DB}

Solving R_{DB} from Figure 12.13:

ancestor	n	n^1	$(1+F_a)$	
B	2	0	—	$=(\frac{1}{2})^2 = 25\%$
B	2	0	—	$=(\frac{1}{2})^2 = 25\%$
J	3	1	—	$=(\frac{1}{2})^4 = 6.2\%$
				$= 56.25\%$
				$R_{DB} = 56.25\%$

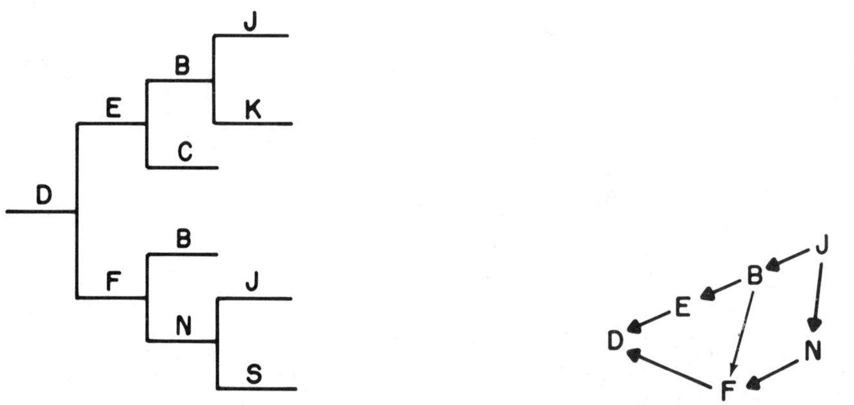

BRACKET PEDIGREE FOR ANIMAL D

ARROW PEDIGREE TO COMPUTE R_{DB}

Figure 12.13. Drawn by CSU Graphics.

Table 12.6. Statistics Helpful in Calculating Relationship and Inbreeding Coefficients

One-half to Various Powers	Number of Times ½ⁿ Occurs in Pathways				
	1	2	3	4	5
$(\frac{1}{2})^1$	.500	1.000			
$(\frac{1}{2})^2$	.250	.500	.750	1.000	
$(\frac{1}{2})^3$	.125	.250	.375	.500	.625
$(\frac{1}{2})^4$	.062	.125	.1882	.250	.312
$(\frac{1}{2})^5$	.031	.062	.094	.125	.016
$(\frac{1}{2})^6$	.0162	.031	.057	.062	.078

Table 12.6 gives calculations which are useful in calculating relationship and inbreeding coefficients. For example, if $(\frac{1}{2})^4$ occurred four times in the calculation of a problem, the table can be used to identify the answer—0.250, rather than compute $(\frac{1}{2})^4$ separately four times then add them together.

INBREEDING

Now that relationship is more clearly defined, the concept of inbreeding (mating of *related* animals) should be more easily understood. Inbreeding increases homozygosity of pairs of genes of the resulting inbred animal. Knowing how to calculate inbreeding gives a more vivid concept of changes in homozygosity.

Computation of Inbreeding

The formula for computing inbreeding is:

$$F_X = \Sigma(\tfrac{1}{2})^{n+n^1+1}(1 + F_a)$$

where F_X = inbreeding coefficient of individual X or the animal in question. Note that inbreeding is computed on one animal where relationship involves two animals.

Σ = sum of all independent paths which connect the sire and dam of X.

n = number of generations (arrows) from ancestor responsible for the inbreeding of X, to the sire of X.

n^1 = number of generations (arrows) from the ancestor, responsible for the inbreeding of X, to the dam of X.

1 = the 1 in the inbreeding formula computes one-half of the relationship of the sire and dam. The relationship of the sire and dam of X, divided by 2, will give the inbreeding of X.

$(1 + F_a)$ = 1 plus the inbreeding of the ancestor common to the sire and dam of X. If the ancestor is also inbred, there will be an increased homozygosity of X.

Example: Inbreeding computation for full brother-sister mating (F_A)

Observation of the arrow pedigree shows inbreeding when arrows pass from an ancestor or ancestors (D and E) through the sire and dam and converge on the animal in question (A).

Solving F_A from Figure 12.14:

ancestor	n	n^1	1	$(1+F_a)$	
D	1	1	1	—	$=(\frac{1}{2})^3 = 12.5\%$
E	1	1	1	—	$=(\frac{1}{2})^3 = 12.5\%$
					$F_A = 25.0\%$

This answer means that by mating Sire B to Dam C, the resulting individual A has 25% *more* of its genes homozygous as compared to a non-inbred animal in the same population. Note that it is an increase in homozygosity or a decrease in heterozygosity.

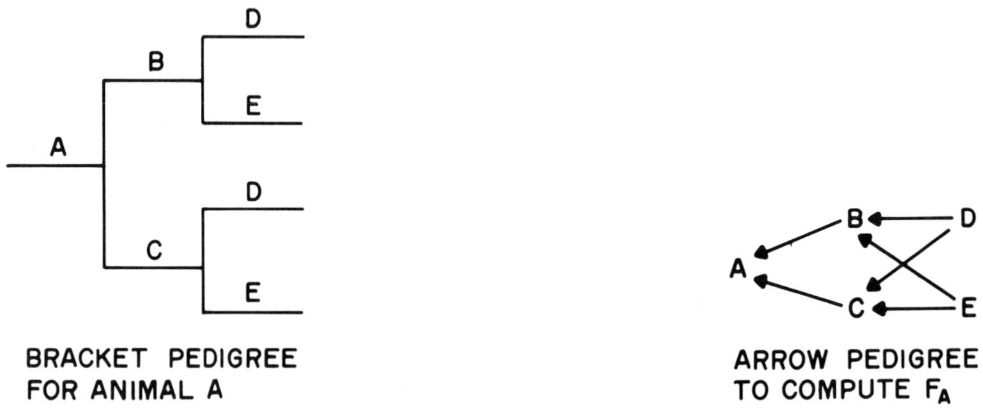

Figure 12.14. Drawn by CSU Graphics.

Example: Inbreeding computation for a sire-daughter mating (F_X)

Solving F_X from Figure 12.15:

ancestor	n	n^1	1	$(1+F_a)$
A	0	1	1	—

$= (1/2)^2 = 25\%$

$F_X = 25\%$

Figure 12.15. Drawn by CSU Graphics.

Example: Pedigree showing linebreeding

Solving F_X from Figure 12.16:

ancestor	n	n^1	1	$(1+F_a)$	
B	2	2	1	—	$=(1/2)^5 = 3.12\%$
B	2	2	1	—	$=(1/2)^5 = 3.12\%$
B	2	2	1	—	$=(1/2)^5 = 3.12\%$
B	2	2	1	—	$=(1/2)^5 = 3.12\%$
					$F_X = 12.5\ \%$

Solving R_{XB} from Figure 12.16:

$R_{XB} =$ ancestor	n	n^1	$(1+F_a)$	
B	3	0	—	$=(1/2)^3 = 12.5\%$
B	3	0	—	$=(1/2)^3 = 12.5\%$
B	3	0	—	$=(1/2)^3 = 12.5\%$
B	3	0	—	$=(1/2)^3 = 12.5\%$
				$R_{XB} = 50.0\%$

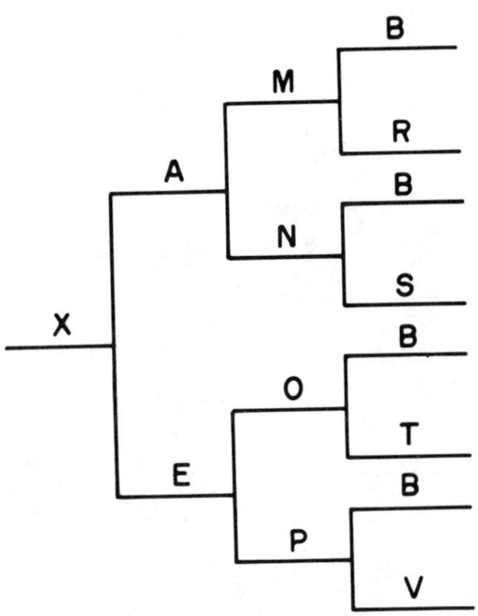

BRACKET PEDIGREE FOR ANIMAL X

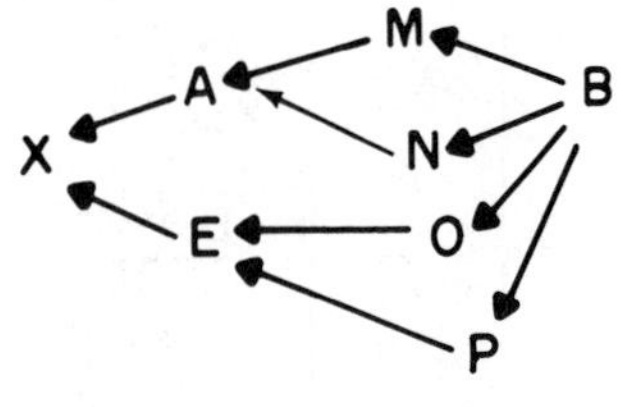

ARROW PEDIGREE TO COMPUTE F_X AND R_{XB}

Figure 12.16. Drawn by CSU Graphics.

If B occurred only once in X's pedigree, the relationship between X and B would be only 12.5%. Due to inbreeding, however, the relationship of X and B is 50% and would be approximately the same as if B would have been the sire of X. This pedigree demonstrates linebreeding where the extent of inbreeding is relatively low (12.5%), while maintaining a high genetic relationship (50%) to an ancestor. The primary value of linebreeding in this example is if B was an outstanding ancestor. There are many other ways where linebreeding can be accomplished other than the one shown here.

Example: Pedigree where $(1 + F_a)$ and the denominator in relationship formula are used.

Computation of R_{XS} from Figure 12.17:

X has resulted from a sire-daughter mating and S has resulted from a half-brother-sister mating.

Solving R_{XS} from Figure 12.17:

ancestor	n	n^1	$(1 + F_a)$
S	1	0	$(1.125) = (\frac{1}{2})^1(1.125) = 56.25$
S	2	0	$(1.125) = (\frac{1}{2})^2(1.125) = 28.12$
			$R_{XS} = 84.37\%$

Both X and S, however, are inbred which means that the denominator part of the formula must be used.

$$R_{XS} = \frac{.8437}{\sqrt{(1 + F_X)(1 + F_S)}}$$

$$R_{XS} = \frac{}{\sqrt{(1.281)(1.125)}}$$

$$R_{XS} = .703 = 70\%$$

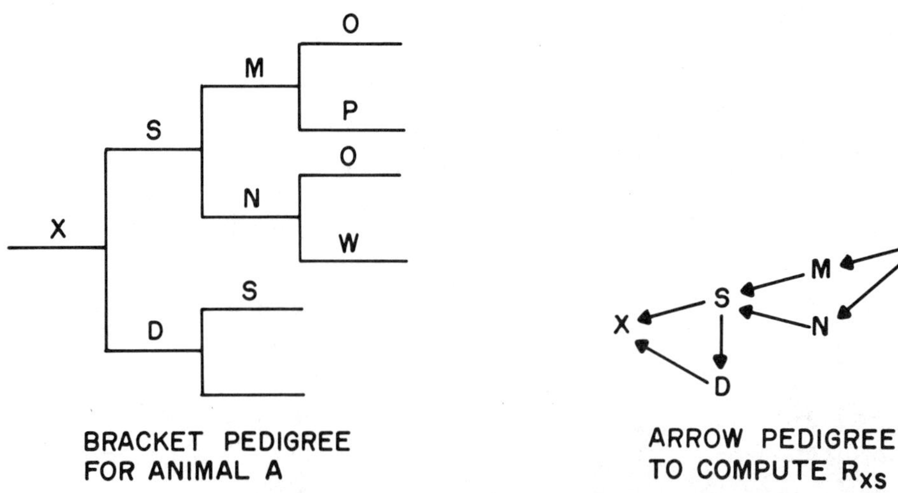

Figure 12.17. Drawn by CSU Graphics.

The reduction in relationship from 84% to 70% occurs because X and S are likely homozygous for different pairs of genes which would decrease their genetic similarity.

Development of Inbred Lines

There are numerous, genetically different inbred lines that can be produced in a given population such as a breed. The number of different inbred lines is 2^n where n is the number of heterozygous gene pairs. For example, with 2 pairs of heterozygous gene pairs, there are 2^2 or 4 different inbred lines that are possible. For example, with AaBb genes, the different, completely homozygous lines which can result are AABB, AAbb, aaBB, and aabb.

There have been research projects with cattle where inbred lines have been produced, anticipating that the crossing of these inbred lines would produce results similar to hybrid corn (Figure 12.18). The cattle breeding project, over the past 30 years, at the San Juan Basin Research Center (SJBRC) at Hesperus, Colorado, has developed some of the most highly inbred cattle in the world. Figure 12.19 shows the arrow pedigree of one inbred bull produced in the Royal line of Hereford cattle.

The inbreeding of Royal 4160 (Figure 12.19) is difficult to compute by the hand method previously described because of the large combination of pathways that connect his sire and dam with common ancestors. The computer is utilized for easily computing relationship and inbreeding on complex pedigrees like this one.

There have been several cattle produced at the SJBRC in the Royal, Brae Arden, Prospector and Tarrington lines of inbred Herefords whose inbreeding coefficients have ranged from the 40s to the high 60s.

12–15

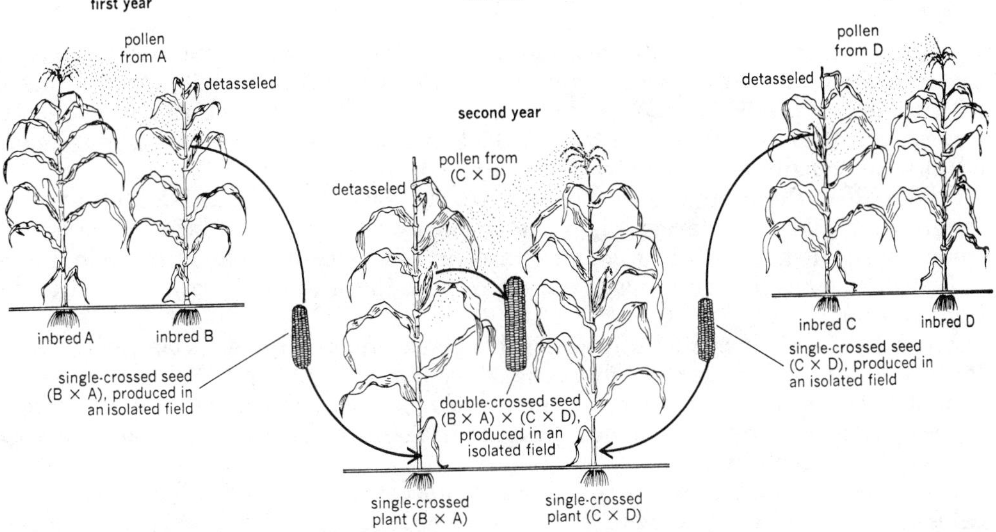

Figure 12.18. Crossing of inbred lines of corn to eventually produce the double-crossing hybrid corn seed utilized widely today. Note the relative size of the corn ears from the inbred lines compared to the resulting hybrid cross. Courtesy of USDA.

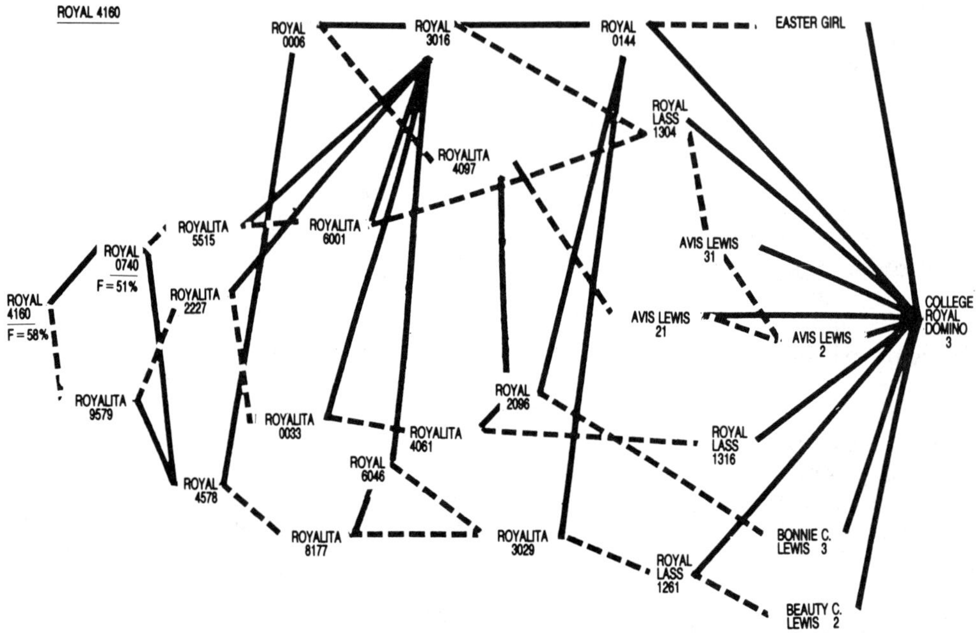

Figure 12.19. Arrow pedigree of Royal 4160 which has an inbreeding coefficient of 58%. Solid lines represent the genetic contribution of the bulls, while the cows' contribution is represented by a broken line. Courtesy of the CSU Expt. (Sta. San Juan Basin Research Center) General Series 982.

Information obtained from these lines and other inbreeding studies with cattle demonstrates the following results and observations:

1. Increased inbreeding is usually detrimental to reproductive performance, pre-weaning and post-weaning growth. Also, the inbred cattle are more susceptible to environmental stresses. While 60% to 70% of the inbred lines show the detrimental effects to increased inbreeding, there are 30% to 40% of the lines which show no detrimental effect with some lines demonstrating improved productivity.
2. At the SJBRC, the inbred lines showed a yearly genetic increase of 2.6 lb in weaning weight over a 26-year period while the line crosses made a 4.6 lb increase over the same time period. Heterosis is demonstrated in the line crosses and the 4.6 lb increase is typical of what breeders might expect from using intense selection in an outbred herd.
3. Inbreeding quickly identifies some desirable genes and also undesirable genes, particularly the hidden, serious recessive genes.
4. Inbred bulls with superior performance are most likely to have superior breeding values which will result in more uniform progeny with high levels of genetically influenced productivity.
5. Crossing of inbred lines results in heterosis; however, in most cases heterosis compensates for inbreeding depression.

It is not logical for breeders to develop their own lines of highly inbred cattle. Inbreeding depression will usually affect economics of the operation. Both purebred and commercial producers can take advantage of highly productive inbred lines by topcrossing these bulls on unrelated cows in their herds.

Inbreeding such as sire-daughter matings are logical ways to test for undesirable recessive genes. Also, seedstock producers can use inbreeding by having well-planned linebreeding programs. Consideration should be given to implementing linebreeding when breeders have difficulty introducing sires from other herds which are genetically superior to those they are raising.

Outbreeding Mating Systems

Species crossing is rarely used except to possibly bring in new sources of genetic material in development of new breeds.

Outcrossing is the most widely used mating system by both commercial and seedstock producers. Genetic material stays primarily in a heterozygous condition even though there are small increases in homozygosity. This is because animals being mated are somewhat genetically related because they are members of the same breed. Usefulness of outcrossing is primarily dependent on effectiveness of selection (selection differential × heritability).

Crossbreeding

There are two primary reasons for using crossbreeding in cattle, (1) breed complementation and (2) heterosis. Breed complementation implies using breeds in a crossbreeding program where their strengths and weaknesses complement one another. There is no one breed that is superior in all desired production characteristics; therefore, planned crossbreeding programs, using breed complementation, can significantly increase herd productivity.

Crossbreeding, if properly managed, allows for the effective use of heterosis (hybrid vigor) which has had a marked effect on productivity in the swine and corn industries. Heterosis is defined as the increase in productivity in the crossbred progeny over and above the average of breeds that go into the cross. The calculation of heterosis for weaning weight is shown in Table 12.7.

Table 12.7 also shows the importance of using highly productive cattle from breeds

Table 12.7. The Computation of Heterosis for Weaning Weights

	Heterosis for Weaning Weight	
	Low Breed Productivity	**High Breed Productivity**
Angus	410	510
Hereford	*390*	*490*
Average of the two breeds	400	500
Average of crossbreds	420	525
Percent heterosis	20/400 = 5%	25/500 = 5%

being utilized in the crossbreeding program. Highly productive crossbred cattle result from crossing highly productive purebred cattle, usually through purebred bulls. The utilization of crossbreeding in commercial breeding programs is discussed in Chapter 13.

SELECTED REFERENCES

Publications

Brinks, J. S. and Knapp, B. W., 1975. *Effects of Inbreeding on Performance Traits of Beef Cattle in the Western Region.* Colo. Expt. Sta. Tech. Bul. 123.

Cundiff, L. W. and Gregory, K. E., 1977. *Beef Cattle Breeding.* Agric. Info. Bulletin No. 286. Washington, D.C.: USDA.

Lasley, J. F., 1978. *Genetics of Livestock Improvement.* Englewood Cliffs: Prentice-Hall, Inc.

Warwick, E. J. and Legates, J. E., 1979. *Breeding and Improvement of Farm Animals.* New York: McGraw-Hill.

Visuals

Cattle abnormalities, 8861, 26 min. (16mm color film). Bureau of Audio-Visual Instruction, Univ. of Wisc. Ext., Box 2093, Madison, WI 53701.

13

Beef Cattle Breeds and Breeding

A breed of cattle is defined as a race or variety, related by descent and similarity in certain distinguishable characteristics. More than 250 breeds of cattle are recognized throughout the world, and several hundred other varieties and types have not been identified with a breed name.

Some of the oldest recognized breeds in the United States were officially recognized as breeds during the middle to late 1800s. Most of these breeds originated from crossing and combining existing strains of cattle. When a breeder or group of breeders decided to establish a breed, distinguishing that breed from other breeds was of paramount importance. Major emphasis was placed on readily distinguishable visual characteristics, such as color, color pattern, polled or horned condition, and rather extreme differences in form and shape.

New cattle breeds, such as Brangus, Santa Gertrudis, and Beefmaster, have come into existence in the United States in recent years. These breeds have been developed by attempting to combine the desirable characteristics of several existing breeds. In most cases, however, the same identifying characteristics as already mentioned have been emphasized in breeding programs to give new breeds visual identity.

Soon after some of the first breeds were developed, the word purebred was attached to them. Herd books and registry associations were established to assure the "purity" of each breed and to promote and improve each breed. Purebred refers to purity of ancestry, established by the pedigree, which shows that only animals recorded in that particular breed have been mated to produce the animal in question. Purebreds, therefore, are cattle within various breeds that have pedigrees recorded in their respective breed registry associations.

When viewing a herd of purebred Angus, Charolais, or other breed, one notes uniformity, particularly uniformity of color or color pattern. Because of this uniformity of one or two characteristics, the word purebred has come to imply genetic uniformity (homozygosity) of all characteristics. Cattle within the same breed are not highly homozygous because high levels of homozygosity occur only after many generations of close inbreeding. This close inbreeding has not occurred in the cattle breeds. If breeds were uniform genetically, they could not be improved or changed even if changes were desired.

WHAT IS A BREED?

The genetic basis of cattle breeds is not well understood by most livestock people. Often the statement is made, "There is more variation within a breed than there is between breeds." The validity of this statement needs to be carefully examined. Considerable variation does exist within a breed for most of the economically important traits. This variation is depicted in Figure 13.1, which shows the number of calves of a particular breed, that fall within certain weight categories at 205 days of age. A bell-shaped curve is formed by connecting the high points of each bar. The breed average is represented by the solid line that separates the bell curve into equal halves. Most calves are near the breed average; however, at the outer edge of the bell curve are high- and low-weaning weight calves. Note that they are fewer in number at these extremes.

Figure 13.2 allows us to hypothetically compare three breeds of cattle in terms of weaning weight. The breed averages are different; however, the variation within each breed is comparable among all three. The statement, "There is more variation within a breed than between breed averages," is more correct than the statement, "There is more variation within a breed than there is between breeds."

Figures 13.1 and 13.2 are hypothetical examples; however, they are based on realistic samples of data obtained from the various breeds of cattle. Figure 13.3 shows differences in percent of fat in milk from groups of cattle from the Holstein and Jersey breeds. It is obvious that by selecting certain Holsteins with high percent fat and Jerseys with low percent fat, a conclusion could be made that Holsteins have a higher percent milk fat average than Jerseys. Random sample selections of cows from each breed, however, would show that Jerseys, on the average, do have a higher percent of fat in their milk than Holsteins.

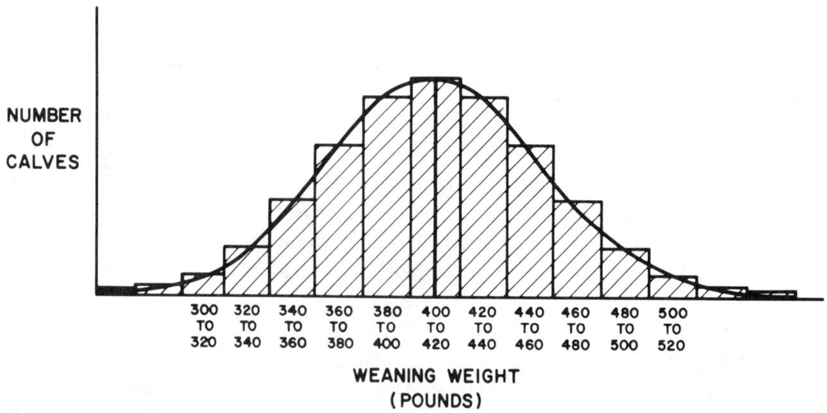

Figure 13.1. Variation or differences in weaning weight in beef cattle. The variation shown by the bell-shaped curve could be representative of a breed or a large herd. The dark vertical line in the center is the average or the mean and, in this example, is 410 pounds. Drawn by CSU Graphics.

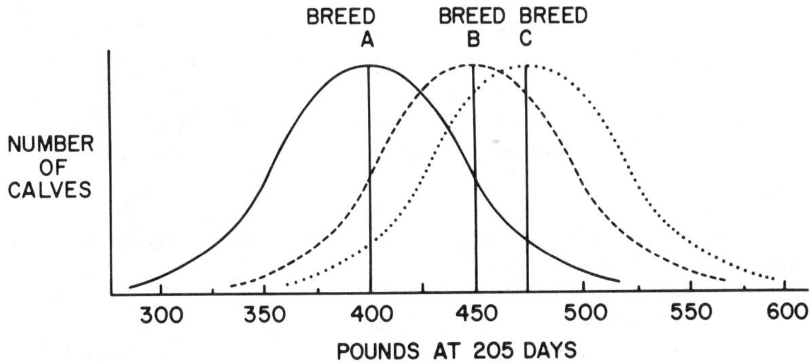

Figure 13.2. Comparison of breed averages and the variation within each breed for three breeds of beef cattle. The vertical lines are the breed averages. Note that some individual animals in Breed B and Breed C can be lower in weaning weight than the average of Breed A. Drawn by CSU Graphics.

Occasionally a bull may seem to have a high level of productivity in a certain trait for which the information comes from selected data and not comparisons based on random samples. Only breed comparisons based on random samples are valid.

MAJOR BEEF BREEDS IN THE UNITED STATES

Until the 1960s and 1970s, the number of cattle breeds in the United States had remained relatively stable at around 15 to 20. Today, more than 60 breeds of cattle are available to United States beef producers. Why the large importation of the different breeds from several countries? Following are several possible reasons: (1) Feeding more cattle larger amounts of grain, a practice started in the 1940s resulted in many overfat cattle—cattle that had been previously selected to fatten on forage rations. Therefore, a need was established for grain-fed cattle that could produce a higher percentage of lean to fat at the desired slaughter weight, (2) Economic pressures to produce more weight in a shorter period of time demonstrated a need for cattle with more milk and more growth, and (3) An opportunity was available for some promoters to capitalize

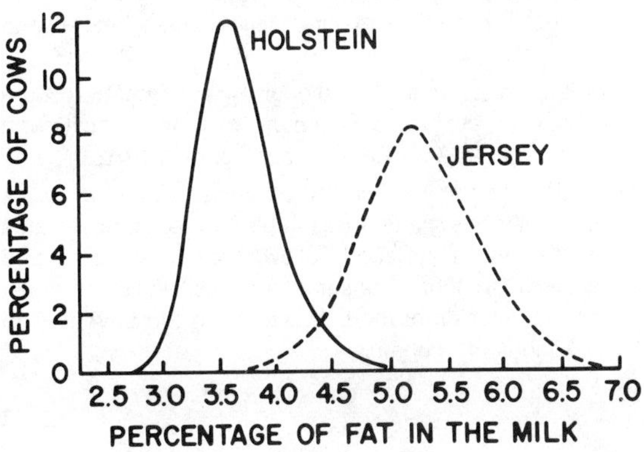

Figure 13.3. Breed differences in percentage of fat in the milk of Holstein and Jersey cows. Data summarized by G. E. Shook, University of Wisconsin, from DHIA records. Courtesy of *Animal Agriculture*, First Edition, edited by H. H. Cole and Magnar Ronning. W. H. Freeman and Company. Copyright © 1974, p. 271.

on merchandising a certain breed as being the ultimate in all production traits. This opportunity could be merchandised easily because there was little comparative information on breeds.

Color Plates A, B, C, and D identify the most numerous breeds of beef cattle in the United States. Table 13.1 gives some distinguishing characteristics and brief background

Table 13.1. Background and Distinguishing Characteristics of the Most Numerous Beef Breeds in the United States, Several of Which are Shown in Color Plates A, B, C, and D

Breed	Distinguishing Characteristics	Brief Background
Angus	Black color and polled.	Originated in Aberdeenshire and Angushire of Scotland. Imported into the United States in 1873 to cross with Longhorn.
Beefmaster	Various colors.	Developed in the United States from Brahman, Hereford and Shorthorn breeds. Selected for its ability to reproduce, produce milk and grow under range conditions.
Blonde d'Aquitaine	Fawn colored, sometimes with a reddish tinge. Heavily muscled.	Originated in France and live cattle imported in the United States in 1973. Semen imported prior to that time.
Brahman	Various colors, with gray predominant. They are one of the Zebu breeds which have the hump over the top of the shoulder. Most Zebu breeds also have large, drooping ears and excess skin in the throat and dewlap.	Major importations to the United States from India and Brazil. Largest introductions in early 1900s. These cattle are heat tolerant and well adapted to the harsh conditions of the Gulf Coast region.
Brangus	Black and polled predominate, although there are Red Brangus. There are separate breed associations for the Brangus and Red Brangus.	Breed developed around 1912 in the United States to become three-eighths Brahman and five-eighths Angus.
Charolais	White color with heavy muscling, horned or polled.	One of the oldest breeds in France. Brought into the United States soon after World War I, but its most rapid expansion occurred in the 1960s.
Chianina	White color with black eyes and nose. Extremely tall cattle.	An old breed originating in Italy. Acknowledged to be the largest breed, with mature bulls weighing more than 3000 pounds.
Devon	Dark red (North Devon) to light red or brown (South Devon).	North Devon is an old breed originating in England. South Devon is more of a dual-purpose breed. The Devon first came to America with the Pilgrims in 1623.
Galloway	Polled with majority solid black in color, some dun-colored, and others white with black noses, ears, and feet. The Belted Galloway is black with distinctive white belt.	An old breed from the Scottish province of Galloway. Long, burly hair has made it adaptable to the harsh climates of the north. The Belted Galloway has a separate breed association. It has the same origin as the Galloway but had an infusion of the belted cattle in the Seventeenth or Eighteenth century.

(cont.)

Table 13.1. (*Continued*)

Breed	Distinguishing Characteristics	Brief Background
Gelbvieh	Golden colored.	Originated in Austria and West Germany. Developed as a dual-purpose breed (draft, milk and meat).
Hereford	Red body with white face and horned.	Introduced in the United States in 1817 by Henry Clay. Followed the Longhorn in becoming the traditionally known range cattle.
Limousin	Golden color with marked expression of muscling.	Introduced into the United States in 1969, primarily from France.
Longhorn	Multi-colored with characteristically long horns.	Came to West Indies with Columbus. Brought to the U.S. through Mexico by the Spanish explorers. Longhorns were the noted trail-drive cattle from Texas into the Plains States.
Maine-Anjou	Red and white spotted.	A large breed developed in France and introduced into Canada in late 1968.
Murray Grey	Silver gray and polled.	Developed in Australia from Shorthorns crossed on Angus.
Pinzgauer	Reddish chestnut color with white markings on rump, back and belly.	A hardy breed developed in the Pinza Valley of Austria and areas of Germany and Italy. Introduced into North America in 1972.
Polled Hereford	Red body with white face and polled.	Bred in 1901 in Iowa by Warren Gammon who accumulated several naturally polled cattle from horned Hereford herds.
Red Angus	Red color and polled.	Founded as a performance breed in 1954 by sorting the genetic recessives from black Angus herds.
Red Poll	Red color and polled.	Introduced into the United States in the late 1800s. Originally a dual-pupose breed but now considered a beef breed.
Santa Gertrudis	Red color and horned.	First United States breed of cattle, developed on the King Ranch in Texas. Combination of five-eighths Shorthorn and three-eighths Brahman.
Scotch Highland	Golden color with long, shaggy hair.	Bred in the highlands of Scotland. Imported into the United States in 1894.
Shorthorn (includes polled and milking)	Red, white, or roan in color. Both horned and polled.	Introduced into the United States in 1783 under the name "Durham." Most prominent in the United States around 1920. Illawarra is an Australian breed of dairy cattle similar to the Milking Shorthorn. Illawarra and Milking Shorthorn have been crossed with U.S. Shorthorn in recent years.
Simmental	Yellow to red and white color pattern. Both polled and horned.	A prominent breed in parts of Switzerland and France. First bull arrived in Canada in 1967. Originally selected as a dual-purpose milk and meat breed.
Tarentaise	Solid wheat-colored hair ranging from a light cherry to dark blond.	Mountain cattle derived from an ancient Alpine strain in France. Originally a dairy breed in which maternal traits have been emphasized.

information for each of these breeds. Other cattle breeds which contribute to the beef supply are described in Table 13.2.

The relative importance of the various breeds' contributions to the total beef industry is best estimated by the registration numbers of the breeds (Table 13.3). These eleven breeds would be considered the major breeds of beef cattle in the United States since they would contribute the major part of the genetics to the commercial cow–calf segment.

Although registration numbers are for purebred animals, they reflect commercial cow–calf producers' demand for different breeds. Registration numbers show that Angus, Hereford, and Polled Hereford are currently the most important breeds of the beef cattle industry in the United States. Numbers of cattle belonging to various breeds in this country have changed over the past years as shown in Figure 13.4. It is reasonable to expect that in future years, some breeds will become more numerous, while other breeds will decrease significantly in numbers. These changes will be influenced by economic conditions and by genetic changes being made in the economically important traits in the breeds.

Table 13.2. Additional Beef Breeds, Dairy Breeds, Synthetics (Composites), and Buffalo (Bison) Breeds in the U.S. Which Contribute to the Beef Supply

Breed	Distinguishing Characteristics	Brief Background
Amerifax	Polled. Solid red or black in color.	Originated in the United States and formally introduced in 1977. Purebreds are five-eighths Angus and three-eighths Beef Friesan.
Ankina	Black.	Originated from crossing Angus and Chianina breeds.
Ankole Watusi	Claimed as the world's longest, largest horned cattle.	Originated in Africa. Approximately 120 purebreds in the United States Registry formed in 1983.
Barzona	Dark reddish-brown. May be polled or horned.	Developed by the Bard Family in 1942 in the intermountain desert area of Arizona. Selected for hardiness traits under rigorous range conditions. Foundation stock came from Angus, Africander, Hereford and Santa Gertrudis breeds.
Beefalo (also referred to as Cattalo)	Various colors.	Originated in California where Mr. Basalo crossed bison with domestic cattle. Term "Cattalo" originated in Canada where similar crosses were made earlier.
Beef Friesan	Similar black and white markings as Holstein-Friesans, although more black color.	Imported to the United States from Ireland in 1971. Selected for more thickness of muscling than Holstein Friesans in the United States.
Braford	White-face with various hair colors.	Originated in the United States in late 1940s by crossing the Brahman and Hereford breeds to achieve five-eighths Hereford and three-eighths Brahman.
Brahmasin	Various colors.	Cross of Brahman and Limousin. Originated in the United States.
Brahmental	Various colors.	Originated in U.S. by crossing Brahman and Simmental. Must have a minimum of one-fourth Brahman and a minimum of three-eighths Simmental.
Buffalo (Bison)	—Not typically considered a breed, however, there are two associations registering buffalo at the present time.	

(cont.)

Table 13.2 (*Continued*)

Breed	Distinguishing Characteristics	Brief Background
Charbray	Various colors.	Cross between Charolais and Brahman.
Chargrey	Various colors.	Developed in Australia by crossing Charolais and Murray Grey breeds.
Charswiss	Various colors.	Cross between Brahman and Brown Swiss.
Dairy Breeds	—Holstein-Friesan, Jersey, Guernsey, Brown Swiss, Ayrshire, Milking Shorthorn, Dutch Belted and Red and White. Of the approximate 11 million dairy cows, the Holstein is most numerous.)	
Dexter	Claimed as smallest breed in the United States (cows, 650 lbs; bulls, 850 lbs).	Originated in Ireland. Introduced into the United States in 1912.
Gelbray	Various colors.	Originated in the United States by crossing Gelbvieh and Brahman breed. Gelbvieh (max. three-fourths; min. five-eighths and Brahman (max. three-eighths; min. one-fourth. Registered in American Gelbvieh Association.
Hays Converter	Various colors.	Developed in the 1950s by Harry Hays, former Canadian Minister of Agriculture by crossing Hereford, Holstein and Brown Swiss breeds.
Marchigiana	Large but refined in bone. Light gray to almost white in color. Tail switch is dark colored.	Originated in the Marchi and surrounding areas of Italy. Resembles Chianina in color and conformation which implies possible intermixing with them.
Normande	Coat colors may be yellowish or blond, very dark brown or almost black, and white. Colors occur in spots and speckles.	Originated in France. First seven-eighths Normande calf was born in the United States in 1976.
Piedmontese	Extremely heavily muscled. Color varies.	Imported from Italy.
Ranger	Various colors.	Developed in the 1940s, 1950s, and 1960s in several Western United States cattle herds. Crosses of Hereford, Red Angus, Shorthorn, Scotch Highland and Brahman are the primary breeds represented.
Romagnola	Similar to Marchigiana	
Salers	Uniform mahogany red with medium to long hair.	Raised in the mountainous area of France where they were selected for milk, meat and work.
Senepol	Mostly polled. Color varies from light tan to dark red.	Developed in the Virgin Islands in 1900s by crossing the Red Poll and N'Dama cattle.
Simbrah	Various colors.	Purebreds are five-eighths Simmental and three-eighths Brahman. Registered in the American Simmental Association.
Synthetics (Composites)	Various colors. May be polled or horned.	There are several different Synthetics or Composites currently being developed (for example, MARC I, MARC II, MARC III, CASH and others). They are not considered breeds because a breed association is not currently associated with them. Several of the other breeds listed (Brahman crosses) could be considered Synthetics or Composites.
White Park	White color with characteristic black ears and muzzle.	Imported from England in 1940. Breed association formed in United States in 1975.

Table 13.3. Beef Breeds with Registration Numbers of More than 10,000 Per Year[a]

Breed	Registration Numbers (1980)
Angus	257,578
Beefmaster	30,000
Brahman	36,450
Brangus	24,500
Charolais	26,907
Hereford	201,495
Limousin	13,793[a]
Polled Hereford	151,658
Santa Gertrudis	31,211
Shorthorn	19,426
Simmental	28,853[a]

[a] Reported by breed associations as purebred only. Registration of percentage cattle are not included in these numbers. Percentage refers to cattle that have other breeds represented in their breeding and that are not recognized as purebred by their respective breed associations. Total registrations, including percentage cattle, for the Limousin and Simmental breeds are 31,937 and 66,109 respectively.
Source: National Society of Livestock Record Associations.

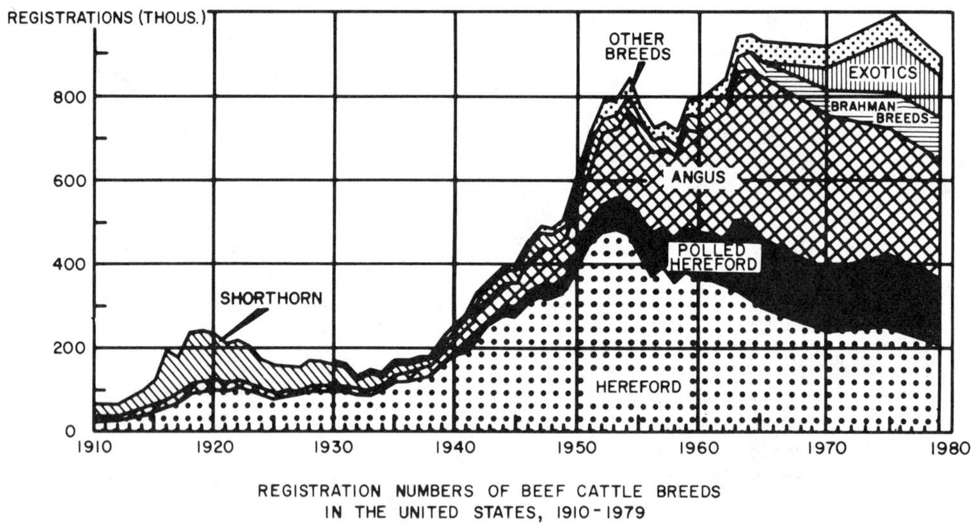

REGISTRATION NUMBERS OF BEEF CATTLE BREEDS IN THE UNITED STATES, 1910-1979

Prior to 1963—Some Polled Herefords were registered only in the American Hereford Association. Others were registered in the American Polled Hereford Association.

Prior to 1965—Brahman breeds were included in other breeds. Very few exotics registered.

After 1965—Shorthorn ends, included in other breeds. Brahman breeds (Beefmaster, Brahman, Brangus, Santa Gertrudis). Exotics (Charolais, Chianina, Limousin, Maine Anjou, Simmental).

Figure 13.4. Registration numbers of beef cattle in the United States, 1910 through 1979. Adapted from the USDA Farmer's Bulletin No. 2228, *Beef Cattle Breeds.*

Angus

Beefmaster

Brahman

Brangus

Charolais

Chianina

Plate A

Devon

Galloway

Gelbvieh

Hereford

Limousin

Longhorn

Plate B

Maine Anjou

Murray Grey

Pinzgauer

Polled Hereford

Red Angus

Red Poll

Plate C

Salers

Santa Gertrudis

Scotch Highland

Shorthorn

Simmental

Tarentaise

Plate D

IMPROVEMENT OF BEEF CATTLE THROUGH BREEDING METHODS

Genetic improvement in beef cattle can occur by selection and also by using a particular mating system. Significant improvement by selection results when selected animals are superior to the herd average, and heritabilities of traits are relatively high (40% and higher).

Factors affecting the rate of genetic improvement from selection include heritability, selection differential, and generation interval.

Variation

Most economically important traits in beef cattle are under the influence of many pairs of genes. This genetic variation along with different environmental effects causes considerable variation in these traits. Reports of yearling weights of bulls in excess of 1700 lbs, rib-eye areas larger than 20 square inches, and 205-day weaning weights over 800 lbs demonstrate that high levels of productivity are possible with exceptional genetic and environmental conditions. Whether these extremely high levels of productivity are economically feasible or practical is another question. Differences in productivity are noticeable even under similar environmental conditions (Figure 13.5).

It is not uncommon for a commercial cow–calf producer to find weaning weight differences of over 100 lbs in calves within the same herd.

Figure 13.5. These two cows had the same productive opportunity being in the same herd for their entire lifetimes. The cow on the left produced 11 calves with an average weaning weight of 533 lbs, while the cow on the right had an average weaning weight of 384 lbs on 11 calves. With calf prices at 75 cents per pound, the cow on the left had a higher gross return of $1229 compared to the other cow. These differences currently exist in most herds today. Courtesy of Dwight F. Stephens.

Table 13.4. Heritability Estimates

Trait	Heritability[a] (%)
calving interval	10
age at puberty (heifer)	40
first service conception rate	25
percent assisted births	20
calving ease score	30
scrotal circumference	50
birth weight	40
weaning weight	30
post weaning ADG (feedlot)	45
post weaning ADG (pasture)	30
efficiency of feedlot gain	45
yearling weight	40
yearling hip height	40
carcass quality grade	40
fat thickness	45
yield grade	30
tenderness	60
cancer eye susceptibility	30
pink eye susceptibility	25

[a] Heritabilities below 20% are considered low, 20–39% are medium in magnitude, while heritabilities over 40% reflect highly heritable traits.

Heritability

A heritability estimate is a percentage figure which shows what proportion of variation in a trait is due to heredity. Heritability identifies that portion of variation that is passed on from parent to offspring. Table 13.4 lists heritability estimates for some economically important traits for beef cattle.

Selection Differential (Reach)

Selection differential is sometimes called *reach* because it shows how superior (or inferior) the average of selected individuals is to the average of the group from which they were selected. An example of a selection differential will be shown in the next section in the calculation of genetic change.

Predicting Genetic change

$$\text{Genetic change per year} = \frac{\textit{heritability} \times \textit{selection differential}}{\text{generation interval}}$$

BULLS

selected bulls average	545 lbs
of bulls in herd	400 lbs
selection differential =	145 lbs
heritability	.30
total genetic superiority =	43.5 lbs
only half passed on (43.5 ÷ 2) =	21.7 lbs

HEIFERS

selected heifers	460 lbs
average of heifers in herd	400 lbs
selection differential =	60 lbs
heritability	.30
total genetic superiority =	18.0 lbs
only half passed on = 18 ÷ 2 =	9.0 lbs

Fertilization combines the genetic superiority of both parents 21.7 lbs + 9.0 lbs = approx. 30 lbs

The selected heifers only represent approximately 15% of the total cow herd, so the 30 lbs is for one generation. This much selection would have to be practiced over 6 lbs ÷ 6 years = 5 lbs per year. Thus, the generation interval influences the yearly genetic change.

This represents genetic change if selection was for only one trait. If selection is practiced for more than one trait, genetic change is $1/\sqrt{n}$ where n is the number of traits in the selection program. If four traits were in the selection program, the genetic change per trait would be $1/\sqrt{4} = ½$. This means that only ½ the progress would be made for any one trait compared to giving all the selection to one trait. This reduction in genetic change per trait should not discourage producers from multiple trait selection. Maximizing genetic progress in a single trait usually is associated with lowering productivity in other economically important traits.

Sometimes traits are genetically correlated. A genetic correlation between two or more traits means that some of the same genes affect each trait. Genetic correlations, varying from a +1.0 to −1.0, may help or hinder a genetic improvement program. Genetic correlation between rate of gain and feed per pound of gain (from similar beginning weights to similar end weights) is negative. This is desirable from a genetic improvement standpoint, because animals that gain faster will require less feed (primarily less feed for maintenance). This relationship is also desirable because rate of gain is easily measured while feed efficiency is an expensive trait to measure. Yearling weight or mature weight is positively correlated with birth weight, which means that as yearling weight increases, birth weight will increase also. This may pose a potential problem since birth weight reflects calving difficulty to a large extent.

Genetic Improvement Through Artificial Insemination

Figure 13.6 shows the rate of genetic improvement for yearling weight under different selection schemes. The most significant points are:

1. marked difference between Herd A (no AI, males and females selected for yearling weight) and Herd D (bulls obtained from a herd where no selection is practiced for yearling weight).
2. the small difference between Herd B (selects bulls from herd A on yearling weight and selects for yearling weight in his females) and herd C (selects bulls from Herd A) but practices no selection for yearling weight on females. Effective sire selection will account for 80% to 90% of the genetic change in a herd.
3. greatest change in Herd F where the top one percent of the sons of the top one percent of the bulls were used to build generation upon generation of AI selection. The rate of improvement in Herd F is 1.5 that of Herd A with natural service. The primary reasons for the greater improvement in Herd F is a larger selection differential.

Types of Selection

The three types of selection are (a) tandem, (2) independent culling level, and (3) selection index based on an index of net merit.

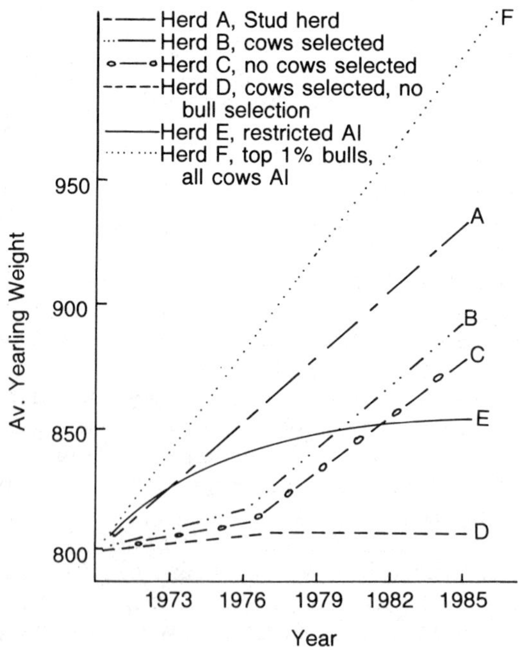

Figure 13.6. Expected genetic change in different types of beef herds (Adapted from Magee, Mich. State Univ. Qtr. Bul. 48:4).

Tandem is selection for one trait at a time. When the desired level is achieved in one trait, then selection is practiced for the second trait. It is not recommended as it is the least effective of the three methods.

Independent culling level establishes minimum culling levels for each trait in the selection program. Even though it is the second most effective type, it is the most prevalent one used. Its major disadvantage is that an animal slightly below the minimum culling level in one trait but highly superior in other traits would be culled.

The selection index is the most effective selection type when selecting for two or more traits. Caution was mentioned earlier about genetic change when many traits are included in the selection program. Selection indexes are not easily constructed as they require highly involved statistical methods in putting the heritabilities, economic values, genetic correlations, and variabilities of several traits into a single formula.

Selection Methods

The primary methods of selection, (1) pedigree, (2) individual appearance or performance, and (3) progeny testing are shown in Figure 13.7.

Pedigree information is most useful before animals have expressed their own individual performance or their progeny's performance is known. Pedigree is also useful for assessing genetic abnormalities, traits expressed much later in life (longevity), and selecting for traits expressed only in one trait (bull selection for maternal characteristics). In evaluating pedigree information, only the information from the animal's closest relatives should be used because of the genetic relationship involved (see Chapter 12).

Selection on individual appearance and performance should be practiced on traits which are economically important and have high heritabilities. One primary advantage of selection on individual performance is that it permits rapid generation turnover and thus shortens the general interval.

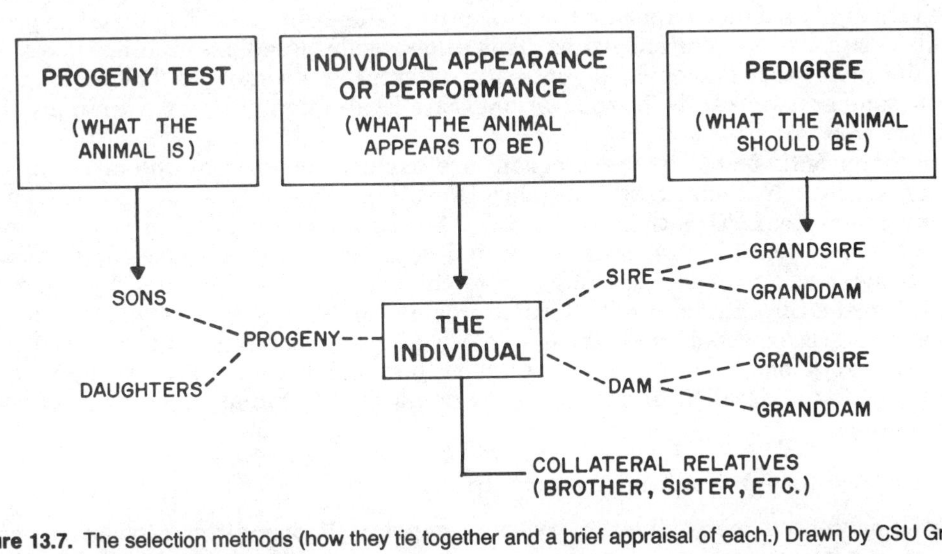

Figure 13.7. The selection methods (how they tie together and a brief appraisal of each.) Drawn by CSU Graphics.

Progeny testing is more accurate than the other two methods, providing the progeny test is accurate and extensive. Progeny tests are particularly useful in selecting for carcass traits (where good carcass indicators are not available in the live animal), sex-limited traits such as mothering ability and traits with low heritabilities.

Disadvantages of progeny testing are that too few animals can be progeny tested, the longer generation interval required to obtain progeny information, and decreased accuracy in poorly conducted tests. Progeny testing is usually limited to bulls because a sufficient number of calves cannot be obtained on a cow to give an adequate progeny test.

In a proper progeny test, bulls are assigned cows at random after the cows have been classified by age, line of breeding, and known performance. A bull bred to a selected group of cows can give biased information which does not accurately evaluate his breeding ability. An adequate progeny test involves a minimum number of progeny, typically 20 head per sire through weaning and 10 head through postweaning and carcass evaluation. The more numbers, the greater the validity of the test.

Progeny test data will be biased and accuracy greatly reduced if cows and calves are not fed and managed uniformly until all data are collected.

All selection methods discussed can be effectively used in a beef cattle breeding program. Outstanding young bulls can best be identified by using individual performance and pedigree information on the sire, dam, and half sibs. These outstanding young sires then need to genetically prove themselves based on a valid progeny testing program. Results of the progeny tests will determine which sires to use extensively in later years.

National Sire Evaluation

Previously, progeny testing was considered valid only when done on an within-herd basis. Artificial insemination and some refined data analysis allows comparison of bulls between herds. Several breed associations now have National Sire Evaluation Programs which follow the guidelines established by the Beef Improvement Federation. These guidelines are based on comparing the progeny of bulls being tested against the progeny of bulls designated as reference sires. This allows widespread use of reference sires in many different herds and obviously in many different environments. The breeding plan and the comparative results for the National Sire Evaluation program are simply shown in Figure 13.8.

Results on bulls tested in the program are usually reported by the different breed associations in a National Sire Summary. Results are usually shown as an expected progeny difference (EPD) rather than a ratio. An example of an EPD for yearling weight on a bull might be +15.5 lbs. This bull would be expected to sire calves approximately 15 lbs heavier at a year of age compared to the average of calves sired by other bulls included in the Sire Summary. This EPD provides an estimate on how future progeny by this sire are expected to perform relative to a set of reference sires used in the breed. Accuracy figures are provided to show how much confidence can be placed in the EPD. A more detailed discussion of National Sire Summary records is presented in Chapter 8.

Progeny Testing for Genetic Abnormalities.

There have been many different types of genetic abnormalities reported in cattle. Most of this genetic trash expresses itself at a relatively low incidence even though

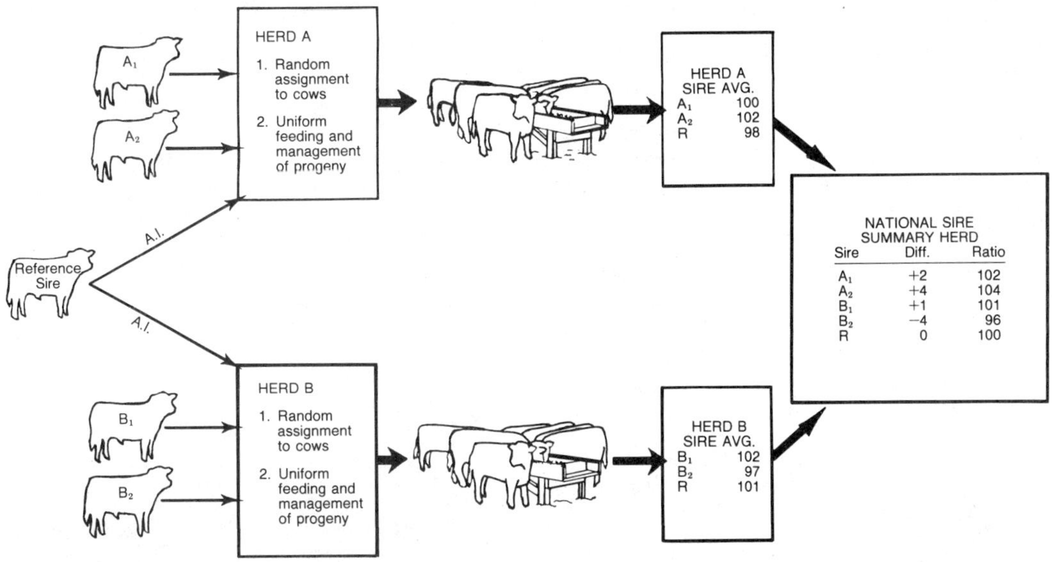

Figure 13.8. Basic breeding and data analysis plan for National Sire Evaluation. Courtesy of USDA.

all breeds and most herds have some present. Most genetic abnormalities remain hidden as recessive genes. A genetic abnormality expresses itself when the homozygous recessive condition occurs from mating a carrier bull and a carrier cow.

When an abnormal calf is born, the producer should first determine if the abnormality is genetic. There are some abnormalities that are environmentally caused such as the *Crooked Calf Disease.* The crooked legged condition makes one suspicious of a genetic abnormality; however, this abnormality is caused by the cow eating the lupine plant during a certain stage of pregnancy (see Chapter 20).

A producer should not panic if a genetically abnormal calf is born in the herd. A commercial producer might consider selling the cow and replacing the bull. A seedstock producer may need to take a more serious look at the problem and take more steps to eliminate or reduce the incidence of the genetic problem.

Some genetic abnormalities causing the greatest alarm in recent years are identified later in this chapter. Several of these abnormalities have reached some unexpected levels of occurrence due to a selection preference for the carrier animal or a noted bull being used extensively through AI. In the latter situation, a carrier bull can sire several thousand calves before the abnormality expresses itself. Obviously one-half of the bull's calves would be carrying the undesirable recessive. This would not be known until one of his carrier daughters was mated to another animal carrying a similar undesirable gene. Even then, the chance of an abnormal calf occurring is one chance in four for each carrier x carrier mating.

Occasionally it is deemed desirable and necessary to progeny test a bull from unknown or known carrier parentage. The results of such a progeny test for *mule foot* were recently reported (*Angus Journal,* Sept. 1979, p. 114). It has been generally accepted

Table 13.5. Probability of Detecting a Bull as a Carrier of an Undesirable Recessive Gene, Using Three Kinds of Test Matings.

	Mating Bull to		
No. Matings	Double Recessive	Known Heterozygote	Own Daughter
1	0.5	0.25	0.125
2	0.75	0.44	0.235
3	0.87	0.58	0.330
4	0.94	0.68	0.414
5	0.97	0.76	0.487
6	0.98	0.82	0.551
7	0.99	0.87	0.607
8	(0.996)	0.90	0.656
9	0.998	0.925	0.699
10	0.999	0.944	0.737
15		(0.987)	0.865
20		0.997	0.931
25		0.9993	0.965
30			(0.982)
35			0.991
40			0.995
50			0.999

that the level of testing should be at the one percent level of probability. At this probability level, there is only one chance in a hundred that a bull can be tested this extensively and still be carrying the undesirable recessive gene.

Using the mule foot abnormality (Table 12.2, Chapter 12), a bull (S?) could be mated to eight females showing the mule foot condition (ss) or 16 known carriers (Ss) or 32 of his own daughters (½ SS and ½ Ss). If no mule foot calves were identified, the bull would be tested to the one percent level of probability (Table 13.5, probabilities are circled for each of the three mating groups).

TRAITS AND THEIR MEASUREMENT

It is important that traits included in a breeding program be of economic importance and that the traits be measured as objectively as possible. Most economically important traits of beef cattle can be classified under (1) reproductive performance, (2) weaning weight, (3) postweaning growth, (4) feed efficiency, (5) carcass merit, (6) longevity, (7) conformation, and (8) freedom from genetic defects.

Reproductive Performance

This trait has the highest economic importance when compared to other traits. Most cow–calf producers have a goal for percent calf crop weaned (number of calves weaned compared to number of cows in the breeding herd) of 90% or higher. Beef producers also desire each cow to calve every 365 days or less and have a calving season for the entire herd of less than 90 days. All of these are good objective measures of reproductive performance. Heritability of fertility in beef cattle is quite low (less than 20%), so little genetic progress can be made through selection. The most effective way to improve reproductive performance is to improve the environment through, for example, adequate nutrition and good herd health practices.

Reproductive performance can be improved through breeding methods by crossbreeding to obtain heterosis for percent calf crop weaned, using bulls with relatively light birth weights (heritability of birth weight is 40%) which decreases calving difficulty, and by selecting bulls that have a relatively large scrotal circumference. Scrotal circumference has a high heritability (50%), and bulls with a larger scrotal size (over 30 cm for yearling bulls) produce a larger volume of semen and have half-sister heifers that reach puberty at earlier ages than heifers related to bulls with a smaller scrotal size.

Weaning Weight

Weaning weight, as measured objectively by scales, reflects milking and mothering ability of the cow and the preweaning growth rate of the calf. Weaning weight is commonly expressed as adjusted 205-day weight, where weaning weight is adjusted for age of calf and age of dam. This adjustment puts all weaning weight records on a comparable basis since older calves will weigh more than younger calves, and mature cows (5 to 9 years of age) will milk heavier than when they are young cows (2 to 4 years of age).

Weaning weights of calves are usually compared by dividing the calf's adjusted weight by the average weight of other calves in the herd. This is expressed as a ratio. For example, a calf with a weaning weight of 440 lbs, where the herd average is 400 lbs, has a ratio of 110. This calf's weaning weight ratio is 10% above the herd average. Ratios should be used primarily for selecting cattle within the same herd. Comparing ratios between herds is misleading from a genetic standpoint because most differences between herds are caused by differences in environment. Weaning weight will respond to selection because it has a heritability of 30%.

Postweaning Growth

This trait measures growth from weaning to a weight that approaches slaughter weight. Postweaning growth may take place on a pasture or in a feedlot. Usually animals with relatively high postweaning gains make efficient gains at a relatively low cost to the producer.

Postweaning gain in cattle is usually measured in pounds gained per day after a calf has been on a feed test for 140 to 160 days. Weaning weight and postweaning gain are usually combined into a single trait, namely, adjusted 365-day weight (yearling weight). It is computed as follows:

$$\text{Adjusted 365-day weight} = (160 \times \text{average daily gain}) + \text{adjusted 205-day weight.}$$

Average daily gain for 140 days and adjusted 365-day weight both have high heritabilities (40%), so genetic improvement can be quite rapid when selection is practiced on postweaning growth or yearling weight.

Feed Efficiency

Feed efficiency is usually measured by pounds of feed required per pound of liveweight gain. Specific records for feed efficiency can only be obtained by keeping records on the amount of feed consumed by individual animals. With the possibile exception of some bull testing programs, determining feed efficiency on an individual animal basis is seldom economically feasible.

The interpretation of feed efficiency records can be rather confusing depending on the endpoint to which animals are fed. Feeding endpoint can be a certain number of days on feed (e.g., 140 days), to a specified slaughter weight (e.g., 1200 lbs), or to a carcass compositional endpoint (e.g., low Choice quality grade or 0.30 inches of fat over the rib eye). Most differences shown by individual animals in feed efficiency are related to pounds of body weight maintained through feeding periods and daily rate of gain or feed intake of each animal. Cattle fed from a similar initial feedlot weight (e.g., 600 lbs) to a similar slaughter weight (e.g., 1100 lbs) will demonstrate a high correlation between rate of gain and efficiency of gain. In this situation, cattle that gain faster will require fewer pounds of feed per pound of gain. Thus, a breeder can select for rate of gain and thereby make genetic improvement in feed efficiency. When cattle, however, are fed to the same compositional endpoint (approximately the same carcass fat), the difference in amount of feed required per pound of gain is relatively small. This is true for different sizes and shapes of cattle and even for various breeds that vary greatly in skeletal size and weight.

The heritability of feed efficiency is high (45%), so selection for more efficient cattle can be effective. It seems logical to use the genetic correlation between gain and efficiency where possible because of the expense in obtaining individual feed efficiency records.

Carcass Merit

Carcass merit is presently measured by quality grade and yield grade, which are discussed in detail in Chapter 16. Many cattle breeding programs have goals to produce cattle that will grade Choice and have yield grades from 2.0 to 3.5. Objective measurements of backfat thickness probes on the live animal and measurements of the height of the skeleton at the hip can assist in predicting the yield grade at certain slaughter weights. Visual appraisal, which is more subjective but accurate with proper training, can be used to predict amount of fat or predisposition to fat and skeletal size. These visual estimates can be relatively accurate in identifying actual yield grades if cattle differ by as much as two yield grades (see Plates E, F, G, and H in Chapter 17).

Because quality grade cannot be evaluated accurately in the live animal, it is necessary to evaluate the carcass. Steer and heifer progeny of different bulls are slaughtered to best identify genetic superiority or inferiority of bulls for both quality grade and yield grade. Heritabilities of most beef carcass traits are high (over 40%), so selection can result in marked genetic improvement for these traits.

Longevity

Longevity measures length of productive life and it is an important trait, particularly for cows. Bulls are usually kept in a herd for only a few years, or inbreeding may occur. Some highly productive cows remain in the herd until 15 years of age or older, while other highly productive cows have been culled from the herd prior to reaching 8 or 9 years of age. These latter cows may have been culled because of problems such as skeletal unsoundness, poor udders, unhealthy eyes or unhealthy teeth. Little selection opportunity for longevity exists because few cows remain highly productive past the age of 10 years. Most producers need to improve their average herd performance as rapidly as possible rather than to improve longevity. In this situation, a relatively rapid turnover of cows is needed.

Some selection for longevity occurs because highly productive cows, that stay in the herd for a long period of time, leave more numbers of potential replacement heifers. Some beef producers attempt to identify bulls that have highly productive, relatively old dams. Certain conformation traits, such as skeletal soundness and udder soundness, may be evaluated to extend the longevity of production.

Conformation

Conformation is the form, shape and visual appearance of an animal. How much emphasis to put on conformation in a beef cattle selection program has been and continues to be controversial. Some producers feel that putting a productive animal into an attractive package contributes to additional economic returns. It is more logical, however, to place more selection emphasis on traits that will produce additional numbers of calves and pounds of lean growth for a given number of cows. Placing some emphasis on conformation traits such as skeletal (particularly feet and legs), udder, eye and teeth soundness is justified (Figures 13.9 and 13.10). Conformation differences, such as fat accumulation or predisposition to fat, can be used effectively to make meaningful genetic improvement in carcass composition.

Most conformation traits are medium to high (30% to 60%) in heritability, so selection for these traits will result in genetic improvement.

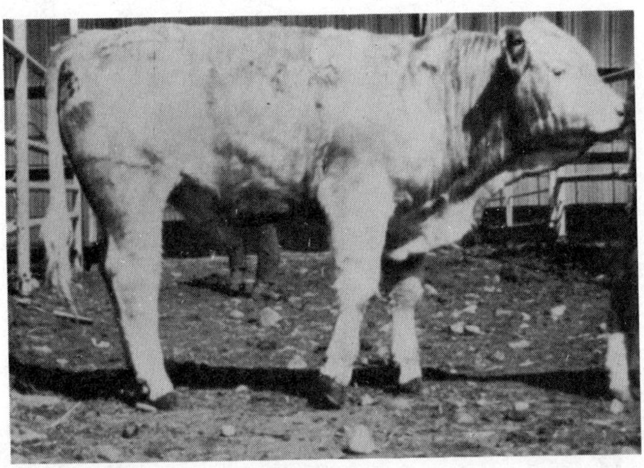

Figure 13.9. Bulls which are post-legged (hind legs too straight) are likely to become stifled. This affects their longevity as breeding animals since they would have to be culled from the herd. Courtesy of CSU.

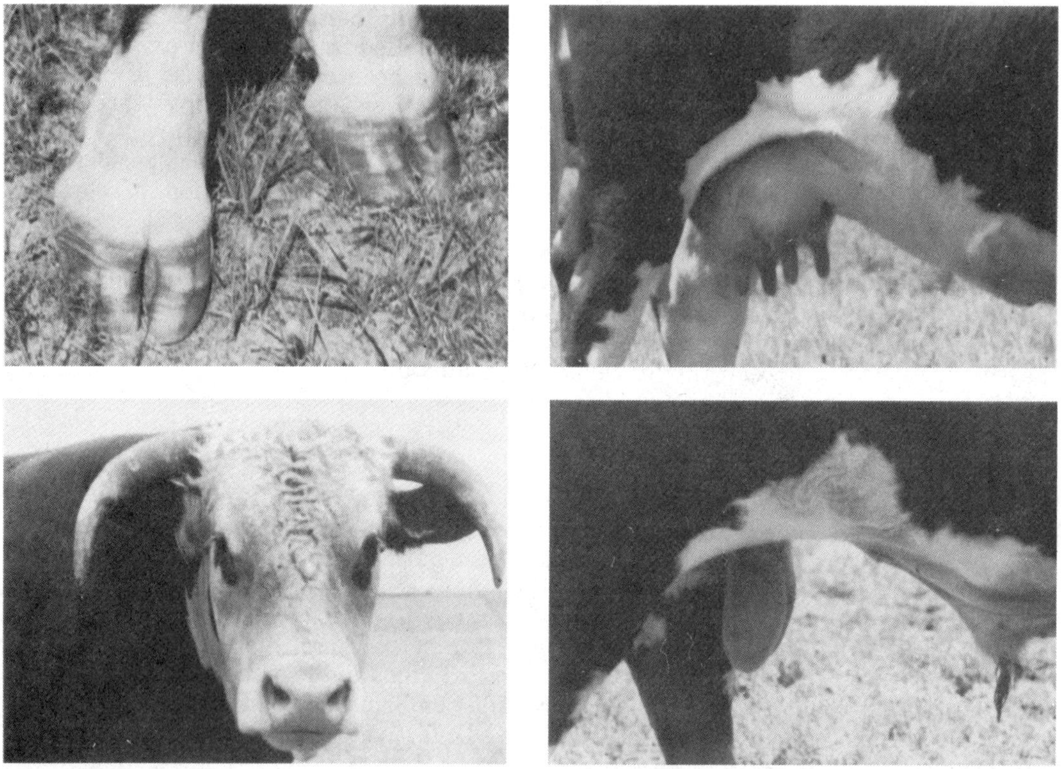

Figure 13.10. Soundness of feet and legs, eyes, and testicles can be evaluated visually. Reproductive soundness, including scrotal circumference, is best measured more objectively. Courtesy of Richard Sell, Booker, Texas.

Genetic Defects

Genetic defects, other than those previously identified under longevity and conformation, need to be considered in breeding productive beef cattle. Cattle have numerous known hereditary defects. Most of them, however, occur very infrequently and are of minor concern. Some defects increase in their frequency and selection needs to be directed against them. Most of these defects are determined by a single pair of genes that are usually recessive. When one of these hereditary defects occur, it is a logical practice to cull both the cow and the bull.

Some of the most common occurring genetic defects in cattle today are double muscling, syndactly (mule foot), cryptorchidism, arthrogryposis (palate-pastern syndrome), osteopetrosis (marble bone disease), hydrocephalus, and dwarfism.

Double muscling is evidenced by an enlargement of the muscles with large grooves between the muscle systems especially noticeable in the hind leg. Double-muscled cattle usually grow slowly, and their fat deposition in and on the carcass is much less than that of normal beef animals.

Syndactyly (mule foot) is a condition in which one or more of the hooves are solid in structure rather than cloven. Mortality rate is high in calves with syndactly.

Cryptorchidism is the retention of one or both testicles in the body cavity.

Arthrogryposis (palate-pastern syndrome) is a defect in which the pastern tendons are contracted, and the upper part of the mouth has not properly fused together (cleft palate).

Osteopetrosis (marble bone disease) is characterized by the marrow cavity of the long bones being filled with bone tissue. All calves having osteopetrosis have short lower jaws, protruding tongue, and impacted molar teeth.

A bulging forehead where fluid has accumulated in the brain area is typical of the defect of *hydrocephalus.* Calves with arthrogryposis, hydrocephalus, or osteopetrosis usually die shortly after birth.

The most common type of *dwarfism* is snorter dwarfism, in which the skeleton is quite small and the forehead has a slight bulge. Some snorter dwarfs exhibit a heavy, labored breathing sound. This defect was most common in the 1950s, and it has decreased significantly since that time.

Disposition

Disposition is the calmness of cattle or their level of excitability. It is best evaluated when producers are near cattle during calving or working them through various kinds

Figure 13.11. Disposition in cattle has a genetic basis, however, the major influence on disposition is how the cattle are handled. Courtesy of J. Carter Thomas, Cuero, Texas.

of handling and restraining facilities. Obvious differences in behavior of cattle exist which affect their disposition (see Chapter 16).

There is evidence that both genetic makeup of cattle and environmental conditions to which cattle are exposed determine their disposition. There are breed differences, line of breeding differences within a breed and individual differences in disposition. There is justification in culling cattle with very poor dispositions to prevent human injury or to decrease repair costs of fences and other facilities. Some cattle have very calm dispositions depending on how producers treat their cattle (Figure 13.11).

BREED EVALUATION FOR COMMERCIAL PRODUCERS

Most commercial beef producers use a crossbreeding program because they can take advantage of the heterosis in addition to genetic improvement from selection. A crossbreeding system should be determined according to which breeds are available and adapted to the commercial producers' feed supply, market demands, and other environmental conditions (Figure 13.12). A good example of adaptability is the Brahman breed, which is more heat and insect resistant than most other breeds. Because of this higher resistance, the level of productivity (in the southern and Gulf regions of the United States) is much higher for the Brahman and other breeds that include Brahman.

Most commercial producers travel less than 150 miles to purchase their bulls. Therefore, a producer should assess the breeders with excellent breeding programs in a 150-mile radius, as well as which breeds are available. This assessment, in most cases, should have a greater priority than determining which breeds to use and in what combination to use them.

Breeds should be chosen for a crossbreeding system based on how well the breeds complement each other. Table 13.6 gives some comparative ranking of major beef breeds for productive characteristics. Although the information in Table 13.6 is useful, it should not be considered the final answer for decisions on breeds to use. First, a producer needs to recognize that this information reflects breed averages; therefore, there are individual animals and herds of the same breed that are much higher or lower than the ranking given. Producers need to use some of the previously described methods to identify superior animals within the breed (see bull and female selection, Chapter 8). Second, it should be recognized that these average breed rankings can change with time, depending on the improvement programs used by the leading breeders within the same breed. Those traits that have high heritabilities would be expected to change most rapidly, assuming the same selection pressure for each trait. A careful analysis of the information in Table 13.6 shows that no one breed is superior in all important productive characteristics. This gives an advantage to commercial producers using a crossbreeding program if they select breeds whose superior traits complement each other. An excellent example of breed complementation is shown by the Angus and Charolais breeds which complement each other for both quality grade and yield grade.

During the past several years there has been an extensive comparative evaluation of the major breeds of cattle which represent diverse biological types. This germ plasm evaluation has been conducted at the Meat Animal Research Center (MARC) at Clay Center, Nebraska. The following background information and Tables 13.7 through 13.12 were taken from the MARC publication, *Beef Research Program,* Progress Report No. 1, April 1982. These tables show comparative performance data of various breed crosses

Figure 13.12. Cattle are expected to produce under some harsh environments—in this situation a limited supply of forage. Some cattle are more adaptable than others. Courtesy of the American Hereford Association.

where all animals have been raised under similar environmental conditions. The ratios reflect a comparison of breed crosses to Hereford-Angus crosses, where the latter ratio is 100.

In 1969, MARC implemented a program to characterize a broad range of biological types of cattle as represented by breeds that differ widely in characteristics such as milk level, growth rate, carcass composition and mature size. The primary objective was to characterize breeds representing diverse biological types for the full spectrum of traits relating to beef production. The germ plasm evaluation program at MARC has included three cycles of sire breeds that were bred by artificial insemination (AI) to Hereford and Angus dams. Hereford-Angus reciprocal crosses were repeated in each cycle as controls and to make possible the comparison of breeds included in different cycles. The first cycle involved breeding Hereford, Angus, Jersey, Limousin, South Devon,

Table 13.6. Breed Evaluation for the Productive Characteristics[a]

Breed	Age at Puberty	Weight at Puberty	Birth Weight	Wean Weight	Pounds of Calf Weaned Per Cow Exposed	Post-Wean Gain	Feed Efficiency				Marbling	Yield Grade		Palatability
							Equal Age	Equal Weight	Choice Grade	Equal Fat	Equal Age	Equal Age	Equal Ribeye Fat	Grain Fed and Equal Age
Angus	①[b]	②	②	④	③	④	③	③	①	①	①	④	④	①
Beefmaster	3	3	4	2	2	2	—	3	—	—	3	3	3	2
Blonde d'Aquitaine	4	4	4	3	2	2	—	2	4	4	4	2	3	1
Brahman	⑤	⑤	④	①	①	③	④	④	④	②	⑤	③	③	③
Brangus	3	4	3	3	3	3	—	3	3	—	4	3	—	2
Charolais	④	⑤	⑤	①	④	①	①	①	②	④	④	①	①	①
Chianina	④	⑤	⑤	①	③	①	②	②	⑤	⑤	⑤	①	①	①
Devon	②	③	③	④	④	②	②	③	①	①	③	④	③	①
Galloway	3	3	3	4	4	3	—	3	—	—	—	3	3	1
Gelbvieh	①	②	④	①	②	②	①	①	②	③	④	③	③	①
Hereford	③	③	③	④	③	③	③	③	①	①	④	④	④	①
Limousin	④	④	④	③	④	③	①	②	④	④	⑤	①	①	①
Longhorn	2	2	1	5	5	5	5	5	5	5	—	4	3	1
Maine-Anjou	③	④	⑤	①	③	①	①	①	②	④	④	②	②	①
Murray Grey	2	2	3	3	3	4	—	3	—	—	3	4	4	1
Pinzgauer	①	②	④	②	②	②	③	③	①	②	④	③	③	①
Polled Hereford	③	③	③	④	③	③	3	③	1	1	④	④	④	①
Red Angus	1	2	2	4	3	3	3	3	1	1	1	4	4	1
Red Poll	①	①	②	④	③	③	⑤	⑤	④	③	④	④	③	①
Santa Gertrudis	3	5	4	2	4	3	4	3	—	—	4	4	3	2
Scotch Highland	3	3	2	4	4	4	—	—	—	—	3	3	3	1
Shorthorn	3	3	3	4	3	3	3	3	1	1	4	5	5	1
Simmental	②	④	⑤	①	③	①	②	②	③	④	④	②	①	①
Tarentaise	①	②	③	②	②	③	④	④	③	③	④	③	③	②

[a] Ranking based on 1 (most desirable) through 5 (least desirable).

[b] Circled numbers are based primarily on extensive breed comparison data from the Meat Animal Research Center (MARC), Clay Center, Nebraska. Most of the MARC data are based on the various breeds of bulls being bred to Angus and Hereford cows. Rankings are not made where there are insufficient comparative data. Uncircled numbers are judgments based on less extensive, comparative data.

Table 13.7. Breed Group Means for Birth and Weaning Traits of Calves Produced in Cycles I, II, III

Breed Group	Number	Gestation Length (Days)	Calving Difficulty[a,b] (%)	Calf Crop Weaned[a,c] (%)	Birth Weight[a] (lb)	ADG[d] (lb)	200-day Weight (lb)	200-day Weight Ratio[e] (%)	Weight Weaned Per Cow Calving 200 Days (lb)	Weight Weaned Per Cow Calving Ratio[e] (%)
Jersey-X	302	282.9	2.9	92.3	68.6	1.70	406	94	375	90
Hereford-Angus-X	962	284.0	2.9	97.3	78.7	1.76	430	100	418	100
Red Poll-X	214	285.2	3.7	97.8	78.7	1.74	426	99	417	100
South Devon-X	232	286.7	11.9	92.9	83.1	1.74	430	100	399	95
Tarentaise-X	202	287.1	6.0	94.8	82.7	1.81	443	103	420	100
Pinzgauer-X	376	286.0	6.3	95.2	86.4	1.76	439	102	418	100
Sahiwal-X	325	294.0	6.2	94.3	83.8	1.74	432	100	407	97
Brahman-X	349	291.7	10.0	93.5	90.2	1.83	456	106	426	102
Brown Swiss-X	263	285.0	8.4	97.2	85.6	1.83	452	105	439	105
Gelbvieh-X	213	286.3	8.0	91.5	86.0	1.87	461	107	422	101
Simmental-X	399	287.3	14.9	89.1	88.9	1.83	452	105	403	96
Maine-Anjou-X	222	285.4	20.4	90.8	90.6	1.81	454	106	412	98
Limousin-X	371	289.2	9.4	91.7	85.8	1.76	437	102	401	96
Charolais-X	382	287.0	18.4	86.5	90.6	1.85	459	107	397	95
Chianina-X	238	287.5	11.8	91.1	89.3	1.83	456	106	415	99
Hereford dams[f]	2440	288.6	10.1	93.0	86.2	1.70	423		393	
Angus dams[f]	3129	285.4	6.5	94.2	80.9	1.87	454		428	

[a] Calving difficulty (%), and birth weight data are from cows calving at 4 years of age or older.
[b] Calving difficulty (%) is percentage of normal presentations requiring calf puller or C-section.
[c] Calf crop weaned is expressed as a percentage of all dams calving.
[d] ADG = average daily gain.
[e] Ratio relative to Hereford-Angus crosses.
[f] Includes straightbred progeny from Cycles I and II but does not include Hereford-Angus cross progeny from Cycle III in this and subsequent tables.
Source: MARC.

Table 13.8. Breed Group Means of Females Produced in Cycles I, II, and III

Breed	Number	200-day Weight (lb)	200-day Weight Ratio[a]	400-day Weight (lb)	400-day Weight Ratio[a]	550-day Weight (lb)	550-day Weight Ratio[a]	550-day Hip Height[b] (in.)	Weight at Puberty (lb)	Weight at Puberty Ratio	Age at Puberty (Days)	Age at Puberty Ratio	Pregnant 550 Days (%)
Jersey-X	117	392	94	613	90	712	92	0	518	83	308	86	80.6
Hereford-Angus-X	322	419	100	681	100	772	100	46.1	622	100	357	100	87.2
Red Poll-X	95	410	98	655	96	750	97	46.4	580	93	337	94	84.8
South Devon-X	120	428	102	703	103	798	103	0	639	103	350	98	79.3
Tarentaise-X	85	432	103	701	103	796	103	47.6	622	100	349	98	96.2
Pinzgauer-X	114	437	104	703	103	798	103	48.0	611	98	334	94	99.2
Sahiwal-X	87	419	100	653	96	758	98	48.0	642	103	414	116	103.3[c]
Brahman-X	103	450	107	703	103	831	108	49.6	712	114	429	120	98.2
Brown Swiss-X	126	439	105	701	103	805	104	48.4	615	99	332	93	93.0
Gelbvieh-X	81	448	107	714	105	822	106	48.4	626	101	326	91	93.2
Simmental-X	157	441	105	723	106	825	107	0	666	107	358	100	80.4
Maine-Anjou-X	89	441	105	734	108	838	108	48.4	672	108	357	100	94.2
Limousin-X	161	423	101	692	102	770	100	0	679	109	384	108	76.2
Charolais-X	132	448	107	723	106	814	105	0	703	113	384	108	74.8
Chianina-X	92	443	106	714	105	833	108	50.8	699	112	384	108	85.6
Hereford dams	997	415	0	675	0	787	0	48.4	642	0	379	0	90.4
Angus dams	1103	441	0	695	0	787	0	48.0	635	0	351	0	87.7

[a] Ratio relative to Hereford-Angus crosses.
[b] 550-day hip height data were not collected on the Cycle I females.
[c] In Cycle III, only 82.0 percent of Angus-Hereford crosses were pregnant at 550 days whereas 98 percent of the Sahiwal crosses were pregnant at 550 days.
Source; MARC.

Table 13.9. Breed Group Means for Postweaning Growth and Feed Efficiency to Time, Weight, Grade, and Fat-trim End Points

					Feed efficiency (Mcal ME/lb gain)			
		Post-weaning ADG	452-day Weight	452-day Weight	Time	Weight	USDA Choice	Fat Trim
Breed Group	Number	(lb)	(lb)	Ratio[a]	0 to 238 Days	545 to 1,036 lb	0 Days to Small Marbling	0 Days to 18.9% Fat Trim
Jersey-X	132	2.23	981	94	10.90	11.76	11.06	10.43
Hereford-Angus-X	508	2.40	1046	100	10.49	10.67	10.54	10.07
Red Poll-X	111	2.20	995	95	11.36	11.49	11.54	11.34
South Devon-X	94	2.58	1082	103	10.35	10.40	10.72	10.29
Tarentaise-X	103	2.38	1054	101	10.75	10.97	11.39	11.11
Pinzgauer-X	176	2.45	1058	101	10.44	10.60	10.77	10.84
Sahiwal-X	154	2.20	999	96	11.04	11.80	11.55	10.98
Brahman-X	153	2.40	1068	102	10.83	11.00	11.81	10.98
Brown Swiss-X	154	2.47	1087	104	10.36	9.67	10.84	11.20
Gelbvieh-X	111	2.56	1115	107	10.17	9.45	10.92	11.11
Simmental-X	176	2.69	1141	109	10.36	9.95	11.20	11.47
Maine-Anjou-X	109	2.65	1133	108	10.22	9.35	11.01	11.84
Limousin-X	173	2.32	1035	99	10.15	10.25	11.74	11.70
Charolais-X	176	2.67	1143	109	10.02	9.46	10.94	11.61
Chianina-X	119	2.49	1099	105	10.39	9.86	12.13	13.29

[a] Ratio relative to Hereford-Angus crosses.
Source: MARC.

Simmental, and Charolais sires (20 to 35 sires per breed) by AI to Hereford and Angus dams (ranging from 2 to 7 years old at calving) to produce three calf crops in March and April of 1970, 1971, and 1972. In Cycle II, Hereford and Angus dams (ranging from 4 to 9 years old at calving) used in Cycle I were bred by AI to Hereford, Angus, Red Poll, Brown Swiss (predominantly European), Gelbvieh, Maine-Anjou, and Chianina sires to produce two calf crops in 1973 and 1974. Cycle III involved the same or comparable Hereford and Angus dams (ranging from 4 to 11 years old at calving) mated by AI to Hereford, Angus, Tarentaise, Pinzgauer, Sahiwal, and Brahman sires to produce two calf crops in 1975 and 1976. The same Hereford and Angus sires were used in all three cycles of the program to ensure a more stable control population of Hereford-Angus reciprocal crosses as a basis for comparison of breeds included in different cycles of the program. In Cycles I and II, but not in Cycle III, Hereford and Angus sires were also mated to produce straightbred progeny of each of the two breeds.

Most of the heterosis achieved in cattle as a result of crossbreeding is expressed by weaning time. The cumulative effect of heterosis on pounds of calf weaned per cow exposed is shown in Figure 13.13, in which one sees that maximum heterosis is obtained when crossbred calves are obtained from crossbred cows. Traits that express an approximate heterosis of a 20% increase in pounds of calf weaned per cow exposed are early puberty of crossbred heifers, high conception rates in crossbred females, high survival

Table 13.10. Breed Group Means for Carcass Traits—Cycles I, II, and III

Breed Group	Number	Shrunk Live Weight (lb)	Carcass Weight (lb)	Fat Thickness (in.)	Marbling Score[a]	Retail Product (%)	Fat Trim (%)	Bone (%)	Kidney Fat (%)	Live Weight (lb)	Carcass Weight (lb)	Fat Thickness (in.)	Marbling Score[a]	Retail Product (%)	Bone (%)	Kidney Fat (%)
		Adjusted to a Constant Age of 458 Days								Adjusted to 19 Percent Fat Trim						
Jersey-X	134	958	593	0.46	13.3	65.5	22.1	12.4	6.2	879	537	0.37	11.8	68.0	13.0	5.6
Hereford-Angus-X	472	1008	637	0.64	11.3	66.3	21.7	12.0	3.9	935	584	0.53	10.2	68.4	12.6	3.6
Red Poll-X	106	979	618	0.49	11.2	66.6	21.0	12.4	5.1	928	581	0.43	10.4	68.2	12.8	4.8
South Devon-X	94	1031	655	0.48	11.3	67.7	20.0	12.3	4.7	1010	639	0.46	10.9	68.5	12.5	4.5
Tarentaise-X	102	1010	638	0.44	10.1	69.8	17.7	12.5	4.9	1016	645	0.44	10.4	68.6	12.4	5.1
Pinzgauer-X	130	1017	629	0.46	10.8	69.4	17.5	13.1	4.4	1028	640	0.46	11.2	68.0	13.0	4.6
Sahiwal-X	141	962	611	0.54	9.7	69.1	18.4	12.4	3.9	951	606	0.52	9.7	68.5	12.5	4.0
Brahman-X	128	1033	663	0.56	9.3	69.4	18.0	12.6	4.1	1032	665	0.55	9.5	68.4	12.6	4.3
Brown Swiss-X	120	1076	677	0.39	10.4	69.1	17.6	13.3	4.0	1144	723	0.46	11.2	68.1	12.9	4.1
Gelbvieh-X	108	1090	687	0.37	9.7	69.8	17.4	12.8	4.5	1169	741	0.45	10.6	68.7	12.3	4.7
Simmental-X	175	1079	673	0.39	9.9	71.0	15.6	13.4	4.3	1251	791	0.55	11.9	68.6	12.4	4.8
Maine-Anjou-X	109	1103	704	0.37	10.2	70.2	16.5	13.3	4.1	1223	787	0.48	11.5	68.4	12.6	4.4
Limousin-X	177	1021	652	0.41	8.9	72.4	15.1	12.5	4.3	1214	787	0.60	11.0	69.7	11.3	4.9
Charolais-X	177	1093	691	0.38	10.3	71.8	15.2	13.0	4.2	1291	828	0.55	12.6	69.1	11.9	4.8
Chianina-X	112	1077	690	0.32	8.5	73.0	13.0	14.0	3.8	1390	910	0.57	11.7	68.8	12.2	4.6
Hereford dams[b]	1044	1025	643	0.43	9.8	69.7	17.3	13.0	4.1	1082	683	0.50	10.6	68.4	12.6	4.4
Angus dams[b]	1353	1024	651	0.51	11.1	68.3	19.4	12.3	4.5	1015	645	0.50	10.9	68.6	12.4	4.5

[a] Marbling scores: traces = 4, 5, 6; slight = 7, 8, 9; small = 10, 11, 12; modest = 13, 14, 15.
[b] Includes straightbred progeny from Cycles I and II but does not include Hereford-Angus cross progeny from Cycle III.
Source: MARC.

Table 13.11. Percentage of Carcass in Whole Cuts and Percentage of Total Retail Product, Fat Trim, or Bone in Each Cut at a Constant Carcass Weight

Item	Round	Chuck	Loin and Rib	Minor Cuts	Kidney Knob	Round	Chuck	Loin and Rib	Minor Cuts	Kidney Knob
			Whole Cut					***Fat Trim***		
Jersey-X	20.99	26.18	21.43	25.03	6.33	7.59	11.76	14.91	39.31	26.43
Hereford-Angus-X	22.72	26.35	21.90	25.08	3.94	9.64	12.95	18.33	41.12	17.96
Red Poll-X	22.27	26.49	21.58	24.33	5.34	7.46	12.57	15.72	39.49	24.76
South Devon-X	23.17	26.10	21.87	24.23	4.58	8.96	11.43	16.36	39.84	23.41
Tarentaise-X	22.73	26.36	21.99	23.94	4.97	6.77	10.83	15.92	39.92	26.56
Pinzgauer-X	22.92	26.56	21.88	24.16	4.47	7.36	11.80	16.16	40.76	23.92
Sahiwal-X	22.98	25.68	22.28	24.67	4.37	8.01	10.63	17.11	43.37	20.88
Brahman-X	23.68	26.25	21.72	24.22	4.12	7.94	8.67	16.26	43.93	23.20
Brown Swiss-X	23.99	26.94	21.87	23.32	3.88	7.74	10.14	16.12	41.77	24.23
Gelbvieh-X	23.27	26.89	21.76	23.76	4.32	6.19	8.48	15.37	42.84	27.12
Simmental-X	24.19	26.72	21.50	23.40	4.18	7.09	10.49	14.55	41.05	26.82
Maine-Anjou-X	24.32	26.91	21.60	23.39	3.78	5.05	6.53	15.18	44.73	28.51
Limousin-X	24.19	26.47	22.05	23.15	4.18	7.02	8.84	16.18	40.82	27.14
Charolais-X	24.76	26.39	21.72	23.21	3.93	8.01	9.66	14.55	40.73	27.05
Chianina-X	25.00	27.43	21.80	22.25	3.52	3.23	4.80	14.88	45.04	32.05
Average	23.41	26.52	21.80	23.88	4.39	7.20	9.97	15.84	41.65	25.34
			Retail Product					***Bone***		
Jersey-X	24.56	30.72	24.74	19.98	—	28.67	30.64	17.07	23.63	—
Hereford-Angus-X	25.80	30.02	24.05	20.14	—	29.76	30.72	16.52	23.01	—
Red Poll-X	25.81	30.23	24.35	19.62	—	29.49	31.00	16.87	22.65	—
South Devon-X	25.84	26.46	24.33	20.45	—	30.07	30.19	16.62	23.13	—
Tarentaise-X	25.89	29.82	24.49	19.81	—	29.46	30.51	17.11	22.93	—
Pinzgauer-X	25.88	29.85	24.43	19.85	—	29.91	30.41	16.43	23.26	—
Sahiwal-X	26.40	29.65	24.71	19.25	—	30.21	29.82	17.63	22.35	—
Brahman-X	26.54	30.02	24.02	19.43	—	30.17	30.11	16.53	23.20	—
Brown Swiss-X	26.54	30.12	24.39	18.96	—	30.81	30.75	15.88	22.57	—
Gelbvieh-X	26.02	30.39	24.25	19.35	—	30.03	30.87	16.41	22.70	—
Simmental-X	26.56	29.61	23.99	19.85	—	30.62	30.19	16.29	22.91	—
Maine-Anjou-X	26.79	30.06	23.93	19.23	—	30.84	30.46	16.01	22.70	—
Limousin-X	26.66	29.50	24.21	19.64	—	30.11	30.49	16.54	22.87	—
Charolais-X	26.81	29.02	24.13	20.04	—	30.98	30.22	16.07	22.74	—
Chianina-X	27.27	30.37	24.01	18.36	—	31.17	30.26	16.28	22.30	—
Average	26.22	29.92	24.27	19.60	—	30.15	30.44	16.55	22.86	—

Source: MARC.

Table 13.12. Breed Group Means for Factors Identified With Meat Quality—Cycles I, II, and III

Breed Crosses	Marbling[a]	Percent Choice	Warner-Bratzler Shear (lb)	Flavor[a]	Juiciness[1]	Tenderness[1]
Chianina-X	8.3	24	7.9	7.3	7.2	6.9
Limousin-X	9.0	37	7.7	7.4	7.3	6.9
Brahman-X	9.3	40	8.4	7.2	6.9	6.5
Gelbvieh-X	9.6	43	7.8	7.4	7.2	6.9
Sahiwal-X	9.7	44	9.1	7.1	7.0	5.8
Simmental-X	9.9	60	7.8	7.3	7.3	6.8
Maine-Anjou-X	10.1	54	7.5	7.3	7.2	7.1
Tarentaise-X	10.2	60	8.1	7.3	7.0	6.7
Charolais-X	10.3	63	7.2	7.4	7.3	7.3
Brown Swiss-X	10.4	61	7.7	7.4	7.2	7.2
Pinzgauer-X	10.8	60	7.4	7.4	7.2	7.1
South Devon-X	11.3	76	6.8	7.3	7.4	7.4
Hereford-Angus-X	11.3	76	7.3	7.3	7.3	7.3
Red Poll-X	11.5	68	7.4	7.4	7.1	7.3
Jersey-X	13.2	85	6.8	7.5	7.5	7.4

[a] Marbling: 5 = traces, 8 = slight, 11 = small, 14 = modest, 17 = moderate. Taste panel scores: 2 = undesirable, 5 = acceptable, 7 = moderately desirable, 9 = extremely desirable.
Source: MARC.

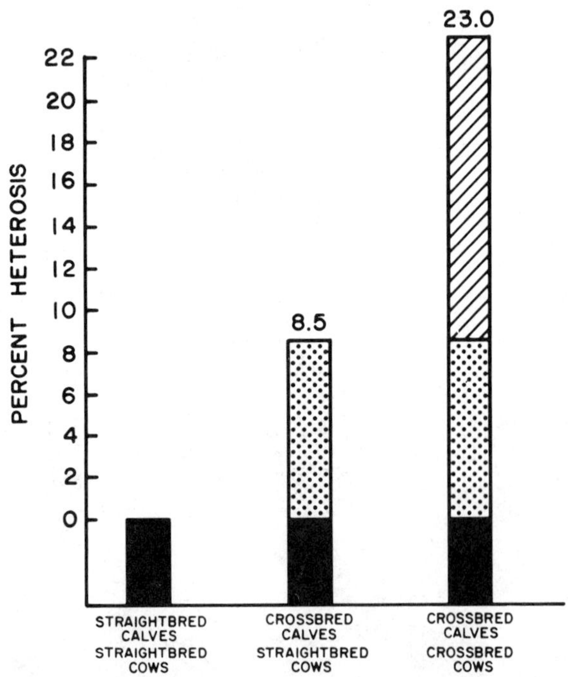

Figure 13.13. Heterosis, resulting from crossbreeding, for pounds of weaned calf per cow exposed to breeding. Courtesy of USDA.

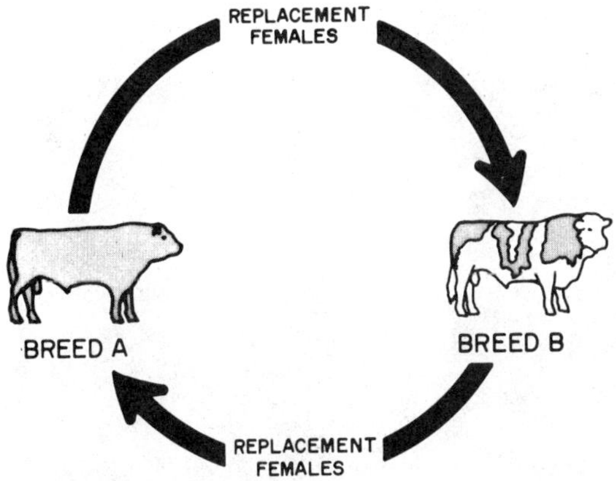

Figure 13.14. Two-breed rotation cross. Females sired by Breed A are mated to Breed B bulls and heifers sired by Breed B are mated to Breed A bulls. This will increase the pounds of calf weaned per cow exposed by approximately 15%. Drawn by CSU Graphics.

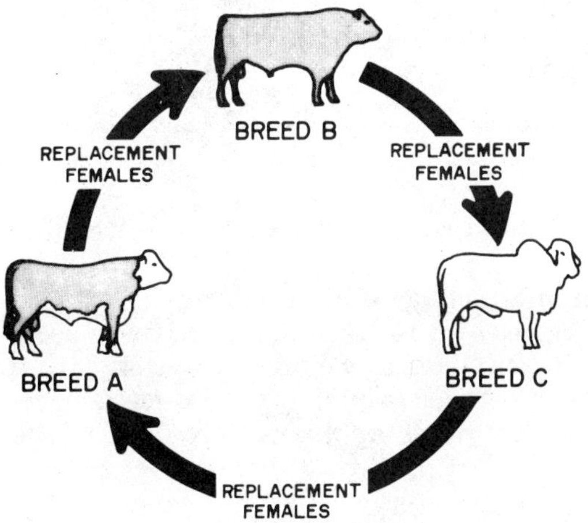

Figure 13.15. Three-breed rotation cross. Females sired by a specific breed are bred to the breed of the next bull in rotation. This will increase the pounds of calf weaned per cow exposed by approximately 20%. Drawn by CSU Graphics.

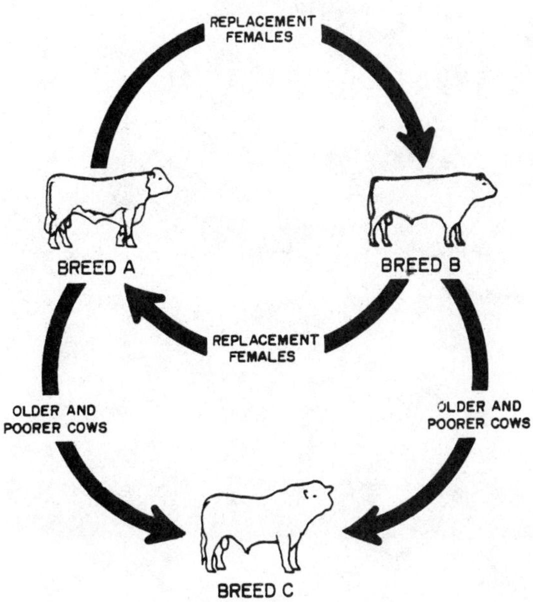

Figure 13.16. Two-breed rotation and terminal sire crossbreeding system. Sires are used in the two-breed rotation primarily to produce replacement heifers. Terminal cross sires are mated to the less productive females. This system will increase the pounds of calf weaned per cow exposed by approximately 20% to 25%. Drawn by CSU Graphics.

Figure 13.17. An example of a terminal crossbreeding program. The smaller sized Angus x Hereford cow, selected for maternal traits is mated to the Simmental bull. Growth and carcass traits have been emphasized in the selection of the bull. Courtesy of the American Simmental Association.

rate of calves, increased milk production of crossbred cows, and a higher preweaning growth rate of crossbred calves.

Consistent high levels of heterosis can be maintained generation after generation if crossbreeding systems such as those shown in Figures 13.14, 13.15, and 13.16 are used. The crossbreeding system shown in Figure 13.16 combines a two-breed rotation with a terminal cross. In this system, the two-breed rotation is used primarily to produce replacement females for the entire cow herd. In most cow herds approximately 50% of the

Figure 13.18. The Brahman-Hereford crossbred cow (pictured) is a popular cow in several southern states. These cows are bred to several breeds of bulls. Pictured is an Angus sired calf. Courtesy of L. I. Smart, Louisiana State Univ.

Table 13.13. Heritability and Heterosis for the Major Beef Cattle Traits

Traits	Heritability	Heterosis
reproduction	low	high
growth	medium	medium
carcass	high	low

cows are bred to sires to produce replacement females with the remaining 50% being bred to terminal cross sires. All terminal cross calves are sold. This crossbreeding system maintains heterosis as high as the three-breed rotation system.

In rotational crossing, breeds with maternal trait superiority (high conception, calving ease, and milking ability) would be selected. The terminal cross sire could come from a larger breed where growth rate and carcass cutability are emphasized (Figure 13.17). A primary advantage of the rotational-terminal cross system is that smaller or medium-sized breeds can be used in the rotational crossing and a larger breed could be used in terminal crossing. Some producers purchase all their maternal cross females rather than raise any replacements.

Currently the most popular crossbreeding system is the three-breed rotational cross, previously shown in Figure 13.15. The selection criteria for each of the three breeds used in this system would be similar. Extremes in size would be avoided with emphasis being placed on optimum combinations of maternal, growth, adaptability, and carcass traits (Figure 13.18).

Table 13.13 shows the heritability and heterosis for the major groups of beef cattle traits. This information demonstrates how selection and crossbreeding complement one another in a commercial breeding program.

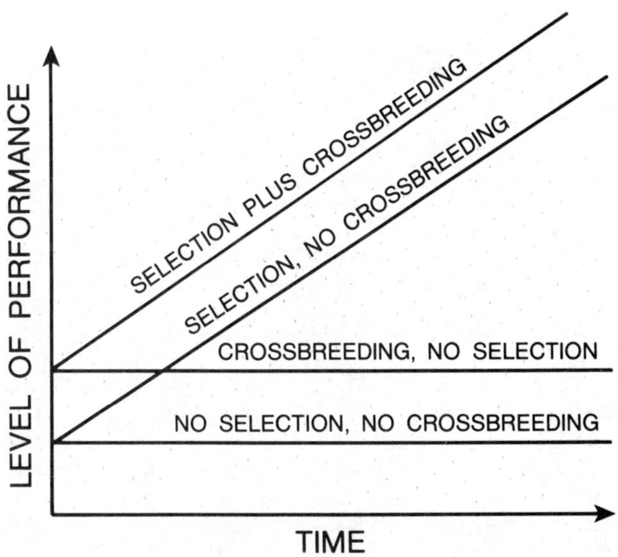

Figure 13.19. Improvement in performance with various combinations of selection and crossbreeding. Drawn by CSU Graphics.

Figure 13.20. Example of 14 years of selection and crossbreeding. These two steers, shown here prior to being slaughtered at 15 months of age, are descendents of the same straightbred Hereford herd. The unselected, straight-bred Hereford steer weighed 874 lb; the selected 4-breed crossbred steer weighed 1525 lb. Both steers would be considered too extreme for current market weight standards. Courtesy of W. T. Magee, Michigan Agric. Expt. Sta.

crossbreeding if they maintain breed purity. Therefore commercial producers have an advantage over purebred breeders in being able to utilize more of the available breeding methods to make genetic improvement (Figure 13.19).

Traits with a low heritability respond very little to genetic selection, but they show a marked improvement in a sound crossbreeding program. Carcass traits associated with growth rate (carcass weight per day of age, fat thickness, and retail product weight at the same age) will show some heterosis. The commercial producer needs to select sires carefully to improve the traits with a high heritability and use a well-planned crossbreeding program to effectively use heterosis (Figure 13.20).

SELECTED REFERENCES

Publications

Beef Research Program, Progress Report No. 1. U.S. Meat Anim. Res. Ctr. ARM-NC-21, Apr. 1982.

Briggs, H. M. and Briggs, D. M., 1980. *Modern Breeds of Livestock.* New York: The Macmillan Company.

Cundiff, L. V. and Gregory, K. E., 1977. *Beef Cattle Breeding.* USDA Agric. Info. Bul. No. 286.

Dickerson, G. E. and Willham, R. L., 1983. *Quantitative engineering of more efficient animal production.* J. Anim. Sci. 57:248 (Suppl. 2).

Gregory, K. E. and Cundiff, L. V., 1980. *Crossbreeding in beef cattle: evaluation of systems.* J. of Anim. Sci. 51:1224.

Guidelines For Uniform Beef Improvement Programs, 1981. USDA Ext. Serv. Program Aid 1020.

Lasley, J. F., 1978. *Genetics of Livestock Improvement.* Englewood Cliffs, N.J.: Prentice-Hall, Inc.

Long, C. R., 1980. *Crossbreeding for beef production: experimental results.* J. of Anim. Sci. 51:1197.

Mason, I. L., 1969. *A World Dictionary of Livestock Breeds, Types and Varieties.* Commonwealth Agric. Bureaux, Farnham Royal, Bucks, England.

Rouse, J. E., 1970. *World Cattle, I: Cattle of Europe, South America, Australia, and New Zealand.* Norman: Univ. of Okla. Press.

Rouse, J. E., 1970. *World Cattle, II: Cattle of Europe and Asia.* Norman: Univ. of Okla. Press.
Rouse, J. E., 1973. *World Cattle, III: Cattle of North America.* Norman: Univ. of Okla. Press.
Warwick, E. J. and Legates, J. E., 1979. *Breeding and Improvement of Farm Animals.* New York: McGraw-Hill.

Visuals

Yearling Bull Test. Sound filmstrip. Vocational Education Productions. Calif. Polytechnic State Univ., San Luis Obispo, CA 93407.

14
Beef Cattle Nutrition

Beef cattle nutrition involves chemical and physiological processes by which cattle utilize available feeds to maintain body functions, reproduce, milk, grow, and produce a desirable end product. Cattle have inherited certain genetic potentials; however, how well they develop those genetic potentials, in expressing the economically important traits, is highly dependent on the environment to which cattle are exposed. Nutrition is a significant part of this environment.

NUTRIENTS

A nutrient is any feed constituent which functions in the support of life. Beef cattle therefore need nutrients to live and feeds they consume provide these nutrients. Nutrients are composed of at least 20 of the more than 100 known chemical elements. These 20 elements and their chemical symbols are calcium (Ca), carbon (C), chlorine (Cl), cobalt (Co), copper (Cu), fluorine (F), hydrogen (H), iodine (I), iron (Fe), magnesium (Mg), manganese (Mn), molybdenum (Mo), nitrogen (N), oxygen (O), phosphorus (P), potassium (K), selenium (Se), sodium (Na), sulphur (S), and zinc (Zn).

The six basic classes of nutrients (water, carbohydrates, fats, proteins, vitamins and minerals) are found in varying amounts in animal feeds. These nutrient classes are briefly described.

Water

Water contains hydrogen and oxygen. The terms water and moisture are used interchangeably. Typically water refers to drinking water, while moisture is used in reference to the amount of water in a given feed or ration. The remainder of the feed, after accounting for moisture, is referred to as dry matter. Moisture is found in all feeds ranging from 10% in air-dry feeds to over 80% in fresh green forage. Cattle will consume several times more water than dry matter each day and will die from lack of water more quickly than from lack of any other nutrient. Water in feed is no more valuable than water from any other source. This is important to understand if proper assessment is to be made with feeds varying in their moisture content.

Water has important body functions because it enters into most of the metabolic reactions, assists in transporting other nutrients, helps maintain normal body temperature, and gives the body its physical shape because water is the major component within cells.

Carbohydrates

Carbohydrates contain carbon, hydrogen, and oxygen in either simple or complex forms. The more simple carbohydrates, such as starch, supply the major energy source for cattle rations, particularly feedlot rations. The more complex carbohydrates, such as cellulose, are the major components of the cell walls of plants. These complex carbohydrates are not as easily digested as simple carbohydrates.

Fats

Fats and oils, also referred to as lipids, contain carbon, hydrogen, and oxygen, although there is more carbon and hydrogen in proportion to oxygen than with carbohydrates. Fats are solid and oils are liquid at room temperature. Fats contain 2.25 times as much energy per pound as do carbohydrates.

Fats are comprised of fatty acids and glycerol.

Example: $3C_{17}H_{35}COOH + C_3H_5(OH_3) = C_{57}H_{110}O_6 + 3H_2O$
Stearic acid (a fatty acid) + glycerol = Stearin (a fat) + water

There are saturated and unsaturated fats depending on their particular chemical composition. Saturated fatty acids have single bonds tying the carbon atoms together (e.g., —C—C—C—C—) while unsaturated fatty acids have one or more double bonds (e.g., —C=C=C—C—): The term polyunsaturated fatty acids is applied to those having more than one double bond. While more than 100 fatty acids have been identified, only 3 to 4 have been determined to be dietary essentials. The two apparent functions of the essential fatty acids are: (1) precursors of prostaglandins and (2) structural components of cells.

Proteins

Proteins always contain carbon, hydrogen, oxygen and nitrogen and sometimes iron, phosphorus or sulphur or both. Protein is the only nutrient class which contains nitrogen. Proteins in feeds, on the average, contain 16% nitrogen. This is why feeds are analyzed for the percent nitrogen in the feed, with the percent multiplied by 6.25 to convert it to percent protein.

Proteins are composed of various combinations of some 25 amino acids. Amino acids are called the building blocks of the animal's body because they make up the muscle mass plus other parts. Amino acids carry the amino group (NH_2) in each of their chemical structures. There are many different combinations of amino acids that can be structured together.

Minerals

Chemical elements, other than carbon, hydrogen, oxygen, and nitrogen are called minerals. They are inorganic because they contain no carbon where organic nutrients do contain carbon. Some minerals are referred to as macro (required in larger amounts) and others are micro or trace minerals (required in smaller amounts).

Minerals have many body functions among which are bone structure, body water balance, oxygen transport and transfer, and metabolic reactions.

Vitamins

Vitamins are organic substances containing carbon, hydrogen and oxygen. There are 16 known vitamins which function in animal nutrition. Vitamins are required by beef cattle in very small amounts and they function in regulating certain body processes toward normal health, growth, and reproduction.

PROXIMATE ANALYSIS OF FEEDS

The nutrient composition of a feed cannot be determined accurately by visual inspection. A system has been devised where the value of a feed can be approximated. Proximate analysis separates feed components into groups according to their feeding value. This analysis is based on a feed sample and analysis therefore is no more accurate than how representative the sample is of entire feed source.

The groups resulting from a proximate analysis are: water, crude protein, crude fat (sometimes referred to as ether extract), crude fiber, nitrogen-free extract, and ash (minerals). Figure 14.1 shows these components resulting from a feed that would have a laboratory analysis of dry matter, 88%; protein, 13%; fat, 4%; crude fiber, 10%; and nitrogen-free extract (NFE) of 56% on a *natural or air-dry basis.* The analysis might be reported on a dry matter basis of moisture, 12%; protein, 14.8%; fat, 4.5%; crude fiber 11.4%; and nitrogen-free extract, 63.6%. Therefore caution needs to be exercised in interpreting the proximate analysis results because different laboratories report their analytical values in different ways.

The proximate analysis for the six basic nutrients does not distinguish the various components of a nutrient. For example, ash content of a feed does not tell the amount of calcium, phosphorus, or other specific minerals. Figure 14.2 gives the chemical analysis for organic and inorganic nutrients. There are specific chemical analyses for each of these nutrients in a feed if such an analysis is needed.

THE RUMINANT DIGESTIVE SYSTEM

Beef cattle are ruminants which means they ruminate or chew a bolus of feed called a cud. Cattle eat the feed and form it into boluses which pass to the stomach. After eating for awhile, cattle will usually lie down, regurgitate the boluses and proceed to chew the feed, breaking it into smaller pieces. Beef cows will normally ruminate approxi-

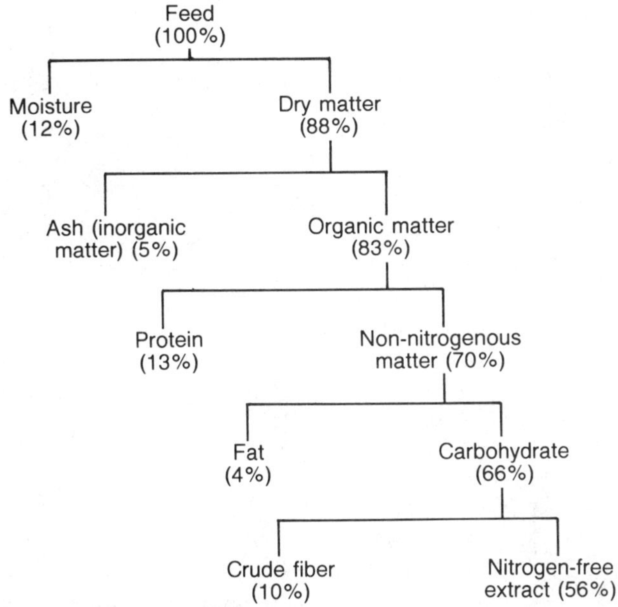

Figure 14.1. Proximate analysis showing the inorganic and organic components of a feed on a natural or air-dry basis. Courtesy of CSU.

mately eight hours each day. As the animal ruminates, large amounts of saliva (15 to 20 gallons per day) is secreted from the salivary glands in the mouth. Saliva is highly alkaline and it neutralizes (buffers) the large amount of acid produced in the ruminant stomach during digestion.

The digestive tract, sometimes referred to as the alimentary canal or tract, are those organs from the mouth to the anus through which the feed passes (Figure 14.3).

Table 14.1 identifies the primary parts of the digestive tract and some of their capacities.

The process of digestion is basically twofold: (1) to physically break the feed into smaller pieces and (2) to chemically break more complex nutrients into simpler forms which can be absorbed from the digestive tract and utilized on the cellular level by the animal. Simply stated, proteins need to be converted to amino acids; starch and other polysaccharides to simple sugars; fiber to organic acids and salts of organic acids; fats to fatty acids and glycerol; and the vitamins and minerals to many soluble forms.

Much of the gross physical breakdown of feed occurs at the time of chewing and particularly during rumination. The omasum is lined with many folds or plies (manyplies) of tissue which are believed to exert a squeezing and grinding action on the feed. There are several digestive juices and enzymes throughout the digestive tract which chemically break feed into more readily usable nutrients.

It is most important to understand some functions of the ruminant stomach since these functions make ration formulation quite unique. The large capacity of the rumen

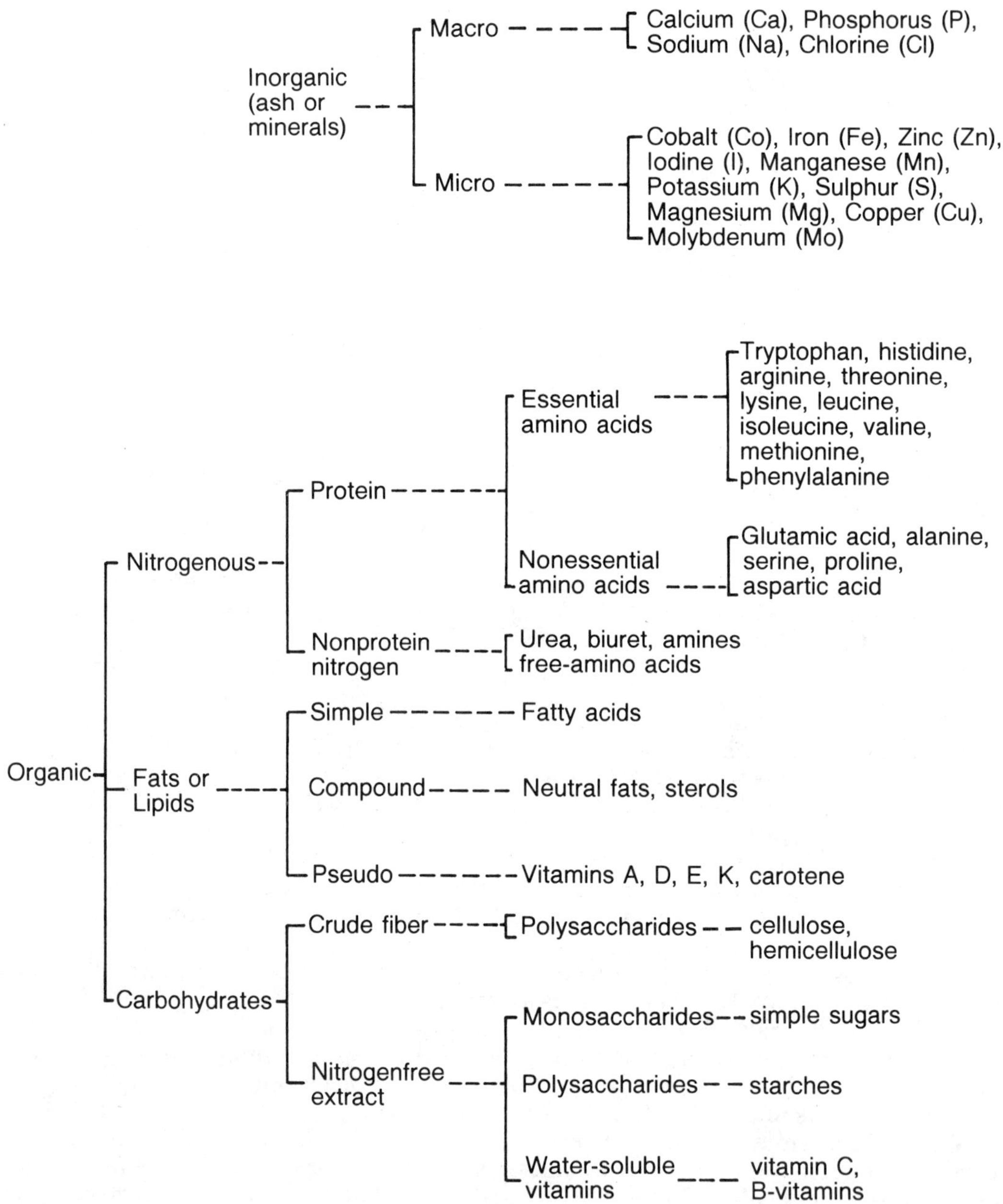

Figure 14.2. Chemical analysis scheme of inorganic and organic nutrients. Courtesy of CSU.

(Table 14.1) permits a large storage area which functions as a fermentation vat. Living in the rumen fluid are billions of microorganisms such as protozoa, bacteria, and yeasts. The kinds and numbers of these microorganisms change depending on the type of feed the animal is eating. This is why there is an animal adjustment when feeds are changed, for example, moving an animal from a high hay or grass ration to a high grain ration.

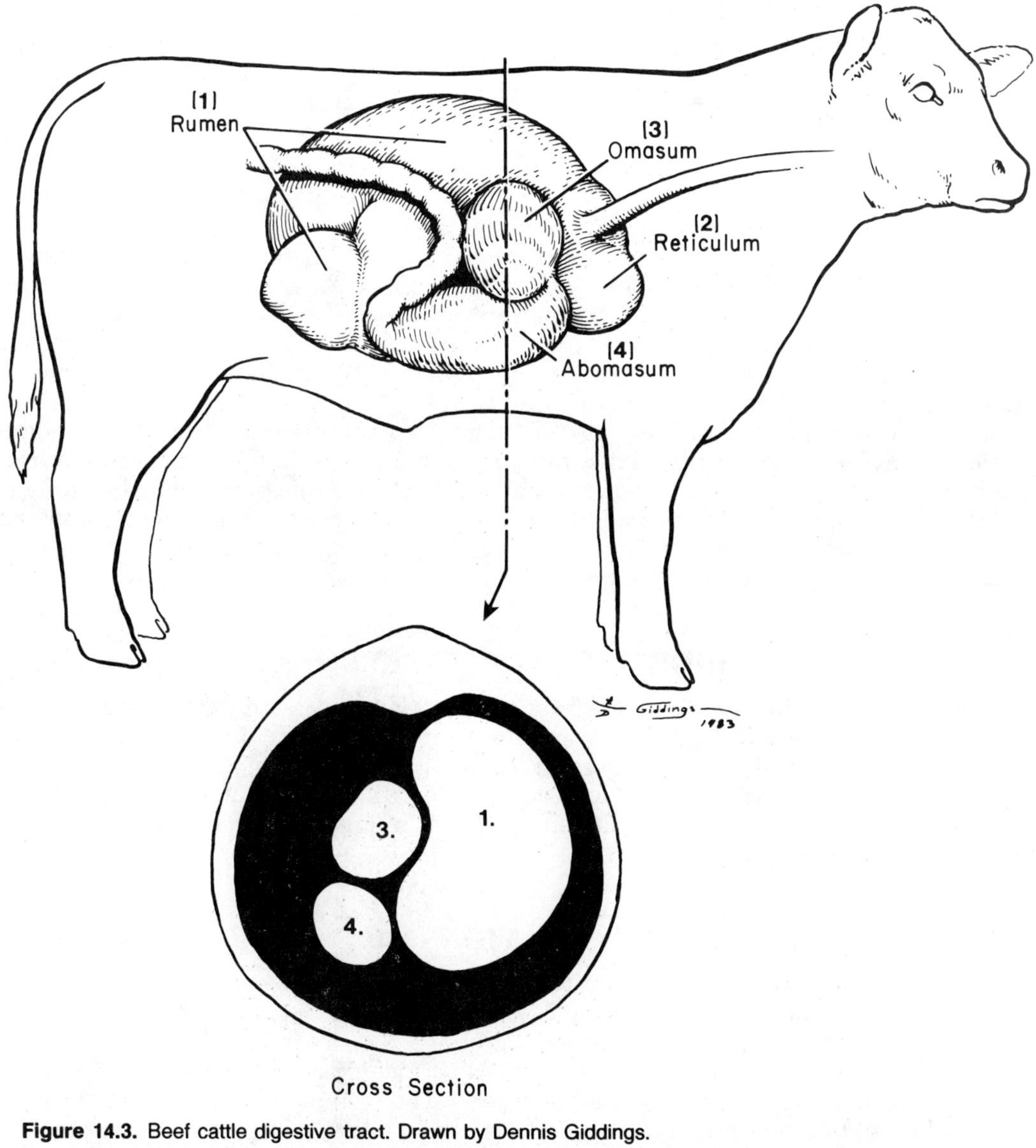

Figure 14.3. Beef cattle digestive tract. Drawn by Dennis Giddings.

Table 14.1. Parts and Capacity of the Digestive Tract of a Cow

Parts	Capacity (gallons)
Mouth	—
Esophagus	—
Stomach (four compartments)	50
Rumen (pauch)	(40)
Reticulum (honeycomb)	(2)
Omasum (manyplies)	(4)
Abomasum (true stomach)	(4)
Small intestine	15
Large intestine	10
Anus	—

Microorganisms manufacture many essential amino acids if there is an adequate supply of nitrogen and simple carbohydrates in the ration. The beef animal then digests the microorganisms, which become a protein source to the animal. Microorganisms also make many B vitamins which therefore do not have to be critically assessed in beef cattle ration formulation. Microorganisms also assist in breaking down complex carbohydrates (crude fiber) which permit the ruminant to consume large amounts of grass and hay.

The rumen fermentation process also provides for microorganisms to convert some simple carbohydrates (starch) to volatile fatty acids such as acetic, propionic, and butyric. In this process, methane gas is produced which the animal releases primarily through belching. Occasionally the gas-releasing mechanism does not function properly and gas accumulates in the rumen causing a condition called bloating to occur. Death will result if gas pressure builds to a high level and interferes with adequate respiration.

DIGESTIBILITY OF FEEDS

Digestibility refers to the amount of various nutrients in a feed that are absorbed from the digestive tract. Different feeds and nutrients vary greatly in their digestibility. Many feeds have been subjected to digestion trials, where feeds of known nutrient composition have been fed to cattle. Feces have been collected and the nutrients in the feces analyzed. The difference between nutrients fed and nutrients excreted in the feces is the apparent digestibility of the feed.

ENERGY EVALUATIONS OF FEEDS

Energy needs of cattle account for the largest portion of feed consumed. Several systems have been devised to evaluate feedstuffs for their energy content. The most commonly used energy system is Total Digestible Nutrients (TDN). It is typically expressed in pounds, kilograms, or percent. TDN is calculated after obtaining the proximate analysis and digestibility figures for a feed. The formula for calculating TDN is TDN = (digestible crude protein) + (digestible crude fiber) + (digestible nitrogen-free extract) + (digestible crude fat $\times$ 2.25). TDN is roughly comparable to Digestible Energy

(DE) but it is expressed in different units. TDN and DE both tend to overvalue roughages compared to the Net Energy system.

Even though there are some apparent shortcomings in using TDN as an energy measurement of feeds, it works well in balancing rations for cows and growing cattle. There is more precision in the energy measurement of feeds by using the Net Energy (NE) system. This system usually measures energy values in megacalories per pound or kilograms of feed. The calorie basis, which measures the heat content of feed, is as follows:

Calorie (*Cal*)—amount of energy or heat required to raise one gram of water 1° C.
Kilocalorie (*Kcal*)—amount of energy or heat required to raise one kilogram of water 1° C.
Megacalorie (*Mcal*)—equal to 1000 kilocalories or 1,000,000 calories.

Figure 14.4. shows various ways that the energy of feeds is utilized by cattle. Also shown are various energy measurements of feeds which reflect the types of energy

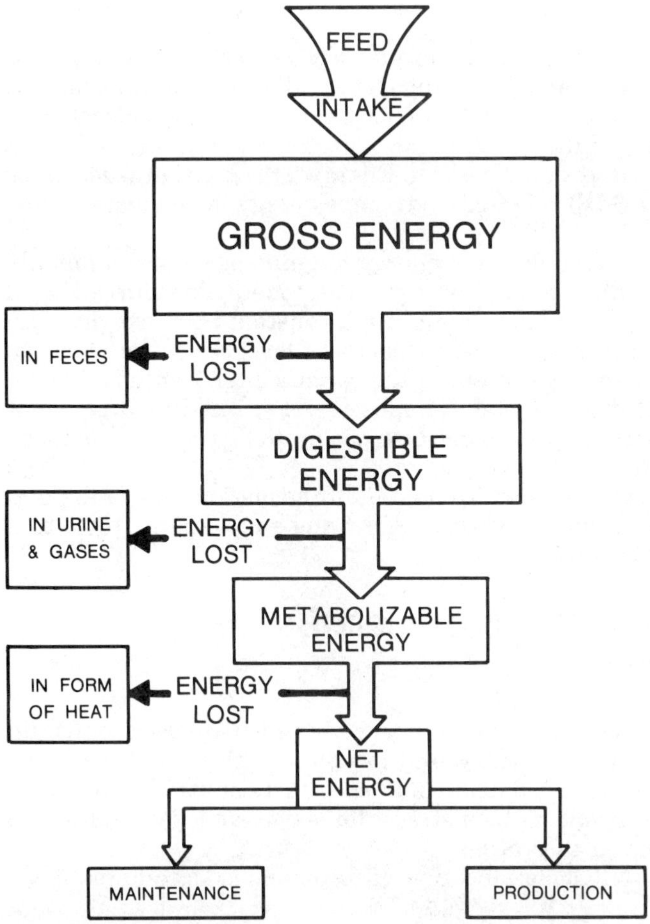

Figure 14.4. Measures of energy and energy utilization by cattle. Drawn by CSU Graphics.

Table 14.2. Comparison of TDN and NE Systems for a Yearling Steer With Varying Rates of Gain

	Rate of Gain (lbs)		
Energy System	**0**	**2**	**2.9**
TDN (lbs)	6.40	12.80	15.00
NEm (Mcal)	6.24	6.24	6.24
NEg (Mcal)		4.29	6.48
Total lbs feed			
TDN basis	8.2	16.5	19.3
NE basis	7.2	15.0	19.0
Feed per lb gain (NE)		7.5	6.6
Feed per lb gain (TDN)		8.2	6.6

utilization by the beef animal. Gross Energy (GE) is the quantity of heat (calories) released from the complete burning of the feed sample in an apparatus called a bomb calorimeter. GE has little practical value in evaluating feeds for cattle because the animal does not metabolize feeds in the same manner as a bomb calorimeter. For example, oat straw has the same GE value as corn grain. Digestible Energy (DE) is GE of feed minus fecal energy. Metabolizable Energy (ME) is GE of feed minus energy in the feces, urine, and gaseous products of digestion.

NEm (net energy for maintenance) and NEg (net energy for gain) are more commonly used for formulating rations for feedlot cattle than any other energy system. NEm in feedlot cattle is the amount of energy needed to maintain a constant body weight. Animals of known weight, fed for zero energy gain, have a constant level of heat production.

The NEg measures the increased energy content of the carcass after feeding a known quantity of feed energy. All feed fed above maintenance is not utilized at a constant level of efficiency. Higher rates of gain require more feed per unit of gain since composition of gain varies with rate of gain.

The TDN and NE systems are compared in Table 14.2. Information in the table is based on feeding a simple ration of ground ear corn (90%) and supplement (10%) to a yearling steer (772 lbs) for different rates of gain.

FEEDS AND FEED COMPOSITION

Classification of Feeds

Feeds are naturally occurring ingredients in rations of cattle which are used to sustain life. The terms feeds and feedstuffs are generally used interchangeably; however, feedstuffs is a more inclusive term. Feedstuffs can include certain nonnutritive products such as additives to promote growth and reduce stress, for flavor and palatability, to add bulk, or to preserve other feeds in the ration.

The National Research Council (NRC) classification of feedstuffs is as follows:

1. Dry roughages and forages
   - Hay (legume and nonlegume)
   - Straw
   - Fodder
   - Stover
   - Other feeds with greater than 18% fiber (hulls and shells)
2. Range, pasture plants, and green forages
3. Silages (corn, legume, and grass)
4. Energy feeds (cereal grains, mill by-products, fruits, nuts, and roots)
5. Protein supplements (animal, marine, avian, and plant)
6. Mineral supplements
7. Vitamin supplements
8. Nonnutritive additives (antibiotics, coloring materials, flavors, hormones, and medicants)

Roughages and forages are used interchangeably although roughage usually implies a bulkier, coarser feed. In the dry state, roughages have more than 18% crude fiber. The crude fiber is primarily a component of cell walls which is not very digestible. Roughages are also relatively low in TDN, although there are exceptions, for example, corn silage which has over 18% crude fiber and approximately 70% TDN.

Feedstuffs that contain 20% or more protein are classified as protein supplements, such as soybean meal and cottonseed meal. Feedstuffs having less than 18% crude fiber and less than 20% protein are classified as energy feeds. Concentrates, such as the cereal grains, are typical energy feeds which is reflected by their high TDN values.

Nutrient Composition of Feeds

Feeds are analyzed for their nutrient composition. The ultimate goal of nutrient analysis of feeds is to predict the productive response of cattle when they are fed rations of a given composition.

The nutrient composition of some of the more common feeds utilized by cattle is shown in Table 14.3. The information in Table 14.3 represents averages of numerous feed samples. Feeds are not constant in composition and an actual analysis should be obtained whenever economically feasible. The actual analysis is not always feasible or possible because of lack of available laboratories and insufficient time to obtain the analysis. Therefore, feed analysis tables become the next best source of reliable information on nutrient composition of feeds. It is not uncommon to expect the following deviations of actual feed analysis from the table values for several feed constituents: crude protein (± 15%), energy values (± 10%), and minerals (± 30%).

Digestible protein is included in many feed composition tables, but because of the large contribution of body protein to the apparent protein in the feces, digestible protein is more misleading than crude protein. For this reason, crude protein is more commonly used in formulating rations.

Digestible protein (DP) can be calculated from crude protein (CP) content by using the following equation (% DP and % CP are on a dry matter basis):

$$\% \text{ DP} = 0.9 (\% \text{ CP}) - 3$$

Table 14.3. Nutrient Composition of Selected Feeds Commonly Used in Beef Cattle Rations

Feed	Dry Matter Percent	On a Dry Matter Basis (Moisture-free)							
		TDN Percent	ME Mcal/lb	NEm Mcal/lb	NEg Mcal/lb	Crude Protein Percent	Vit. A (Carotene) mg/lb	Calcium Percent	Phosphorus Percent
Alfalfa hay (early bloom)	90.0	57.0	0.94	0.55	0.25	18.4	57.8	1.25	0.23
Alfalfa (dehydrated)	92.7	62.0	1.02	0.60	0.33	19.2	78.8	1.43	0.26
Alfalfa (grazed)	27.2	57.0	0.94	0.55	0.25	19.3	90.4	1.72	0.31
Barley	89.0	83.0	1.36	0.97	0.64	13.0	—	0.09	0.47
Beet tops (ensiled)	20.7	54.0	0.89	0.52	0.19	12.7	—	2.32	0.20
Beet pulp (with molasses)	77.0	89.0	1.22	0.93	0.62	8.7	—	0.21	0.04
Bermudagrass (grazed, Coastal	31.0	66.0	—	0.67	0.38	14.6	150.2	0.49	0.27
Bermudagrass, hay, Coastal	91.5	49.0	0.80	0.48	0.10	9.5	—	0.46	0.18
Birdsfoot Trefoil (grazed)	25.0	63.0	1.23	0.63	0.35	20.2	—	1.76	0.20
Bluegrass (grazed, Kentucky)	30.6	58.0	0.95	0.56	0.27	17.0	86.4	0.39	0.39
Bone meal (steamed)	95.0	16.0	—	—	—	12.7	—	30.51	14.31
Brome (grazed, early vegetation)	32.5	63.0	1.04	0.62	0.34	20.3	208.8	0.59	0.37
Brome (grazed, mature)	56.1	50.0	0.82	0.48	0.12	6.4	—	0.30	0.26
Buffalo grass	47.7	56.0	0.92	0.54	0.23	9.2	42.6	0.52	0.16
Clover, Red (Hay)	87.7	59.0	0.97	0.57	0.28	14.9	16.7	1.61	0.22
Corn (No. 2 dent)	89.0	91.0	1.50	1.04	0.67	10.0	0.91	0.02	0.35
Corn (ground ear)	87.0	90.0	1.48	1.01	0.63	9.3	—	0.05	0.31
Corn silage	27.9	70.0	1.15	0.71	0.45	8.4	—	0.28	0.21
Corn stover (without ears and husks	87.2	59.0	0.97	0.55	0.25	5.9	—	0.49	0.09
Cottonseed meal	93.5	74.0	1.22	.75	0.49	42.4	—	0.20	1.09
Dicalcium phosphate	96.0	—	—	—	—	—	—	23.1	18.65
Grama (early vegetation)	41.0	64.0	1.05	0.63	0.36	13.1	—	0.54	0.19
Grama (mature)	63.0	58.0	0.95	0.56	0.27	6.5	13.8	0.34	0.12

Lespedeza (grazed, early bloom)	25.0	60.0	0.99	0.59	0.30	16.4	—	1.35	0.21
Lespedza (hay)	93.0	57.0	0.94	0.54	0.25	15.7	—	1.19	0.26
Limestone (ground)	100.0	—	—	—	—	—	—	33.84	0.02
Meadow hay (native, intermountain)	92.9	51.0	0.84	0.50	0.14	9.1	—	0.57	0.17
Milk	12.0	130.0	2.14	2.09	0.91	25.8	—	0.93	0.75
Milo	89.0	80.0	1.31	0.84	0.56	12.4	—	0.04	0.33
Molasses (cane)	75.0	72.0	1.25	2.27	1.48	4.3	—	1.19	0.11
Oats (grain)	89.0	76.0	1.25	0.79	0.52	13.2	—	0.11	0.39
Oats (silage)	31.7	59.0	0.97	0.58	0.29	9.7	54.3	0.37	0.30
Prairie hay (midbloom, midwest)	91.0	51.0	0.84	0.50	0.14	8.1	9.1	0.34	0.21
Reed Canarygrass (grazed)	25.8	56.0	0.92	0.54	0.23	13.2	—	0.40	0.30
Sagebrush (grazed)[a]	75.0	60.0	0.91	0.45	0.18	8.5	7.2	0.60	0.20
Saltbrush, Shadscale	80.0	—	0.40	0.30	0.00	7.7	8.9	2.53	0.09
Sorghum Stover (without heads)	85.1	57.0	0.94	0.55	0.25	5.3	—	0.40	0.11
Sorghum, Sudan grass (grazed, midbloom)	22.7	63.0	1.04	0.62	0.35	8.7	—	—	—
Soybean meal (solvent)	89.0	81.0	1.33	0.88	0.59	51.5	—	0.36	0.75
Soybean (straw)	87.6	38.0	0.62	0.85	0.00	5.5	—	1.59	0.06
Sweet Clover (hay)	87.2	57.0	0.94	0.55	0.25	16.3	56.5	1.77	0.26
Timothy (hay, mid-bloom)	88.4	57.0	0.90	0.56	0.26	8.5	24.3	0.41	0.19
Wheat (hard, red winter)	89.1	88.0	1.45	0.98	0.64	14.6	—	0.06	0.57
Wheat (grazed, early vegetative)	21.5	73.0	1.20	0.75	0.49	28.6	236.4	0.42	0.40
Wheat (straw)	90.1	48.0	0.79	0.47	0.09	3.6	1.0	0.17	0.08
Wheatgrass, crested (early vegetation)	30.8	67.0	1.10	0.66	0.40	23.6	197.1	0.46	0.35
Wheatgrass, crested (post ripe)	80.0	49.0	0.80	0.47	0.10	3.1	0.09	0.27	0.07
Winterfat (grazed)	76.0	—	0.54	0.35	0.00	11.0	7.6	2.14	0.12

[a] Average of several species.

Source: National Research Council, 1976. *Nutrient Requirements of Beef Cattle,* and Preston, 1984.

Four measures of energy values (TDN, ME, NEm and NEg) are shown in Table 14.3. TDN is shown because there are more TDN values for feeds and because TDN has been a standard system of expressing the energy value of feeds for cattle. Some individuals desire to have ME (metabolizable energy) values for feed because the values are in calories rather than pounds. NEm and NEg values are needed primarily to formulate feedlot ration as these values offset the major problem associated with the TDN energy system.

Management for High Nutrient Content of Feeds

How feeds are grown, harvested, and preserved can have a significant effect on total nutrients produced per unit of land, nutrient content of feeds, and safety of the feeds at the time they are being fed to cattle. Some major management practices utilized to enhance nutrient content of feeds will be discussed here.

Hays

Hays are harvested from legumes, grasses, and cereal crops. Alfalfa hay comprises nearly 60% of more than 60 million acres of hayland in the United States.

The primary objective in haymaking is to reduce moisture content of green forage, which ranges from 65% to 85%, down to approximately 20%. Hay which is stored with excessive moisture can be destroyed by fire resulting from spontaneous combustion or it will mold making it unsuitable for feeding purposes. Hay which is harvested too dry will likely lose many leaves which are much higher in protein, carotene, and other

Table 14.4. Proposed Market Grades for Legume Hay (USA Hay Marketing Task Force)

		Description				Typical Composition (dry basis)	
Grade	Stage of Maturity	Leaves %	Color	Foreign Matter %	Dust, Mold, Odor	Protein %	ADF[a] %
1	*Late vegetative* bud to first flower	40–50	Green	0–5	OK	>19	<31
2	*Early bloom* Initiation of bloom to one-half bloom	35–40	Light Green	<10	OK	17–19	31–35
3	*Mid-bloom* (mid-to full bloom); more than one-half bloom	25–40	Yellowgreen to Green	<15	OK	13–16	36–41
4	*Full bloom* (full and beyond)	<30	Brown to Green	<20	Slight	<13	>41
6	Contains toxic weeds or hardward, or has bad odor, or is heat damaged, hot, wet musty moldy, badly weathered, dusty, extremely overripe, or contains more than 20% foreign material or less than 80% dry matter.						

[a] Acid detergent fiber (a measure of fibrous and indigestible matter).

nutrients compared to the stems. Also hay which is bleached by the sun has a reduced carotene content. Bleaching of hay can be most extensive on hay left too long in the windrow and it also occurs with hay which is stored outside.

Maturity at harvest has one of the most pronounced effects on nutrient composition. Crude protein content of alfalfa shows the following values on a dry matter basis: immature, 21.5%; prebloom, 19.4%; early bloom, 18.4%; midbloom, 17.1%; full bloom, 15.9%; and mature, 13.6%.

Hay exposed to moisture during the haymaking process can be subject to extensive leaching of nutrients from the plant material. Studies have shown alfalfa hay exposed to 0.8 inch of moisture in the swath can have the dry matter yield reduced 10% and cellular proteins and carbohydrates reduced from 10% to 20%. The leaching losses from this amount of moisture were not as large as allowing the maturity of plants to advance 7 days.

The USDA has established grading standards for hay based on subjective evaluations of leafiness, color, and amount of foreign material (weeds, stubble, sticks, dirt, and rocks) in the hay. These standards have minimum use in officially grading hay, however, the descriptive terms are commonly used by producers in evaluating hay quality. Table 14.4 shows the new proposed market grades for legume hay which would be based on additional, objective measurements. Market hay grades have also been proposed for grass and grass-legume hays (*J. Anim. Sci.* 47:747), as shown in Tables 14.4 and 14.5.

It is not uncommon to have a 30% to 50% loss of dry matter from the combination of losses from harvesting, storing, and feeding. Small bales of hay can be stored in well constructed stacks which can make significant reductions in nutrient storage losses (see Appendix). Losses from large round hay bales can be much greater than losses from smaller, square bales unless managed properly. A study in a 19 inch rainfall area showed the following losses in storing large round bales: pyramid, 10.3%; individu-

Table 14.5. Proposed Market Hay Grades for Grass Hay (USA Hay Marketing Task Force)

		Description				Typical Composition (Dry Basis)	
Grade	**Stage of Maturity**	**Leaves %**	**Color**	**Foreign Matter %**	**Dust, Mold, Odor**	**Protein %**	**ADF[a] %**
2	*Late vegetative* grass heads in boot	>50	Green	<5	OK	>18	<33
3	*Early head (early bloom)* less than one-half headed	>40	Light Green to Green	<10	OK	13–18	33–38
4	*Headed (late bloom)* seeds begin to form	>30	Yellowgreen to Green	<15	Slight	8–12	39–41
5	*Post head (dough stage)* and beyond)	>20	Brown to Green	<20	Slight	<8	>41
6	Contains toxic or noxious weeds, has objectionable odor, heat damaged, wet, moldy or badly weathered.						

[a] Acid detergent fiber (a measure of fibrous and indigestible matter).

Figure 14.5. Large, round hay bales can be satisfactorily fed to reduce the amount of hay wasted. Courtesy of Elanco Products Company.

ally separated, 4.0%; and end-to-end arrangement, 0.8%. It is extremely important to use feeding systems which reduce the amount of hay which is trampled and wasted (Figure 14.5).

Crop Residues

Numerous kinds of crop residues can provide a substantial amount of roughage and forage for beef cattle. Most common of these crop aftermath feeds which can be harvested or grazed are corn stalks, straw, grain, or sorghum stubble and soybean residue. It is estimated that the 75 million acres of harvested corn and 35 million acres of harvested grain sorghum would produce approximately 200 million tons of crop residue each year. This could supply the winter energy needs for 35 to 40 million cows. An additional 75 million tons of cereal grain residue (primarily straw) is available in the United States.

Crop residues, because of their relatively low availability of metabolizable energy, are most effectively utilized in maintenance rations for gestating cows when the energy requirement is less than during lactation. Typically the crop residues are low in protein and phosphorus, marginal in calcium, and high in fiber and lignin.

Grazing crop aftermath is usually the most economical method to utilize feed; however, harvested material can be used if supplemented properly (Figure 14.6). Snow and mud interfere with effective utilization of grazed feed. During the past several years, more crop residues have been used for cattle feed due to the combined efforts of producers, researchers, and equipment manufacturers.

Silages

Silages are produced from green forage crops which are compressed and stored under anaerobic (without oxygen) conditions in upright or horizontal silos. The feed is preserved through a process of acid fermentation which prevents mold and other spoilage from occurring.

The most common crops used for silage are corn, sorghums, grasses, legumes, and

Figure 14.6. Corn stalk fields are available to cattle after the grain has been harvested. Courtesy of BEEF Magazine.

some small grains. Haylage is a forage which is intermediate between hay and silage in moisture content (40% to 60%), while the moisture content of silage is usually 60% to 75%.

Silage comprises a large part of the feeding program of many beef cattle programs primarily because amount TDN and beef produced per acre is greatest with silage. Caution should be exercised, however, because silage is one of the most variable feedstuffs in nutrient content and storage loss.

Nine factors which interact together in affecting silage quality are type of silage crop, weather, stage of maturity, moisture, additives, fineness of chopping, packing, and filling the silo, silo structure, and feed-out of the silage.

Moisture content of silage is the single most important factor affecting silage quality. Ensiling above 72% to 74% moisture can produce seepage, undesirable butyric acid, high fermentation losses and reduced intake of silage by cattle. Ensiling below 50% to 55% moisture creates problems in eliminating air (poor compaction) and achieving sufficient fermentation and low pH.

An optimum stage of maturity should be selected which will make the best compromise between increased dry matter yield and decreased digestibility. Table 14.6 gives recommended maturity stage for several silage crops.

A fine chop, one-fourth to one-half inch, is recommended for most crops. Hollow-strawed cereals need to be chopped finer than solid-stemmed corn and sorghum. Crops that are chopped too coarse will have more air trapped when ensiled and more silage will be refused by cattle when it is fed.

Type of silo can affect quality of silage as losses can be 30% or higher (Table 14.7). Silo construction cost and silage losses need to be compared.

Table 14.6. Stage of Maturity Recomendations for Several Silages

Crop	Recommended Stage of Maturity
Alfalfa	Late bud to 1/10th bloom
Cereal grains (barley, oats and wheat)	Soft-dough kernels
Corn	Kernels fully dented
Grasses (summer and perennial)	Prior to head emergence
Sorghum	Soft-dough kernels

Packing and filling the silo should be done to eliminate air pockets, exposing a small surface area. After filling a trench or bunker silo the surface should be sealed to keep out air, rain, and snow thus reducing spoilage loss.

Producers should grow those silage crops which are best suited agronomically for their area, selecting those crops which can economically produce the greatest amount of TDN per acre.

Feed-out should allow silage to be fresh and not exposed to air for extended periods of time. This is especially critical when the weather is warm which stimulates mold growth and other spoilage organisms.

There are four general categories of silage additives: (1) acids for direct acidification, (2) preservatives (sterilants and fatty acids), (3) feedstuffs (molasses, NPN, grain, and so forth), and (4) fermentation aids (enzymes, innoculants and antioxidants). Before using any additive, costs and returns should be carefully weighed. Claims made for additives should be based on well-documented research, not on emotion. Additives should not be used as a substitute for poor management.

Amount of grain and moisture in silage have major influences on its feed value and both should be used to determine the dollar value of silage. Value of corn silage as a standing crop can be determined in the following three ways: (1) leave a few rows for later grain harvest to determine grain yield per acre, (2) harvest by hand a small plot (1/100 acre) and determine grain yield, or (3) use relationship of one ton of 30% dry matter silage for each harvested foot of plant, not counting the tassel. The latter assumes an average stand so adjustments would have to be made for light or heavy stands.

Corn grain yield estimates in silage can be used to estimate value of the silage. For example, if it is estimated there are 6 bushels of number two corn in a ton of wet

Table 14.7. Silo Type and Silage Losses

Type of Silo	Percent Dry Matter Loss
Oxygen-limiting	3 to 8
Concrete upright	5 to 15
Trench or bunker	12 to 25
Open stack	20 to 40

silage (30% to 35% dry matter), then the cash corn price times the 6 bushel plus harvesting costs would give a reasonable estimate on value of the silage (e.g., corn selling for $3 per bu × 6 bu corn per ton = $18 + $4 harvest cost = $22 for 35% dry matter silage).

If the amount of grain in silage is not known, a rough estimate of silage value is 8 times the market value of the corn grain. If silage is already in storage, the silage is worth about 10 times the market price of corn.

Sorghum silage should be priced similar to corn of the same moisture content. Forage sorghums with fairly high grain yield usually have 80% to 90% the value of corn silage per unit of dry matter. Sudan and Sudan-sorghum crosses or varieties, with low grain yields, may have only 65% to 80% the value of corn silage per unit of dry matter.

Grains and Concentrates

Contentrates are high-energy feeds, mostly feed grains and their by-products, which contain less than 20% protein and less than 18% fiber.

Corn and sorghum (milo) are the most common feed grains in cattle rations with barley, oats, wheat and other concentrates being of lesser importance.

Table 14.8 shows the relative feeding values of various grains.

The relative value, using the information in Table 14.8, is calculated in the following manner: Relative value of grain × corn cost per lb = relative cost of grain per lb. For example, to determine the price one could pay for barley relative to corn if corn was $3.10 per bu or 5.5¢ per lb (price of corn) = 4.95¢ per lb or 48 lb per bu × 4.95¢ per lb = $2.38 per bu for barley. One would purchase barley if the price was less than $2.38 per bu compared to corn at $3.10 per bu.

The maximum level recommendations for the grains listed in Table 14.8 are for high concentrate rations. They would be higher for oats, rye, and wheat if grain was being limit-fed in the ration.

Feedlot cattle consume most feed grains, although concentrates are frequently fed to cows during 1 to 3 months of the calving-breeding seasons, when energy requirements are highest.

Table 14.8. Relative Feeding Value of Various Grains for Cattle

Grain	Relative Value, Percent (lb for lb Basis)	Maximum Level, Percent	Bushel wt (lbs)
Barley	88–90	100	48
Corn	100	100	56
Corn and cob meal	85–95	100	70
Milo	85–95	100	56
Oats	70–90	25	32
Rye	80–85	20	56
Wheat	100–105	50	60

Feed grains are harvested and stored in a relatively dry condition (less than 20%); otherwise the grain will spoil and lose its nutrients and feeding acceptability to cattle.

Most grains are processed in one of the following more commonly used methods: (1) dry process (ground or rolled), (2) steam flaking, (3) oxygen-limited reconstitution (adding of water, then storing in air-tight area), (4) early harvested—ground and ensiled in trench or air-tight silos, (5) early harvest with acid treatment of grain, and (6) pelleted. Corn is the only grain which can be fed satisfactorily in whole form to feedlot cattle. Whole corn is usually fed to feedlot cattle on high (80% to 90%) concentrate rations. It is essential that milo be processed to improve its digestibility. Other advantages in processing grains are improved gain and efficiency of gain and increased feed intake by cattle. Using high-moisture grains reduces field losses during time of harvesting.

Increased animal performance must be tempered with energy costs in processing. Currently, steam flaking is the most intensive grain processing alternative; however, it is feasible only in larger sized feedlots (20,000 head and above).

Caution should be exercised in keeping wheat below 40% of the total ration. It can cause compaction if ground too fine and acidosis if cattle are not accustomed to it.

Protein Supplements

Protein supplements are dry or liquid feedstuffs which contain 20% or more protein. They may be of plant origin (soybean meal, cottonseed meal, linseed meal, corn gluten meal, and others) or animal origin (tankage, meat and bone scraps, blood meal, dried skim milk, and others). Molasses is commonly used as the base for liquid supplements.

Soybean meal is the most widely used protein supplement in the United States. Cottonseed meal, however, is commonly used in the southern United States since most of the cotton is grown in that area.

Urea is not a protein supplement but it is a source of nitrogen (42% to 45%) for protein synthesis by rumen bacteria. Urea has a protein equivalent of 262% to 281% since 1 lb of urea contains as much nitrogen as 2.62 lb to 2.81 lb of protein. Urea works well with other plant proteins to lower the cost of protein in the ration. Urea works best in feedlot rations which are relatively high in energy since rumen organisms need a readily available energy source to convert the urea nitrogen into microbial protein.

General guidelines for using urea or other sources of NPN (non-protein nitrogen) are: (1) maximum of one-third of total nitrogen in the ration from urea, (2) urea to be not over 10% to 15% of a typical protein supplement, (3) not over 1% of diet or 3% of concentrate mixture should be urea, and (4) not over 5% of supplement if urea is to be used with low grade roughages. Urea poisoning can occur if urea level in the ration is too high or is not mixed properly.

Vitamin Supplements

The only vitamin of general practical importance is vitamin A value (vitamin A and carotene). Vitamin A or carotene content in feeds depends largely on maturity, harvest conditions, and length and conditions of storage. Where roughages being fed contain a bright green color or are being fed as immature fresh forage (e.g. pasture),

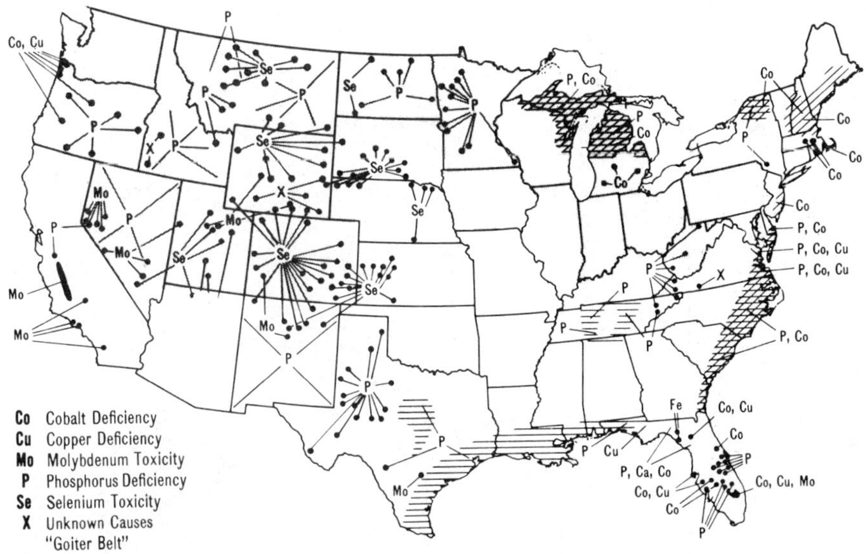

Figure 14.7. Areas of mineral deficiency or toxicities (USDA).

there will probably be sufficient vitamin A value in these roughages to meet the animal's requirement. If there is any serious question about vitamin A availability, periodic vitamin A shots are relatively cheap or a feed analysis may be needed.

Mineral Supplements

Mineral supplements can be very simple or quite complex, especially in the latter situation where most of the trace minerals are included. In most areas of the United States, mineral supplementation of salt, phosphorus, and calcium prove to be adequate. The level of many trace minerals in feeds is largely determined by the level in the soil on which feeds are grown or to some other environmental factors. There are known areas where mineral deficiencies or toxicities do exist (Figure 14.7). In these areas, additional mineral supplememtation should be included and methods to reduce toxicity effects should be employed.

For the cow herd, dicalcium phosphate and steam bone meal are two calcium-phosphorus sources that are readily available. Mixtures of two parts salt with either calcium-phorphorus source are recommended for most areas and times of the year. If phosphorus is very deficient, then equal parts of salt and a phosphorus source is recommended.

Trace minerals, if necessary, are best supplied in a trace mineralized salt. The salt and calcium-phosphorus mixture would be the same as those just previously mentioned.

Implants and Additives

Some commonly used implants and feed additives used for calves and growing-finishing cattle are shown in Table 14.9.

Table 14.9. Selected Implants and Feed Additives for Growing and Finishing Cattle

Common or Trade Name	Name or Content	Claims	Recommended Level	Withdrawal	% Improvement in Gain	% Improvement in Efficiency
Implants						
Compudose	Estradiol–17B	Stimulates pituitary to produce higher levels of growth hormone. Effective over a 200 day period.	Each dose contains 24mg of Estradiol–17B. For steers only.	None	Suckling calves 5 Growing cattle 9 Feedlot 17	7
Ralgro	Zeranol	Stimulates pituitary to produce higher levels of growth hormone.	36 mg can be given at birth and every 65 to 100 days thereafter. Not for breeding cattle.	No implants within 65 days of slaughter.	10–18	6–9
Synovex-H	200 mg testosterone Propionate and 20 mg Estradiol Benzoate	Stimulates pituitary to produce higher levels of growth hormone.	One dose for heifers weighing 400 to 800 lbs Reimplant every 60 to 100 days.	None	10–18	6–9
Synovex-S	200 mg progesterone (per dose) 20 mg Estradiol Benzoate (per dose)		One dose for steers weighing 400 to 1000 lbs Reimplant every 60 to 100 days.			
Steeroid	Same as Synovex-S				10–18	6–9

Table 14.9. (*Continued*)

Common or Trade Name	Name or Content	Claims	Recommended Level	Withdrawal	% Improvement in Gain	% Improvement in Efficiency
Additives						
Aureomycin	Chlortetracycline	Improve gain, feed efficiency and reduces liver abscesses.	0.1 mg to 0.5 mg per lb of body wt. per day (or 70 mg/hd/day).		3–5	3–5
Bovatec	Lasalocid Sodium	An antibiotic which increases the proportion of propionic acid in the rumen. Decreases occurrence of acidosis and bloat.	30 gm per ton of feed.	None	5	8
MGA	Melengestrol Acetate	A synthetic progesterone which inhibits estrus.	0.25 to 0.50 mg per day.	48 hrs before slaughter.	5–11	5
Rumensin	Monensin Sodium	An antibiotic which increases the proportion of propionic acid in the rumen. Decreases occurrence of acidosis.	Feedlot—25 gms per ton air dry feed (200 to 300 mg per head per day. Not for breeding cattle. Pasture—max 400 gm per ton of complete feed.	None	0	5–12
Terramycin	Oxytetracycline	Improve gain, feed efficiency and reduce liver abscesses.	0.1 mg to 0.5 mg per lb body wt. per day.		3–5	3–5
Tylan	Tylosin		60 to 90 mg per head per day.		3–5	3–5

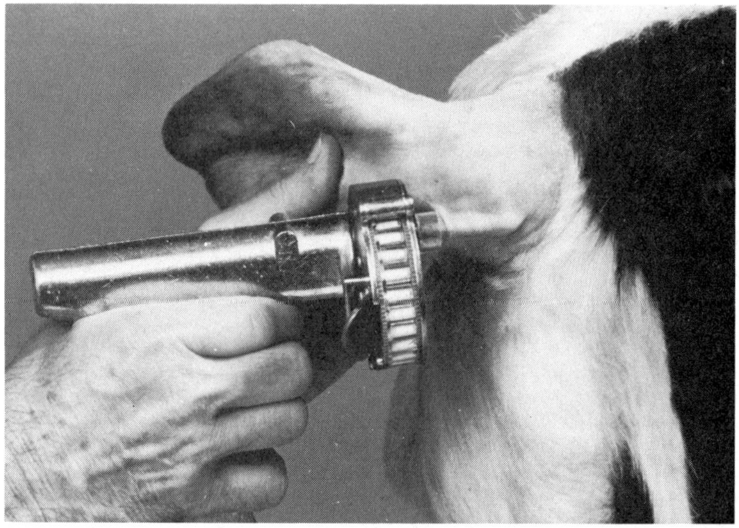

Figure 14.8. Proper position of gun and needle for implanting Ralgro® implants. Hair has been removed from ear for better demonstration. Courtesy of Ralgro® Implants, International Minerals and Chemical Corp.

For implants to be effective, they must be implanted properly. Following these steps will ensure a good job of implanting. First, the animal must be properly restrained so that movement of the head is minimal during the implanting process. Determine where the implant will be located (Figures 14.8, 14.9, 14.10, and 14.11), then insert the needle the proper distance away from that point, push it to the hilt and then back the needle off slightly before commencing to squeeze the trigger and withdraw needle. This prevents possible crushing of the implant, which might lead to decreased effectiveness or possible

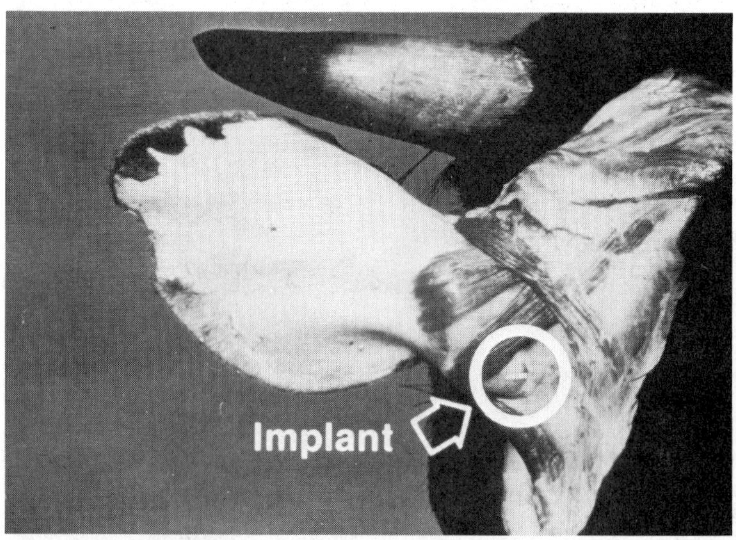

Figure 14.9. Proper location of Ralgro® implants are shown at the base of ear in circled area. Skin has been removed. Courtesy of Ralgro® Implants, International Minerals and Chemical Corp.

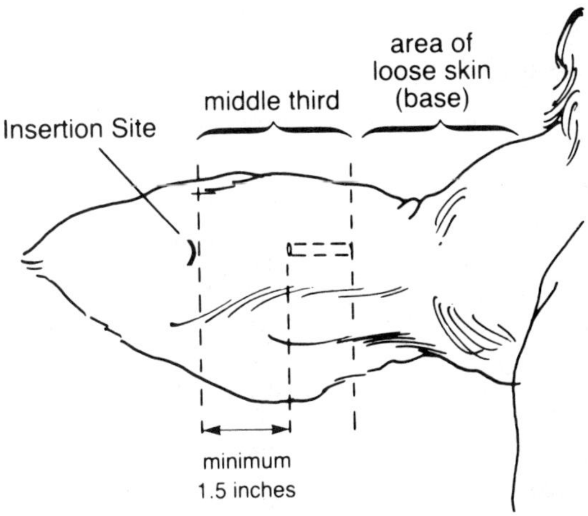

Figure 14.10. Insertion site and location of Compudose® implant (see Figure 7.9) centered in middle one-third of ear. Location of Synovex® implants should be in same position. Courtesy of Elanco Products Co.

problems due to too rapid absorption. The needle must be sharp or a bigger hole may be put in the skin which invites bacteria into the wound causing infection and poor absorption of the pellets.

Once the needle has been removed from the ear, it is a good practice to press the injection opening between the thumb and finger so the opening tends to seal. This practice is not done to prevent the implants from coming out but to prevent infection of an open wound.

NUTRIENT REQUIREMENTS OF BEEF CATTLE

Table 14.10 and 14.11 shows the nutrient requirements of several sexes of cattle within a few selected weights and levels of productivity. More extensive tables of requirements and feed composition can be found in the National Research Council (NRC), *Nutrient Requirements of Beef Cattle.*

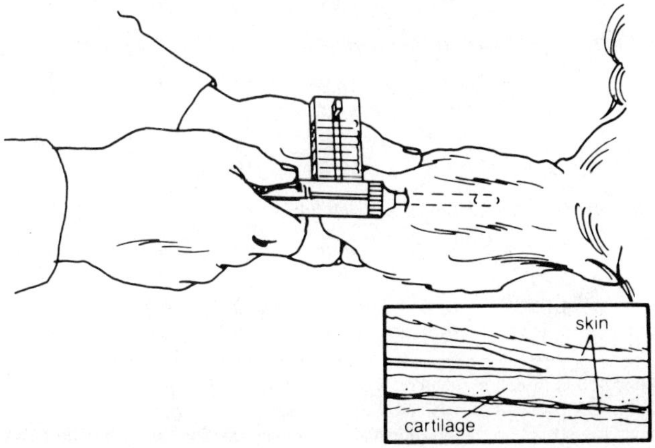

Figure 14.11. Needle insertion site for Compudose® and Synovex® implants. All implants should be between the skin and cartilage (note diagram in box). Courtesy of Elanco Products Co.

Table 14.10. Daily Nutrient Requirements (NRC) for Heifers, Cows, and Bulls

Weight (lb)	Daily Gain (lb)	Minimum Dry Matter Consumpt. (lb)	Roughage (%)	(Crude) Total Prot. (lb)	TDN (lb)	ME (Mcal)	Ca (gm)	P (gm)	Vitamin A (1000 IU)
			Growing Heifer Calves and Yearlings						
330	1.1	9.0	70–80	0.99	5.7	9.4	14	12	9
330	2.0	8.8	25–30	1.19	6.8	11.3	23	17	9
440	0	7.7	100	0.66	4.2	7.0	6	6	8
440	1.1	13.2	80–90	1.28	7.7	12.7	14	13	13
550	0.7	14.1	100	1.25	7.8	12.8	12	12	14
550	1.1	14.3	80–90	1.36	8.6	14.2	13	13	14
660	0	10.4	100	0.88	5.7	9.5	9	9	10
660	0.7	16.3	100	1.39	8.4	14.5	13	13	16
770	0.7	18.1	100	1.52	10.0	16.5	15	15	18
880	0.7	20.0	100	1.67	11.0	18.2	16	16	19
990	0	14.1	100	1.21	7.9	12.9	12	12	14
990	0.4	19.2	100	1.63	10.6	17.4	16	16	19
			Pregnant Heifers, Last 3–4 Months of Pregnancy						
720	0.9	14.5	100	1.28	7.7	12.6	15	15	19
770	0.9	15.2	100	1.34	8.1	13.2	15	15	19
830	0.9	15.9	100	1.39	8.4	13.7	15	15	20
940	0.9	17.2	100	1.52	9.0	14.8	16	16	22
			Cows Nursing Calves—Average Milking Ability[a]—First 3–4 Months Postpartum						
880	—	19.4	100	1.78	10.4[d]	17.0	25	25	21
1100	—	21.6	100	1.98	11.7[d]	19.2	27	27	24
1320	—	24.2	100	2.22	13.0[d]	21.3	28	28	27
			Cows Nursing Calves—Superior Milking Ability[b]—First 3–4 Months Postpartum						
880	—	23.8	100[c]	2.57	13.5[d]	22.1	45	41	34
1100	—	26.0	100[c]	2.84	14.8[d]	24.3	46	43	38
1320	—	28.4	100[c]	3.10	16.1[d]	26.4	46	44	43
			Dry Pregnant Mature Cows—Middle Third of Pregnancy						
880	—	13.4	100	0.79	7.3	11.9	11	11	17
1100	—	15.9	100	0.92	8.6	14.1	13	13	20
1320	—	18.3	100	1.08	9.7	16.1	15	15	23
			Dry Pregnant Cows—Last Third of Pregnancy						
880	—	15.4	100	0.97	8.8	14.3	14	14	21
990	—	16.5	100	1.06	9.4	15.4	15	15	23
1210	—	19.0	100	1.19	10.7	17.5	16	16	26
1430	—	22.4	100	1.32	11.9	19.6	18	18	29
			Bulls, Growth and Maintenance (Moderate Activity)						
880	2.0	24.2	70–75	2.27	15.4	25.3	23	23	43
1320	1.1	26.4	80–85	2.24	16.1	26.4	22	22	48
1760	0	23.1	100	1.96	12.8	21.0	19	19	41
2200	0	27.3	100	2.40	15.2	24.8	22	22	48

[a] 10 pounds of milk per day (Equivalent of approximately 450 lb of calf at weaning if there is adequate forage).
[b] 20 pounds of milk per day (Equivalent of approximately 650 lb of calf at weaning if there is adequate forage).
[c] Would need to be high-quality roughage. Some concentrates may be needed for high pregnancy rates in heavy milking cows.
[d] There is evidence that the TDN requirement should be approximately 10% higher for satisfactory reproductive performance under variable environmental conditions.
Source: National Research Council, 1976. Nutrient Requirements of Beef Cattle.

Table 14.11. Daily Nutrient Requirements for Growing-Finishing Heifers and Steers

Weight (lb)	Daily Gain (lb)	Minimum Dry Matter Consumpt. (lb)	Roughage (%)	Crude Protein (lb)	NEm (Mcal)	NEg (Mcal)	TDN (lb)	Ca (gm)	P (gm)	Vitamin A (1000 IU)
			Growing-Finishing Heifer Calves and Yearlings							
330	1.1	9.0	70–80	0.99	3.30	1.34	5.7	14	12	9
330	1.5	8.8	50–60	1.10	3.30	1.95	6.2	18	14	9
330	2.0	8.8	25–30	1.19	3.30	2.60	6.8	23	17	9
330	2.4	8.8	15	1.32	3.30	3.30	7.5	28	20	9
440	1.1	13.2	80–90	1.28	4.10	1.66	7.7	14	13	13
440	2.0	11.7	35–45	1.36	4.10	3.23	8.8	22	17	13
440	2.4	11.0	<15	1.41	4.10	4.09	9.5	25	19	13
550	1.1	14.3	80–90	1.36	4.84	1.96	8.6	13	13	14
550	2.2	13.0	35–45	1.43	4.84	3.81	10.1	21	17	14
550	2.6	13.9	<15	1.65	4.84	5.37	11.9	27	21	14
660	1.1	16.3	80–90	1.47	5.55	2.25	9.9	14	14	16
660	2.0	15.0	35–45	1.54	5.55	4.37	11.5	19	17	16
660	2.6	15.9	<15	1.74	5.55	6.16	13.7	24	20	16
770	1.1	18.3	80–90	1.61	6.24	2.52	11.2	15	15	18
770	2.2	17.9	35–45	1.69	6.24	4.91	13.2	17	17	18
770	2.6	17.9	<15	1.78	6.24	6.91	15.2	21	20	18
880	1.1	18.7	70–80	1.72	6.89	2.79	11.9	15	15	19
880	2.0	18.5	20–25	1.74	6.89	5.43	14.3	17	17	19
880	2.4	18.3	<15	1.78	6.89	6.88	15.9	19	18	19
990	1.1	20.5	70–80	1.76	7.52	3.05	13.0	17	17	20
990	1.8	20.1	35–45	1.80	7.52	5.17	15.0	16	16	20
990	2.2	18.7	<15	1.83	7.52	6.71	16.3	19	19	20
			Growing-Finishing Steer Calves and Yearlings							
440	1.1	12.8	80–90	1.25	4.10	1.49	7.5	14	13	12
440	1.5	12.6	70–80	1.34	4.10	2.14	7.9	18	16	13
440	2.0	10.8	35–45	1.34	4.10	2.82	8.2	23	18	13
440	2.4	10.1	15	1.39	4.10	3.52	8.6	27	20	13
550	1.5	12.8	55–65	1.36	4.84	2.53	8.8	18	16	14
550	2.0	13.7	45–50	1.52	4.84	3.33	9.9	22	19	14
550	2.4	13.2	20–25	1.61	4.84	4.17	10.4	26	21	14
550	2.9	13.2	15	1.67	4.84	5.04	11.5	30	23	14
660	2.0	17.9	55–65	1.78	5.55	3.82	11.9	22	19	16
660	2.4	16.8	20–25	1.80	5.55	4.78	12.3	25	22	16
660	3.1[b]	16.1	15	1.91	5.55	6.29	13.7	31	25	16
770	2.0	17.6	45–55	1.76	6.24	4.29	12.8	20	18	18
770	2.4	17.6	20–25	2.20	6.24	5.36	13.7	23	20	18
770	3.1	18.1	15	3.10	6.24	7.06	15.4	28	24	18

Table 14.11. (*Continued*)

Weight (lb)	Daily Gain (lb)	Minimum Dry Matter Consumpt. (lb)	Roughage (%)	Crude Protein (lb)	NEm (Mcal)	Neg (Mcal)	TDN (lb)	Ca (gm)	P (gm)	Vitamin A (1000 IU)
880	2.2	20.7	45–55	1.91	6.89	5.33	15.0	21	20	19
880	2.6	18.7	20–25	1.91	6.89	6.54	15.4	23	21	19
880	3.1[b]	19.8	15	2.07	6.89	7.80	17.0	26	23	19
990	2.2	22.7	45–55	2.11	7.52	5.82	16.3	20	20	20
990	2.6	22.5	20–25	2.13	7.52	7.14	17.4	23	22	20
990	3.1	21.6	15	2.16	7.52	8.52	18.5	25	23	20
1100	2.0	23.1	45–55	2.09	8.14	5.60	16.5	19	19	23
1100	2.6	21.2	15	2.11	8.14	7.73	18.1	21	21	23
1100	2.9[b]	22.0	15	2.13	8.14	8.47	19.2	22	22	23

[a] Average weight for a feeding period.
[b] Most steers of the weight indicated, and not exhibiting compensatory growth, will fail to sustain the energy intake necessary to maintain this rate of gain for an extended period.
Source: National Research Council, 1976. Nutrient Requirements of Beef Cattle.

RATION FORMULATION

Rations should be formulated or balanced to meet the nutrient requirements of animals in their particular stages of production. Also to be considered is the cost of the ration with an assurance that it is palatable and will not cause any serious digestive disturbances or toxic effects.

Rations should be balanced for energy (using TDN, ME or NEm, and NEg), crude protein, vitamin A, calcium and phosphorus. Salt is usually provided free choice or mixed in the ration. An assessment should be made to determine if any trace minerals are needed.

Balancing Cow Rations Using TDN

TDN is a satisfactory basis for balancing rations for cows. Table 14.12 shows an example of balancing a cow's ration where silage is the primary source of feed.

The calculations used in Table 14.12 are as follows:

Line a: TDN requirement from Table 14.10.
Line b: Calculate pounds of corn silage dry matter (DM) needed by dividing pounds of TDN required by percent TDN in corn silage ($11.7 \div 0.7 = 16.7$).
Line c: Calculate pounds of "wet" corn silage needed by dividing pounds of corn silage

Table 14.12. Beef Cow Ration With Corn Silage and Protein Supplement

1100 lb Cow (Ave. Milk–Early Lactation)	Amt.
a. TDN required, lb/day	11.7
Corn silage needed:[a]	
b. lb of silage, dry matter (DM)/day	16.7
c. lb of wet silage/day	59.9
Crude Protein (CP) suppl. needed:[b]	
d. CP required, lb/day	1.98
e. CP supplied by corn silage, lb/day	1.40
f. CP deficiency, lb/day	0.58
g. lb of protein suppl./day	1.66

[a] Composition of corn silage taken from Table 14.3; Dry Matter 27.9%, TDN 70% and Crude Protein 8.4%.

[b] Assume protein supplement 35% CP.

DM by percent DM in corn silage ($16.7 \div 0.279 = 59.9$ lbs).

Line d: Crude protein requirement from Table 14.10.

Line e: Calculate lbs of CP supplied by corn silage by multiplying pounds of corn silage DM fed (line b) by percent CP in corn silage ($16.7 \times 0.084 = 1.40$).

Line f: Calculate CP deficiency by subtracting pounds of CP supplied on line e from the CP requirement on line d ($1.98 - 1.40 = 0.58$).

Line g: Calculate pounds of CP supplement needed by dividing pounds of CP deficiency by percent CP in the supplement ($0.58 \div .35 = 1.66$).

This ration is now balanced for TDN and CP, however, it does not meet the minimum dry matter requirement of 21.6 lbs (Table 14.10). To meet the dry matter requirement, one would increase the silage fed to approximately 70 lbs. Fulfilling the dry matter requirement is not a major concern in feeding cows as long as energy, protein, mineral, and vitamin requirements are met. Salt, phosphorus, and possibly vitamin A, would be supplied on a free choice basis to meet the cow's nutrient requirements.

It is not uncommon to have dry matter vary in the same silage crop. Amount of wet silage fed should be adjusted as moisture content changes. If moisture changes are not considered, then critical nutrients will be over-supplied or under-supplied.

Using the same cow requirements as shown in Table 14.12, the CP deficiency may be filled using a legume such as alfalfa hay. The Pearson Square Method could be used to determine the amount of alfalfa hay needed, along with corn silage, to meet the dry matter, and crude protein requirement. The example of using Pearson's Square is as follows:

corn silage (8.4% CP)		9.2 parts corn silage
	9.2%	
alfalfa hay (18.4% CP)		0.8 parts alfalfa hay
		10.0 total parts

On the left-hand side of the square are the crude protein percentages of corn silage and alfalfa hay taken from Table 14.3. The 9.2% in the middle of the square is the crude protein requirement (expressed in %) for the 1100 lb cow. This was calculated:

$$\frac{\text{1.98 lb CP requirement}}{\text{21.6 lb dry matter requirement}} = 9.2\%$$

On the right-hand side of the square are parts of corn silage and parts of alfalfa hay of the total ration. These parts are obtained by subtracting diagonally, the smallest % from the largest %. Table 14.13 permits a check of the calculations.

Least cost rations can best be calculated using a computer program which can analyze many feed sources, their nutrient composition, and cost. Some rough approximations of least cost rations can be obtained by calculating energy and protein costs on several readily available feeds. Since energy and protein costs make up the largest part of the total nutrient costs, using information in Tables 14.14 and 14.15 can be helpful. The information in these tables is for example purposes only. Individual producers should substitute feeds common to their area, energy and protein composition of the feeds and the current feed cost.

Formulating Feedlot Rations

The net energy system is typically used when formulating feedlot rations.

In example No. 1, a ration will be evaluated as to its adequacy in net energy and crude protein for a 650 pound steer expected to gain approximately 3.0 pounds per day during a 150-day finishing period. The requirements for this steer (Table 14.11) are 16.1 lbs dry matter, 1.91 lbs (11.9% of ration) crude protein, 5.55 Mcal NEm and

Table 14.13. Checking Pearson's Square Calculations

	Part of Ration		Protein	
Feed	**Percent**	**lbs**	**Percent**	**lbs**
Corn silage	92	19.8	8.4	1.66
Alfalfa hay	8	1.7	18.4	.31
Total	100	21.5	9.2	1.97

Table 14.14. Comparative Energy Costs Using TDN and NEg

Feedstuff	Cost/ton $	Percent TDN	Cost/lb TDN $	NEg	Cost/Mcal NEg $
Alfalfa hay[a]	70	57	0.061	0.25	$0.140
Barley	130	83	0.078	0.64	0.102
Corn	115	91	0.063	0.67	0.086
Corn silage (28% dry matter)	20	70	0.051	0.45	0.079
Cottonseed meal	160	74	0.108	0.49	0.163
Dehydrated alfalfa	130	61	0.107	0.31	0.210
Milo	110	80	0.069	0.56	0.098
Soybean meal	185	81	0.114	0.59	0.157
Wheat	130	88	0.074	0.64	0.102

[a] Example of calculation: Alfalfa hay at
$\frac{\$70/\text{ton}}{2000} = \$0.035/\text{lb}$; $\text{TDN} = \frac{0.035}{0.57} = \$0.061/\text{lb}$; $\text{NEg} = \frac{0.035}{0.25} = \$0.14/\text{Mcal}$.

6.29 Mcal NEg. Table 14.16 shows pounds of feeds being fed and their conversion from an "as fed" to a dry matter basis. This conversion is done by multiplying the pounds "as fed" by its dry matter content (Table 14.3).

Table 14.17 shows calculations for determining the megacalories for maintenance and gain and amount of protein in the ration (Example for corn silage: NEm $= 3.35 \times 0.71 = 2.38$; NEg $= 3.35 \times 0.45 = 1.51$; CP $= 3.35 \times 0.084 = 0.28$). Values are obtained from Table 14.3.

The questions posed in example No. 1 are answered in the following steps:

STEP 1—Determine NEm and NEg per pound of feed:

From Table 14.17, the NEm per lb ration $= \frac{14.39}{16.1} = 0.90$

Table 14.15. Calculation of Value of Protein Supplements on Basis of Crude Protein

Supplement	Percent Crude Protein	Cost/ Ton ($)	Cost/lb Crude Protein
Alfalfa hay	15	70	$0.233
Cottonseed oil meal	41	160	0.195
Dehydrated alfalfa	17	130	0.382
Soybean oil meal[a]	44	185	0.210

[a] Example of calculation: soybean oil
meal at $\frac{\$185/\text{ton}}{2000} = \$0.0925/\text{lb}$ CP cost $= \frac{0.0925}{0.44} = \$0.21/\text{lb}$.

Table 14.16. Feed Being Fed to 650 lb Steer

Feed	Lbs Fed	Percent Dry Matter	Lbs Dry Matter
Corn silage	12.0	27.9	3.35
Corn	10.0	89.0	8.90
Chopped alfalfa hay	2.0	90.0	1.80
Beet pulp	1.5	77.0	1.16
32% protein supplement	1.0	90.0	0.90
Total			16.11

From Table 14.17, the NEg per lb ration $= \frac{9.09}{16.1} = 0.56$

STEP 2—Pounds of feed needed to meet the maintenance requirement of the 650 lb steer:

$$5.55 \text{ Mcal} \div 0.90 = 6.17 \text{ lbs}$$

STEP 3—Pounds of feed needed to meet the requirement for 3.0 lb daily gain:

$$6.29 \text{ Mcal} \div 0.56 = 11.2 \text{ lbs}$$

STEP 4—Total lbs of feed the steer must eat to gain 3.0 lb per day:

$$6.17 \text{ lbs} + 11.2 \text{ lbs} = 17.3 \text{ lbs}$$

Conclusion: If the steer has a capacity to eat only 16.1 lbs of dry matter of the ration fed, then 3.0 lbs per day gain will not be realized. The ration could be made more dense in energy by substituting corn grain for corn silage on a lb for lb basis until the desired gain could be realized.

The 1.89 lb of protein supplied in the 16.1 lb of dry matter is sufficiently close to the 1.91 lb protein requirement.

Example No. 2—How can the net energy system be used to predict the amount of daily gain if the feed consumption is known? Using the content of example No. 1, what

Table 14.17. Calculating the NEm and NEg in the Ration

Feed	Lbs Dry Matter	NEm (Mcal)	NEg (Mcal)	Lb Protein
Corn silage	3.35	2.38	1.51	0.28
Corn	8.90	9.26	5.96	0.89
Chopped alfalfa hay	1.80	0.99	0.45	0.33
Beet pulp	1.16	1.08	0.72	0.10
32% protein suppl.[a]	0.90	0.68	0.45	0.29
Total	16.11	14.39	9.09	1.89

[a] 32% protein suppl. contains 0.75 Mcal/lb (NEm) and 0.0 Mcal/lb (NEg).

would be the expected gain if the 650 lb steer consumed 16.1 lbs of the ration in Table 14.16?

STEP 1—Pounds of feed needed to meet daily maintenance requirement = 6.17 lbs (STEP 2 in example No. 1).

STEP 2—Pounds of feed remaining for gain:

$$16.1 \text{ lbs} - 6.17 = 9.93 \text{ lbs}$$

Table 14.18 Net Energy Requirement Tables for Growing and Finishing Cattle

Body Wt., Lb.	**300**		**400**		**500**		**600**	
NE_m Required	**3.10**		**3.85**		**4.55**		**5.21**	
	NE_g		**NE_g**		**NE_g**		**NE_g**	
Daily Gain Lb	**Steers**	**Heifers**	**Steers**	**Heifers —Mcal/day—**	**Steers**	**Heifers**	**Steers**	**Heifers**
.5	0.47	0.52	0.59	0.64	0.72	0.78	0.83	0.90
.6	0.59	0.64	0.73	0.80	0.87	0.95	1.00	1.09
.7	0.68	0.74	0.84	0.92	1.02	1.12	1.17	1.28
.8	0.79	0.87	0.98	1.08	1.17	1.29	1.34	1.48
.9	0.88	0.97	1.10	1.21	1.33	1.46	1.52	1.68
1.0	1.00	1.11	1.24	1.38	1.48	1.64	1.70	1.88
1.1	1.12	1.24	1.39	1.54	1.64	1.82	1.88	2.09
1.2	1.21	1.35	1.51	1.68	1.80	2.00	2.06	2.30
1.3	1.33	1.49	1.66	1.86	1.96	2.19	2.24	2.51
1.4	1.43	1.61	1.78	2.00	2.12	2.38	2.43	2.73
1.5	1.55	1.75	1.93	2.18	2.28	2.57	2.62	2.95
1.6	1.65	1.87	2.06	2.32	2.45	2.77	2.81	3.17
1.7	1.78	2.02	2.21	2.51	2.61	2.97	3.00	3.40
1.8	1.88	2.14	2.34	2.66	2.78	3.17	3.19	3.63
1.9	2.01	2.29	2.50	2.85	2.95	3.37	3.39	3.87
2.0	2.14	2.42	2.63	3.01	3.12	3.58	3.58	4.11
2.1	2.24	2.57	2.79	3.20	3.30	3.79	3.78	4.35
2.2	2.37	2.74	2.95	3.40	3.47	4.01	3.98	4.59
2.3	2.48	2.87	3.08	3.57	3.65	4.22	4.18	4.84
2.4	2.61	3.03	3.25	3.77	3.83	4.44	4.39	5.09
2.5	2.72	3.17	3.38	3.93	4.01	4.66	4.59	5.35
2.6	2.86	3.33	3.55	4.15	4.19	4.89	4.80	5.61
2.7	2.97	3.47	3.69	4.32	4.37	5.12	5.01	5.87
2.8	3.11	3.65	3.86	4.54	4.56	5.35	5.22	6.14
2.9	3.22	3.79	4.00	4.71	4.74	5.59	5.44	6.40
3.0	3.36	3.97	4.18	4.94	4.93	5.82	5.65	6.68
3.1	3.47	4.11	4.32	5.12	5.12	6.06	5.87	6.95
3.2	3.62	4.30	4.50	5.34	5.31	6.31	6.09	7.23
3.3	3.76	4.48	4.68	5.51	5.51	6.56	6.31	7.57
3.4	3.88	4.63	4.83	5.76	5.70	6.81	6.54	7.80

Source: National Research Council, 1976. Nutrient Requirements of Beef Cattle.

Table 14.18. ***(Continued)***

Body Wt., Lb.	700		800		900		1000	
NE_m Required	5.85		6.47		7.06		7.65	
	NE_g		NE_g		NE_g		NE_g	
Daily Gain Lb.	Steers	Heifers	Steers	Heifers —Mcal/day—	Steers	Heifers	Steers	Heifers
.5	0.93	1.01	1.02	1.11	1.12	1.22	1.21	1.32
.6	1.12	1.22	1.24	1.35	1.35	1.47	1.46	1.59
.7	1.31	1.44	1.45	1.59	1.58	1.73	1.71	1.88
.8	1.51	1.66	1.67	1.83	1.82	2.00	1.97	2.17
.9	1.71	1.88	1.89	2.08	2.06	2.27	2.23	2.46
1.0	1.91	2.11	2.11	2.33	2.30	2.55	2.49	2.76
1.1	2.11	2.34	2.33	2.59	2.55	2.83	2.76	3.06
1.2	2.31	2.58	2.56	2.85	2.79	3.11	3.02	3.37
1.3	2.52	2.82	2.78	3.12	3.04	3.40	3.29	3.68
1.4	2.73	3.06	3.01	3.39	3.29	3.70	3.56	4.00
1.5	2.94	3.31	3.25	3.66	3.55	4.00	3.84	4.33
1.6	3.15	3.56	3.48	3.94	3.80	4.30	4.12	4.66
1.7	3.37	3.82	3.72	4.22	4.06	4.61	4.40	4.99
1.8	3.58	4.08	3.96	4.51	4.33	4.92	4.68	5.33
1.9	3.80	4.34	4.20	4.80	4.59	5.24	4.97	5.67
2.0	4.02	4.61	4.45	5.09	4.86	5.57	5.26	6.02
2.1	4.25	4.88	4.69	5.39	5.13	5.89	5.55	6.38
2.2	4.47	5.16	4.94	5.70	5.40	6.23	5.84	6.74
2.3	4.70	5.44	5.19	6.01	5.67	6.56	6.14	7.10
2.4	4.93	5.72	5.45	6.32	5.59	6.90	6.44	7.47
2.5	5.16	6.01	5.70	6.64	6.23	7.25	6.74	7.85
2.6	5.39	6.30	5.96	6.96	6.51	7.60	7.05	8.23
2.7	5.63	6.59	6.22	7.28	6.80	7.96	7.35	8.61
2.8	5.87	6.89	6.48	7.61	7.08	8.32	7.67	9.00
2.9	6.11	7.19	6.75	7.95	7.37	8.68	7.98	9.40
3.0	6.35	7.50	7.02	8.29	7.67	9.05	8.30	9.80
3.1	6.59	7.81	7.29	8.63	7.96	9.43	8.61	10.20
3.2	6.84	8.12	7.56	8.98	8.26	9.81	8.94	10.61
3.3	7.09	8.44	7.83	9.33	8.56	10.19	9.26	11.03
3.4	7.34	8.76	8.11	9.68	8.86	10.58	9.59	11.45

STEP 3—Mcal of NEg supplied from the remaining amount of feed:

9.93 lbs $\times$ 0.56 (NEg per lb in ration) = 5.56 Mcal

STEP 4—Daily gain expected from 5.56 Mcal of NEg for 650 lb steer (see Table 14–18):

5.56 Mcal produces approximately 2.8 lb of gain

Example No. 3—Supoose the genetic ability of the 650 lb steer only allowed the daily consumption of 12.5 lbs of the ration—what would be the expcted gain?

The same steps as in example No. 2 would be followed.

Mcal of NEg for gain = 3.54 Mcal. Expected gain from 3.54 Mcal (Table 14.18) = approximately 1.9 lbs per day.

Most feedlot rations are balanced for vitamin A, calcium and phosphorus. Potassium becomes important as the level of concentrate increases or when non-protein nitrogen is substituted for intact protein. Sulfur also becomes important when the level of non-protein nitrogen increases in the ration. Most purchased protein supplements will have trace minerals added, or a trace mineral salt mixture could be included in the ration if there is a need for trace mineral supplementation.

Adjusting Rations for Weather Changes

Nutrient requirements and predicted gains of cattle are published from research studies where animals are protected from environmental extremes. The most common environmental factor that alters both performance and nutrient requirements is temperature. Thus cattle producers should be aware of critical temperatures which affect performance of their cattle and consider making changes in their feeding programs if economics so dictate.

The thermoneutral zone (TNZ) is the range in effective temperature where rate and efficiency of performance is maximized. Critical temperature is the lower limit of the TNZ and is typified by the ambient temperature below which the performance of cattle begins to decline as temperatures become colder. Figure 14.12 shows that the maintenance energy requirement increases more rapidly during cold weather than does rate

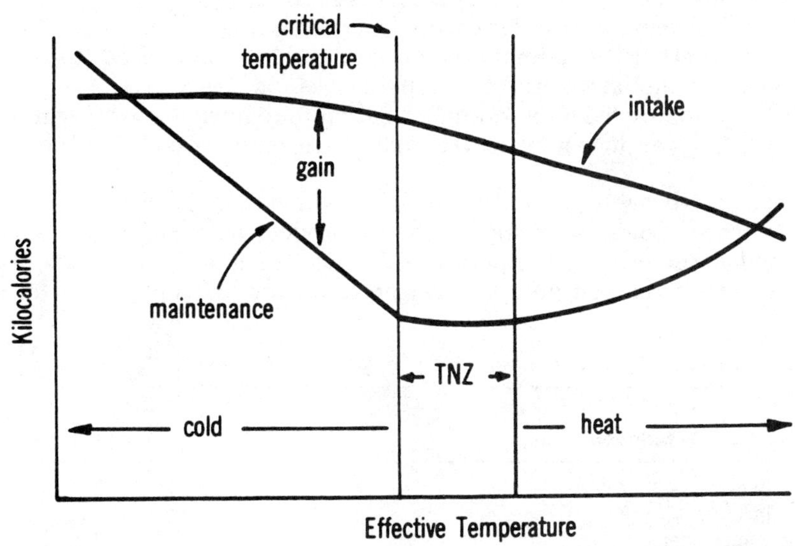

Figure 14.12. Effect of temperature on rate of feed intake, maintenance energy requirement and gain (Ames 1980).

Table 14.19. Wind-Chill Factors for Cattle With Winter Coat

Wind Speed (mph)	Temperature (°F)												
	−10	−5	0	5	10	15	20	25	30	35	40	45	50
Calm	−10	−5	0	5	10	15	20	25	30	35	40	45	50
5	−16	−11	−6	−1	3	8	13	18	23	28	33	38	43
10	−21	−16	−11	−6	−1	3	8	13	18	23	28	33	38
15	−25	−20	−15	−10	−5	0	4	9	14	19	24	29	34
20	−30	−25	−20	−15	−10	−5	0	4	9	14	19	24	29
25	−37	−32	−27	−22	−17	−12	−7	−2	2	7	12	17	22
30	−46	−41	−36	−31	−26	−21	−16	−11	−6	−1	3	8	13
35	−60	−55	−50	−45	−40	−35	−30	−25	−20	−15	−10	−5	0
40	−78	−73	−68	−63	−58	−53	−48	−43	−38	−33	−28	−23	−18

of feed intake. This results in a reduction of gain, more feed required per pound of gain which typically causes cost per pound of gain to be higher.

Effective ambient temperatures below the lower critical temperature (below the TNZ) constitute cold stress and those above the TNZ, heat stress. The term, "effective ambient temperature," is an index of the heating or cooling power of the environment in terms of dry bulb temperature. It would include any environmental factor, such as radiation, wind, humidity, or precipitation that alters environmental heat demand. Not all effective ambient temperatures have been calculated for cattle, although some combined environmental factors are available. One of these, the wind-chill index, is shown in Table 14.19. An example in Table 14.19 shows that with a temperature of 20 °F and a wind speed of 30 mph, the effective ambient temperature is −16 °F.

The lower critical temperature for cattle depends on how much insulation is provided by the hair coat, whether the animal is wet or dry and how much feed the cow consumes. Table 14.20 shows some lower critical temperatures for beef cattle.
For example, a cow being fed a maintenance ration may have a low critical temperature of 32 °F if dry, but the lower critical temperature may be 60 °F if the same cow is wet.

Coldness of a specific environment is the value which must be considered when adjusting rations for cows. Coldness is simply the differences between effective temperature (wind-chill) and lower critical temperature. Using this definition for coldness instead of using temperature read on an ordinary thermometer helps explain why wet, windy

Table 14.20. Estimated Lower Critical Temperatures for Beef Cattle

Coat Description	Critical Temperature
Summer Coat or Wet	59°F
Fall Coat	45°F
Winter Coat	32°F
Heavy Winter Coat	18°F

days in March may be colder for a cow than extremely cold but dry, calm days of January.

The major effect of cold on nutrient requirements of cows is increased need for energy which usually means the total amount of daily feed will need to be increased.

Feeding tables recommend that a 1200 lb cow receive 16.5 lb of good mixed hay to supply energy needs during the last one-third of pregnancy. How much feed should the cow receive if she is dry and has a winter hair coat but the temperature is 20 °F with a 15 mph wind? The coldness is calculated by subtracting the wind-chill or effective temperature (4 °F) from the cow's lower critical temperature (32 °F). Thus the magnitude of coldness is 28 °F. A rule of thumb (more detailed tables are available) is to increase the amount of feed 1% for each degree of coldness. A 28% increase of the 16.5 lb (original requirement) would mean that 21.1 lbs of feed must be fed to compensate for the coldness.

This example is typical of many feeding situations; however, if the cow was wet, the same increase in feed would be required at 31 °F wind-chill (28 °F of coldness).

RANGE COW–CALF NUTRITION

Following is a short summary of general recommendations for feeding commercial range cow herds in most of the Plains and Western States. The basic goal is to produce the most pounds of calf per unit of land available at the least cost over a period of several years. Individual operations are unique situations and these recommendations should be adjusted to meet the needs of each unique operation.

Feeding the Brood Cow Herd

Energy

Most pregnant cows should gain 100 lbs to 150 lbs during winter months. This weight gain represents primarily the weight of calf, fluids, and membranes. More weight gain may be necessary if the cow enters the winter period in poor body condition. During the time between weaning calves and the winter period, cows usually need only a salt-phosphorus supplement if they have sufficient feed to maintain their body weight. During the last two months of pregnancy, however, cows should gain 1 lb to 2 lbs per day unless they are in an above average body condition. Most cows gaining less than one pound per day will not provide adequate nutrition for the developing calf. Some producers attempt to reduce the birth weight of the calf and subsequent calving problems by reducing feed intake during late pregnancy. This practice usually affects calf birth weight only by 2 lbs to 4 lbs, but more importantly, these cows will produce less colostrum and will have a longer post-partum interval. Calves which receive an inadequate level of colostrum are more susceptible to disease. One of the more important nutritional management practices is to see that calves receive adequate quantities of colostrum within a few hours after birth.

The first 90 days after calving is the most important nutritional period for the brood cow. Most calves are born during late winter or early spring months when most pastures are at their lowest point nutritionally. Supplemental feed should be provided at this time to allow cows to produce an adequate supply of milk and still gain in body weight.

Cows that do not gain weight prior to breeding and during the breeding season are most likely to have poor reproductive performance. Producers may want to weigh a sample of their cows during this period to assess that weight gains are occurring and possibly adjust the amount of feed being provided.

A logical management practice is to sort cows into groups prior to calving through part of the breeding season, to provide levels of feed consistent with cow age and body condition. First and second calf females and thin cows need to be separated from other cows to assure feeding levels that will assure high reproductive performance.

Cow size should be adjusted to fit existing environmental conditions. Moderate sized cows appear to function better in most Western, Plains, and Corn Belt States, where higher levels of feed exist. No doubt this would be advantageous for cows of a larger mature size also.

Summer forage, water, and salt will provide all of the cow's nutrient requirements if these nutrient sources are available in ample supply. Winter pasture will provide the cow's energy needs if it is present in adequate amounts and not covered with snow. When adequate amounts of feed are not present, the cheapest source of supplemental energy should be provided. Some common sources include hay, grain, and silage. Supplementing a high-protein concentrate when protein is not needed as a supplement is a common practice, and it is not economically sound.

Protein

Natural Protein

The following graph (Figure 14.13) illustrates that during spring and early summer, protein is adequate in most range and pasture grasses.

The pregnant cow needs 6.5% protein in her ration, 9% for a lactating cow, and 11% for a growing calf or yearling. During the winter period, protein in range grass frequently decreases to 4% to 5%.

Protein is usually more than adequate when feeding alfalfa hay. When grass hay is fed, if a question arises as to its meeting the cow's requirements, a feed analysis should

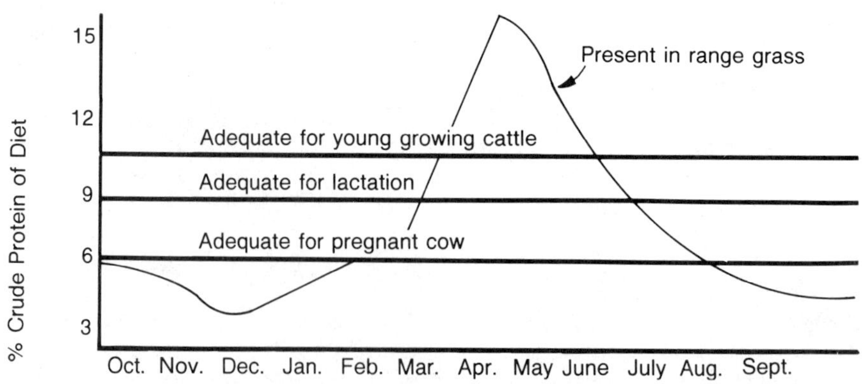

Figure 14.13. Crude protein content in grazing cow's diet. Courtesy of CSU.

Table 14.21. Recommended Levels of Protein to Pregnant and Lactating Cows and Heifers

Item	Percent Crude Protein Needed in Supplement					
	50	40	30	20	15	10
Supplement fed to cows last 2 months of pregnancy (lbs/cow/day)	0.6	1.0	1.5	2	3.5	7
Supplement fed during lactation and to pregnant heifers (lbs/cow/day)	1.3	1.8	2.3	4	11	20

be made. Corn stalks, straw, and other aftermath feeds provide a cheap source of energy, but as a rule, need to be supplemented with protein, especially during the latter part of the grazing period. Under-feeding protein is common and results in decreased performance. Over-feeding protein will not harm animals, but the practice is extremely uneconomical. When a protein supplement is necessary, the cheapest source available should be fed. Often, cottonseed cake, alfalfa hay or a feed grain is a much more economical route than a commercially prepared supplement. Supplements are available, however, in loose, block, and range cube forms. Table 14.21 gives a general idea of how much supplement of varying percentages of protein to feed.

Non-Protein-Nitrogen

Many cow–calf producers and yearling operators make an error in feeding non-protein nitrogen (NPN) as the supplemental source of protein when cattle are on an all forage ration. NPN in the form of urea, nucleic acids, biuret, or IBDU must be fed along with a readily available source of carbohydrates to be adequately utilized. This readily available carbohydrate source is not provided in adequate amounts with range forage alone. If adequate energy is not available, cattle show little benefit from feeding NPN, and money spent on NPN is wasted. Regardless of source of NPN, it is of questionable benefit in meeting the beef cow's protein needs while grazing dry forage.

Minerals

Salt

Salt, in either a loose or block form, should be available free choice to range cattle at all times. The cheapest source should be utilized.

Calcium

Calcium, as shown in Figure 14.14, is present in adequate amounts throughout the year in range and pasture forages. Supplementation is unnecessary.

Phosphorus

Figure 14.14 shows that phosphorus is adequate during summer months but deficient for more than half the year. Inadequate phosphorus causes reduced growth rates, de-

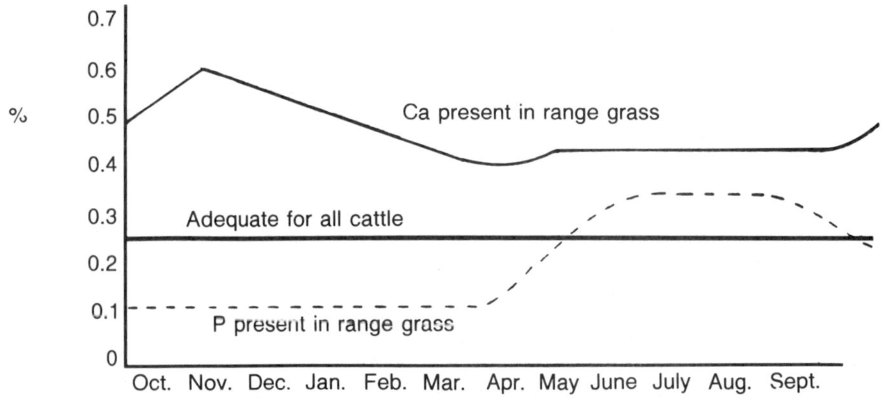

Figure 14.14. Calcium and phosphorus in grazing cow's diet. Courtesy of CSU.

creased milk production, and delayed conception. The latter effect can be seen in the research results shown in Table 14.22.

Phosphorus can be supplemented by mixing a phosphorus source with free-choice salt. The most common ration is one part salt to one part phosphorus source. Some sources of phosphorus include dicalcium phosphate and steamed bone meal. Producers usually find that mixing their own salt-phosphorus supplement and providing it free choice is significantly cheaper than using a commercially prepared supplement.

Magnesium

On some pastures grass tetany is a problem (see Chapter 20). This condition is most commonly observed early in the spring on fresh, green, rapidly growing grasses. Affected animals are often seen staggering. The best prevention is to mix magnesium oxide (MgO) in the salt supplement. This should be started several weeks ahead of the potential problem time and continued throughout the grazing period. If palatability is a problem, MgO can be mixed with soybean meal or grain. Usually no benefit is seen from supplementation of Mg when tetany is not a problem.

Table 14.22. Effect of Phosphorus (P) Supplementation on Conception in Lactating Range Cows

Period	Percent Conception When Total Ration Contained	
	0.2% P	*0.3% P*
First breeding	51	67
Second breeding	23	29
Third breeding	18	4
Fourth breeding	4	
Total	96	100

Trace Minerals.

Except for a few isolated local deficiencies, which in general are not common, all other essential minerals in the range cow diet are present in adequate amounts in the forage. Individual herd and area problems should be considered. As a general rule, cafeteria style mineral feeding or the feeding of complex mineral mixtures is not necessary or beneficial.

Vitamins

Vitamin A

Generally, if the forage being consumed contains green color, it may supply sufficient carotene to meet vitamin A requirements. In forages, beta carotene is present, which is the precursor of vitamin A. Vitamin A is stored in the liver primarily during the spring and summer months, and this storage usually supplies the animals' needs during the winter. Vitamin A deficiency in cows occurs infrequently. Some pastures during periods of extended dryness, however, may not supply adequate amounts of vitamin A. In these cases, 3 to 6 months hepatic supply is sufficient to meet the cow's requirements until the last few months before calving, at which time the cow may become deficient. She needs it primarily at this time, both to secrete colostrum for the calf and for her own needs. Often a cow has ample vitamin A available, but the calf is born deficient. Colostrum must provide the vitamin A to make up the deficiency. Deficiencies make calves more susceptible to disease, and the deficient cow is more susceptible to retained placentas and infections. The calf's deficiency is supplied only if the colostrum contains adequate vitamin A and the calf receives an adequate amount of colostrum. Most veterinarians and nutritionists recommend an intramuscular injection of vitamin A for the calf at birth, considering this an inexpensive insurance policy. If producers are concerned that their brood cows are deficient, they may have them tested. Current recommendations suggest that blood from five to seven randomly selected cows be analyzed for plasma vitamin A levels. Currently, however, the analysis is approximately $5.00 per cow, and it may be more economical to inject all cows with vitamin A. In some cases, vitamin A can be fed in a supplement form.

Other Vitamins

In the cow, B vitamins are synthesized by rumen microorganisms. Other vitamins are usually found in more than adequate amounts in the range cow diet. There are several commerical products available for supplementation of these vitamins. Their use is questionable from a nutritional and economical standpoint.

Water

Clean, fresh water should be available free choice at all times. If water is of questionable quality, it should be tested at a local laboratory. Often when range cattle are brought into drylot, they do not recognize water in automatic waterers. By running the water fast enough to make noise the first few days, cattle become more easily adapted to drylot conditions.

DEVELOPING RELACEMENT HEIFERS

Replacement heifers are generally weaned in the fall, bred the following summer, and calve the following spring at two years of age. From an economic standpoint, this is the best management scheme for most operations. Replacement heifers usually should not be creep fed because this apparently deposits mammary fat, which later is detrimental to milk production. Puberty is a function of age and weight. Most heifers are old enough to breed at 14 to 16 months of age. The critical constraint often is weight. Heifers should be fed to attain 65% of their mature weight at the beginning of the breeding season. To do this, most heifers must gain 1 lb to 1½ lbs per day. For these gains to occur, heifers will need all the good quality forage they can eat plus a small amount of concentrate, with possibly some additional protein and phosphorus. Heifers of 500 lbs to 700 lbs need a 4% higher level of protein in their diet than do mature animals. The concentrate fed should contain enough protein, phosphorus, additional energy, and vitamin A to meet requirements of the heifer. Heifers should reach 85% of their mature weight by calving time. Again it should be stressed that "starving out" calving difficulty is not effective. After calving, heifers should be separated from cows and should be fed to meet additional requirements and gain weight. Implants and growth stimulants should not be used on replacement heifers.

The following calculations should be applied using the weights and days that fit each producer's program and goals.

Example:			
	Expected mature weight		1000 lbs
		×	.65
	Weight desired at breeding		650 lbs
	Present weaning weight	−	450 lbs
	Gain needed		200 lbs
	Days until breeding season		180
			1.1 lbs per day

These 450 lb heifers must gain 1.1 lbs per day for most of them to reach puberty prior to the breeding season. To accomplish this, they must consume about 13 lbs of dry matter a day that is 80% to 90% roughage, and it must contain about 7.7 lbs of TDN. A ration this dense in TDN will require some concentrate to be supplemented in addition to the forage. Refer to Table 14.10 for further protein, vitamin A, and phosphorus requirements. Some protein, phosphorus, and vitamin A may be necessary also, depending on the forage being consumed.

STOCKER-YEARLING NUTRITION

Stocker-yearling cattle are those typically being grown on high roughage rations. Most of these cattle would be from 5 months to 25 months of age weighing from 300 lbs to 1000 lbs. These cattle could be on grazed forage or be in confinement (e.g. a feedlot), in the latter situation referred to as background cattle.

Because rations to stocker-yearling cattle are high roughage with little or no grain, the TDN system of formulating rations is most generally used. Rations can be formulated using the same procedure as discussed for balancing cow rations.

Successful nutritional programs for stocker-yearling cattle must be carefully coordinated with a well planned health program. Many of the newly received cattle are stressed having been sold through several markets, exposed to diseases, and shipped long distances through varying kinds of weather without feed and water. How these cattle adjust to the new environment will be largely determined by complementary nutritional and health programs. Herd health programs for stocker-yearling cattle are discussed in Chapter 19.

Low feed intake and water consumption are the two biggest nutritional challenges for newly arrived stocker-yearling cattle. These challenges exist for the first three to four weeks. It is important that the cattle consume feed before filling-up on water. During the first week, a milled concentrate ration [with steam flaked or whole grains along with long stemmed oat, millet, alfalfa or grass hay (listed in order of preference)] should be fed. Silage, haylage or high-moisture grains should be avoided because fermented feeds tend to reduce feed intake. Cottonseed hulls and ground ear corn have been reported as good ingredients in receiving rations. Feeds which contribute dust to the ration, for example, ground grains, ground, or chopped hay, should be avoided since dust contributes to respiratory problems.

Table 14.23 shows some additional nutritional recommendations for stocker-yearling cattle which have been shown to increase feed intake and contribute to the nutritional well-being of cattle during their initial adjustment period. Concentrate level for heavier cattle is less because their feed consumption is usually higher compared to lighter calves. Protein should be supplemented on a lbs per head basis rather than percent of ration because of the variability in feed consumption. Natural protein is recommended in the ration with minimal amounts of non-protein nitrogen. Bypass protein may also be beneficial. Potassium additions to receiving diets have proven to improve performance and health status of cattle.

Water should be provided three to four hours after the cattle's arrival. Water tanks should be clean with the cattle (especially small cattle) being able to reach the water without being frightened. Allowing water to run into the tank or trough may be an encouraging stimulus to attract the cattle to water.

FEEDING FEEDLOT CATTLE

There are numerous feeding programs utilized by cattle feeders. Individual feeding programs depend on (1) age, weight, and sex of cattle, (2) whether feeds are raised or

Table 14.23. Nutritional Recommendations for Stocker-Yearling Cattle of Varying Weights

	Weight	
Item	Less than 500 lbs	More than 500 lbs
Level of concentrate	55 to 75	Less than 50
Protein	16	12
Potassium	1.5	1.0

purchased, (3) relative costs of available feedstuffs, (4) feeding one group of cattle per year or feeding more than two groups of cattle per year through the same facilities, (5) length of feeding time and others.

Simple feeding programs are those feeding silage with an added protein-mineral-vitamin supplement or self-feeding a ration where both rations stay relatively constant for the entire feeding period. Complex feeding programs combine the altering of concentrate to roughage ratios several times (3 to 5) with formulating least-cost (both ingredient and gain) rations. In the more complex feeding programs, feed intake is maximized by feeding a ration several times a day while carefully monitoring feedbunks and digestive patterns of the cattle.

Calves are usually grown on higher roughage rations before being fed high-grain, finishing rations; otherwise, cattle finish at lighter weights than the preferred slaughter

Table 14.24. Returns Per Head for Yearlings in One-Lot/Year and Continuous Feedings Systems[a]

	Corn:Alfalfa							
Feed Prices	**20:80**		**40:60**		**60:40**		**90:10**	
	one lot/yr.	cont. feed	one lot/yr.	cont. feed	one lot/yr.	cont. feed	one lot/yr.	cont. feed
Corn at $2.50/bu								
Hay $30/ton	$ 48	$116	$ 53	$158	$53	$192	$55	$241
$60/ton	−17	5	11	74	30	138	50	227
$90/ton	−82	−105	−29	−11	8	84	46	216
Corn at $3.00/bu								
Hay $30/ton	39	100	36	126	33	144	30	171
$60/ton	−26	−10	−4	42	10	89	25	157
$90/ton	−91	−121	−45	−44	−12	36	21	146
Corn at $3.50/bu								
Hay $30/ton	29	83	20	92	13	96	5	101
$60/ton	−36	−27	−20	8	−10	41	1	90
$90/ton	−101	−138	−60	−76	−33	−14	−4	76
Corn at $4.00/bu								
Hay $30/ton	19	68	4	59	−8	46	−19	34
$60/ton	−46	−42	−36	−25	−30	−7	−24	20
$90/ton	−111	−155	−76	−109	−53	−62	−29	6
Corn at $4.50/bu								
Hay $30/ton	10	−51	−12	25	−28	−2	−44	−36
$60/ton	−55	−60	−52	−59	−51	−58	−49	−50
$90/ton	−120	−170	−92	−143	−73	−110	−53	−62
Corn at $5.00/bu								
Hay $30/ton	0	34	−28	−8	−48	−51	−69	−106
$60/ton	−65	−77	−68	−92	−71	−105	−73	−119
$90/ton	−130	−187	−108	−177	−93	−160	−78	−131

[a] Returns calculated for selling price $64/100 lb for 1100 lb steers, and purchase price of $60/100 lb for 700 lb feeder steer.
Source: Meiske, 1983.

weights. Other large-framed calves, weaning 500 to 700 lbs at weaning and possessing a high genetic potential for gain, are usually phased onto high-energy rations as quickly as possible. The latter feeding program would be similar for 700 lb to 900 lb thin to moderately fleshed, yearling cattle.

Concentrate to roughage ratios can vary from 100% roughage to 100% concentrate to any combination in between the two extremes. More conventional rations will range from 40% roughage-60% concentrate to 10% roughage-90% concentrate. The most commonly used sources of roughage are silage (essentially 50% grain and 50% forage on a dry matter basis), hay, and corn cobs, while the most common concentrates or grains are corn in the Midwest and Northern Great Plains, milo in the Southern Great Plains and Southwest, and barley in the northwestern part of the United States. The supplement mixture of protein, minerals, vitamins, and possibly additives, usually supplied at 1 to 2 lbs per day, would also be considered concentrates.

Selecting the concentrate to roughage ratio to be fed is largely based on economics, assuming the feeder has the flexibility of making these choices. Table 14.24 gives an example of economically evaluating different concentrate to roughage choices at various feed prices.

Proper ration mixing and delivering correct amounts of feed are probably the most important concern in feeding cattle. Where high daily rates of gain are economically feasible, a high level of feed intake of a well-balanced, energy-dense ration is essential. Large amounts of feed consumed on a daily basis by each animal in the feedlot is largely determined by the appetite of the animal. Appetite is influenced both by genetic and environmental factors. Genetic selection for gain appears to be primarily a selection preference for cattle which have large appetites. Previous level of nutrition which accounts for compensatory gain, health of cattle, weather, kinds of feed, and condition of feed in the feedbunk all influence the appetite of cattle.

Individuals monitoring feedbunks and observing recent fecal material assess the appetite of cattle and condition of feed presented to the cattle. This individual assessment is critical to proper management of the daily feeding program in the feedlot.

SELECTED REFERENCES

Publications

Ames, D. R. *Feeding beef cows in winter.* Angus Journal, Jan. 1981.

Ames, D. R. *Livestock nutrition in cold weather.* Animal Nutrition and Health, Oct. 1980.

Anderson, D. C. *Use of Cereal Residues in Beef Cattle Production Systems.* J. of Anim. Sci. 46:849.

Church, D. C., 1976. *Digestive Physiology and Nutrition of Ruminants Vol 1. Digestive Physiology.* Corvallis: O & B Books.

Church, D. C., 1979. *Digestive Physiology and Nutrition of Ruminants Vol 2. Nutrition.* Corvallis: O & B Books.

Church, D. C., 1980. *Digestive Physiology and Nutrition of Ruminants Vol 3. Practical Nutrition.* Corvallis: O & B Books.

Church, D. C. and Pond, W. G., 1982. *Basic Animal Nutrition and Feeding.* New York: John Wiley and Sons.

Ensminger, M. E. and Olentine, C. G., Jr., 1978. *Feeds and Nutrition.* Clovis, Ca.: The Ensminger Publishing Company.

Great Plains Beef Cattle Feeding Handbook. Coop. Ext. Serv. in CO, KS, MT, NE, NM, ND, OK, SD, TX, and WY.

Great Plains Beef Cow–Calf Handbook. Coop. Ext. Serv. in CO, KS, MT, NE, NM, ND, OK, SD, TX, and WY.

Guyer, P. and Henderson, P. *Estimating corn and sorghum silage value.* Great Plains Beef Cattle Handbook, GPE-2401.

Jurgens, M. H., 1982. *Animal Feeding and Nutrition.* Dubuque, Ia.: Kendall/Hunt Publishing Co.

Light Cattle Management Seminar, 1982. Colo. State Univ. Vet. Med. and Anim. Sci. Ext. Serv.

Matsushima, J. K., 1979. *Feeding Beef Cattle.* New York: Springer-Verlag.

Meiske, J. *Maximizing ration returns.* Feedlot Management, Feb. 1983.

National Research Council, 1976. *Nutrient Requirements of Beef Cattle.* Washington, D.C.: National Academy Press.

National Research Council, 1981. *Effect of Environment on Nutrient Requirements of Domestic Animals.* Washington, D.C.: National Academy Press.

Perry, T. W., 1980. *Beef Cattle Feeding and Nutrition.* New York: Academic Press.

Preston, R. L. *Typical composition of feeds for cattle and sheep, 1983–1984.* Feedlot Mgt. 1984 Planner Issue.

Schake, L. *The Best Way to Process Grain.* Feedlot Mgt., Sept. 1981.

Thomas, J. W., 1978. *Preservatives for Conserved Forage Crops.* J. of Anim. Sci. 47:721.

Ward, J. K., 1978. *Utilization of Corn and Grain Sorghum Residues in Beef Cow Forage Systems.* J. of Anim. Sci. 46:831.

Visuals

"Basic Beef Cattle Nutrition." Sound filmstrip. Vocational Educational Productions. Calif. Polytechnic State Univ., San Luis Obispo, CA 93407.

"Nutritional Deficiencies and Toxicities of Poultry, Ruminants and Swine." A slide set. Contact F. N. Owens, 168 Vet. Med., Okla, State Univ., Stillwater, OK 74074.

"Supplementing Beef Cattle." Video tape. 27 minutes. National Cottonseed Products Assoc. Inc., P.O. Box 12023, Memphis, TN 38182–0023.

15
The Retail Product of Beef

The major consumable product from beef cattle is the carcass which represents from 45% to 65% of the animal's liveweight. Carcass components of primary interest are lean, fat, and bone, and to a lesser extent, connective tissue. Characteristics of the retail product of beef, preferred by consumers, are a large proportion of wholesome, highly palatable lean, minimum amounts of fat, and little or no bone or connective tissue. Carcasses are evaluated with the intent of identifying these desired characteristics of the retail product of beef, along with preferred palatability characteristics.

CARCASS GRADES

Carcass grades have been developed by the USDA to primarily facilitate trading in the marketplace. Carcass grades were officially accepted on a voluntary basis in 1927. There have been several revisions in grades (see Chapter 24) because of changes in feeding programs, economics of product values, consumer preferences, and other production-management changes.

Carcass grades should not be confused with inspection. Both grading and inspection have marks that are applied to carcasses (Figure 15.1). Inspection is concerned with wholesomeness of meat. USDA inspectors inspect the beef in packing plants during the slaughtering process, where they examine the carcass and internal organs for evidence of disease. Carcasses that pass inspection are given the mark of a circle which contains the abbreviations, U.S. INSP'D & P'S'D (U.S. INSPECTED AND PASSED) and a number which identifies the packing plant where inspection occurred. Carcasses which are condemned have uses other than for human consumption. Inspectors also evaluate packing plant sanitation, assure correct labeling, and prevent adulteration of the carcass or its resulting products.

Figure 15.1. Inspection marks and grade marks which are placed on the beef carcass. USDA choice represents one of several designated quality grades. Yield grade 2 represents one of the five yield grades. Courtesy of USDA.

INSPECTION MARK

GRADE MARKS

Quality Grades

Carcass quality grades are Prime, Choice, Good, Standard, Commercial, Utility, Cutter, and Canner. These quality grades are generally expected to relate to consumer palatability characteristics of tenderness, juiciness, and flavor. Factors used to determine quality grades are (1) bone maturity, (2) marbling, (3) lean color, and (4) firmness and texture of the lean. Bone maturity and marbling are the two most important factors.

Maturity classifications are based on the relationship that as animals increase in age, tenderness of the meat decreases. Bone maturity estimates physiological age of the animal and maturity is determined primarily by bone ossification. After the carcass has been split into two halves, the dorsal processes of the vertebrae are exposed and are evaluated for amount of ossification. Carcasses having A maturity will have the largest amount of cartilage on the tips of the vertebrae processes, while complete ossification has occurred in carcasses with E maturity. Carcasses with C, D, or E maturity are eligible for only the Commercial, Utility, Cutter, and Canner grades.

Color of the lean is evaluated along with bone maturity to determine the final maturity classification. Color of the lean becomes darker with advancing age of the animal. An A maturity carcass is usually a bright, cherry red while an older (C, D, or E) maturity carcass is usually a darker, coarser textured lean. Occasionally an A or B maturity will have dark lean ("dark cutter"), resulting from the animal being stressed shortly before slaughter. Dark cutters, with A or B maturity, are given a quality grade often lower than what the amount of marbling would indicate.

Marbling, flecks of fat within the ribeye muscle, is visually evaluated at the 12th and 13th rib interface. The nine degrees of marbling vary from abundant to practically devoid (Figure 15.2).

Figure 15.3 shows how the final carcass quality grade is determined by using a combination of maturity and marbling, while Figure 15.4 visually depicts several quality grades. Note that a carcass usually has to have more marbling to stay in the same

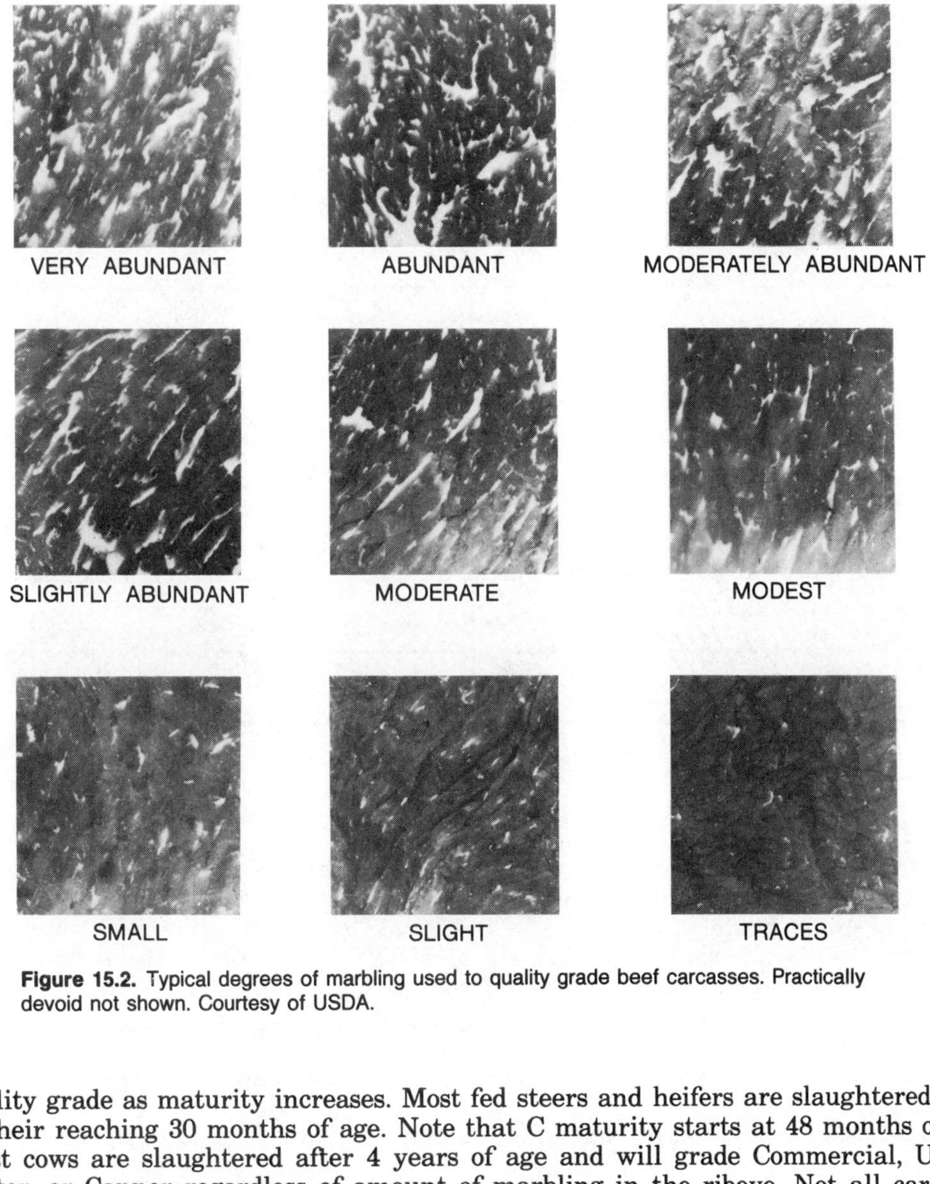

Figure 15.2. Typical degrees of marbling used to quality grade beef carcasses. Practically devoid not shown. Courtesy of USDA.

quality grade as maturity increases. Most fed steers and heifers are slaughtered prior to their reaching 30 months of age. Note that C maturity starts at 48 months of age. Most cows are slaughtered after 4 years of age and will grade Commercial, Utility, Cutter, or Canner regardless of amount of marbling in the ribeye. Not all carcasses are quality graded; however, those that are graded will be identified with the quality grade stamp (Figure 15.5).

Yield Grades

Yield grades (also called cutability) refer to pounds of boneless, closely trimmed, retail cuts (BCTRC) from the round, loin, rib, and chuck. Marketing communications

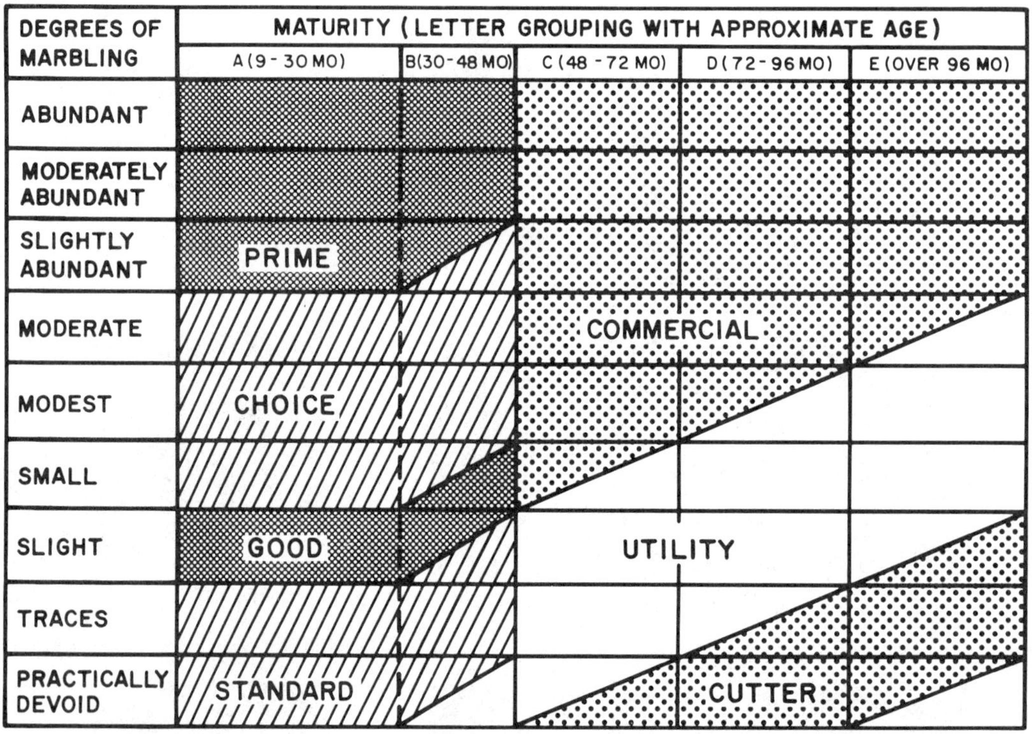

Figure 15.3. Relationship between marbling, maturity, and carcass quality grade. Courtesy of USDA and CSU Graphics.

generally use the yield grades designated numerically 1 through 5. Yield grades, however, will often be reported in tenths in research and carcass contests. A 700 lb carcass with a 3.0 yield grade would have 350 lbs of BCTRC (700 × 0.50), while a 700 lb carcass with a 4.0 yield grade would have 334 lbs of BCTRC (700 × 0.477). Table 15.1 shows yield grades and their respective percentages of BCTRC.

Yield grades are determined from four carcass characteristics:

1. amount of fat, measured in tenths of inches, over the ribeye muscle or longissimus dorsi (Figure 15.6).
2. kidney, pelvic, and heart fat (KPH fat—usually estimated as a percentage of carcass weight).
3. area of ribeye muscle (REA), measured in square inches (Figure 15.7).
4. hot carcass weight. Carcass weight reflects amount of intermuscular fat. Generally as the carcass increases in weight, the amount of intermuscular fat increases also.

Fatness is the primary factor in determining yield grades. The fat measurement, over the ribeye muscle, measures most differences in fatness in the carcass (Figure 15.8).

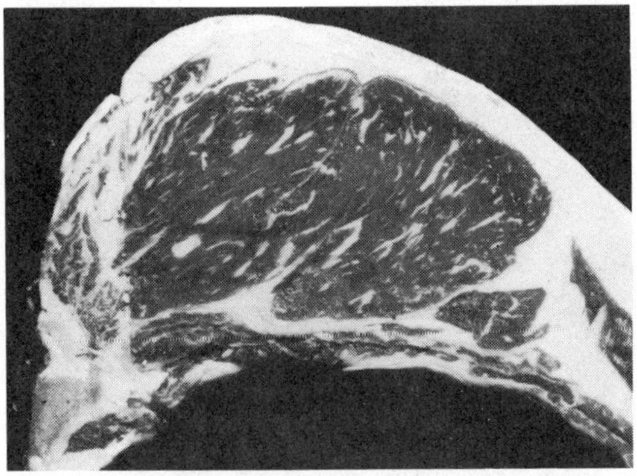

Prime (moderately abundant marbling)

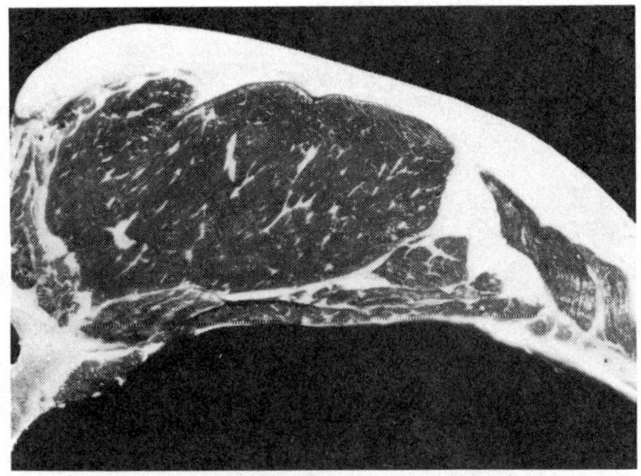

Choice (modest marbling)

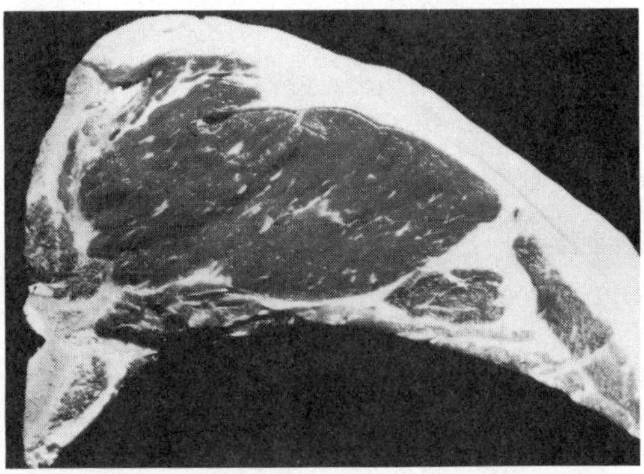

Good (slight marbling)

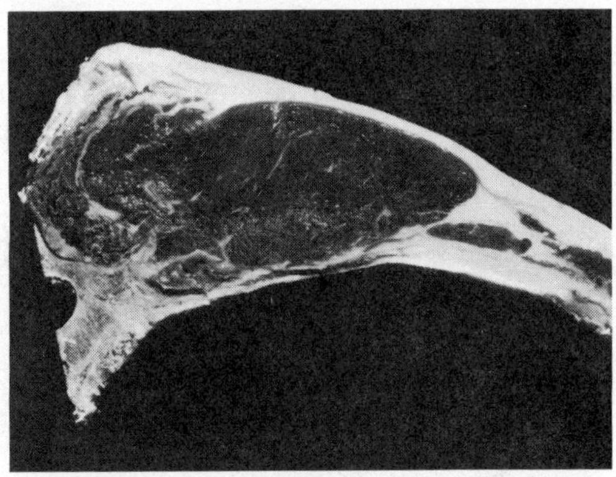

Standard (traces of marbling)

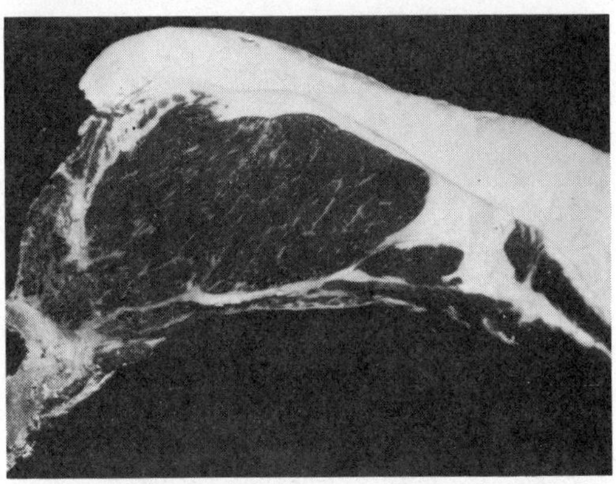

Commercial (moderately abundant marbling; however, E maturity put it into this grade)

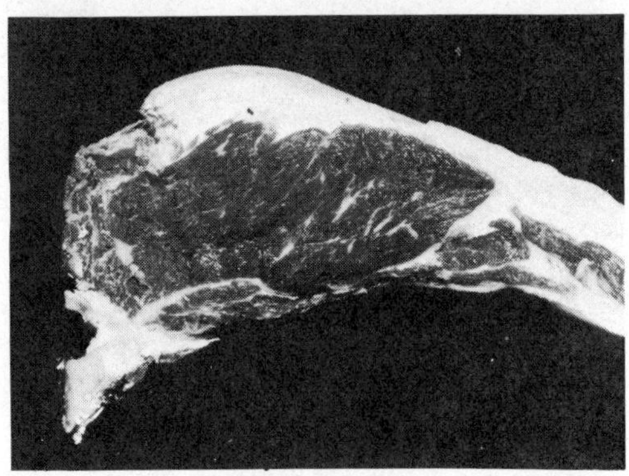

Utility (modest marbling)

Figure 15.4. Exposed ribeye muscles (between twelfth and thirteenth ribs) showing several carcass quality grades resulting from combinations of marbling and maturity. Courtesy of the National Live Stock and Meat Board.

Figure 15.5. Carcasses which are quality graded have the quality grade stamp "rolled" over the entire length of the carcass. Courtesy of the American Simmental Association.

Yield grades have been developed by numerous cut-out evaluations. Actual formulas which are utilized to compute yield grades or cutability are as follows:

$$
\begin{aligned}
\%\ \text{BCTRC} = 51.54 &- 5.784\ (\text{inches of fat at 12th-13th rib}). \\
&- 0.462\ (\%\ \text{kidney, heart, and pelvic fat}) \\
&- 0.0093\ (\text{lbs hot carcass weight}) \\
&+ 0.740\ (\text{square inches of ribeye muscle}) \\
\text{Yield Grade} = 2.50 &+ 2.50\ (\text{inches of fat at 12th-13th rib}) \\
&+ 0.20\ (\%\ \text{kidney, heart, and pelvic fat}) \\
&+ 0.0038\ (\text{lbs hot carcass weight}) \\
&- 0.32\ (\text{square inches of ribeye muscle})
\end{aligned}
$$

Table 15.1. Carcass Yield Grades and the Yield of BCTRC (Boneless, Closely Trimmed Retail Cuts from the Round, Loin, Rib and Chuck)

Yield Grade	% BCTRC	Yield Grade	% BCTRC
1.0	54.6	3.6	48.7
1.2	54.2	3.8	48.2
1.4	53.7	4.0	47.7
1.6	53.3	4.2	47.3
1.8	52.8	4.4	46.8
2.0	52.3	4.6	46.4
2.2	51.9	4.8	45.9
2.4	51.4	5.0	45.4
2.6	51.0	5.2	45.0
2.8	50.5	5.4	44.5
3.0	50.0	5.6	44.1
3.2	49.6	5.8	43.6
3.4	49.1		

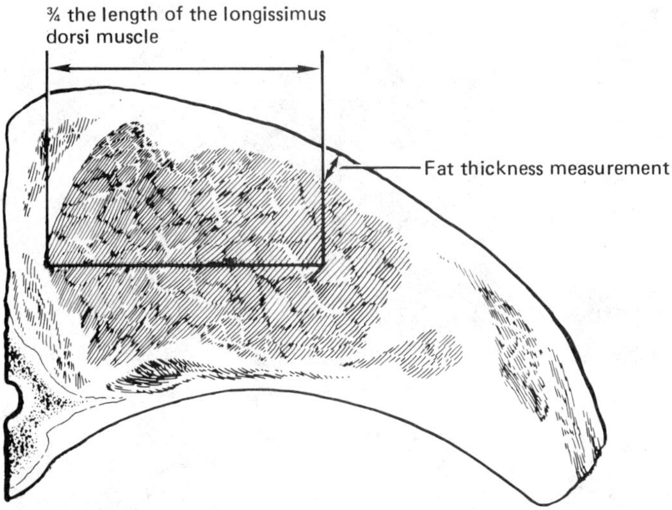

Figure 15.6. Location of fat measurement over the ribeye (longissimus dorsi muscle). Drawn by Dennis Giddings.

Instead of using the formulas, a couple of simplified methods can be used to determine yield grades of carcasses or potential yield grades in live, slaughter cattle.

The first method is used by acquiring a Beef Carcass Yield Grade Finder. These are available from the USDA. This Yield Grade Finder is a 3 inch by 8 inch simplified slide rule which is calibrated to determine the yield grade. The four carcass characteristics are entered in the slide rule and resulting yield grade is read.

The second method uses the increments for the carcass characteristics as shown in Table 15.2.

An example of a carcass with 0.50 in. of fat, REA of 11.30 sq. in., KPH of 3.50%, and a carcass weight of 625 lbs will be used. The starting base is a preliminary yield grade of 2.00. With 0.50 in. of fat ($5 \times 0.25 + 2.00 = 3.25$). A REA of 11.3 ($11.0 - 11.3 =$

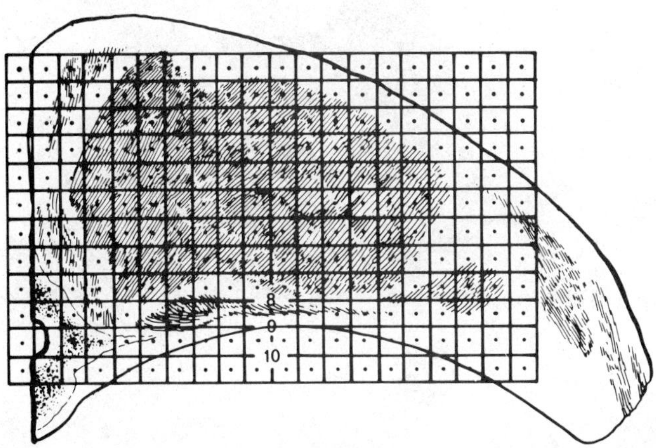

Figure 15.7. Plastic grid is placed over the ribeye muscle to measure area. Each square represents 0.1 square inch. Drawn by Dennis Giddings.

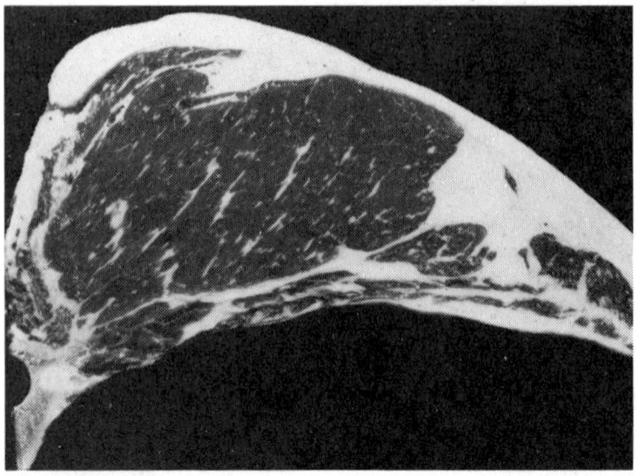

Yield Grade 1 (Fat 0.2 in., ribeye area 13.9 sq in.)

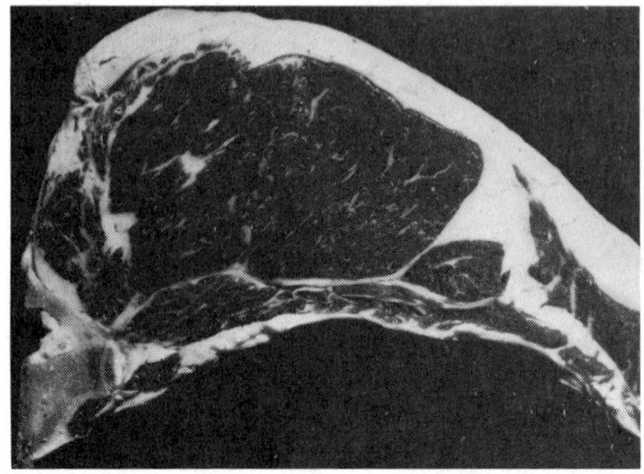

Yield Grade 2 (Fat 0.4 in., ribeye area 12.3 sq in.)

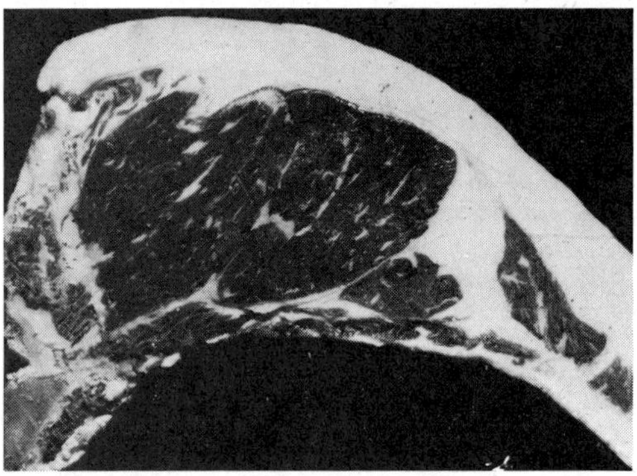

Yield Grade 3 (Fat 0.6 in., ribeye area 11.8 sq in.)

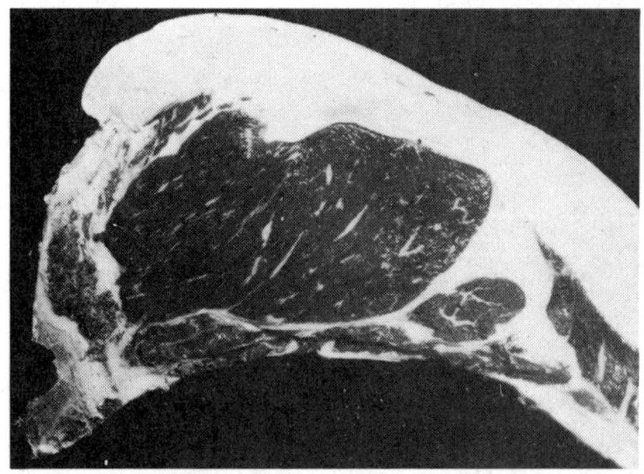

Yield Grade 4 (Fat 0.9 in., ribeye area 10.5 sq in.)

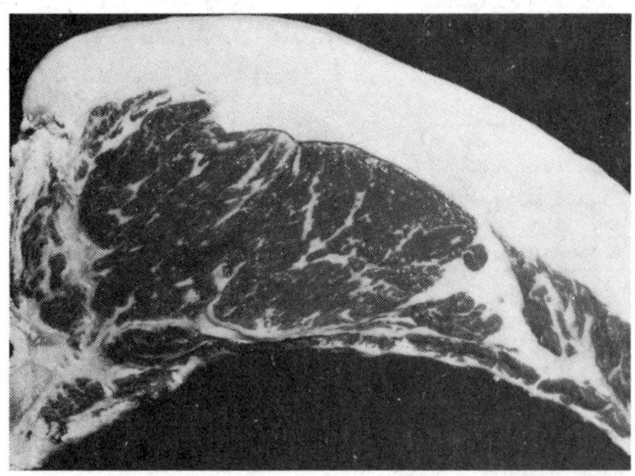

Yield Grade 5 (Fat 1.1 in., ribeye area 10.9 sq in.)

Figure 15.8. The five yield grades of beef shown at twelfth and thirteenth ribs. (Courtesy of the National Live Stock and Meat Board.

Table 15.2. Short-Cut Method to Determine Yield Grade. (Assumes a Starting Yield Grade Base of 2.00)

Carcass Trait	Base	Increment
Fat	0.00 in.	0.10 in. = ±0.25
REA	11.0 sq. in.	1.0 sq. in. = ±0.33
KPH fat	3.51	1.0% = ±0.20
Carcass wt	600 lbs	100 lb = ±0.40

0.30) is ⅓ of 1.0, so the increment would be −0.11 from 3.25 = 3.14. KPH fat is the same as the base, so KPH has no influence on yield grade in this example. The carcass weight is 625 (625 − 600 = +25 which is ¼ of 100 lbs) so the increment is 4 × ¼ = +0.1 added to 3.14 equals a final yield grade of 3.24 or a 3.2 yield grade.

DISTRIBUTION OF BEEF GRADES

Grading of beef by USDA meat graders is a voluntary program on the part of packers. Each packer may have all, none, or part of their beef carcasses graded and marked with grade stamps. Most packers will have part of their beef marked, usually those carcasses that will be in higher yield and quality grades. Since 1976 all beef that is quality graded must be yield graded as well.

Table 15.3 shows the distribution of grades for beef actually graded and stamped during 1981.

VALUE OF THE CURRENT CARCASS GRADES

During the early 1980s, there has been considerable discussion and assessment of current USDA beef carcass grades, with primary attention focused on quality grades.

Table 15.3. Summary of Beef Graded for Calendar Year 1981[a]

Quality Grade	%	Yield Grade	%
Prime	5.9	1	1.9
Choice	89.0		
Good	4.3	2	30.3
Standard	0.2		
Commercial	0.2	3	57.2
Utility	0.4		
Cutter	0.0	4	9.4
Canner	0.0	5	1.2

[a] Represents 12.1 million lbs which is approximately 55% of the beef produced for 1981.
Source: Livestock, Meat, Wool Market News, March 9, 1982.

There has been strong expression from several segments of the beef industry that quality grades need to be revised.

Evidence submitted to support revision of quality grades is based on several factors. The primary one is that too much emphasis is placed on marbling where substantial research results show that only 10% to 15% of beef palatability differences are accounted for by differences in marbling. Yet marbling accounts for 90% of the variation in carcass grades of young cattle.

Another reason is that consumers are desiring leaner beef. Marbling is difficult to identify in the live animal, therefore, the feeder extends feeding time to hopefully produce a higher percent of cattle grading Choice. This supposedly gives a higher economic return to the cattle feeder because of a price differential of Choice over Good. This is a questionable practice, because it dictates a longer feeding program involving higher gain costs, particularly during the last 30 days of the feeding program. Feeding for higher marbling levels encourages overfattening of cattle which in turn produces a less desirable consumer product.

During 1981 and 1982, several proposals were made to the USDA to change quality grades (Figure 15.9). One proposal was to change the Good grade name to Select and narrow the marbling requirements for the Choice grade.

The National Cattlemen's Association's (NCA) proposal recommended the following changes: (1) reduce the minimum amount of marbling required for Prime grade from slightly abundant to moderate, (2) allow carcasses with modest, small, and slight marbling to grade Choice, provided carcasses with slight marbling also have at least 0.30 inches of fat over the 12th rib, and (3) redefine the Good grade to include carcasses with slight marbling and less than 0.30 inches fat thickness and carcasses with less than slight marbling or yellow fat. NCA has also encouraged the USDA to consider changes in grade names for all C, D, and E maturity carcasses since most of this beef goes into the manufacturing trade.

The USDA made proposed changes in quality grades on 30 December 1981 (Figure 15.9). The proposed changes were: (1) The minimum amount of marbling required for steer, heifer, and cow beef in A maturity to qualify for Prime, Choice, and Good would be reduced to minimum moderate, typical slight, and minimum traces, respectively, (2) An increase of two degrees of marbling would be required to offset increases in maturity across the width of B maturity, (3) Beef in A and B maturities failing to meet the minimum requirements for Good would be graded Utility, and (4) The quality grade requirements for bullock beef would be changed to the same as those proposed for steer, heifer, and cow beef in A maturity.

Public hearings on the proposed changes were held at five different locations during February 1982, and written comments were submitted to the USDA 31 March 1982.

Most comments from various beef industry segments appeared to favor some change in quality grade standards. Some felt the proposed grade changes, if implemented, would be only a token change and that further changes would be needed. They cited the already widely accepted no-roll (primarily Good grade) beef being merchandised by Safeway (Safeway Quality) and King Soopers (Generic Beef) which in some of their stores comprised the highest percentage of beef sales. Certain individuals and organizations (National Restaurant Association, National Association of Meat Purveyors, and others) opposed the grade changes. They felt the changes would cause them and consumers

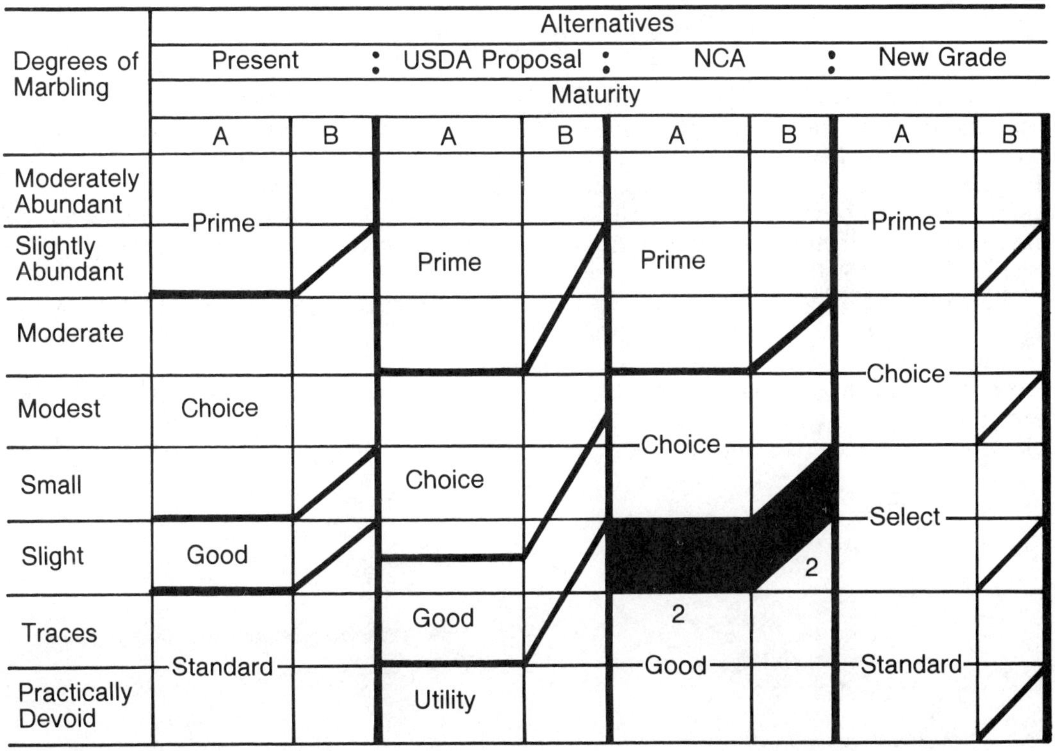

Figure 15.9. Comparison of marbling requirements, maturity, and quality grades for three alternatives to the present quality grades. Courtesy of USDA.

to pay higher prices for a lower quality product. Some midwest cattle feeders opposed the changes because they believed the change would economically favor large commercial feeders because of a shorter feeding time and a larger turnover of cattle fed each year.

The USDA decided in late 1982 to make no changes in current quality grades. They did, however, suggest that the issue was not dead and that further discussion should continue.

Research studies which resulted in the yield grade formula were done in the 1950s. A report by Murphey in 1960 on some of these studies became the basis for USDA yield grade standards for beef which were officially adopted in 1965. Since that time, there have been several studies assessing accuracy of the yield grade equation. Some studies have verified the original yield grade equation (*J. Anim. Sci.* 50:841), while other studies suggest that revisions in cutability calculations are needed (*J. Anim. Sci.* 41:548). The latter may be true because the present cattle population is more diverse in biological types (e.g., Limousin, Charolais, and Longhorn) compared to the 1950s and 1960s.

CARCASS CHARACTERISTICS AFFECTING BEEF PALATABILITY

Tenderness, flavor and juiciness are palatability characteristics of beef that have been identified as being most important to the consumer Numerous research studies have been conducted to identify relationships of carcass traits and processing and handling of carcasses to the desired palatability characteristics.

Tenderness is considered the most important palatability characteristic of meat. To understand the factors affecting tenderness, some knowledge of muscle structure is needed (Figure 15.10).

The three major components of meat tenderness are (1) myofibrillar component, (2) collagen content of connective tissue, and (3) how method of cooking affects the myofibrillar and connective tissue components in different ways.

It is well known that longer sarcomeres in myofibrils will be more tender than shorter sarcomeres. In living muscle, myofibrils shorten during contraction and lengthen during relaxation. After death, muscle fibers contract during the onset of rigor mortis. Rate

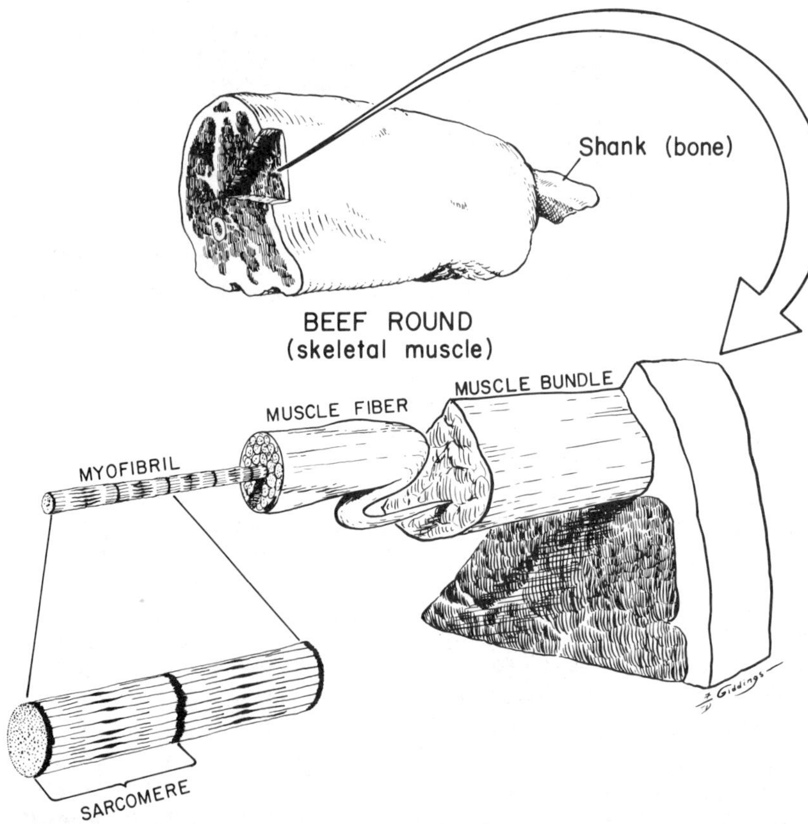

Figure 15.10. The fundamental structure of muscle in the beef carcass. Drawn by Dennis Giddings.

of cooling during the first few hours of postmortem determines sacromere length and thus a portion of the tenderness of the meat.

Rate of cooling and the resulting cold-induced toughening of meat appears to be significantly affected by size of carcass and amount of fat insulation covering the carcass. Evidence is accumulating that intensive feeding improves tenderness by increasing carcass weight and fatness and thereby reducing susceptibility of the carcass to rapid postmortem chilling and cold-induced toughening (Tatum 1981). There is also some evidence that fat cover on the carcass relates to tenderness in ways other than preventing cold shortening (Marsh, Natl. Beef Grading Conf., p. 99).

The other part of myofibril tenderness is determined by the strength of the myofibril. Myofibrils are considered to have a relatively constant strength in healthy, living muscle tissue. After death and rigor development, the myofibril structure is weakened by proteolytic enzymes. This breakdown of protein components of myofibrils increases tenderness of meat. There is some evidence that intensive feeding and rapid gains generate more protein turnover which in turn is associated with a higher myofibril fragmentation and collagen solubility (*J. Anim. Sci.* 52:757).

Aging is holding beef under refrigeration for 7 to 14 days postmortem. Proteolylic weakening of the myofibril structure that occurs during this time counteracts some of the cold-induced sarcomere shortening occurring during rigor. Some methods of aging beef have been accelerated to 2 to 5 days by increasing the temperature (60 °F for 6 to 12 hours; then below 40 °F for 2 to 3 days versus below 40 °F for 7 to 14 days). Proteolytic enzymes are more active at higher temperatures.

Electrical stimulation, administering an electrical shock to the carcass during the slaughter-dressing sequence, appears to speed up rigor reducing cold-shortening and initiates myofibril weakening earlier. There is also some evidence that electrical stimulation causes some structural damage to muscle fibers, thus improving tenderness.

Muscle stretching, hanging the carcass by hip suspension rather than hock suspension, has been shown to lengthen the sacromeres during rigor. This method has not been widely accepted by the industry because of economics, even though tenderness is improved.

The connective tissue component of meat tenderness has several interesting relationships. Collagen is the primary protein in connective tissue. This unique protein forms an extensive network that envelops individual muscle fibers and extends to the attachments to the bones of the skeleton. Since collagen must transmit the contraction of myofibrils to the skeleton, it must be a very tough, resistant tissue.

Muscles used for locomotion require much more collagen to cause extensive skeleton movement. Because limb muscles have considerably more collagen than support muscles, they are less tender. Loin muscles, which are support muscles, have relatively low levels of collagen and hence are the most tender muscles. Differences in collagen levels between different muscles reflect some value differences between different beef cuts. There is no change in amount of collagen in muscles after slaughter. Therefore, tenderness due to collagen content of muscles, is established at time of slaughter. Meat cuts that have high collagen content do not benefit to the same degree from the aging process as do low collagen meats.

In addition to collagen content, there is crosslinking between collagen molecules which affect meat tenderness. As animals grow older, more crosslinkages between collagen molecules are formed. Older animals are less tender than younger animals because crosslinkages between collagen molecules are more numerous and more stable. This is the primary reason for the maturity classifications in USDA quality grades. Collagen crosslinkages are established at time of slaughter and change very little during postmortem aging.

Another aspect of collagen, being evaluated more critically, is its stability or solubility. There is some research work which indicates that steers fed high-concentrate rations, thus having more rapid daily gains, have more tender meat than those fed low-energy rations. This is because the collagen of rapidly gaining steers is more soluble or less stable.

The cooking aspect of beef tenderness can both tenderize and toughen protein components of meat at the same time. Myofibrillar porteins become less tender as cooking temperature increases. This effect is similar to proteins in egg whites that turn from a liquid to solid during cooking. Collagen, however, softens during moist heat cookery. Under steam cooking, collagen will eventually turn to a liquid. If collagen is allowed to dry, however, it will toughen. Therefore, meat with high collagen content must be cooked under moist heat to maximize tenderness, and low collagen meats can be cooked under dry heat methods and still be tender. Cooking recommendations for various retail cuts are shown in Figure 15.11.

Mechanical and enzyme tenderizing can make dramatic changes in tenderness of beef. Both the myofibrillar and connective tissue component of tenderness are modified. The mechanical tenderizer physically disrupts the tissue through penetration of small blades into the meat. Meat quality is not altered with the mechanical tenderizer. The enzyme tenderizing process involves application of plant or fungus enzymes to meat. These enzymes soften the myofibrillar and connective tissue proteins during the cooking process.

Restructured Retail Cuts

Figure 15.11 shows the standard retail cuts typically processed from the beef carcass. Table 2.4 in Chapter 2 identifies very few beef products available to the consumer which are intermediate in value between hamburger, steaks, or roasts. There has been considerable interest in developing intermediate valued retail cuts of beef by restructuring some products which are typically used in the manufacture of ground or processed beef. *Intermediate Value* is a term used to identify products having characteristics, particularly textural, somewhere between ground beef and those of intact muscle cuts. The assumption is made that the more closely these products simulate the intact muscle cuts, the higher their value is perceived to be.

Restructured beef refers primarily to those intermediate valued products which are flaked and formed, sliced and formed, or sectioned and formed. Restructuring requires a particle size reduction or modification, blending, and reforming. Binding together of meat pieces is an essential requirement of restructuring. Currently that adhesion in the uncooked state is achieved by freezing, and in the cooked state is dependent upon heat induced bonding. Binding agents of preference at this time are intracellular proteins, especially myosin. Since the bind must occur at particle interfaces, extraction of intracellular proteins to particle surfaces is necesssary. This is accomplished by stirring-mixing-massaging and is known to be enhanced by the presence of salt and phos-

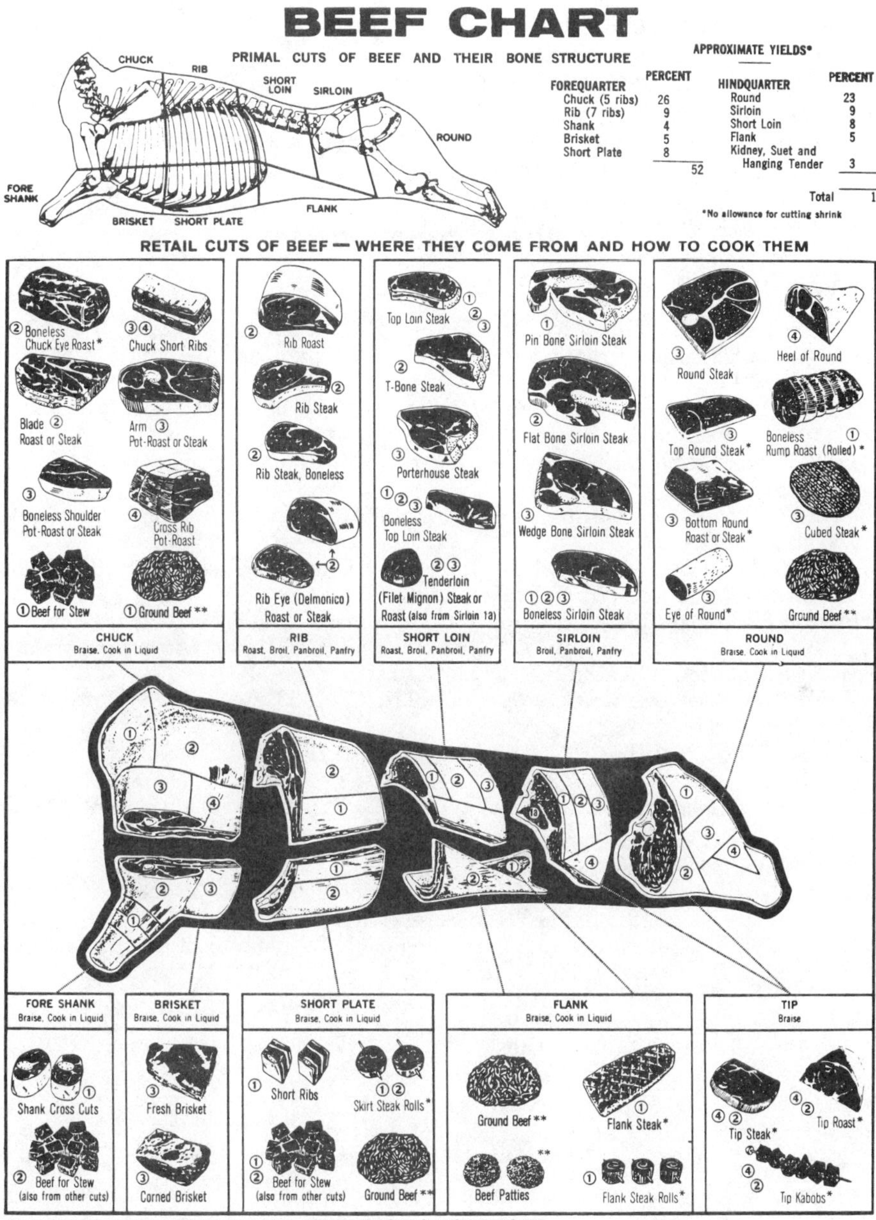

Figure 15.11. The wholesale and retail cuts of beef with the recommended type of cooking. Courtesy of the National Live Stock and Meat Board.

phates. Raw materials typically used for restructuring are trimmings from Choice beef, boneless cow beef, and Choice chucks.

Changing consumer demands, in combination with changes in the type of cattle produced and different beef processing methods, along with economic constraints, dictate the need to continually evaluate the acceptability of the end product being produced. Beef has to compete with other protein sources both in cost and palatability.

SELECTED REFERENCES

Publications

Aberle, E. D., Reeves, E. S., Judge, M. D., Hunsley, R. E., and Perry, T. W., 1981. *Palatability and muscle characteristics of cattle with controlled weight gain: Time on a high energy diet.* J. of Anim. Sci. 52:757.

Breidenstein, B. C., 1982. *Intermediate Value Beef Products* (Restructured Beef Products). Chicago: National Live Stock and Meat Board.

Breidensten, B. C. and Carpenter, Z. L., 1983. *The Red Meat Industry: Product and Consumerism.* J. Anim. Sci. 57:119.

Foods from Animals: Quantity, Quality, and Safety, 1980. Report No. 82. Ames: Council for Agric. Sci. and Tech. (CAST).

Forrest, J. C., Aberle, E. D., Hedrick, H. B., Judge, M. D., and Merkel, R. A., 1975. *Principles of Meat Science.* San Francisco: W. H. Freeman and Company.

Meat Evaluation Handbook, 1977. Chicago: National Livestock and Meat Board.

Murphey, C. E., Hallett, D. K., Tyler, W. E., and Pierce, J. C., 1960. *Estimating yields of retail cuts from beef carcasses.* J. Anim. Sci. 19:1240 (Abstr.).

Proceedings National Beef Grading Conference, 1981. CE-1633. Cooperative Extension Service, Iowa State Univ., Ames, Iowa.

Romans, J. R. and Ziegler, P. T., 1977. *The Meat We Eat.* Danville: The Interstate Printers and Publishers, Inc.

Tatum, J. D., 1981. *Is tenderness nutritionally controlled?* Reciprocal Meat Conf. Proc., vol. 34.

Visuals

"Evaluating Market Beef and Lambs for Yield Grade, 8864;" 28 min. (16 mm color film). Bureau of Audio-Visual Instruction, Univ. of Wisc. Extension, Box 2093, Madison, WI 53701.

"Beef Carcass Evaluation," 82 slides; 32 minute tape using 3 slide projectors. Instructional Media Center, Marketing Division, Michigan State Univ., East Lansign, MI 48824.

16 Growth, Development and Behavior of Beef Cattle

Many aspects of growth, development, and behavior of beef cattle are contained in other chapters of this book. This is because growth and development is defined as an integration of nutrition, genetics, animal breeding, physiology, and meat science. It seems logical, however, to separate and identify some of these specific relationships which allow producers and others to better understand the beef animal, its production and management.

Growth, development, and behavior patterns are not well defined and understood. Most of our information comes from observations (particularly for behavior) and piecing together some limited research data. Most research information comes from studies which were not designed specifically to identify growth and development relationships. With the advent of embryo transfer and embryo-splitting, research techniques are now available to establish some new and unique milestones in understanding growth, development, and behavior relationships. For now, we need to evaluate and synthesize the available information on this subject which can give a reasonably sound basis for more effective production and management decisions.

GROWTH AND DEVELOPMENT

During growth and development, the beef animal changes in form and composition. Growth is defined in several different ways with no one definition being acceptable to all concerned. Generally speaking, growth is an increase in weight until mature size is reached. More specifically, growth implies the production of new biochemical units brought about by cell division, cell enlargement, or incorporation of materials from the environment.

Development has been defined as the directive coordination of all diverse processes until maturity is reached. It involves growth, cellular differentiation, and changes in body shape and form. For the purpose of this discussion, growth and development will be combined basically into one entity.

Beef producers must understand growth and development if they are to optimize the combined traits of reproduction, post-natal growth, efficiency of growth, and carcass composition.

Pre-Natal

The three phases of pre-natal life: the ovum, the embryo, and the fetus are briefly discussed in Chapter 11. The relative size of the fetus changes during gestation with the largest increase in weight occurring during the last 3 months of pregnancy (Figure 16.3).

Embryological development is fascinating in observing a spherical mass of cells differentiating into different shapes of recognizable organs (Figures 16.1 and 16.2.). The endoderm differentiates into the digestive tract, lungs, and bladder; the mesoderm into the skeleton, skeletal muscle, and connective tissue; while the ectoderm differentiates into the skin, hair, brain, and spinal cord. This growth, development and differentiation process, involving primary protein synthesis is directed by DNA chains of chromosomes in the cells and the organizers in the developing embryo.

The fetus undergoes marked changes in shape and form during prenatal growth and development. Early in the prenatal period, the head is much larger than the body. Later, the body and limbs grow more rapidly than other parts. The order of tissue growth follows a sequential trend determined by physiological importance, starting with the central nervous system and progressing to bones, tendons, muscles, intermuscular fat, and subcutaneous fat.

During the first two-thirds of the pre-natal period, most of the increase in muscle weight is due to hyperplasia (increase in number of fibers). During the last 3 months of pregnancy, hypertrophy (increase in size of fibers) represents most of the muscle growth. Individual muscles vary in their rate of growth with larger muscles (those of legs and back) having the greatest rate of postnatal growth. Water content of fetal muscle declines with fetal age, and this decline in water content continues through postnatal growth as well.

Birth

After birth, the number of muscle fibers does not appear to increase significantly, therefore postnatal muscle growth is primarily by hypertrophy. All muscle fibers appear to be red at birth, but shortly thereafter some of them differentiate into white and intermediate muscle types.

At birth, the various body parts have considerably different proportions when compared to mature body size and shape. At birth, the head is relatively large; the legs are long; and the body is small, while in the mature animal, the head is relatively small; the legs relatively short; and the body is relatively large. Birth weight represents approximately 5% to 7% of the mature weight, while leg length at birth is approximately 60%, height at withers approximately 50% of those same measurements at maturity. Hip width and chest width at birth are approximately one-third of the same measure-

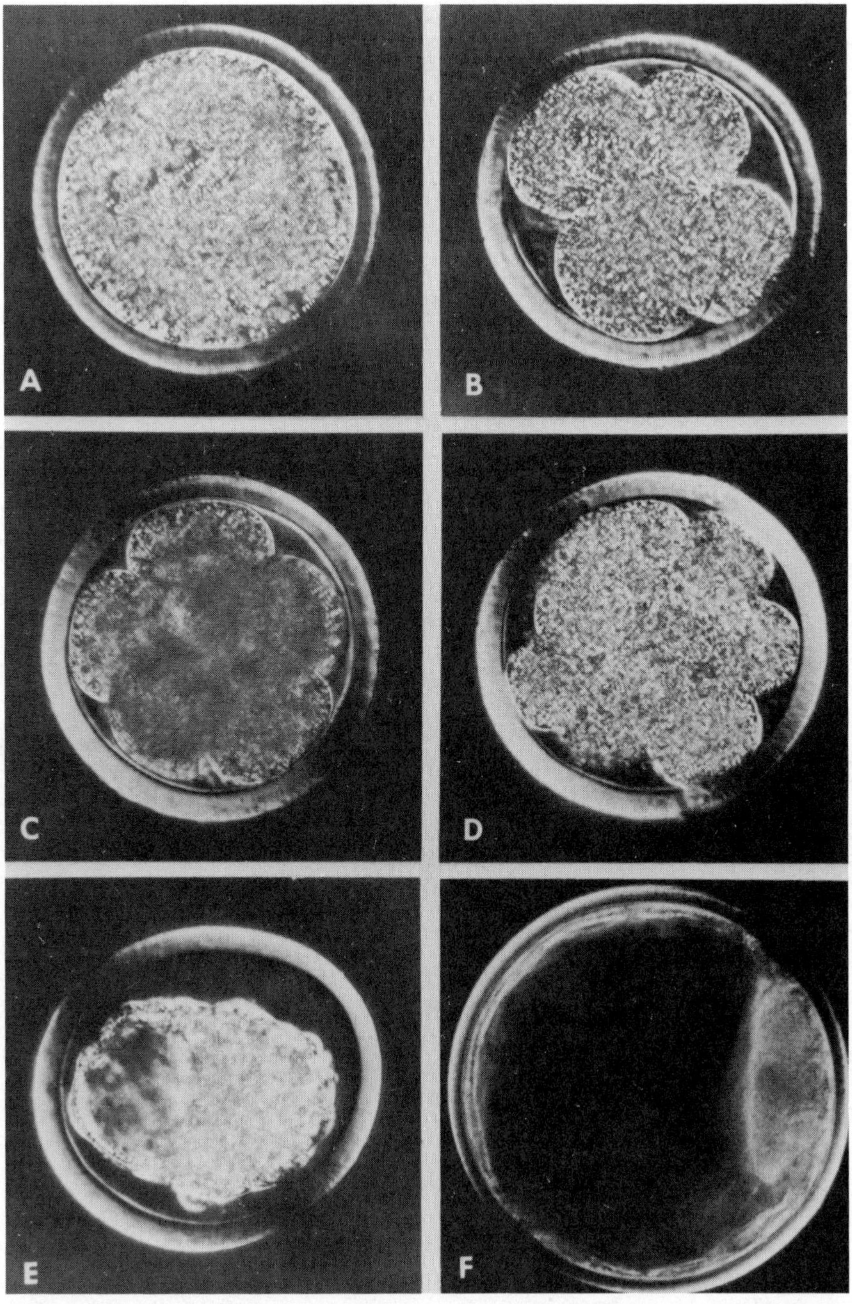

Figure 16.1. Embryonic development during the first 8 days of pregnancy. A. unfertilized egg; B. Four-cell embryo on day 2 of pregnancy (estrus = day 0); C. eight-cell embryo on day 3 of pregnancy; D. an 8- to 16-cell embryo on day 4 of pregnancy; E. very early blastocyst stage (approx. 60 cells); F. expanded blastocyst (> 100 cells) recovered from the uterus on day 8 of pregnancy. Magnification approximately × 300. Courtesy of Seidel, G. E., Jr., 1981. Superovulation and embryo transfer in cattle. Science 211:351. Copyright 1981 by the American Association for the Advancement of Science.

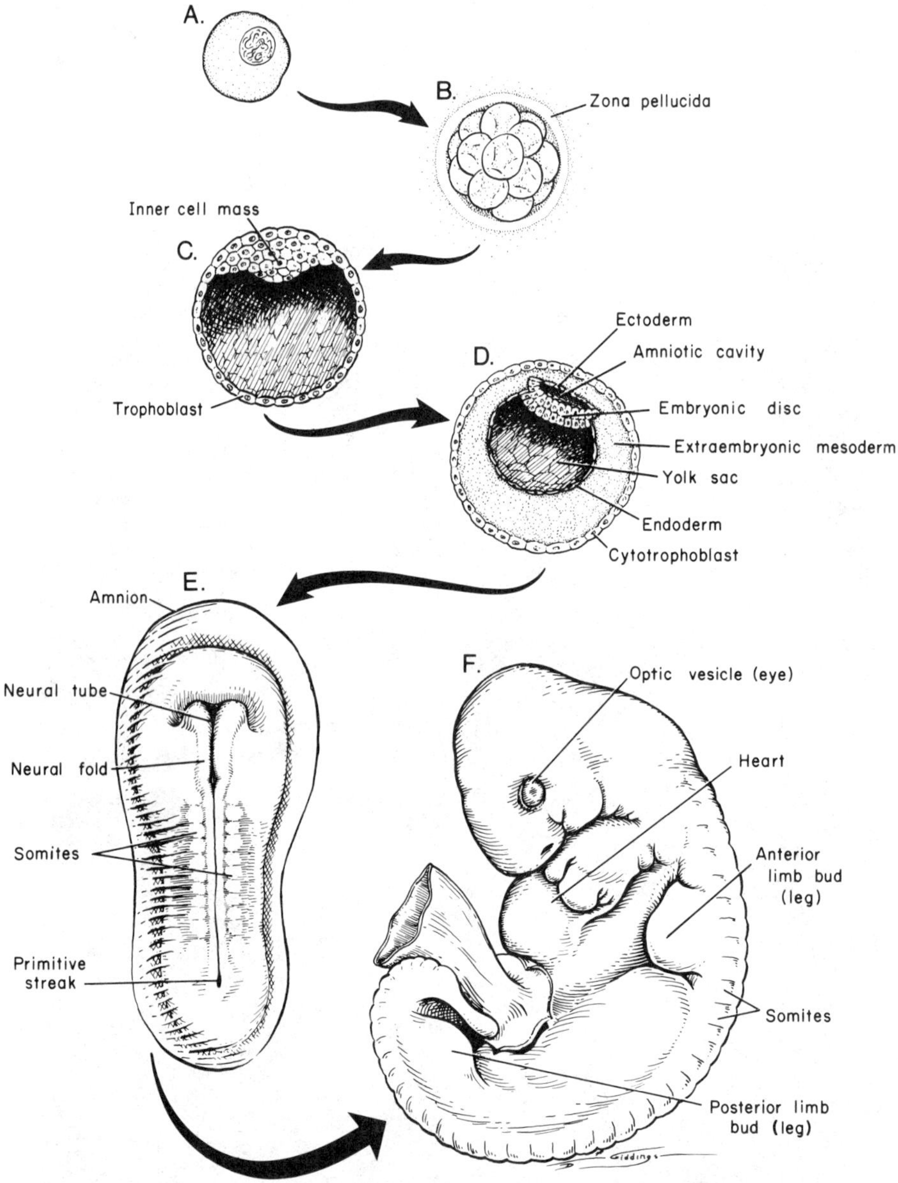

Figure 16.2. The morphogenesis of a single egg cell (A) into a morula (B), then to a blastocyst (C). D. shows the stage at which the two cavities have formed in the inner cell mass; an upper (amniotic) cavity and a lower cavity yolk sac. The embryonic disc containing the ectoderm and endoderm germ layers is located between cavities. E is a cattle embryo showing the neural tube and the stomites. F illustrates the development of a 14 day cattle embryo. Drawn by Dennis Giddings.

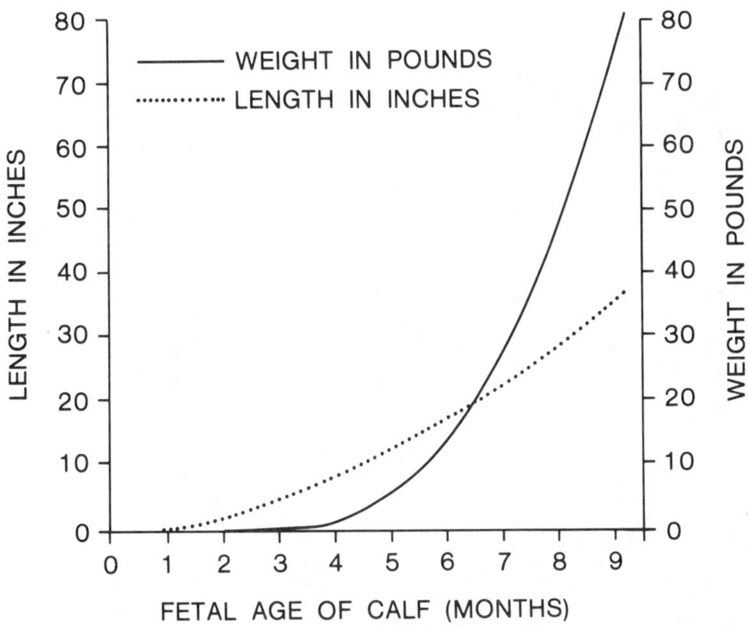

Figure 16.3. Growth of the fetus. Drawn by CSU Graphics.

ments at maturity. This shows that the distal parts (leg and wither height) are developed earlier than the proximal parts (hips and chest).

Liveweight Gain

Beef cattle growth is typically observed or measured by two methods, (1) evaluating body weight changes per unit of time or (2) plotting body weight against age. The first method measures relative growth expressed as average daily gain, weight per day of age, or pounds of lean (muscle) per day of age. The second method results in the expression of growth curves. Growth curves are similar in shape for various breeds and types of cattle although there are noticeable differences when mature size is reached relative to age (Figure 16.4). Growth curves will no doubt receive increased emphasis in future years in anticipation of altering the shape of curves or achieving the best combination of rapid early growth in slaughter cattle and reducing mature size in breeding cattle.

Carcass Composition

A superior carcass is characterized by a low proportion of bone, a high proportion of muscle, and an optimum amount of fat. During the past two to three decades, carcass composition has been improved in beef cattle primarily by reducing the amount of fat in the carcass.

Figure 16.5 shows expected changes in fat, muscle, and bone as cattle increase in liveweight. Fat growth begins rather slowly, then increases geometrically as the animal enters the fattening phase. Bone has a smaller relative growth rate than either fat or

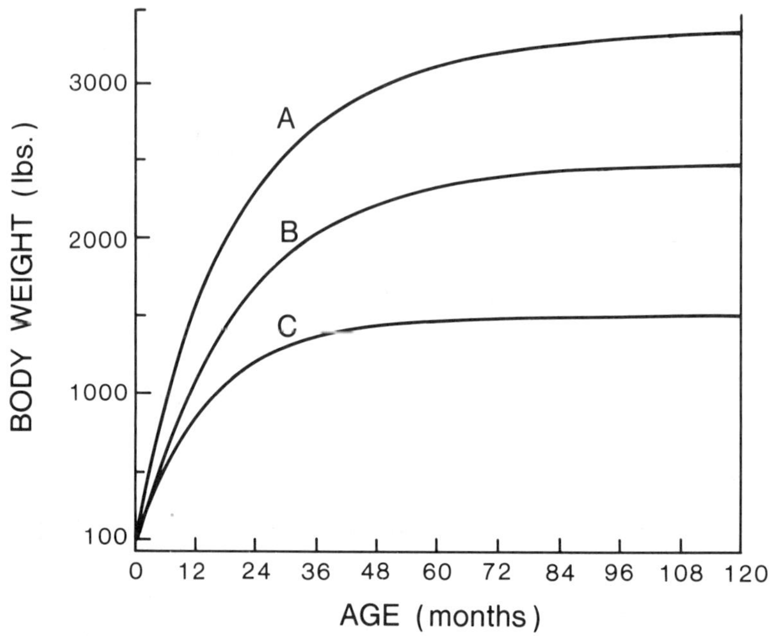

Figure 16.4. Growth curves representative of: A. Bulls of very large frame sizes, B. Bulls of intermediate frame sizes and C. Bulls of small frame sizes. Growth curves for steers and heifers of the three frame sizes would be markedly less than their counterpart bulls. Drawn by CSU Graphics.

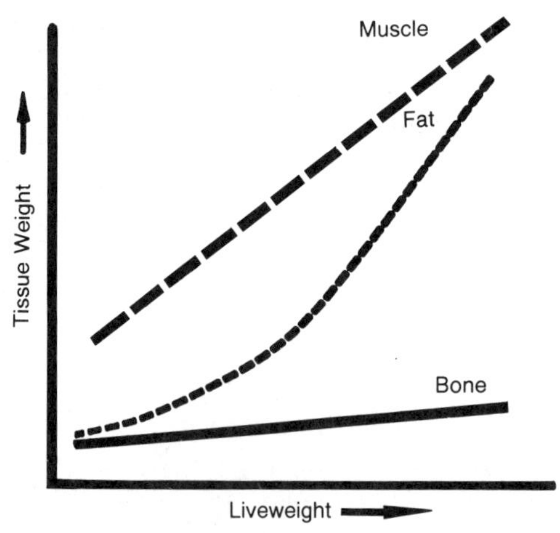

Figure 16.5. Tissue growth relative to liveweight. Courtesy of the American Society of Animal Science, Berg and Walters 1983.

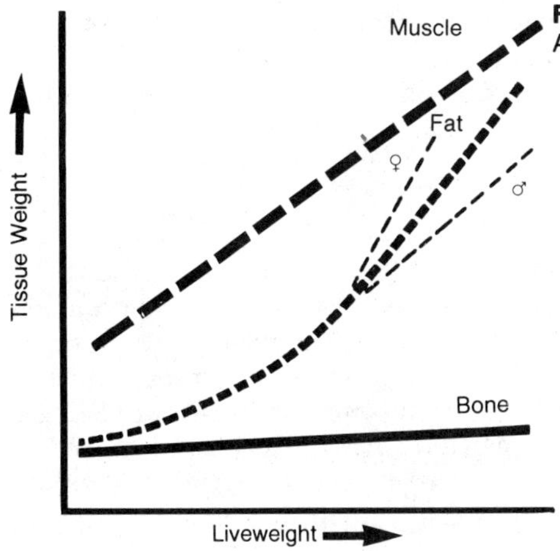

Figure 16.6. Sex effects on carcass composition. Courtesy of the American Society of Animal Science, Berg and Walters 1983.

muscle. Muscle to bone ratio in the young calf will be approximately 2:1, whereas at a slaughter weight of 1100 lbs to 1200 lbs, muscle to bone ratio would likely be in the 4:1 to 5:1 range.

The sex effects on tissue growth is primarily on the fat component (Figure 16.6). Heifers fatten at an earlier weight which is the primary reason why they are slaughtered 100 lbs to 200 lbs less than steers of a similar frame size. Bulls have less fat than steers and heifers at the same slaughter weight.

Genetic differences in carcass composition are shown in Figure 16.7. These different maturity types can be described as breeds or frame sizes. It is not entirely clear how much of the breed differences can be accounted for by frame size alone.

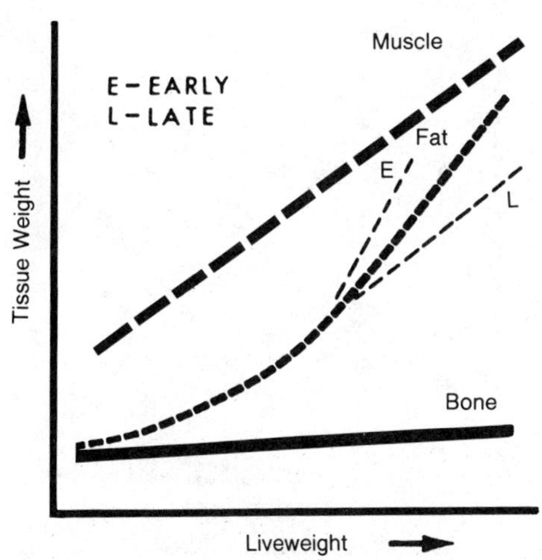

Figure 16.7. Effects of different maturity types on carcass composition. Courtesy of the American Society of Animal Science, Berg and Walters 1983.

There are maturity type differences within a breed which are determined primarily by frame size differences. This is particularly noticeable in several English breeds since selection emphasis in recent years has increased frame size differences from small to large.

Fat Growth and Development

Plane of nutrition or level of energy in the ration changes the relative amount of fat deposition. Cattle on high-energy rations will deposit fat earlier than cattle on lower energy rations, even when the cattle are evaluated at the same liveweights.

Where fat is deposited, as liveweight increases, is of high economic importance to the beef industry. Current market preferences favor an optimum combination of intramuscular fat (marbling) and total fat. There is evidence that intramuscular fat increases in proportion to total fat and at a similar relative rate, thus it is a challenge to produce 1 and 2 yield grades having a Choice quality grade. Level of marbling needed continues to be an important unanswered question for the beef industry.

Fat partitioning among the various fat depots has both genetic and environmental influences. Both genetic and environmental tendencies to earlier fattening result in increased proportion of subcutaneous fat to intermuscular fat, while delayed or decreased fattening has an opposite effect. Kidney and pelvic fat generally grows at the same relative rate as total fat, although there are some breed, sex, and plane of nutrition effects. There is evidence that during the fattening phase, kidney and pelvic fat increases at a faster rate than intermuscular fat. Dairy breeds deposit a higher proportion of internal fat and a lower proportion subcutaneously than traditional beef breeds. Limited data indicate that at equal fatness the fat partitioning in steers, heifers, and bulls is quite similar.

Fat growth gradients are low at the distal limb joints with increased growth gradients centripetally on the limbs, increasing further toward the rib and loin area. Fat growth coefficients in the flank and brisket areas tend to be the highest.

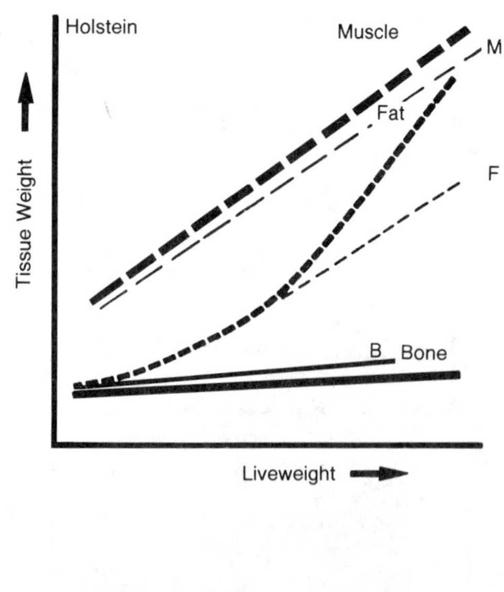

Figure 16.8. Lighter-muscled Holsteins (lighter lines) compared to average cattle. Courtesy of the American Society of Animal Science, Berg and Walters 1983.

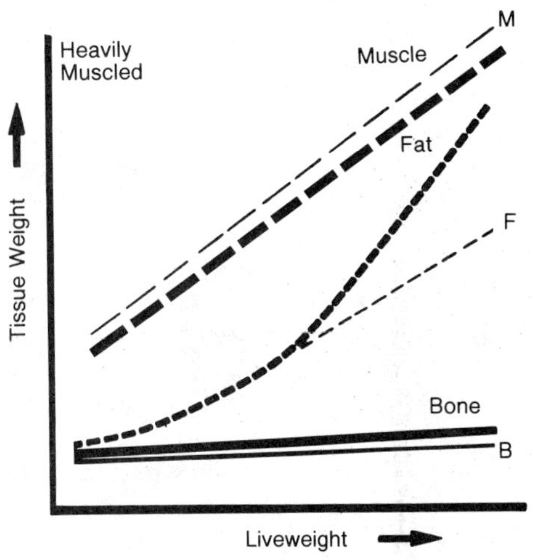

Figure 16.9. Heavy-muscled cattle (lighter lines) compared to average cattle. Courtesy of the American Society of Animal Science, Berg and Walters 1983.

Muscle Growth and Development

The proportion of muscle in the carcass varies indirectly with fat, and a higher proportion of fat is associated with a lower proportion of muscle and vice versa. Muscle has a much faster relative growth rate than bone (Figure 16.5). Muscle weight relative to liveweight or muscle to bone ratio can be used as a valuable measurement of muscle yield.

Relatively large genetic differences exist for muscling in cattle. These muscling differences are shown for lighter-muscled Holsteins (Figure 16.8), heavy-muscled (Figure 16.9), and double-muscled cattle (Figure 16.10).

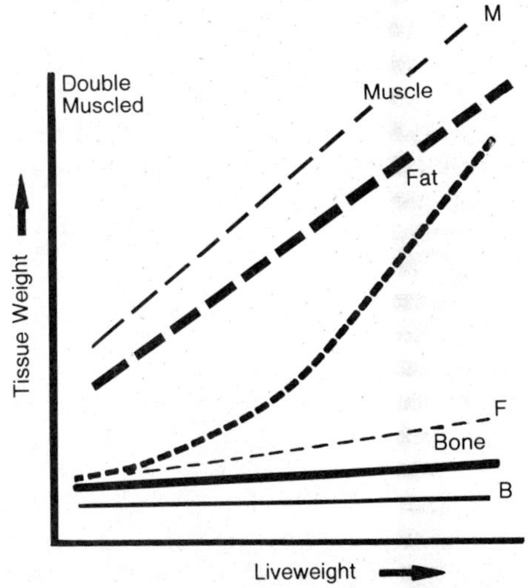

Figure 16.10. Double-muscled cattle (lighter lines) compared to average cattle. Courtesy of the American Society of Animal Science, Berg and Walters 1983.

Most of the current population of cattle in the United States does not have large muscling differences. There are, however, muscling differences in 10% to 15% of the cattle which represent wide differences in muscling. Muscle yield relative to liveweight of more than 50% to 33% (muscle to bone ratio ranges of 3.5:1 to 5.0:1) has been observed in steers and heifers of slaughter weight without experiencing the extremes of double-muscling. These differences are economically important, particularly if more boneless cuts are merchandised.

Extreme muscling as observed in double-muscling types is associated with several problems of fitness such as increased susceptibility to stress and reduced reproductive efficiency in the female. There may be some latitude in increasing muscle to bone ratio in slaughter cattle without encountering negative side effects.

Individual muscles have different relative growth rates. The newborn calf has relatively well developed muscles in the hind legs. As maturity continues, there is a progressive rise in relative growth rate from rump to the neck with the shoulder girdle and neck muscles having the highest growth rates of all muscles. Also, front limb muscles develop more rapidly than hind limb muscles as the weight of the animal shifts progressively forward. Muscle growth rate is relatively high in the abdominal and thoracic areas.

Muscle Distribution

Individual muscle dissection has shown a similarity of proportion of muscles to the total body muscle weight (Berg and Butterfield 1976). Thus it appears that increasing muscle weight in areas of high-priced cuts, at the expense of muscle weight in lower priced cuts is not commercially feasible. Some statistically significant small differences in muscle distribution have been shown, but the differences do not seem to be economically significant for the industry. Double-muscled animals show the greatest evidence of a higher proportion of muscle weight in higher priced muscles.

There is little evidence to support claims that selection used to form the breeds of cattle resulted in muscle distribution differences between the breeds.

Bone Growth and Distribution

Bone growth has the slowest growth rate of the muscle, fat, and bone tissues (Figure 16.5). There appears to be minor differences in the relative growth rates of different bones, although growth of limb bones is somewhat less.

Breed differences in bone distribution have been reported although these differences appear to have little commercial value.

Age and Teeth Relationship

For many years, producers have observed teeth of cattle to estimate their age or the cattle's ability to graze effectively with advancing age. Cattle have also been "mouthed" to classify them into appropriate age groups for specific showring classes. The latter has not always proven exact because of animal variation in teeth condition and the ability of some exhibitors to manipulate the condition of the teeth.

Table 16.1 gives some guidelines used in estimating the age of cattle by observing

Table 16.1. Description of Cattle's Teeth to Estimate Age

Approx. Age	Description of Teeth
Birth	Usually only 1 pair of middle incisors.
One month	All eight, temporary incisors.
1½ to 2 yrs	First pair of permanent (middle incisors).
2½ to 3 yrs	Second pair of permanent incisors.
3¼ to 4 yrs	Third pair of permanent incisors.
4 to 4½ yrs	Fourth (corner) pair of permanent incisors.
5 to 6 yrs	Middle pair of incisors begins to level off from wear. Corner teeth may also show some wear.
7 to 8 yrs	Both middle and second pair of incisors show wear.
8 to 9 yrs	Middle, second and third pair of incisors show wear.
10 yrs and over	All eight incisors show wear.

their teeth. The incisors (front cutting teeth) are the teeth used. There are no upper incisors and the molars (back teeth) are not commonly used to determine age.

Figure 16.11 shows pictures of incisor teeth of cattle of different ages. If grazing area is sandy and the grass is short, the wearing of the teeth will progress faster than that previously described. Also some cows may be "broken-mouth" which means some of their permanent incisor teeth have been lost.

BEHAVIOR

Producers who have raised cattle for a period of time have observed behavior patterns of cattle, many of which have a direct relationship to cattle productivity. Disposition, especially when cattle are being handled, and social behavior when feeding are some specific behavior areas which producers have evaluated as being economically important.

Scientific information on cattle behavior is limited with most of it becoming available in recent years. It is anticipated that additional scientific information on this subject will become available in future years with greater rapidity than in the past.

It is important for producers to understand how cattle behave and how they can influence cattle behavior through their care, training, selection, and the type of social and physical environment provided. With this understanding, producers can make management decisions which will improve cattle productivity and profitability.

General Disposition

Cattle producers consider the disposition or temperament of cattle from a range of docile to being very wild or "high-strung." This evaluation is usually made when cattle

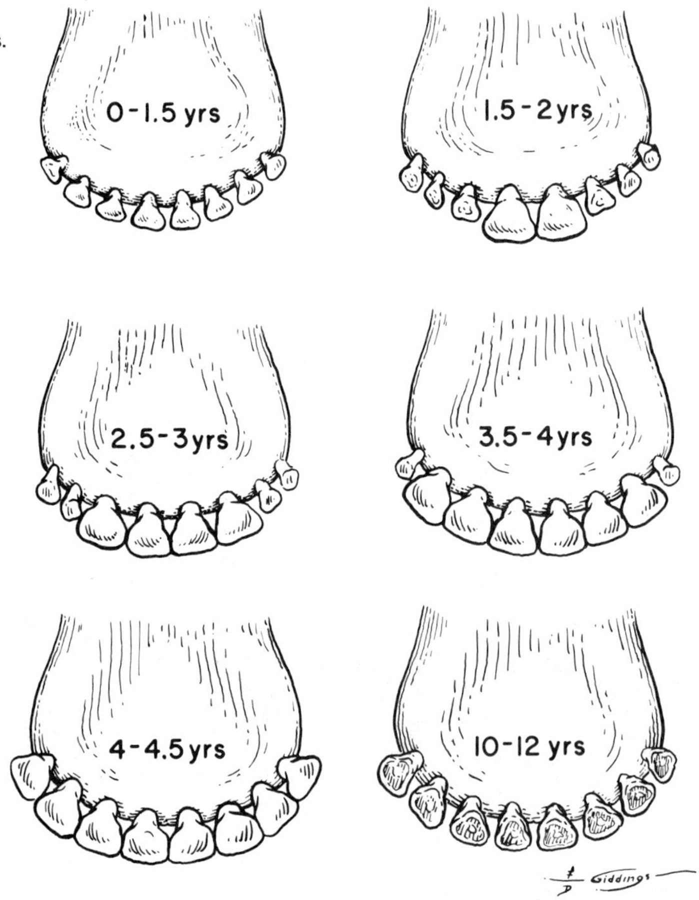

Figure 16.11. Incisor teeth of cattle of different ages. Drawn by Dennis Giddings.

are handled through various types of corrals, pens, chutes, and other working facilities. The typical behavior exhibited by cattle with poor disposition is one of fear of man or aggressive fighting behavior.

There is evidence that cattle develop good or poor dispositions from the way they have been treated, or the behavior tendency has been inherited. The few heritability estimates for disposition are in the medium to high categories indicating that the trait would respond to selection. Some producers cull cattle with poor dispositions because of economics losses due to broken fences, other facilities repair, risk of injury, and the desire to reduce excitability of other animals.

Communication

Communication exists when some type of information is exchanged between individuals. This may occur with the transfer of information through any of the senses.

Figure 16.12. A calf, separated from its dam, has a distinct bawl which communicates distress or dissatisfaction. Courtesy of the American Hereford Association.

A distress call, involving a different type of sound, occurs from either the cow or her calf when they become separated (Figure 16–12). The cow and calf may recognize each other's vocal sounds, however, it appears the most effective way the cow recognizes her calf is by smell. Cows will more easily adopt other calves through transfer of the odor of one calf to another.

Cattle learn to respond to vocal calls or whistles of the producer who wants the cattle to come to feed. Cattle soon learn that the stimulus of the sound is related to being fed.

The bull vocally communicates his aggressive behavior challenge to other bulls and intruders to his area through a deep bellow. This bellow and aggressive behavior is under the control of the male hormone (testosterone) as the castrated male seldom exhibits similar behavior.

The bull also issues calls to cows and heifers, especially when he is separated from them, but they are within his sight.

Cattle are especially perceptive in their sight since they have 310 to 360 degree vision. This affects their behavior in many ways, for example, as they are approached from different angles and as they are handled through various types of facilities.

Social Behavior

Some cows will withdraw from the group to find a secluded spot just prior to calving. Cattle sometimes withdraw from the group if they are sick.

Early and continuous association of calves is associated with greater social tolerance, delayed onset of aggressive behavior, and relatively slow formation of social hierarchies.

Status and social rank typically exist in a herd of cows with certain individuals dominating others who become submissive. The presence or absence of horns are important in determining social rank, especially when strange cows are mixed together. Also, horned cows will usually outrank polled or dehorned cows where close contact is encountered, such as at feedbunks or on the feed ground.

Large differences in age, size, strength, genetic background, and previous experience have powerful effects in determining social rank. Once the rank is established in the cow herd, it tends to be consistent from one year to the next. There is evidence that genetic differences exist for social rank. Wagnon et al. (1966) determined that even though Hereford cows were larger than Angus and Shorthorn cows in the same herd, they were lowest in dominance status.

There is evidence that animals fed together will consume more feed than when fed individually. This competitive environment evidently is a stimulus for greater feed consumption (Kidwell 1954). Dairy calves, separated from their dams at birth, appear to gain equally well whether fed milk in a group or kept separate. There is, however, evidence that they learn to eat grain earlier when group fed compared to being individually fed (Warnick et al. 1977). Cattle individually fed in metabolism stalls will consume only 50% to 60% the amount of feed they would eat if the animals were group fed.

When fed in a group of older cows, two-year-old heifers have difficulty getting their share of supplemental feed (Table 16.2). There is, however, evidence that two-year-old heifers and three-year-old cows can be fed together without there being a significant age effect in competition for supplemental feed (Wagnon 1965). These behavior differences no doubt explain some of the nutrition, weight gains, and post-partum internal relationships which are age related when cows of all ages compete for the same supplemental feed.

Table 16.2. Weight Changes of Two-Year-Old Heifers Fed Separately as a Group or Together with Older Cows

Treatment	Weight Change (Lbs)
Pastured and fed with older cows	25 lb loss
Pastured and fed separately	46 lb gain

Source: Wagnon 1965.

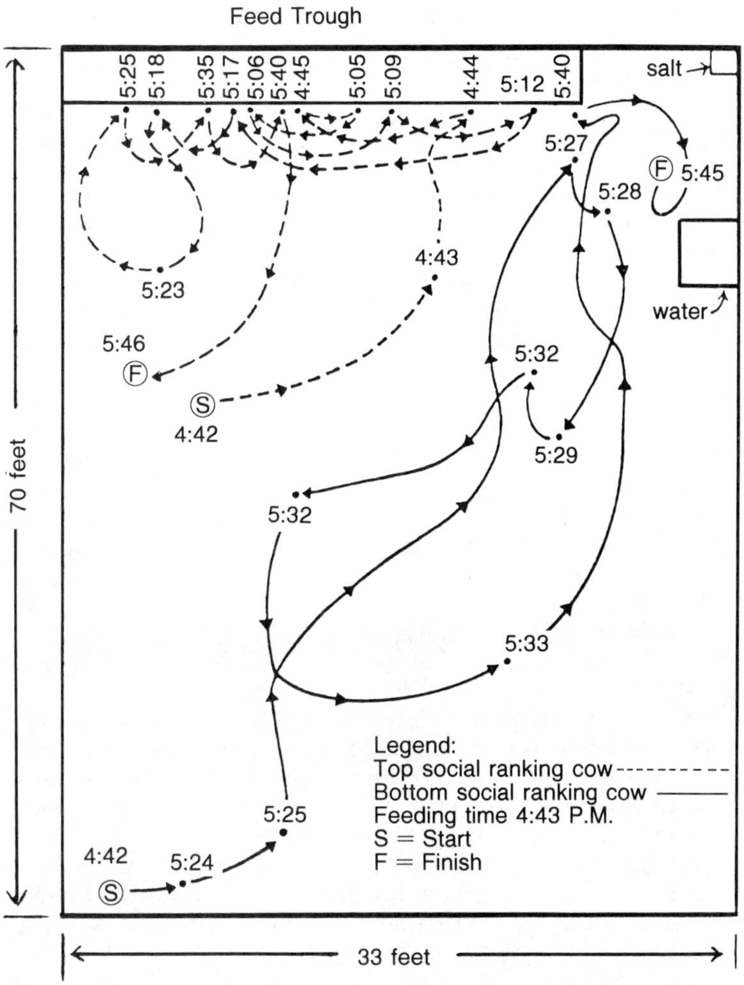

Figure 16.13. One hour feeding pattern for two cows of different social rankings (Schake and Riggs 1972).

Dominant cows, raised in a confinement operation, will usually consume more feed and wean heavier calves. More submissive cows will wean lighter calves (25%), and few of them will be pregnant compared to more aggressive cows (Schake and Riggs 1972). Figure 16.13 shows a one-hour feeding pattern for two cows of different social rankings.

Reproductive Behavior

Some profound behavior patterns are associated with sex of the animal and changes resulting from castration. This verifies the importance of hormonal directed expression of behavior. Bulls have more aggressive behavior, while steers are more docile after losing their source of male hormone.

The bull will frequently curl his nose and upper lip (known as an olfactory reflex or *flehmen*) after sniffing the urine or genitals of females. This behavior is associated with inhalation through the upper respiratory tract. The bull, through the use of the vomeronasal organ, is attempting to identify pheromones which are chemical substances secreted in the urine of the proestrus or estrus female. It appears that in a sexually active group of cows, the bull is attracted to a cow in heat most often by visual means (observing cow-to-cow mounting) rather than by olfactory clues.

When females are sexually receptive, they will usually seek out a bull if mating has previously not occurred. Primarily through hormonal influence, the female in estrus will stand for the bull when he attempts to mount her. Cows in heat demonstrate male behavior by mounting and being mounted by other cows. A thrusting action is common in the cows as they terminate the mounting action.

The bull may guard a female which he has determined is approaching estrus. His success in guarding the female or actually mating with her is dependent on his rank of dominance in a multiple-sire herd. During the mating process, the bull will usually playfully nudge the female, lick various parts of her body with his tongue, lick her side and back prior to mounting and rest his chin on her rump. After mating, most bulls temporarily lose interest in the cow. Some bulls leave the cow and do not make repeated matings, while other bulls may breed the same cow numerous times even in the presence of other cows in estrus. These different behavior patterns may well explain the wide variations in the success of individual bulls with varying bull to cows ratios.

Recent studies show that many individual bulls have sufficient sex drive and mating ability to fertilize more females than are commonly allotted to them. An excessive number of males, however, are used in multiple-sire herds to offset the few which are poor breeders and to cover for the social dominance which exists among several bulls running with the same herd of cows. If low fertility exists in the dominant bull or bulls, then calf crop percent will be seriously affected even in multiple-sire herds.

Tests have been developed to measure libido and mating ability differences in young bulls. While behavioral differences are evident between different bulls, tests, based on pregnancy rates, have not proven accurate for use by the beef industry.

Some cows will eat or chew the afterbirth (placental membranes), and most cows will begin licking the calf after it is born. Exceptions to the latter occur when the cow does not feel well after a difficult birth or some other type of illness. The licking of the calf provides an olfactory response to identifying the calf in the future. The calf usually nurses with its rear end towards the cows head. This allows the cow to smell the calf and decide to accept or reject it. A rejected calf is usually bunted with the head of the cow and kicked with the rear legs when it attempts to nurse.

Cows have fostered several calves, by removing their own calves at birth, then smearing the foster calves with amniotic fluid previously collected from the second "water bag." Where foster calves have been raised on a cow, it has been shown that foster calves suckled less frequently and for shorter periods and weighed less than the cow's own calf.

Certain cows become very aggressive in protecting their calves shortly after calving. Serious injury can occur to producers who do not use caution with these cows.

Bulls, being raised with other bulls, commonly mount one another, have a penal erection and occasionally ejaculate. Individual bulls can be observed arching their back, thrusting their penis toward their front legs and ejaculating.

Bulls can be easily trained to mount objects which provide the stimulus for them to experience ejaculation. AI studs commonly use restrained steers for collection of semen. Bulls soon respond to the artificial vagina, when mounting steers, which provides them with a sensual reward.

Mating behavior has an apparent genetic base as there is evidence of more frequent mountings in hybrid or crossbred animals.

There is evidence that more cows calve during periods of darkness than during daylight hours. The calving pattern, however, can be changed by when cows are fed. Cows which are fed during late evening will have a higher percentage of their calves during daylight hours.

Suckling Behavior

Animals have certain instinctive behavior patterns as compared to learned behavior. One of these instinctive behaviors is the newborn calf's searching movement of head and neck, sucking of protruding objects, and swallowing of fluids. Obviously the function of this behavior is to locate the mammary gland and suckle the teats for milk.

The number and duration of suckling incidences have been observed in range beef calves (Odde 1983). In this study, the average time spent suckling during each of several 24-hour periods was approximately 45 minutes with an average of 5 suckles per calf. Peaks in suckling activity occurred from 5:00 AM to 7:00 AM, 10:00 AM to 1:00 PM, and 5:00 PM to 9:00 PM. The most suckles in a single hour occurred between 5:00 AM and 6:00 AM and least between 10:00 PM and 11:00 PM. Cows that gave more milk nursed less frequently, and heavier calves sucked less frequently. Age, breed (Polled Hereford and Simmental) and sex of calf did not influence the number or duration of suckles.

The calf is usually born away from the herd in a secluded spot if one is available. The calf remains secluded while the cow grazes or feeds elsewhere, usually with the herd. The cow will return to the calf several times each day to allow the calf to nurse. In a few days, the cow will lead the calf to the herd where both will remain.

Cross-nursing is more frequent when cows have been forced to accept a second calf compared to a cow raising only a single calf. In the latter situation, some cross-nursing may be observed, and occasionally a cow will nurse another cow. These behaviors pose problems to the validity of weaning weight comparisons.

Grazing Behavior

Cattle have or develop palatability preferences for certain plants and may have difficulty changing over from one type or types of plants to other types. Also, cattle have preference to graze lower areas, especially if they are near water. These grazing behaviors tend to cause overgrazing in certain areas of the pasture and reduce weight gains.

Age and weather affect the behavior of cows grazing native range during the winter. At the Range Research Station at Miles City, Montana, cows grazed less as temperatures dropped below 20 degrees Fahrenheit, and at a minus 30 degrees, three-year-olds grazed approximately 2 hours less than six-year-olds. Also, the colder the temperature, the longer cows wait before starting to graze in the morning. At 30 degrees, cows started grazing between 6:30 and 7:00 in the morning; but at minus 30 degrees, they waited until about 10 AM.

Table 16.3. Activities of Cows Grazing on Winter Range

Activity	Hours
Grazing	9.45
Ruminating	
Standing	0.63
Lying	8.30
Idle	
Standing	1.11
Lying	3.93
Traveling	0.58
Total	24.00

In Northern Utah, observations were made on cows grazing open, treeless winter range (crested wheatgrass) from mid-November to mid-January. Bitter cold temperatures, strong winds and snow accumulations up to 10 inches are common in this area. The average time spent in various activities is shown in Table 16.3.

On the average, cows walked 3.6 miles per day in the course of eating and moving to and from water. When temperatures lowered and wind speeds were in excess of 2 mph, travel was restricted to as little as 1.5 miles per day, and the cows spent less time grazing.

Feedlot Behavior

Cattle kept in barns have been observed to be awake approximately 20 hours per day, and they sleep the remaining 4 hours with 40% of the time spent standing. Cattle tend to sleep less when on pasture compared to being in familiar buildings.

The "buller-steer" problem with steers in a feedlot is a unique behavior problem with serious economic significance to feedlot operators. Certain steers are singled out, probably by an olfactory identification through the vomeronasal organ, and ridden continually by other steers. While the exact cause of the "buller-steer" syndrome is not known, there is evidence that the feeding and implanting of hormones (especially estrogenic ones) have increased the incidence. Other factors affecting the incidence appear to be keeping large numbers in a group and introducing new cattle into a pen. This behavior pattern reduces feedlot gains and usually the "buller-steer" becomes crippled, stressed to where other serious health problems occur, including death. This abnormal mating behavior differs from that seen in all-female groups, since "buller-steers" are unlikely to mount others that also stand.

Bulls can become very aggressive toward one another in riding and fighting, especially when group size becomes large (usually over 50 head). The group of bulls become extremely aggressive toward another bull which may have been separated from the group for a few days, then introduced back into the group.

Feedlot steers have different patterns of water consumption throughout the day. Some of these patterns are significantly affected by season of the year (Figure 16.14).

Feeding behavior of steers in confinement on concrete floors appears to be slightly different from steers in dry lot, dirt pens. A higher percentage of steers on concrete eat during the night with fewer eating during the middle of the day (10 AM to 3 PM).

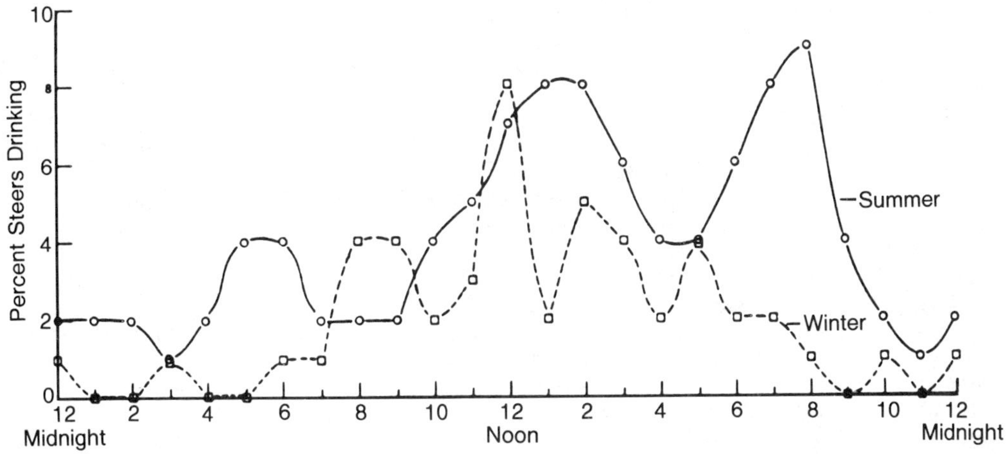

Figure 16.14. Seasonal influence on diurnal activity of yearling steers in the feedlot (Ray and Roubicek 1971).

Behavior During Handling and Restraint

Most cattle are handled and restrained several times during their lifetime. Ease of handling will depend largely on their temperament, size, previous experience, and design of the handling facilities. Knowledge of animal behavior can assist in preventing injury, undue stress, and physical exertion for both cattle and producers.

Blood odor appears to be offensive to cattle; therefore, the reduction or elimination of such odors may well encourage cattle to move through handling facilities with greater ease. Cattle are easily disturbed by loud or unusual noises such as motors, pumps, and compressed air.

With their 310 to 360 degree vision, cattle are sensitive to shadows and unusual movements observed at the end of the chute or outside the chutes. For these reasons, cattle will move with more ease through curved chutes with solid sides.

Round pens, having an absence of square corners, will handle cattle which are more excitable with less injury occurring to them.

Head catches are almost a necessity to immobilize cattle of large size. After the head is caught, nose rings or nose pinchers can assist in controlling large, aggressive bulls. A machine which uses an electrical current from a source connected to the lip and tail is a unique way to immobilize an animal and immediately control behavior.

Some breeders claim that AI facilities should permit cows to be handled quietly and carefully, not using facilities where cows have previously felt pain. They state that pregnancy rates will be higher. This sounds logical, but there is little research data to substantiate these claims.

SELECTED REFERENCES

Publications

Berg, R. T., 1979. *Growth and development of carcass components in cattle.* Arkansas Agric. Expt. Sta. Spl. Rpt. 12.

Berg, R. T. and Butterfield, R. M., 1976. *New Concepts of Cattle Growth.* Sydney, Australia: Sydney University Press.

Berg, R. T. and Walters, L. E., 1983. *The meat animal: Changes and challenges.* J. Anim. Sci. 57:133 (Suppl. 2).

Blockey, M. A. de B., 1976. *Sexual behaviour of bulls at pasture: A review.* Theriogenology 6:387.

Brown, C. J., Brown, A. H., Jr., and Johnson, Z., 1983. *Studies in body dimensions of beef cattle.* .Arkansas Agric. Expt. Sta. Bull. 863.

Craig, J. V., 1981. *Domestic Animal Behavior.* Englewood Cliffs, N.J.: Prentice-Hall.

Curtis, S. E. and Haupt, K. A., 1983. *Animal ethology: Its emergence in animal science.* J. Anim. Sci. 57:234 (Suppl. 2).

Dickson, D. P., Barr, G. R., Johnson, L. P., and Wieckert, D. A., 1970. *Social dominance and temperament of Holstein cows.* J. of Dairy Sci. 53:904.

Grandin, T., 1977. *Practical applications of behavioral principles to the design of cattle handling facilities.* Amer. Society Animal Science Meetings, Madison, Wisc. (Abstract).

Hammond, J., Jr., Robinson, T., and Bowman, J., 1983. *Hammond's Farm Animals.* Baltimore: Edward Arnold Univ. Park Press.

Hafez, E. S. E. and Bouissou, M. F., 1975. *The Behaviour of Domestic Animals.* Baltimore: Williams and Wilkins Co.

Kidwell, J. F., Bohman, V. R., and Hunter, J. E., 1954. *Individual and group feeding of experimental cattle as influenced by hay maturity.* J. of Anim. Sci. 13:543.

Kilgour, R. and Dalton, C., 1983. *Livestock Behaviour: A practical guide.* Boulder, Colo.: Westview Press.

Odde, K. G., 1983. *The postpartum cow: effect of early weaning and limited suckling on cow and calf performance, and suckling behavior in range calves.* Ph.D Thesis. Kansas State Univ.

Pierson, R. E., Jensen, R., Brady, P. M., Horton, D. A., and Christie, R. M., 1976. *Bulling among yearling feedlot steers.* J. Amer. Vet. Med. Assoc. 169:521.

Prior, R. L. and Lasater, D. B., 1979. *Development of the bovine fetus.* J. of Anim. Sci. 48:1546.

Ray, D. E. and Roubicek, C. B., 1971. *Behavior of feedlot cattle during two seasons.* J. Anim. Sci. 33:72.

Rile, R. W., MacNeil, M. D., Jenkins, T. G., and Koong, L. J., 1982. *A simulation model of grazing behavior in cattle.* Proc. West. Sect., Amer. Soc. Anim. Sci.

Rupp, G. P., Ball, L., Shoop, M. C., and Chenoweth, P. J., 1977. *Reproductive efficiency of bulls in natural service: effects of male to female ratio and single vs. multiple sire breeding groups.* J. Amer. Vet. Med. Assoc. 171:639.

Schake, L. M. and Riggs, J. K., 1972. *Behavior of beef cattle in confinement.* Texas Agric. Exp. Sta. Tech. Rpt. 27.

Schake, L. M. and Riggs, J. K., 1970. *Activities of beef calves reared in confinement.* J. of Anim. Sci. 31:414.

Sherry, C. J., Klemm, W. R., Sis, R. F., and Schake, L, M., 1982. *Reproductive and feedlot behavior: The role of the vomeronasal organ.* Beef Cattle Research in Texas. PR 3923.

Trenkle, A. H. and Marple, D. N., 1983. *Growth and development of meat animals.* J. Anim. Sci. 57:273 (Suppl. 2).

Swatland, H. J. 1984. Structure and Development of Meat Animals. Englewood Cliffs: Prentice-Hall, Inc.

Wagnon, K. A., 1965. *Social dominance in range cows and its effects on supplemental feeding.* Calif. Agric. Exp. Sta. Bull. 819.

Wagnon, K. A., Loy, R. G., Rollins, W. C., and Carroll, F. D., 1966. *Social dominance in a herd of Angus, Hereford and Shorthorn Cows.* Anim. Behavior 14:474.

Warnick, V. D., Arave, C. W., and Mickelsen, C. H., 1977. *Effects of group, individual and isolation rearing of calves in weight gain and behavior.* J. of Dairy Sci. 60:947.

Wyatt, R. D., Gould, M. B., and Totusek, R., 1977. *Effects of single vs. simulated twin rearing on cow and calf performance.* J. of Anim. Sci. 45:1409.

17

Beef Type, Form and Function

Type has been defined as an ideal or standard of perfection combining all characteristics that contribute to an animal's usefulness for a specific purpose. Conformation has been used interchangeably with type and broadly refers to the form and shape of the animal. Both type and conformation involve external characteristics of the animal which can be evaluated visually or measured more objectively. Most general descriptive terms have involved length, height, thickness, and general symmetry. More specifically, body parts are given description both individually and how they blend together.

One of the first known writings on livestock judging was from Bourgelat of Lyon who developed his classic book, *Traite de la Conformation Exterieure du Cheval* (1768). Since that time, and no doubt before, there has been an extensive interest in visual appraisal of cattle. Conformation or form and shape has been widely used in evaluating beef cattle and carcasses.

There has been and continues to be wide differences of opinion as to the relationship of form and function in beef cattle. The role of visual appraisal and measurements of body form are currently being discussed as to their value in breeding programs, market grades, and defining so-called "ideal types" of cattle. It is important to separate true relationships from opinion, to the best of one's ability, particularly in the area of beef form and function which continues to be controversial.

Obvious differences in cattle form and function come from a broad evaluation of the various uses of cattle. Development of cattle, over time, has brought several divergent types and forms into existence. Where cattle have been used as work animals, the form generally has been one of large, massive size. Cattle selected for milk production only have developed into more angular animals. Cattle developed for maximum meat production express more thickness and total volume of form. Dual-purpose cattle, where emphasis has been placed on meat, milk, and possibly work, typically reflect some combinations of body form unique to each of those functions being combined.

Man has directed the change in cattle form under certain environmental conditions and thus is responsible for most differences in cattle types which are observed today (Figure 17.1). Man has attempted to appraise form and its relationship to function (growth, carcass, reproductive ability, and longevity).

While some relationships have been observed and validated, man has confused the issue. Some cattle producers decided that certain animals were more attractive or "more pleasing to the eye." Some of these characteristics had little, if any, relationship to productivity. Strong opinions, however, were verbalized that these relationships did exist. Those who could verbalize with great conviction, and particularly those who became recognized authorities in beef evaluation, have greatly influenced the opinions of many others. Thus, there has been perpetuation of both valid and invalid opinions as to the value of beef type and conformation.

Part of the beef production goal should be to put a highly productive animal into a merchandisable package which will command the highest market price. Productivity of breeding and slaughter cattle is best identified by using meaningful performance records and effective visual appraisal.

It has been well demonstrated that performance records are much more effective than visual appraisal in improving beef cattle productivity. Visual appraisal, however, does have importance in evaluating productivity, primarily in identifying reproductive and skeletal soundness and health status. Visual appraisal also has some importance in evaluating productive differences in carcass composition. Both performance records (such as backfat probes and sonoray readings for meatiness) and visual appraisal are important in evaluating slaughter animals. Visual appraisal of the slaughter cattle can be used effectively to predict carcass composition (fat, lean, and bone) when relatively large differences exist.

It should be recognized that most slaughter cattle are purchased at market time based on a visual appraisal of their apparent carcass merit.

PARTS OF THE BEEF ANIMAL

Figure 17.2 shows the parts of the beef animal. Figures 17.3 and 17.4 show some additional descriptions of conformation which are used to identify body dimensions. Knowledge of beef animal parts and their description is important for effective communication in any phase of the beef industry involving the live animal.

CARCASS CONFORMATION

When the USDA beef carcass quality grades were established in 1927, conformation was a factor in grade standards and remained a factor until 1975, at which time it was removed. The reason for eliminating conformation from quality grade standards was that research studies had shown that conformation had no positive effect on beef palatability.

Fat thickness, carcass weight and area of the ribeye are carcass characteristics related to carcass conformation which are presently included in yield grade standards. Research reports disagree as to the importance of carcass conformation, as described in the USDA quality grade standards prior to 1975, and yield of retail cuts. Reports range from a slightly positive relationship between conformation and retail cut yield to no relationship.

Figure 17.1. Body type changes, reflecting man's preferences, have been dramatic over the past 100 years—even within the same breed. A. Angus steer, Black Prince imported from Scotland to United States in 1883. Champion steer, weighing approximately 2400 lbs at 1883 Chicago Fat Stock Show. B. Champion steer at 1912 Chicago International (1630 lbs). Also champion at 1913 Chicago International weighing approximately 150 lbs less than 1912. C. Champion steer at 1958 Chicago International weighing less than 900 lbs. D. Preferred type of Angus steer for the 1980s, weighing approximately 1200 lbs. Note the type resemblance of steers A and C; also steers B and D. Courtesy of the American Angus Association.

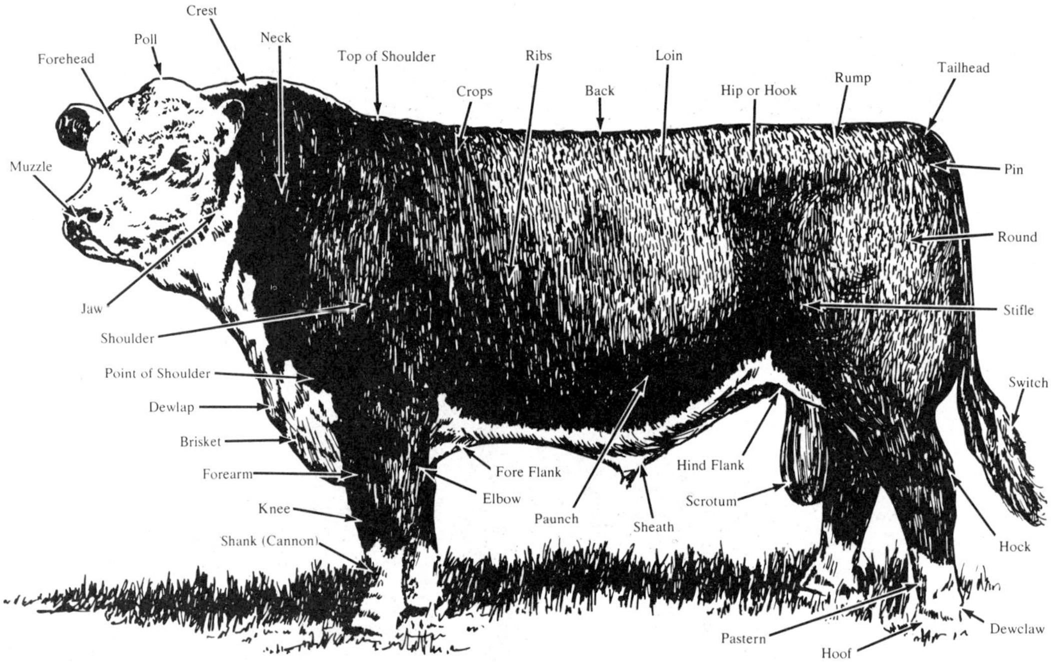

Figure 17.2. Parts of the beef animal. Courtesy of the American Polled Hereford Association.

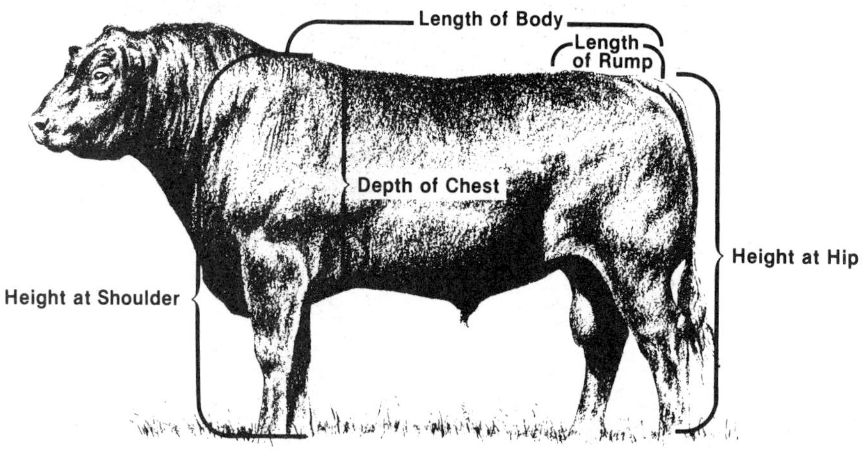

Figure 17.3. Parts of the beef animal used to identify body dimensions (side view). Courtesy of the American Angus Association.

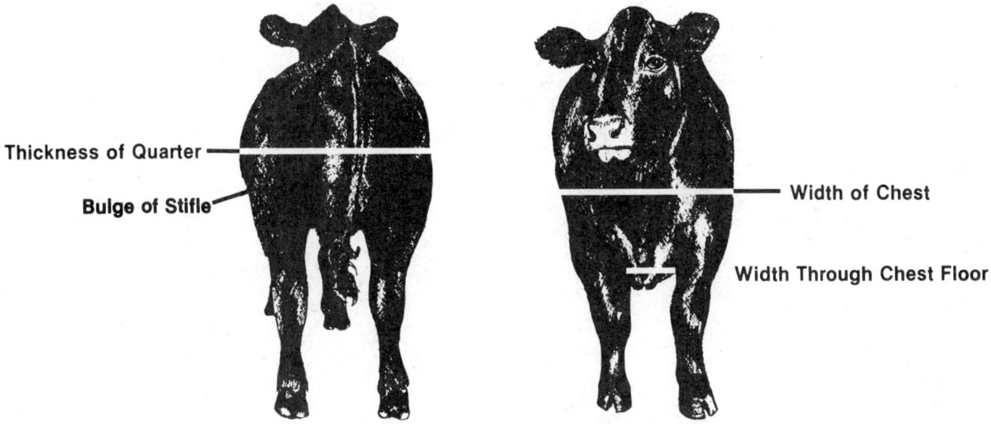

Figure 17.4. Parts of the beef animal used to identify body dimensions (front and rear views). Courtesy of the American Angus Association.

Fat greatly influences form and shape of the carcass. Subcutaneous fat affects conformation the most; however, intermuscular fat also causes muscles to have an appearance of greater fullness and depth. There are large differences in carcass muscle to bone ratios, independent of fat. These widely different ratios can be observed visually by noting muscling areas of the carcass which are not greatly influenced by subcutaneous and intermuscular fat.

Some packers and retailers still consider carcass conformation important from a marketing standpoint. Preference is given to a thicker carcass with a thick, plump round. Most of these shape differences have little value after the carcass is processed into retail cuts.

CONFORMATION OF SLAUGHTER CATTLE

Visual appraisal appears to be more accurate for carcass composition than for the other economically important traits. Accuracy of individual animal estimates for quality grade is low; however, live animal estimates of carcass yield grades have been shown to have useful accuracy (Crouse 1974). This accuracy is dependent on the experience of appraisers and variability in yield grades of cattle.

Effective visual appraisal, after becoming familiar with the external parts of the animal, involves an understanding of where major meat cuts are located in live animals. A carcass is evaluated after an animal has been slaughtered, eviscerated, and split into two halves. When the carcass is evaluated, it is hanging by the hind leg from the rail in the packing plant. This makes it difficult to perceive how the carcass would appear as part of the live animal standing on all four legs. Figure 17.5 which shows the location of wholesale cuts on the slaughter steer, helps to understand the major carcass component parts of the live animal.

The carcass is composed of fat, lean (red meat), and bone. The meat industry goal

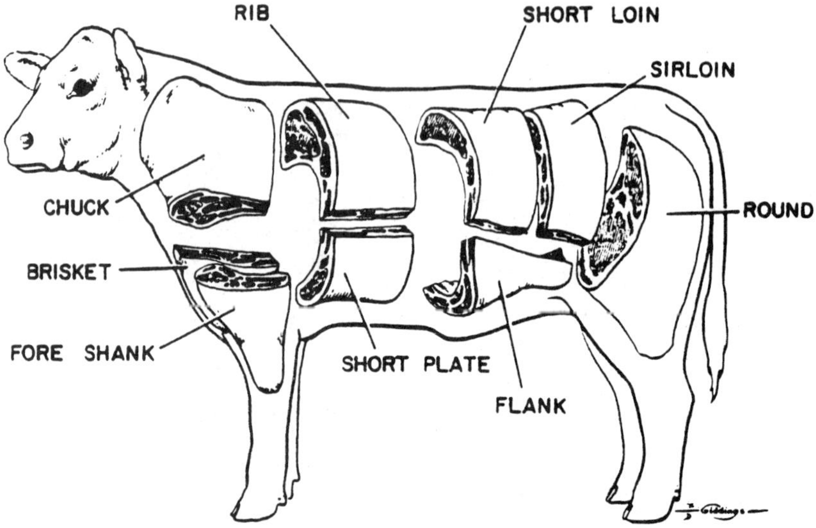

Figure 17.5. Location of the wholesale cuts on the live steer. Drawn by Dennis Giddings.

is to produce large amounts of highly palatable lean and minimal amounts of fat and bone.

Effective visual appraisal of carcass composition requires knowing the body areas of the live animal where fat deposits and muscle growth occur. Color Plates F and G show the fat and lean composition at several cross sections of a yield grade 2 and a yield grade 5 steer. Ribbon on the frozen carcasses shows location of the cross sections. Cross section 3 removes the bulge to the round, cross section 4 is in front of the hip bone down through the flank, cross section 5 is at the twelfth and thirteenth rib, and cross section 6 is at the point of the shoulder down through the brisket. Note the contrasts in fat (18% versus 44%) and lean (66% versus 43%) percentages. Both steers graded Choice and were slaughtered at approximately 1,100 lb.

Color Plates E and H show an additional perspective of fat, lean, and bone composition of slaughter steers. Removing fat and muscle, a layer at a time, allows a better understanding of what is under the hide of slaughter steers.

The conformation characteristics of the two steers (Color Plates F and G), which are primarily influenced by fat deposits and some muscling differences are contrasted in Figures 17.6 and 17.7. Conformation of each of these two steers is also contrasted to the conformation of a slaughter steer that is underfinished and thinly muscled (Figure 17.8). All body parts referred to in these figures can be identified in Figure 17.2, with the exception of the twist. The twist is distance from the top of the tail to where the hind legs separate as observed from a rear view (note No. 4, Color Plate H).

Accuracy in visually appraising slaughter cattle is obtained by making visual estimates of yield grades, then comparing the visual estimates with carcass measurements. Accurate visual appraisal can be used as one tool in producing slaughter cattle with a more desirable carcass composition of lean to fat.

Figure 17.6. Slaughter steer typical of yield grades 4 or 5. Drawn by Dennis Giddings.

Conformation characteristics

1. short, deep body (side view)
2. flat, wide top (rear view)
3. pear shaped (rear view)
4. deep in the **twist** (rear view)
5. deep in rear flank which makes a straight underline (side view)
6. uniform width or wider in middle of back (top view)
7. full dewlap and brisket (front view)
8. filled in behind the shoulders (side and rear view)

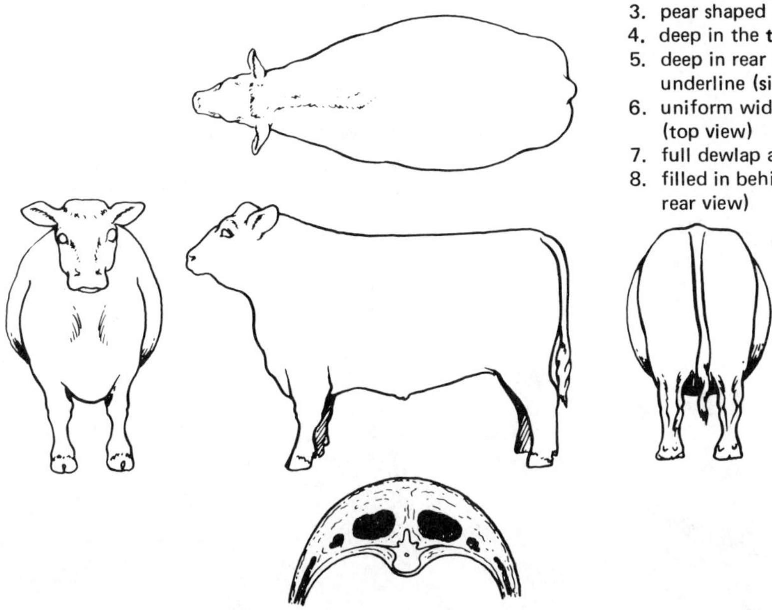

Conformation characteristics

1. relatively long body with moderate body depth (side view)
2. well turned (curved) top (rear view)
3. wide through center of round (rear view)
4. trim in the **twist** (rear view)
5. higher in rear flank than foreflank (side view)
6. wider in the rump than in the middle of the back (top view)
7. trim in dewlap and brisket (front view)
8. slightly dished behind the shoulders (side and rear view)

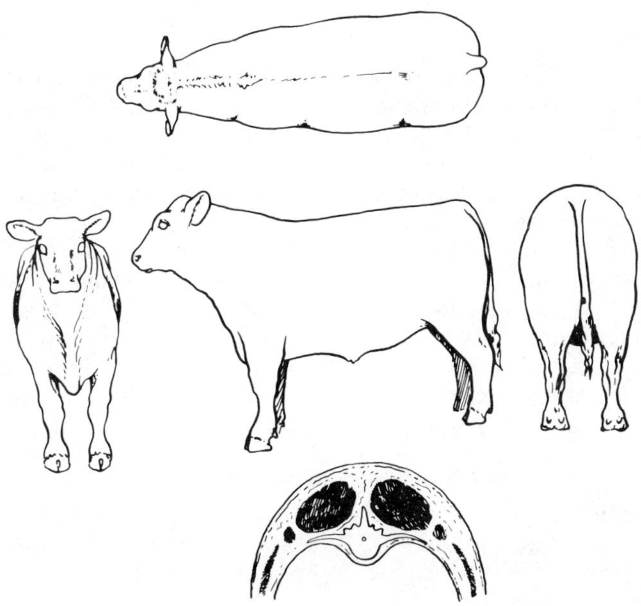

Figure 17.7. Slaughter steer typical of yield grade 2. Drawn by Dennis Giddings.

Figure 17.8. Slaughter steer that is underfinished and thinly muscled. Drawn by Dennis Giddings.

Underfinished (contrast with Figure 17.6)

1. narrow, pleated brisket (front view)
2. relatively shallow body (side view)
3. prominent hip and rib bones (side and rear view)
4. markedly dished behind the shoulders (side and rear view)

Thinly muscled (contrast with Figure 17.7)

1. narrow through center of round (rear view)
2. flat and narrow forearm (front and side view)

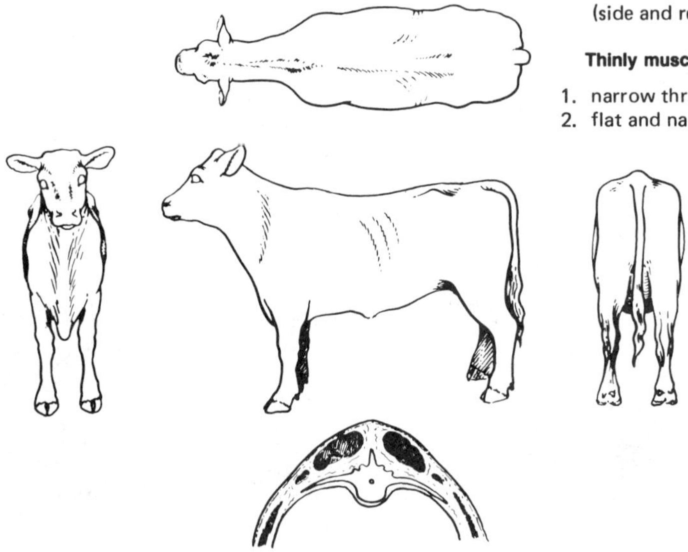

Rear views in Color Plates F and G show thickness or muscling differences that can be appraised visually, even though there are current disagreements about differences in muscling in slaughter cattle. Kauffman, et al. (1973) reported that muscular shaped steers contained two percent more fat-free muscle when compared to nonmuscular shaped steers at similar weight and fat composition. Muscular steers had a muscle/bone ratio of 3.8 while the nonmuscular steers' ratio was 3.0.

Muscling appears best described by muscle to bone ratios where observed differences typically range from 2.5:1 to 4.5:1. The latter would not include double muscled steers which would be in excess of that figure. Muscling differences (2.5 to 3.0:1 vs 4.0 to 4.5:1) can be visually appraised in slaughter cattle with a reasonable degree of accuracy (CSU, unpublished data).

A common belief is that large differences exist in muscle distribution in cattle (e.g., higher percent of carcass weight in hind limb muscle than forelimb muscle and vice versa). There is no evidence of significant differences in muscle distribution in slaughter steers (Kauffman, et al. 1973, Berg and Butterfield 1976, and MARC data).

Under today's marketing stipulations, market steers should have the conformation to: (1) have a 1.5 to 3.0 yield grade carcass, (2) have a 3.8 to 4.2:1 muscle/bone ratio,

(3) produce a 600 to 800 pound carcass with a minimum of 0.25 inch of fat over the twelfth rib.

CONFORMATION OF FEEDER CATTLE

Feedlot performance of most feeder cattle is estimated from visual appraisal. Feeder cattle buyers, for the majority of commercial feedlots, use visual criteria of skeletal size, finish, weight relative to age, and health status. Preference is given to cattle which show evidence of compensatory gain. These cattle are relatively large framed, thin, and old for their weight.

Frame size has become quantified by measuring skeletal height at the shoulder or hip. Reference is occasionally made today to the Madison Type Conferences (Brungardt 1972) where Angus, Charolais cross, and Hereford calves of various frame size (skeletal height and length) were evaluated for feedlot and carcass characteristics. A summary of results of the three-year study is shown in Table 17.1. The calves were born in April and May, fed for approximately 150 days (Charolais crosses fed 190 days), and slaughtered at approximately 14 to 15 months of age. Cattle of larger body types had to be fed to heavier weights to make similar carcass quality grades. At the respective slaughter weights, each body type graded between low Choice and average Choice. When fed to the same quality grade, there was little difference in feed efficiency. At similar weights, cattle that gained faster were more efficient in feed conversion. Profitability generally favored larger framed feeder calves (primarily up through body type 5) when evaluated through the feedlot and the carcass.

Evidence shows that visual estimates of trimness, finish, length, and height in feeder cattle can give fair estimates of cutability at slaughter (Crouse 1974). In the same study, where feeder calves of widely different biological types were evaluated, no combination of feeder calf traits could make a meaningful estimate of carcass quality grades.

Feedlot gains are usually higher for longer, taller cattle when fed to a constant time on feed or weight constant basis (Crouse 1974). Frame size differences appear to have little effect on rate of gain and feed efficiency when cattle are fed to a similar compositional basis (Crickenberger and Black 1976). The latter implies rather large differences in final carcass weights. This could have economic significance depending on market weight specifications which can vary by area of the United States.

Large differences in rate and efficiency of gain do exist in cattle of similar frame size and environmental background. Visual appraisal is not able to detect these differences with any high degree of accuracy.

Frame size, whether appraised visually or measured as hip height, can be used as a management tool in determining the logical slaughter weight of cattle in the feedlot.

Figure 17.9 shows compositional differences of different frame sizes of feeder cattle fed to different slaughter weights. No one frame size is superior to the others. Feed resources, market weight preferences, and other production costs will determine frame size preference for different feeders. A visual evaluation of frame size and condition can be used to sort feeder cattle into separate feeding groups. Days on feed can be projected for each group to reach similar yield grade and quality grade endpoints (Daley, et al. 1983).

Table 17.1. Feedlot and Carcass Characteristics of Various Body Types of Angus (A), Charolais Cross (CX), and Hereford (H) Calves (100 head per breed group)

Visual Body Types as Feeders	Ht at Should. at Slau. (in.)			Slaughter Wt (lbs)			Ave. Daily Gain (lbs)			Fat Cover (in.)		
	A	*CX*	*H*	*A*	*CX*	*H*	*A*	*CX*	*H*	*A*	*CX*	*H*
	42.1	—	42.4	877	—	889	2.7	—	2.7	0.57	—	0.53
	43.8	—	43.4	984	—	958	2.8	—	2.8	0.65	—	0.50
	45.5	45.5	45.3	1024	1029	1055	2.9	2.9	3.0	0.73	0.35	0.66
	45.9	46.8	47.0	1035	1093	1107	2.9	3.0	3.0	0.74	0.34	0.56
	47.1	47.9	48.4	1106	1133	1175	2.9	3.1	3.2	0.80	0.30	0.60
	—	49.2	—	—	1137	—	—	3.0	—	—	0.33	—
	—	49.5	—	—	1211	—	—	3.2	—	—	0.34	—

Source: Brungardt, 1972.

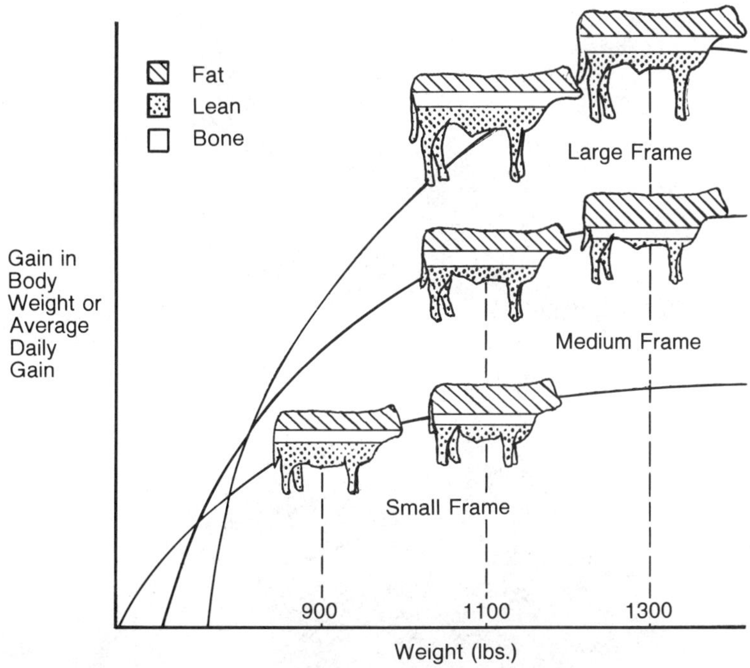

Figure 17.9. The relationship of frame size and weight to carcass composition. Drawn by CSU Graphics.

The new USDA feeder grades are based on visual appraisal of frame size and thickness (Figures 17.10 and 17.11). There are three frame size specifications: Large (steers and heifers would have Choice carcasses at approximately 0.50 inch fat at 1200+ lbs and 1000+ lbs, respectively); Medium (Choice carcasses for steers at 1000 to 1200 lbs liveweight and heifers at 850 to 1000 lbs); Small (steers grading Choice at less than 1000 lbs and heifers less than 850 lbs).

Body types or frame sizes shown in Table 17.1 would relate to the proposed frame size scores as follows: small frame (body type 1, 2, and lower half of 3); medium frame (upper half of 3, 4, and lower two-thirds of 5); large frame (upper third of 5, 6, and 7).

No. 1 thickness would be predominately for cattle with a high proportion of beef breeding. No. 2 thickness grade includes cattle that are narrow through the hindquarter while the No. 3 thickness grade shows very little thickness through the hindquarter. There is some question whether many No. 3s actually exist and if the thickness scores proposed would meaningfully separate the existing population of feeder cattle for muscling.

The last grade of the proposed feeder grades is inferior. It would include unthrifty cattle and also *double-muscled* cattle regardless of their frame size or thickness.

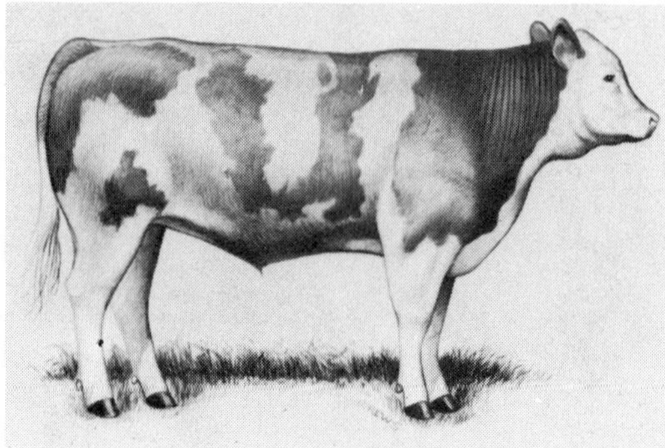

a. Large frame

b. Medium frame

c. Small frame

Figure 17.10. The three frame sizes of the USDA feeder grades for cattle. Courtesy of USDA.

No. 1

No. 2

No. 3

Figure 17.11. The three thickness standards of the USDA feeder grades for cattle. Courtesy of USDA.

CONFORMATION OF BREEDING CATTLE

Carcass Composition

Cattle breeders should study carefully the value of conformation in carcasses, slaughter, and feeder cattle before making decisions on how conformation characteristics could be used in a breeding program. Packers and retailers are forced to accept the basic conformation resulting from previous decisions made by cattle breeders and feeders.

In addition to some of the previous relationships mentioned, the breeder should recognize how sex differences affect conformation, particularly body composition. Figure 17.12 shows compositional differences of heifers, steers, and bulls at different weights. This information tells the feeder why heifers are typically marketed 100 to 150 lbs lighter than the counterpart steers. The breeder needs to understand these relationships to be more effective in sire selection. When the breeder focuses on a particular steer end-point (primarily weight and carcass composition), this dictates that the bull must be much more extreme in composition at the same weight, or in other words, the bull must be approximately 150 lbs heavier than the steer to be similar in carcass composition.

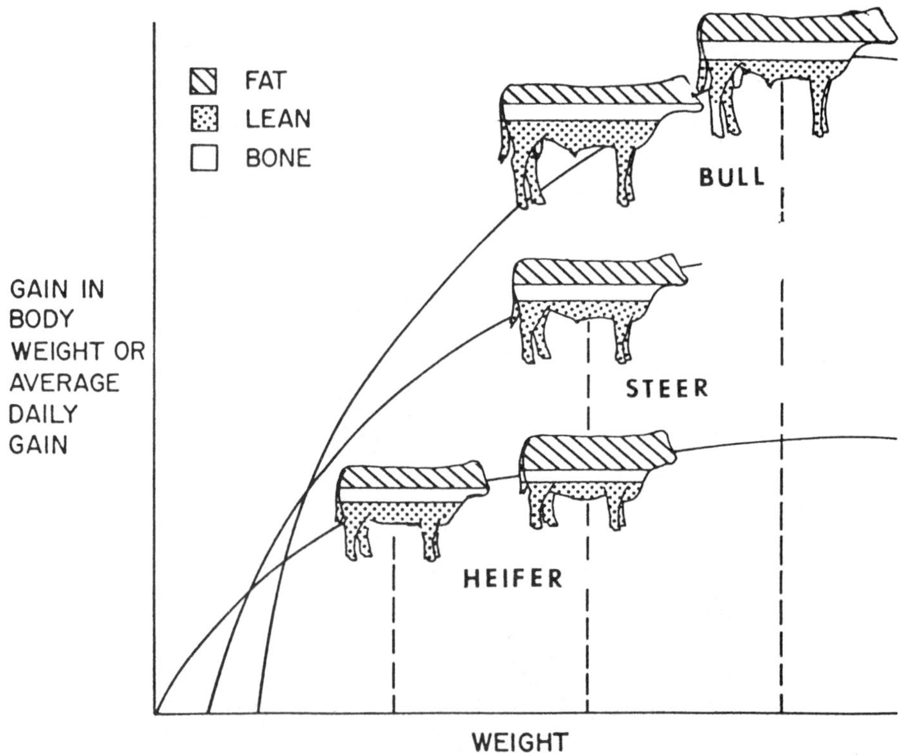

Figure 17.12. The influence of sex on body (carcass) composition at various weights. Drawn by CSU Graphics.

A mechanical backfat probe and visual evaluation can be used to accurately estimate subcutaneous fat. Recent studies show both methods to be similar in accuracy, accounting for approximately two-thirds of the differences in carcass subcutaneous fat thickness (Daley, et al. 1983). Visual appraisal can have added value by evaluating predisposition to fat, primarily in the brisket and flank areas. Validity of this appraisal is based on the fact that fat-cover over the ribeye does not always give an accurate prediction of fat trim as it affects yield grade. The carcass yield grader makes adjustments to the carcass fat measurement by visually appraising fat accumulation in the brisket, plate, and flank areas.

Accuracy of the fat probe or visual estimates on live cattle appears to be related to the experience and technique of the individual doing the probing or visual evaluation. The fat probe, at the twelfth to thirteenth rib, is most useful on yearling bulls that have been fed relatively high energy rations and are probed at weights similar to preferred slaughter steer weights (1050 to 1200 lbs). At these similar weights, approximately 0.20 inch should be added to the backfat probe of a bull to project to a slaughter steer basis.

Linear measurements of skeletal size appear to be most useful with breeding cattle, particularly young bulls 12 to 15 months of age. Height at hip is the most common measurement used today to measure size of skeleton. Body length is a questionable measurement because it is quite proportional to skeletal height and adds little additional information. There is also more difficulty in obtaining accurate body length measurements. Table 17.2 gives the hip height measurements used to identify various frame sizes in cattle.

Table 17.2. Hip Height Measurement (Inches) to Determine Various Frame Sizes of Bulls at Different Ages.[a]

Age	Frame Size							
(mos)	1	2	3	4	5	6	7	8
6	35	37	39	41	43	45	47	49
7	36	38	40	42	44	46	48	50
8	37	39	41	43	45	47	49	51
9	38	40	42	44	46	48	50	52
10	39	41	43	45	47	49	51	53
11	40	42	44	46	48	50	52	54
12	41	43	45	47	49	51	53	55
13	41.5	43.5	45.5	47.5	49.5	51.5	53.5	55.5
14	42.0	44.0	46.0	48.0	50.0	52.0	54.0	56.0
15	42.5	44.5	46.5	48.5	50.5	52.5	54.5	56.5
16	43.0	45.0	47.0	49.0	51.0	53.0	55.0	57.0
17	43.5	45.5	47.5	49.5	51.5	53.5	55.5	57.5
18	44.0	46.0	48.0	50.0	52.0	54.0	56.0	58.0

Data adapted from shoulder height measurements on more than 3000 bulls of the three British breeds at University of Missouri. Sex adjustments are estimated to be: steer (minus 1 inch), heifer (minus 2 inchs). Height at shoulder is approximately two inches less.

The primary usefulness of backfat probes, hip height measurements, and visual appraisal is to estimate carcass cutability at a certain slaughter weight endpoint.

Skeletal Soundness

Visual appraisal is used extensively by beef producers to evaluate skeletal soundness as it relates to longevity and productivity. Figure 17.13 shows some feet and leg descriptions commonly used today. Showring standards for years have dictated that each of the four legs should be straight and set on the corners. Also, the toes should point straight ahead and the animal should move with a long, easy stride.

Opinions are varied as to what constitutes skeletal soundness. Well-documented research findings are rather sparse in this area. The following points appear to have a good basis for consideration:

1. Cattle with hock angles greater than 150 degrees appear to be more predisposed to becoming "stifled." Cattle that are post-legged have more serious problems than those with sickle hocked condition (Woodward 1968).
2. Excessive hoof growth can result in lameness and economic concern if foot trimming is necessary. Hoof growth appears to be highly heritable (Brinks 1979).
3. "Founder," where lameness is typically involved, can be relatively serious under feedlot conditions. This structural soundness condition is usually observed when cattle are on relatively high energy rations. There appears to be a genetic basis to founder (Brown, et al. 1967).
4. Greenough, 1972, states, "Many so-called points of abnormal conformation are more of aesthetic interest than practical importance." Some respected cattle judges have stated that show cattle, which are straight on their legs, are that way primarily because of numerous trips to the hoof-trimming table. They also state that productive cattle are naturally a little splayed in front (toed out) and slightly cow-hocked in back with some set in their hock.
5. At this time, there appears to be little value in selecting for bone size and so-called "bone quality." Bone size appears to be independent of performing ability with the exception that a smaller bone, larger muscle combination will contribute to higher muscle to bone ratios. Also, larger bone may contribute to increase calving difficulty.
6. Further evidence that selection for skeletal soundness may have been overemphasized in the past is the observation of cattle in closed, inbred herds over the past 20 to 30 years. Examples are the cattle at the U.S. Range Station in Miles City, Montana, and the San Juan Basin Research Center at Hesperus, Colorado. There have been marked improvements in productivity where the primary selection criteria have been weaning and yearling weights in linebreeding and inbreeding programs with minor emphasis to selection for skeletal soundness.

More research needs to be done in evaluating the relationship between skeletal soundness and productivity, particularly longevity. Currently there are more questions than answers in determining whether differences in bone structure and joints (Figures 17.13 and 17.14) are more important from an aesthetic standpoint rather than a measure of productivity.

Correct skeletal structure

Sickle-hocked and back at the knees

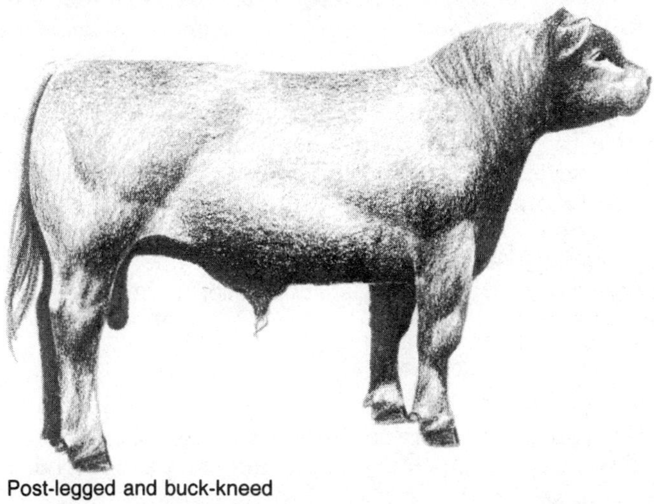

Post-legged and buck-kneed

Figure 17.13. Foot and leg structure as presently associated with structural soundness. Courtesy of R. A. Long, Texas Tech. University.

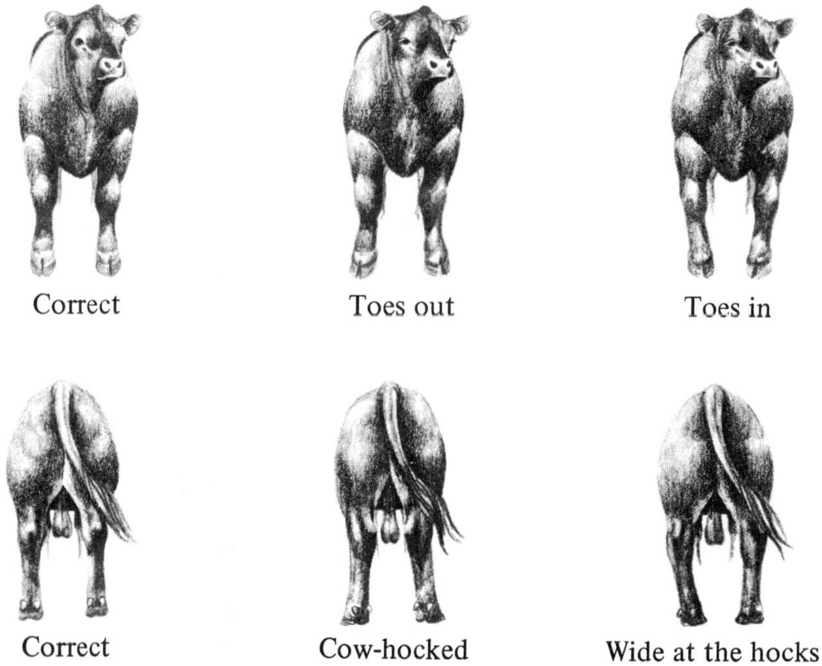

Figure 17.13 (cont.)

Reproduction

Bonsma (1983) has demonstrated that cattle can be selected visually for reproductive performance, although Wilson (1981) found no conclusive evidence in relating feminine features in cows to fertility. Most of Bonsma's demonstrations have been predicting the number of calves produced by each cow. His appraisal appears to reflect the expression of hormone levels or hormone imbalances as they affect specific target locations of neck, shoulders, hair, tailhead, and reproductive organs. This type of appraisal has not proven effective in heifer selection where hormone expression has been limited. Also, the use of good performance records in mature cows is more useful and economically feasible than trying to gain experience to duplicate Bonsma's visual ability.

Visual appraisal of body condition in breeding females is useful in assessing postpartum interval. Visual body scores have been shown to be highly related to carcass fat and carcass energy content ($r = 0.80$) and more useful than weight to height ratios (*J. Anim. Sci.* 57:391, Suppl. 1).

Merchandising

Visual appraisal can be used to put productive animals into attractive packages. Body form and appearance is generally independent of productivity. It does appear to

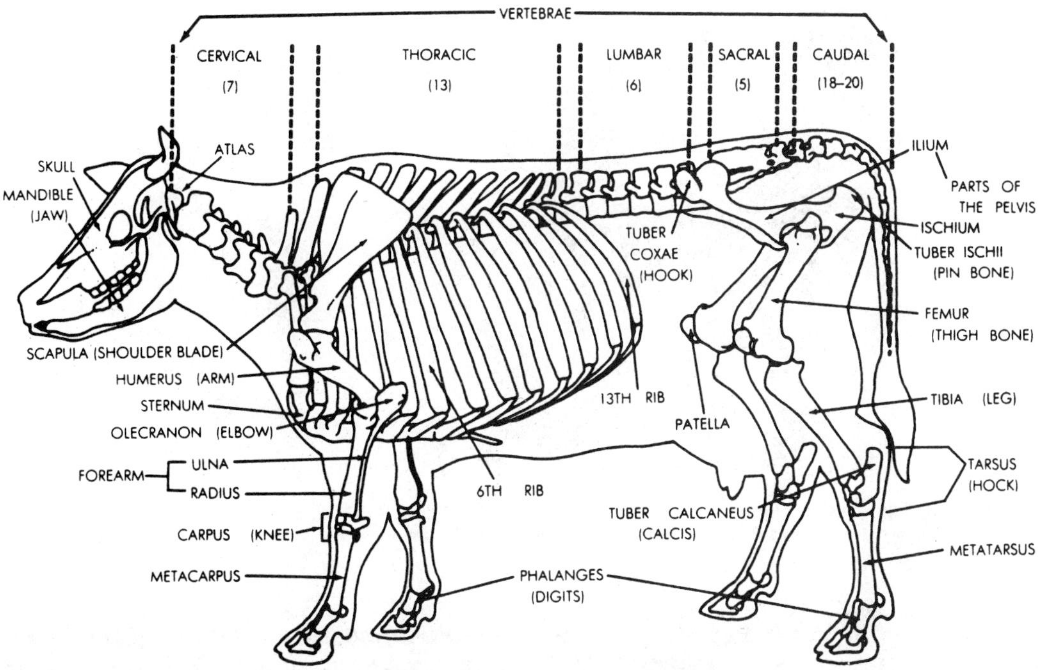

Figure 17.14. The parts of the skeleton of the cow. Courtesy of Battaglia, R. A., and Mayrose, V. B., 1981. *Handbook of Livestock Management Techniques.* Minneapolis: Burgess Publishing Co., copyright ©.

be economically feasible for producers to combine highly productive animals with form traits which are pleasing to the eye (Figure 17.15). This is part of merchandising whether the evaluation is a retail cut, a carcass, a feeder calf, a slaughter steer, or a breeding animal. Caution must be exercised in keeping the major priority on productivity with secondary emphasis on eye appeal.

THE LIVESTOCK SHOW

Competitiveness of the beef cattle show has challenged the beef producer for centuries. Breeding and management programs have been designed with showring placings as their measuring stick.

Heated discussions have ensued as to the value of showrings to beef producers and the beef industry. As emotion began to give way to genetic research in the 1940s and 1950s, it became evident that major differences in showring cattle were caused by environmental differences. Showring placings and high sale prices in many cases proved to be negatively related to productivity as measured by the commercial cattle industry.

Today most beef producers recognize the showring for what it really is—a tool for

Figure 17.15. Alertness, straightness of lines, femininity and blending together of body parts are some of the characteristics considered important in attractiveness of heifers. Courtesy of Barton Charolais Ranch, Manti, Utah.

promotion and merchandising and not a comparative measure of breeding programs. It is not a place to measure the genetic difference between animals or a place where breed standards of productivity should be established.

Cattle Judging

Numerous fairs and shows require judges to visually rank cattle in groups or in individual placings. Usually one judge assumes this task, while on occasion an associate judge is also involved. Several major national beef shows have used the 3-judge system. In this situation, three judges work independently of each other and the final placing of cattle is a composite of either two or three of the judges place. For each class, the three judges are randomly chosen to be Judge A, Judge B, and Referee Judge. The Referee Judge enters into the decision on the placing only when Judge A and B disagree on the placing. The 3-judge system is used to eliminate possible bias that may occur with one judge and to make placings on more than one judgement. Usually the 3-judge system is more costly.

Most beef judges have developed their judging ability through participation in 4-H, FFA, junior college, and university judging contests which are usually associated with fairs and shows. Most land grant universities have courses given in beef cattle and other livestock judging. These courses improve student's abilities to make decisions and communicate more effectively rather than make marked improvement in their abilities to identify more productive cattle. Shanteau (1978) presented evidence that livestock judges have unusual abilities to make complex judgments, and the courses and training programs used to develop these judges have a clear impact on these abilities. Many successful cattlemen and other businessmen attribute their decision-making success to participation in livestock judging programs.

Plate E

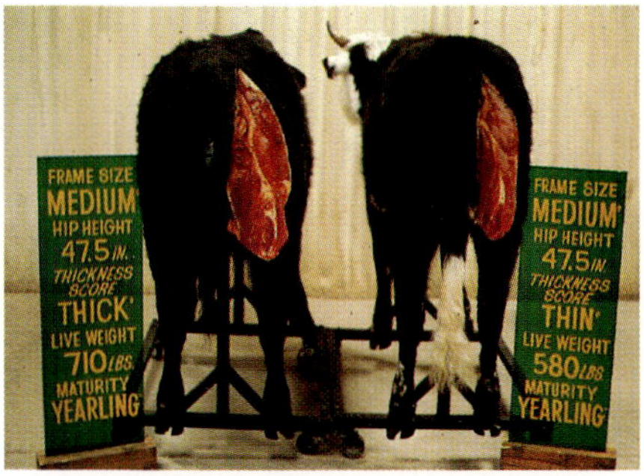

1

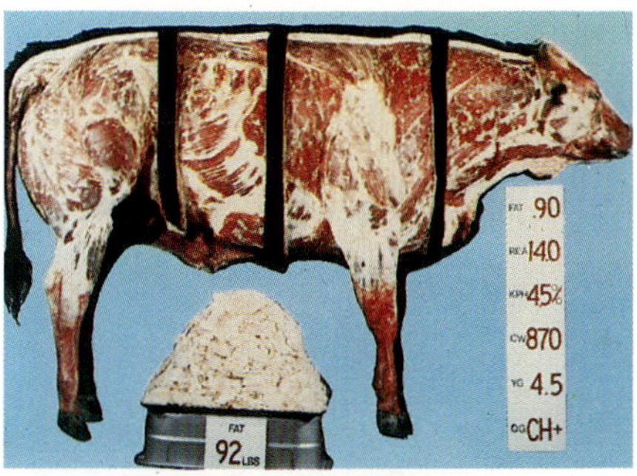

2

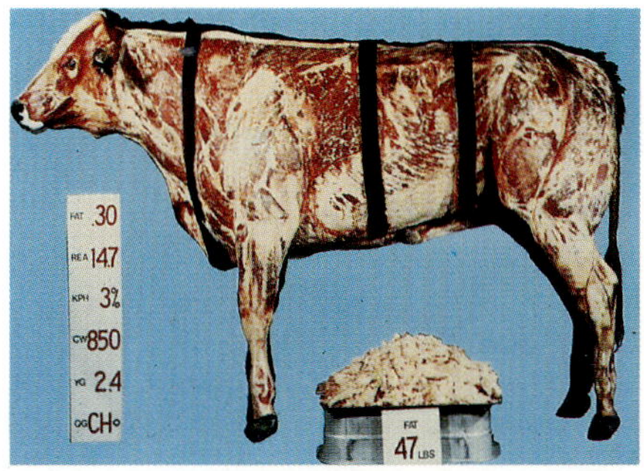

3

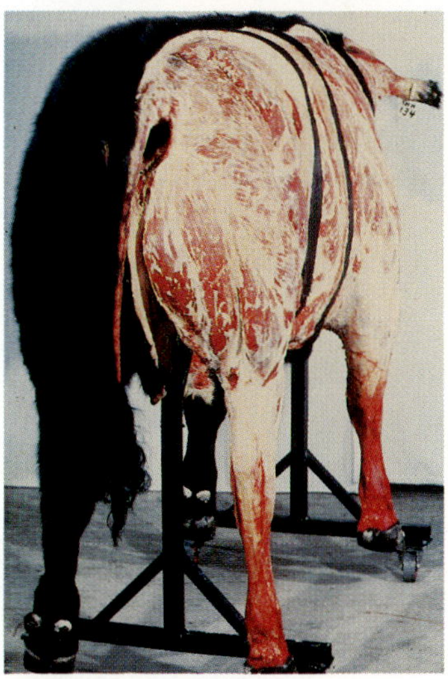

4

5

Plate E compares body conformation of different yield grades in both live animals and their carcasses: (1) thickness of muscling in two medium-framed feeder steers. The steer on the left has more thickness of lean meat in the round than the steer on the right; (2 & 3) pounds of fat trimmed from one-half of the body of a yield grade 4 steer and a yield grade 2 steer. Black stripes are hide and fat which remain at key locations on the body; (4) rear view showing how the fat increases in thickness from the middle of the back to the edge of the loin; (5) two yearling Hereford bulls of approximately the same weight. Breeding cattle can be visually appraised for fat and lean compositional differences. The bull on the left would sire slaughter steers having yield grade 4 or 5 carcasses, while the bull on the right would sire yield grade 1 or 2 steers. This assumes that the bulls would be bred to similar frame-size cows, and the steers slaughtered at approximately 1,150 pounds. Compare to plates F and G.

Plate F

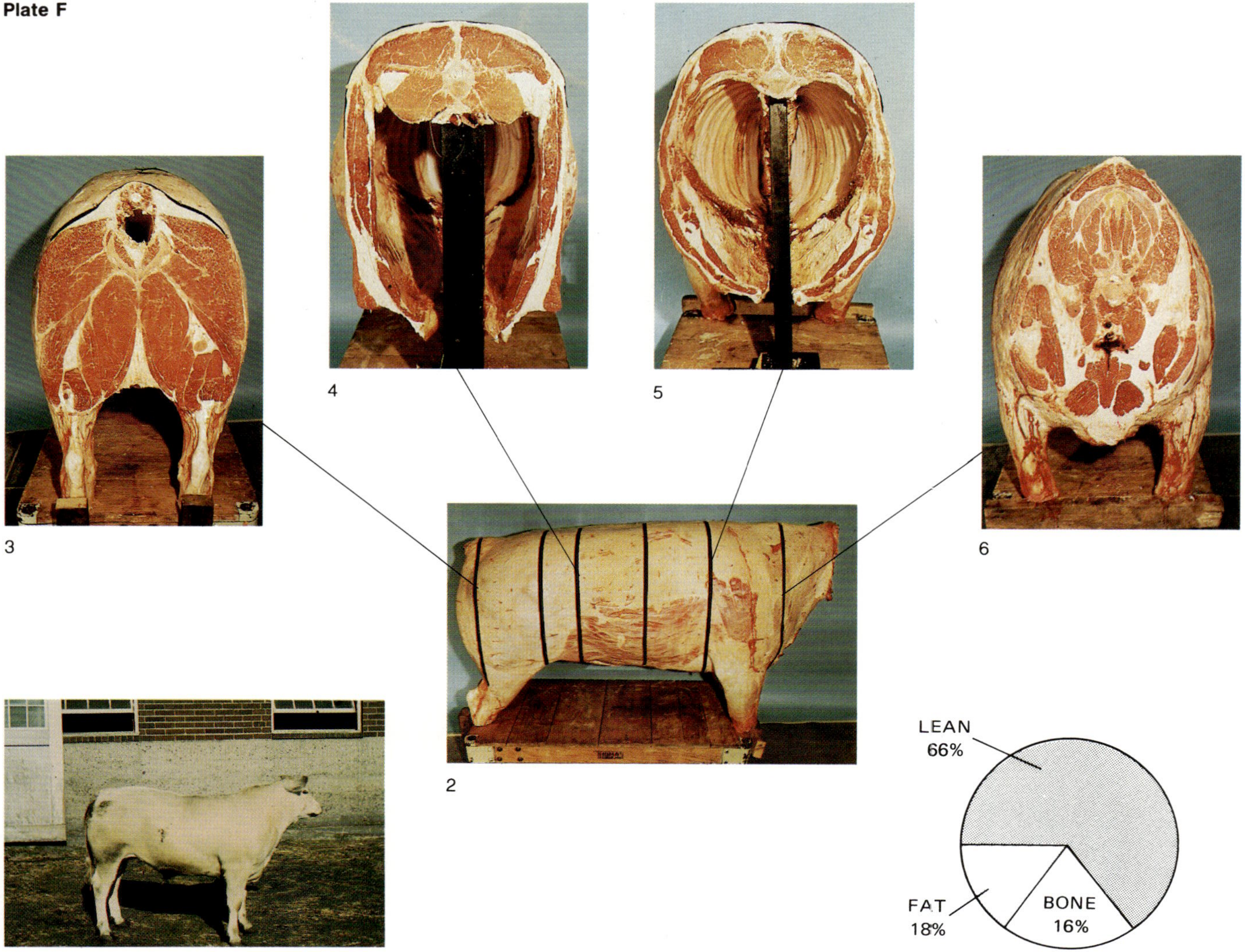

1 2 3 4 5 6 7

Plate G

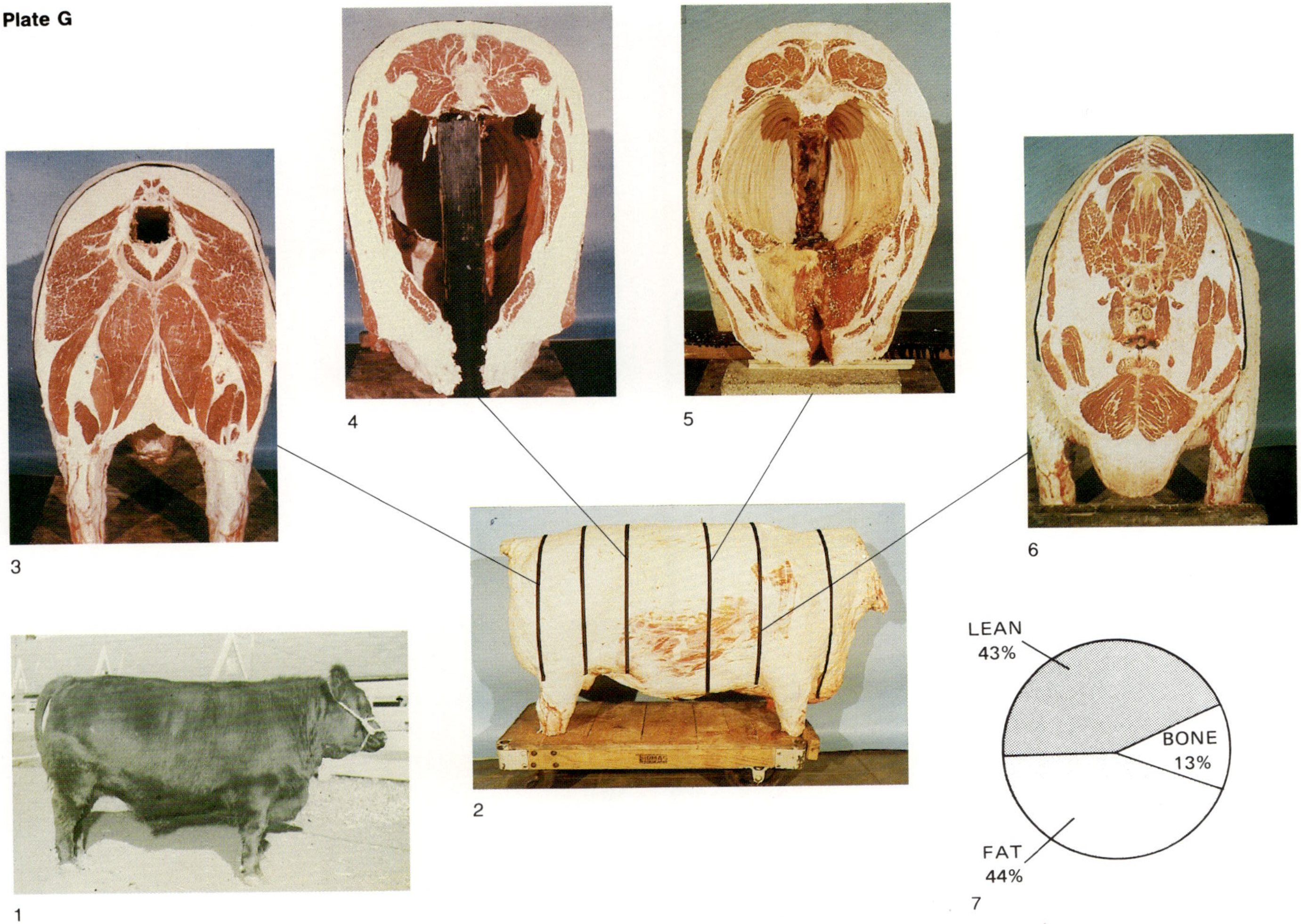

Plates F & G compare fat to lean composition on a yield grade 2 steer (plate F) and a yield grade 5 steer (plate G): (1) live animal—side view; (2) numbered ribbons indicate where following cross sections were made; (3,4,5, & 6) cross sections from rump, hip, mid-carcass, and shoulder; (7) the percentages of fat, lean, and bone found in this carcass.

Plate H

1

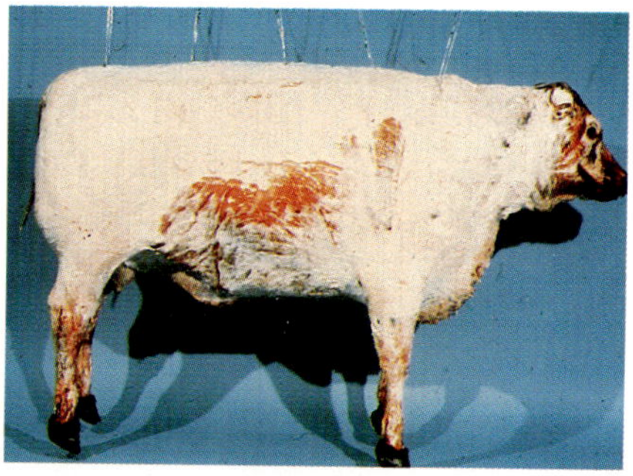

2

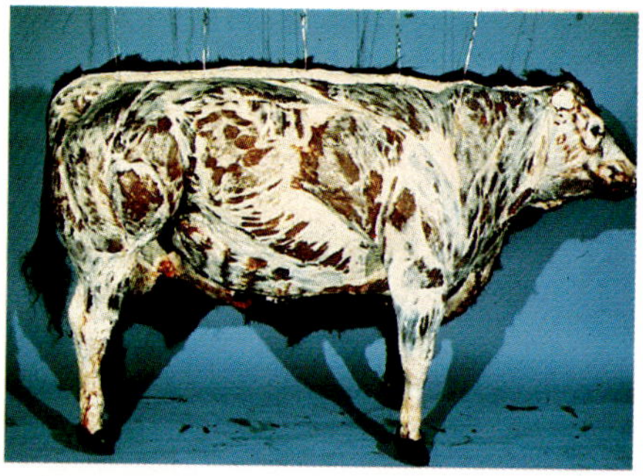

3

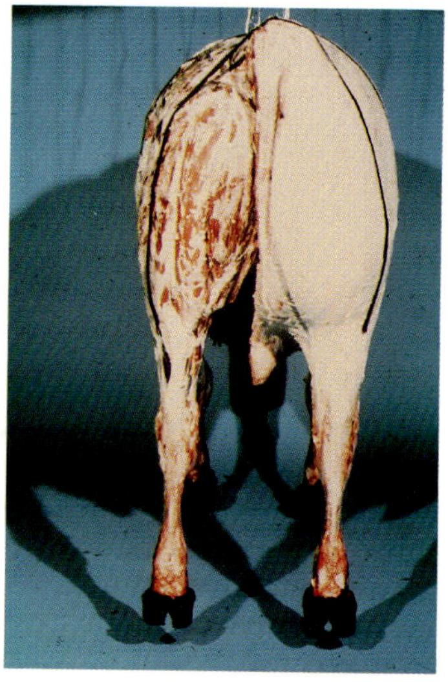

4

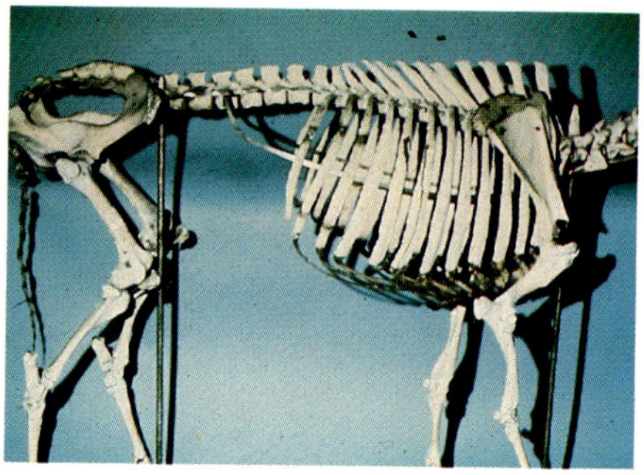

5

Plate H shows the composition and appearance of a steer carcass at various stages of removal of hide, fat, and muscle: (1) live steer with hair clipped from one side; (2) frozen steer with hide removed and fat exposed; (3) fat removed from one-half of steer's body; (4) rear view with fat removed from the left side. Note cod and twist fat on the right side; (5) skeleton of the beef animal after all the muscle has been removed.

Youth Projects

Youth programs (4-H and FFA) have typically used county, state, and national shows as a completion date for the beef project. Steer shows still have considerable participant and audience appeal. This is because the Grand Champion steer receives major publicity, primarily because of a high sale price.

In recent years, several 4-H beef projects have become more consistent with commercial beef production. Carcass and rate of gain evaluations have been added to the live placing. In this situation, steers with the best combination of live placing, gain and carcass are designated as "The Supreme Champions." Great strides in meaningful youth education have been achieved where this type of program receives primary emphasis.

The dimensions of realistic commercial beef marketing is still seriously lacking in many youth beef programs particularly in national shows. It is not uncommon for feeder steers to be purchased by the youth for several times the current commerical price. The motivation is still the live champion steer that sells for several times the commercial market value. One cannot argue the economics for the winner, but low economic return is evident for many steers placing lower than first. It is a great show, particularly for

Figure 17.16. Beef cattle projects give youth experience in assuming responsibility, understanding more about the beef industry and becoming familiar with some of the factors affecting beef cattle productivity. Courtesy of Santa Gertrudis Breeders International.

parents living in their second childhood. It is this dimension of the youth program that causes deep thought after reading the educational objectives of the 4-H program.

Steer shows have received considerable public attention and media appeal. The beef industry is fortunate to have this public interest and can continue to capitalize on this consumer focus on beef.

Junior heifer shows, sponsored by several breed associations, have provided a meaningful stimulus for young people to initiate breeding herds of their own. These have merit in encouraging youth to have beef industry involvement (Figure 17.16).

In recent years, there have been some excellent accomplishments in combining the competition and excitement of the show with the valid performance comparison of animals. Most of these have centered around steers on a competitive feedlot and carcass test or bulls on a common 140-day gain test, then exhibited in competition and sold. This type of show or exhibit, plus other new innovations, will continue to have appeal to the beef producer in the years to come. The cattle show or exhibit will not die or become outmoded as some have predicted. It will be molded to fit the needs of the day. If for no other reason, the gathering of those with common interests and a desire to share friendship and ideas will perpetuate the beef show. And you can be guaranteed that somewhere in the mingling of people, there will be some live beef cattle in the center of the discussion.

SELECTED REFERENCES

Publications

Barton, R. A., 1967. *The relation between live animal conformation and the carcass of cattle.* Anim. Breeding Abstracts 35:1.

Beeson, M. W., Hunsley, R. E., and Nordby, J. E., 1970. *Livestock Judging and Evaluation.* Danville, Illinois: The Interstate Printers and Publishers, Inc.

Berg, R. T. and Butterfield, R. M., 1976. *New Concepts of Cattle Growth.* New York: John Wiley and Sons.

Berg, R. T. and Walters, L. W., 1983. *The meat animal: Changes and challenges.* J. Anim. Sci. 57:133.

Bonsma, J. C., 1983. *Man Must Measure.* Livestock Production. Cody, Wyo.: Agi Books.

Boggs, D. L. and Merkel, R. A., 1979. *Live Animal, Carcass Evaluation, and Selection Manual.* Dubuque, Iowa: Kendall/Hunt Publishing Company.

Briggs, H. M., 1981. *Uses and Abuses of the Show Ring.* Ark. Expt. Sta. Spl. Rpt. 93.

Brinks, J. S., Davis, M. E., Mangus, W. L., and Denham, A. H., 1979. *Genetic aspects of hoof growth in cattle.* CSU Gen. Series 982.

Brown, C. J., Brown, A. H. Jr., and Johnson, Z., 1983. *Studies of body dimensions of beef cattle.* Ark. Agric. Expt. Sta. Bull. 863.

Brown, C. J., Roussel, J. D., and Stallcup, O. T., 1967. *Genetic and other aspects of a hoof anomaly of Hereford bulls.* J. Anim. Sci. 26:201 (Abst.).

Brungardt, V. H., 1972. *Efficiency and profit differences of Angus, Charolais, and Hereford cattle varying in size and growth.* Research Reports R2397-R2401. Univ. of Wis.

Butterfield, R., 1974. *Exposing the myths of show judging.* Calif. Livestock Symposium.

Crickenberger, R. G. and Black, J. R., 1976. *Influence of frame size on performance and economic considerations of feedlot cattle.* Mich. State Agric. Expt. Sta. Res. Rpt. 318.

Crouse, J. D., Dikeman, M. E., and Allen, D. M., 1974. *Prediction of beef carcass composition by live-animal traits.* J. Anim. Sci. 38:264.

Daley, D. A., Tatum, J. D., and Taylor, R. E., 1983 *Practical methodology of feeder cattle sorting.* J. Anim. Sci. 57:390 (Suppl. 1)

Daley, D. A., Tatum, J. D., and Taylor, R. E., 1983. *Accuracy of subjective and objective preslaughter estimates of beef carcass fat thickness.* J. Anim. Sci. 56:1.

Greenough, P. P., MacCallum, F. J., and Weaver, A. D., 1972. *Lameness in Cattle.* Philadelphia: J. B. Lippincott Co.

Hedrick, H. B. 1983. *Methods of estimating live animal and carcass composition.* J. Anim. Sci. 57:1316.

Kauffman, R. G., Grummer, R. H., Smith, R. E., Long, R. A., and Shook, G., 1973. *Does live-animal and carcass shape influence gross composition?* J. Anim. Sci. 37:1112.

Shanteau, J., 1978. *Psychological Abilities of Livestock Judges.* Kansas Agric. Expt. Sta. Bull. 620.

Totusek, R., 1979. *The importance of visual evaluation and the influence of the showring.* International Limousin Journal, Aug. p. 51.

Wilson, G. R., Reef, J. E., Turner, T. B. and Wilson, G. W., 1981. *Relating visual feminine characteristics to reproductive performance in a herd of Angus cows.* Ohio Agric. Res. and Devel. Center Anim. Sci. Series 81–1:25.

Woodward, R., 1968. *The importance of soundness in the selection of breeding animals.* Charolais Banner, July.

Visuals

Beef Conformation (carcass cross-section slides). Iowa State University, Ames, IA 50010.

Fitting and Showing Beef Cattle (filmstrip and tape). Vocational Education Productions. Calif. Polytechnic State Univ. San Luis Obispo, CA 93047

Livestock Judging Slide Sets (35mm color slides). Vocational Education Productions. Calif. Polytechnic State Univ. San Luis Obispo, CA 93407.

Market and Feeder Cattle Selection (filmstrip and tape). Vocational Education Productions. Calif. Polytechnic State Univ. San Luis Obispo, CA 93407.

Shaping Carcass Cutability (16mm film or video tape). Available from Dept. of Animal Science, Land Grant University (see Chapter 22) or Elanco Products Co., P.O. Box 1705, Indianapolis, IN 46206.

18
Marketing Beef Cattle

The primary goal for the beef industry, "the efficient production of a highly palatable product with a profitable return to those who produce it," implies that two primary factors affecting profitable beef production are (1) efficiency of production and (2) price received for the product. There are limits to improving efficiency of production. Even at optimum levels of production efficiency (numbers and weights at least cost), if market price is not high enough to cover these costs, profitability will not be realized. Producers need to understand advantages and disadvantages of different types of available markets and which factors affect supply and demand for beef. These latter factors determine price structure for beef, and producers can identify ways in which beef prices can be predicted and possibly changed.

MARKET CHANNELS

Market channels are pathways through which cattle pass as they move from farm or ranch to slaughter. Figure 18.1 shows marketing channels currently utilized for cattle. The auction market is most important in marketing feeder cattle, slaughter calves, slaughter cows and slaughter bulls. Selling direct (also identified as country) is most important for slaughter steers and heifers. In this situation, slaughter cattle are being marketed directly from the feedlot to the packer. The terminal market, while declining in numbers compared to previous years, is still an important market, particularly to midwestern cattle producers. Figure 18.1 does not show market channels for breeding cattle which are marketed primarily through auctions (private or consignment) and directly to commercial producers through private treaty.

Producers should understand various market channels which differ in availability, marketing costs, and buyer competition.

Auction Market

Approximately 2000 livestock auctions, sometimes called sale barns, are located throughout the United States. Cattle auctions are more numerous in areas of greatest

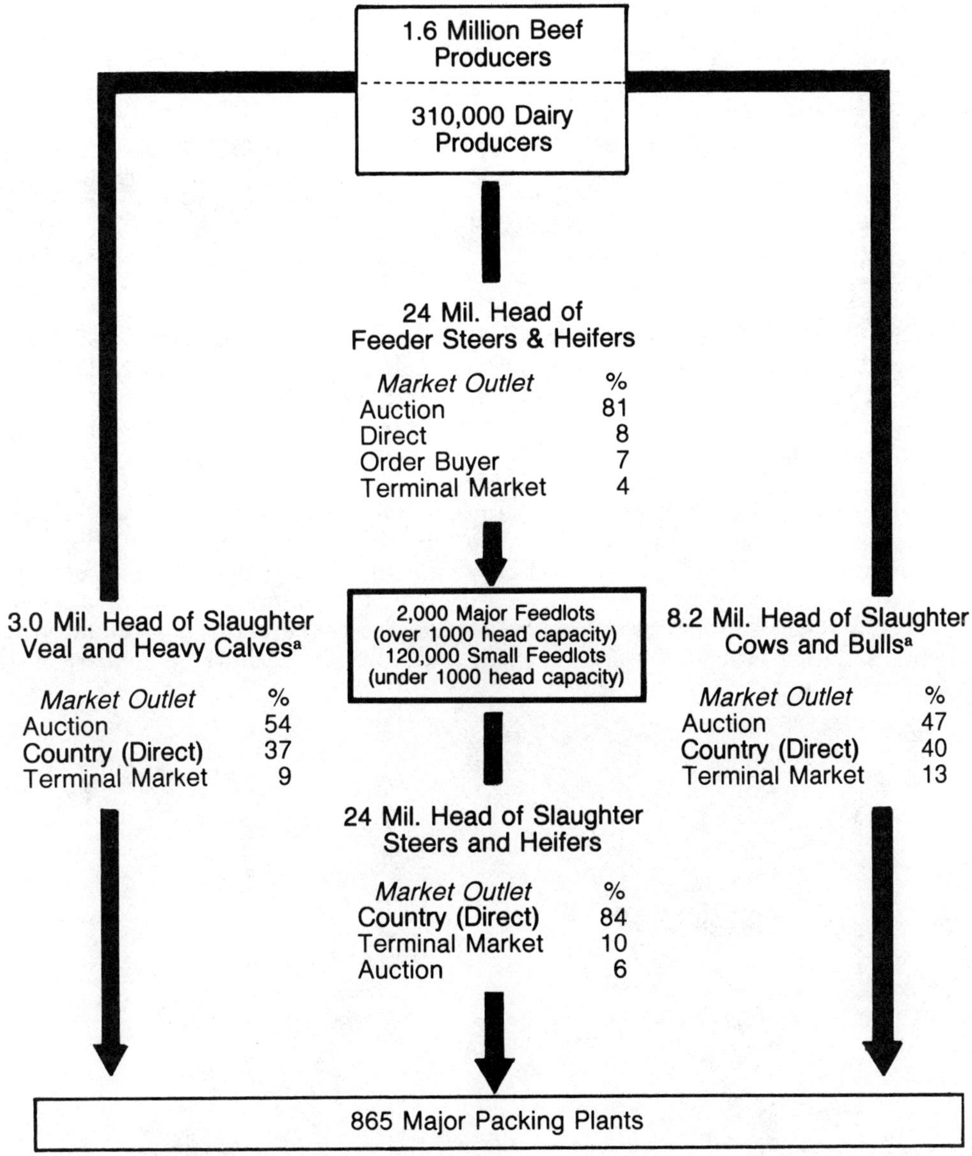

[a] Some go to feedlots for a short feeding period prior to slaughter.
Source: Packers and Stockyards Resume, USDA, 23 Dec. 1979. Personal communication, C. K. Gee, Economic Research Service.

Figure 18.1. Marketing channels for cattle in the United States, 1982. Courtesy of CSU.

cow–calf numbers since primarily feeder cattle, cull cows and bulls are sold through them. Larger auctions are owned by corporations while most smaller auctions are under individual proprietorships. Holding pens, scales, and sale area are under the same ownership.

Figure 18.2. The Blue Grass Stockyards in Lexington, KY has one of the largest annual receipts of cattle numbers compared to other auction markets in the United States. Courtesy of Cy Alexander.

The major cattle auction markets, each with 1980 cattle receipts over 240,000 head, were Amarillo, TX, Springfield, MO, Dodge City, KS, Torrington, WY, Billings, MT, Lexington, KY, LaJunta, CO, San Angelo, TX, and Louisville, KY (Figures 18.2 and 18.3).

Cattle owners consign their livestock to the auction where cattle are sorted into uniform sale groups if there is a considerable amount of variation in type and grade. At sale time, cattle enter a sales ring where the auctioneer accepts incremental bids from the audience until no one advances the bid. Most cattle are sold by the pound, and current weight is typically displayed electronically while the cattle are being sold.

Sales costs are usually by the head as shown in the following costs of a selected auction: feeder steers and heifers ($4.50 to $5.50), slaughter cows ($7.50), slaughter bulls ($6 to $15), breeding bulls ($20), and cow–calf pairs ($12.50). Additional representative

Figure 18.3. Auction day at Livestock Exchange, Inc. in Brush, Colorado. The nation's eighth largest auction market has pen space for more than 12,000 head of cattle. Pens immediately behind the Sales Pavilion (center of photo) contain hay and water for the cattle prior to the sale. Pens to the right are dry, holding pens for the cattle after they sell. Courtesy of Livestock Exchange, Inc., Brush, Colorado.

costs per head are feed (cost plus 25%), brand inspection ($0.20) in some states, health inspection ($0.20), insurance ($0.06), and fees collected for advertising, promotion and research ($0.15 to $0.50) for the National Live Stock and Meat Board and state checkoff program.

Terminal (Public) Market

In 1978 there were 28 terminal markets in operation which is a marked reduction from the 80 terminal markets that existed in the 1930s. Importance of the terminal market declined rapidly when many packing plants moved away from terminal markets and closer to large commercial feedlots. In 1980 the leading terminal markets, based on cattle receipts of over 500,000 head, were Oklahoma City, OK, Omaha, NE, South St. Paul, MN, Sioux City, IA, and Sioux Falls, SD (Figure 18.4 and 18.5).

Terminal markets are similar in their structure and activities. The stockyards company, usually a corporation, owns the physical facilities (land, buildings, pens, and scales). They charge a yardage fee (approximately $2.50 per head), to producers consigning their cattle, for use of the yards and other services such as feed and handling the cattle. Cattle are consigned to a commission firm designated by the producer. The commission firm represents the producer as the seller and charges a fee for services rendered (approximately $2.85 per head). There is a wide range of buyers on the terminal market; however, the primary buyers are packer buyers (salaried by different packing firms), order buyers (purchasing for someone else for a fee), and yard dealers (sometimes known as traders, speculators, or scalpers whose objective is to resell the cattle at higher prices).

Selling price of cattle on the terminal market is determined by private treaty where the commission representative and the buyer (one at a time in the pen of cattle) bargain until a firm price is agreed upon. Some terminal markets have incorporated the auction

Figure 18.4. View of the 12 acre covered cattle division located at the Sioux City, Iowa Terminal Market. Courtesy of the Sioux City Stockyards.

Figure 18.5. Penned cattle ready for sale at the Sioux City, Iowa Terminal Market. Courtesy of the Sioux City Stockyards.

as another marketing alternative; however, it is primarily feeder cattle that are sold through the auction.

Both terminal markets and auction markets, which handle livestock involved in interstate movement, are considered public markets. These public markets are subject to regulations specified by government personnel of Packers and Stockyards (P&S), an agency of the Agricultural Marketing Service of the USDA. The P&S assures that markets are properly bonded, prescribes rules for fair trade and competition, and periodically checks the scales.

Country Market

Country markets include order buyers, commission representatives, and direct marketing. Order buyers and commission representatives are intermediaries between the seller and the ultimate buyer, while direct marketing does not involve the services of a marketing middle person. Examples of direct marketing are where commercial producers sell their feeder cattle directly to cattle feeders and where packer buyers purchase slaughter cattle directly from the feedlot owner. This latter marketing method is the most significant country market since a high percentage of slaughter steers, heifers, and calves are purchased in this manner.

Direct marketing of slaughter steers and heifers is accomplished primarily two ways: (1) cattle are sold on a live weight basis per 100 lbs considering estimated carcass grade basis and live weight, and (2) sale of cattle is based on carcass weight and usually carcass quality grade. The latter is sometimes referred to as "grade and yield," "on the rail," or "in the meat." Approximately 28% of slaughter cattle are currently sold by this method, which has increased from approximately 20% in 1971. Nebraska, Iowa, and Colorado account for about two-thirds of all carcass grade and weight purchases of steers and heifers. Packers in Texas, Iowa, Nebraska, Minnesota, and Wisconsin account for approximately three-fifths of all cows and bulls purchased on a carcass basis.

Some producers prefer carcass grade and weight (yield) marketing, indicating that it is a more accurate method than dressing percentage and carcass grade in live animals. Other producers are fearful of this method of marketing, feeling they lose control of the marketing process when weighing, grading, and occasional condemnation of carcasses occurs behind "closed doors."

Electronic Marketing

Electronic marketing involves some recent innovations in marketing cattle, primarily in feeder cattle. Different types of electronic equipment, such as video tape, telephone hookup, teletype, and computers are used to collect and distribute information to buyers and sellers.

Most electronic marketing systems share the following common characteristics: (1) pricing is determined at a single location, or over a single communication system, (2) the cattle are not moved from the seller's farm until they are sold, and (3) buyers can participate without being at the location where pricing occurs.

While large numbers of feeder cattle have not yet been marketed by electronic means, the video auction has experienced the largest sales volume. Video auctions are conducted by showing buyers a video tape (2 to 4 minutes) taken at the farm or ranch of the cattle being offered for sale. The audio portion of the tape describes weight, location, and background information on the cattle. The seller also states the weighing location and various options of delivery dates. The video tape is shown to buyers assembled in the same room with an auctioneer who sells the cattle to the highest bidder. After the sale, buyer and seller arrange for shipping the cattle.

KEEPING CURRENT ON BEEF PRICES

Table 18.1 lists some comparative cattle and product prices and the source of these prices. It is not uncommon for prices to fluctuate on a daily or even an hourly basis, so it is important to have current prices in making valid marketing decisions to assure maximum dollar return.

Market information on cattle numbers and prices are available on a daily or weekly basis in many local newspapers and radio stations. These sources generally give market

Table 18.1. Comparative Market Prices for Cattle and Cattle Products at a Given Point in Time

Beef Industry Segment	Product	Price[a] (October 1983)	Reference Source[b]
Purebred (Seedstock)	Bulls (14–30 mo.)		
	—to purebred breeders	$1000–$25,000	2, 3
	—to commercial producers	$750–$3000	2, 3
	Semen		
	—registered herds	up to $100/unit	3, 6
	—commercial herds	$4–$8/unit	3, 6
Commercial Cow–Calf	Calves		
	—under 1 month	$35–$90/head	2
	—400–500 lbs (6–10 mo. steers)	$59–$69/cwt	1, 2, 5
	Sl. cows (900–1200 lbs) (good condition)	$35–$38/cwt	1, 2, 5
	Sl. bulls (1200–2000 lbs) (good condition)	$42–$50/cwt	1, 2
	Cow–Calf pairs	$400–$600/pair	2
Yearling	Steers (600–800 lbs)	$56–$62/cwt	1, 2, 5
	Heifers (500–700 lbs)	$50–$55/cwt	1, 2, 5
Feeder	Sl. steers (Choice, YG3, 1050–1250 lbs)	$59–$62/cwt	1, 2, 5
	Sl. heifers (Choice, YG3, 900–1100 lbs)	$56–59/cwt	1, 2, 5
Packer	600–750 lbs Choice steer carcasses (YG 3)	$92–$96/cwt	1, 2, 5
	550–700 lbs Choice heifer carcasses (YG 3)	$90–$94/cwt	1, 2, 5
	Hide and offal	$6.11/cwt (liveweight)	1, 2, 5
Retailer	Ave. price of all retail cuts	$2.33/lb	4
Consumer	Chuck roast	$1.59	4
	Porterhouse steak	$3.53	4
	Round steak	$2.49	4
	Ground beef	$1.54	4

[a]Price shown gives a range which would include most of the cattle or product fitting the particular description.
[b] Sources of market prices.
1. *Livestock Market News* (USDA).
2. Weekly livestock publications, e.g., *Record Stockman, Drovers Journal, Western Livestock Journal.*
3. Breed Journals, e.g., *Angus Journal, Polled Hereford World, Limousin Journal.*
4. *Livestock and Meat Situation* (USDA).
5. Cattle-Fax.
6. AI price lists.

reports on local auctions, direct sales, terminal markets in certain areas, and occasionally report the carcass beef trade. Several major weekly publications, i.e., *Drovers Journal, Record Stockman* and *Western Livestock Journal,* give rather extensive market numbers and price coverage.

Several more extensively used resources for market information are presented.

Cattle-Fax

Cattle-Fax is a market analysis service associated with the National Cattlemen's Association since 1968. Cattle-Fax members can receive up-to-the-minute cattle marketing information from regional analysts on toll free WATS lines. Also provided are written update reports, mailed weekly, giving current, intermediate, and long-term market evaluations. A video communication system called DATA QUOTE IV, available to Cattle-Fax members, provides current information on Cattle-Fax sales and comments, terminal markets, grain information, cattle prices, and commodity markets.

Instant Market News

Instant Market News is an automatic telephone answering service provided by the USDA and sponsored by producer organizations or commercial companies. The reports, updated two to five times daily, give information on livestock and meat numbers, prices, and futures trading pertinent to the area serviced. Table 18.2 gives telephone numbers for several locations throughout the United States.

MAJOR FACTORS AFFECTING CATTLE PRICES

Price of cattle and their resulting products is the most widely discussed topic in the beef industry. The many factors which determine cattle prices are complex and not easily understood. Because of this complexity, many misunderstandings are generated between various beef industry segments which have resulted in serious accusations, lawsuits, price freezes, and consumer boycotts.

Cattle producers are told that cattle prices result from a free enterprise system where supply and demand factors totally determine the price of cattle. Producers find themselves as "price takers" rather than "price makers," yet they purchase most of their goods and services in a reverse pricing structure. Even within the beef industry, the pricing structure changes from the producer essentially saying, "What will you pay me?" to the retailer who establishes a price for beef.

Supply and Demand

Factors affecting the supply and demand of beef and ultimately the price of beef are shown in Figure 18.6.

It is obvious that most of these factors have an effect on pounds of beef produced and the retail price of beef.

Beef Cattle Cycles and Beef Prices

For almost 100 years, there have been several beef cattle cycles with characteristic peaks and valleys in cattle numbers and prices. These periodic changes in cattle numbers have occurred primarily in response to changes in profitability and weather (drought). During the peaks of cattle numbers, the supply of beef was excessive; therefore price of beef dropped, while at the low points of cattle numbers, supply of beef was limited, and this stimulated price increases. These characteristic number and price cycles are shown in Figure 18.7.

Table 18.2. Locations and Telephone Numbers of Cattle Market Information Provided by the USDA.

INSTANT MARKET NEWS

The latest livestock market information is now available from an automatic telephone answering device. Producers and others who need up-to-the minute market news can get this service by dialing a number anytime of the day or night. Most of the machines are sponsored by producers organizations or commercial concerns. The reports are updated from two to five times daily, depending on the area serviced. All reports are not the same, as each one is individually tailored to best serve the needs of the industry in their respective areas. The reports offer a variety of the most current information on livestock and meat prices, Federally inspected slaughter, salable receipts, and futures trading. The following directory lists 100 services in 34 states that are now in operation:

Location	Telephone
Alabama	
Montgomery	
5 PM–8 AM	C-(800) 392–5804
5 PM–8 AM	H-(800) 392–5801
Arkansas	
Ft. Smith	(918) 875–3892
Little Rock	(501) 372–3933
Arizona	
Phoenix	(602) 275–7972
California	
Bell	(213) 268–8020
El Centro	(714) 352–8160
Redding	(916) 246–8480
Stockton	(209) 466–3085
Visalia	(209) 733–3750
Colorado	
Denver	(303) 629–7081
Florida	
Bonifay	(904) 547–2016
Ft. Pierce	(305) 465–6216
Monticello	(904) 997–3081
Orlando	(305) 841–7700
Quincy	(904) 875–2414
Sebring	(813) 385–2472
Tallahassee	(904) 488–0274
Tampa	(813) 272–2254
Trenton	(904) 463–2427
Winter Park	(305) 628–0412
Georgia	
Thomasville	(800) 342–1440
Illinois	
Chicago	(312) 922–1253
Joliet	(815) 423–5026
Peoria	(309) 676–8811
Nat'l Stk. Yds.	(618) 874–1900
Springfield	(217) 525–4019
Idaho	
Burley	(208) 678–2424
Pocatello	(800) 632–9494
Indiana	
Indianapolis	(317) 784–7778
Indianapolis	1–(800) 382–1567 (Indiana only)
Iowa	
Ames	(515) 294–6899
	(515) 294–4347
Des Moines	(515) 282–6870
Durrant	(319) 785–6032
Sioux City	(712) 252–2100
Kansas	
Dodge City	(316) 225–1311
	(316) 227–2306
Kentucky	
Frankfort	(502) 564–4958
Louisville	(502) 584–6617
Louisiana	
Baton Rouge	(504) 291–0054
Michigan	
Lansing	(517) 373–6330
Minnesota	
So. St. Paul	(612) 451–3692
Missouri	
Jefferson City	(314) 636–4203
Joplin	(417) 781–9451
Kansas City	(816) 421–7694
Mexico	(314) 581–6250
N/A 8:30–9:15 & 11:30–12:00AM	
Sikeston	(314) 472–1564
St. Joseph	(816) 238–1203
Springfield	(417) 866–4986
West Plains	(417) 256–9631
Nebraska	
Aurora	(402) 694–3183
Beatrice	(402) 223–5231
Beemer	(402) 528–3654
Columbus	(402) 564–1133
Grand Island	(308) 384–5101
Kearney	(308) 237–5908
Lincoln	(402) 477–3366
Omaha	(402) 731–5355
Tekemah	(402) 374–1667
West Point	(402) 372–5650
New Mexico	
Clovis	(505) 763–3030
New York	
Central	(315) 495–6562
Eastern	(518) 457–6672
Northern	(315) 788–2590
Western	(716) 343–0678
North Dakota	
West Fargo	(701) 282–4593
Ohio	
Columbus	1–(800) 282–7605
Oklahoma	
Okla. City	(405) 236–5491
Tulsa	(918) 437–0740
Oregon	
Corvallis	(503) 754–2037
Pennsylvania	
New Holland	(717) 354–7288
South Carolina	
Columbia	(803) 799–5568
Walterboro	(803) 549 5232
South Dakota	
Rapid City	(605) 342–1833
Sioux Falls	(605) 336–7765
Tennessee	
Jackson	(901) 423–2080
Knoxville	(615) 525–3211
Nashville	(615) 833–4046
Nashville	(800) 342–8206
Texas	
Amarillo	(806) 372–3494
Fort Worth	(817) 624–7451
Corsicana	(214) 872–4001
San Angelo	(915) 655–2358
San Antonio	(512) 223–4100
Sealy	(713) 885–2050
Utah	
No. Salt Lake	(801) 524–5001
5:00 PM–7:30 AM	
Salina	(801) 529–7000
Virginia	
Abingdon	(703) 628–8601
Harrisonburg	(703) 434–5871
Onley	(804) 787–3500
Richmond	(804) 786–8749
Roanoke	(703) 563–5956
Warrenton	(703) 347–1326
Warsaw	(804) 333–3430
Windsor	(804) 242–6978
Washington	
Moses Lake	(509) 765–0311
Yakima	(509) 575–0377
West Virginia	
Charleston	(304) 348–8883
N/A 11:00 AM–2:00 PM	
Wyoming	
Cheyenne	(307) 777–7959
Torrington	(307) 532–7200

Source: Livestock Market News 18 June 1983

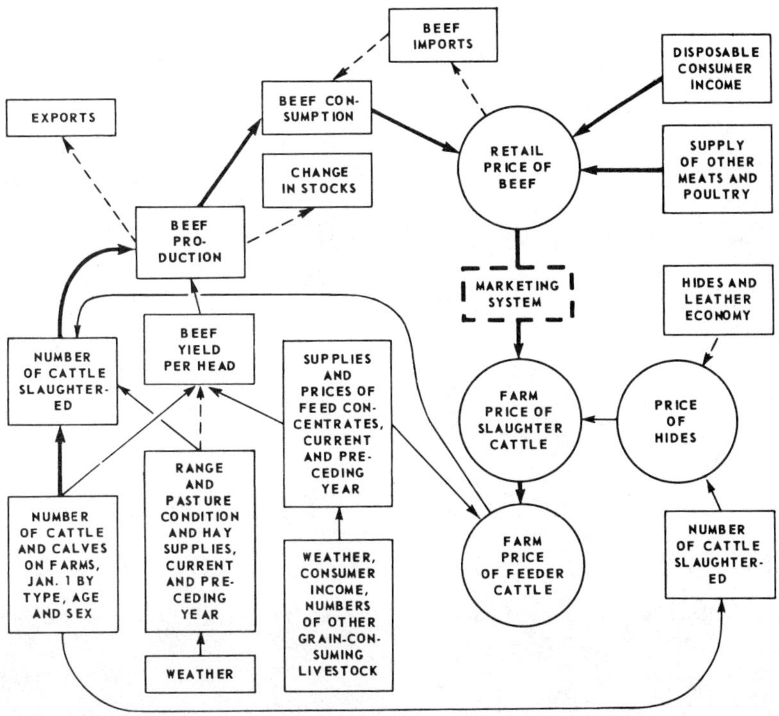

Figure 18.6. The demand and supply structure for beef. Arrows show direction of influence. Heavy arrows indicate major paths of influence. Light, solid arrows indicate less important actors of influence, while dashed arrows are pathways of minor influence. Courtesy of USDA.

These cattle numbers and price effects on slaughter and feeder cattle prices are shown more vividly for the past 40 years in Figure 18.8.

Typically, price cycles lay behind number cycles. This means cattle numbers cannot be changed rapidly even though prices are high. There is a biological time lag to increase cattle numbers (Figure 18.9).

The most apparent problem of beef cattle cycles has been wide price fluctuations, due to wide swings in oversupplying and undersupplying the beef market. This is best demonstrated in Figure 18.10 which shows changes in number of cows per 100 people in relation to feeder cattle price.

Some analysts say that historic cattle cycles will not repeat themselves in the same way. They say that rapid changes of the past have had either cheap cattle, cheap land, cheap feed, or cheap money as a stimulus. During the early 1980s, none of these factors existed. Also the recession put a lid on consumer demand as their buying power was restricted.

There are others who claim beef cattle cycles will occur in the future even though the beef industry has encountered some dramatic changes in recent years. If the cycles do continue, then each cycle will be expected to bring several years of unprofitable beef prices as well as several years of profitable prices.

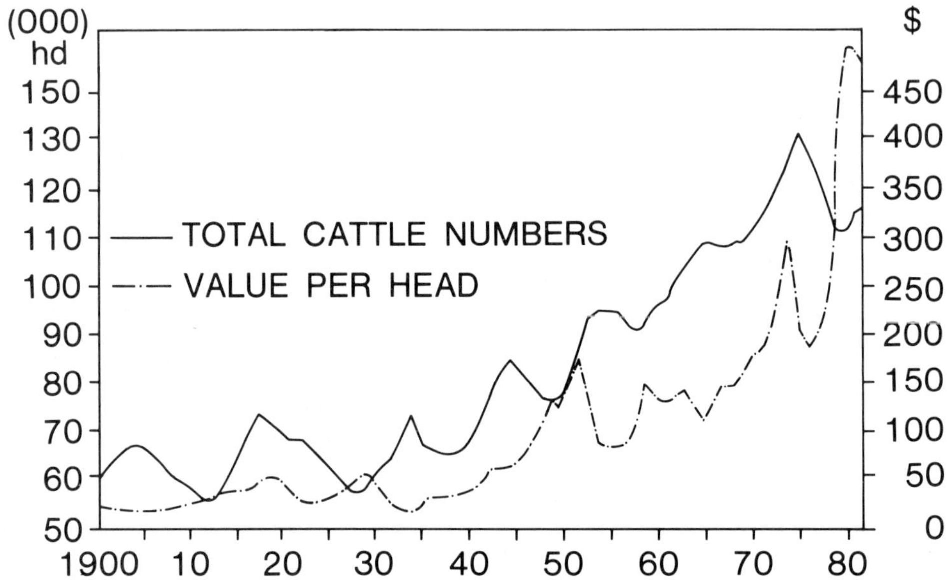

Figure 18.7. Total cattle numbers and farm value per head, 1900–1980. Drawn by CSU Graphics.

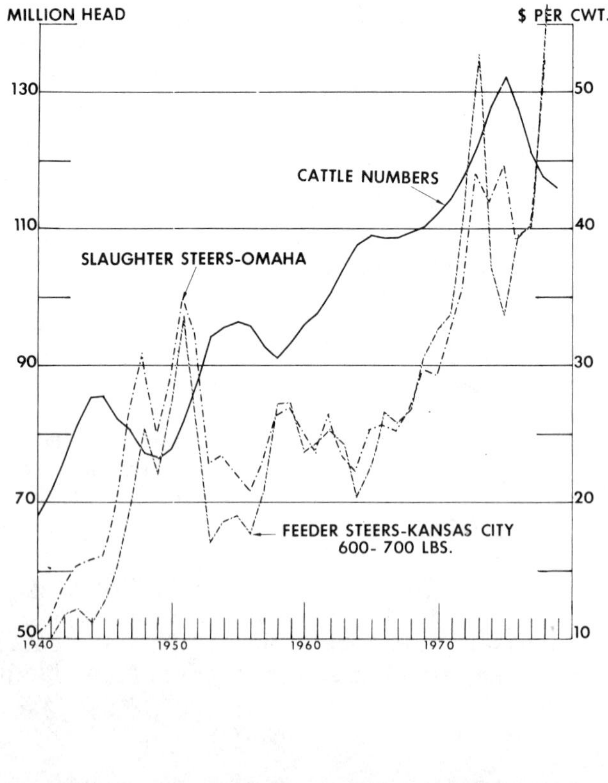

Figure 18.8. Annual numbers and average prices of cattle, 1940 to 1979. Courtesy of the Western Livestock Round-Up.

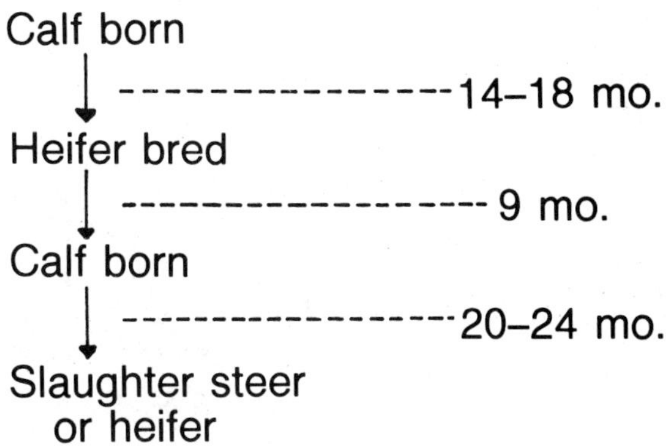

Figure 18.9. Approximately 3 to 4 years is needed before a heifer calf, saved as a breeding animal replacement, can produce another slaughter animal. Courtesy of CSU.

Carcass Weights and Prices

Tonnage of beef produced is a function of the number of head slaughtered and average carcass weight. It is not uncommon for the average carcass weight to vary 25 lbs from one year to the next. Considering 30 million head of slaughter cattle, this carcass weight fluctuation would result in an increase or decrease of 750 million pounds of beef.

There is a relationship between carcass weight and cattle prices (Figure 18.11). Heavy carcass weights do not necessarily cause lower beef prices, but they result from the

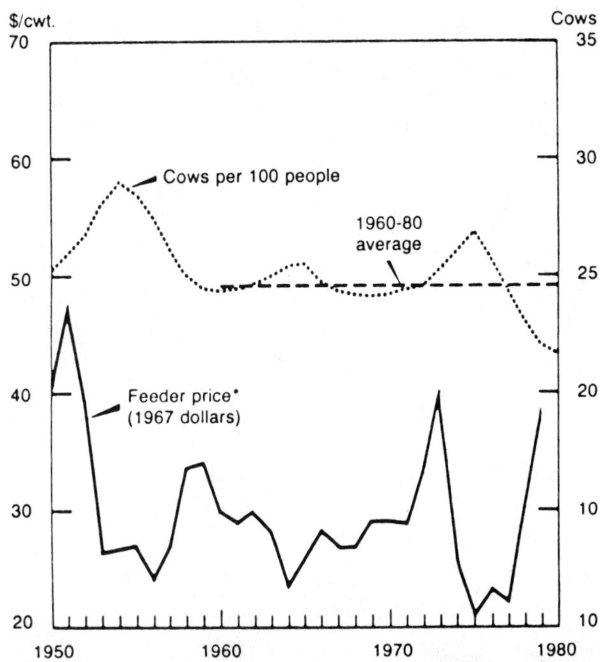

Figure 18.10. Cow to people ratio and its relationship to cattle prices. Courtesy of the National Cattlemen's Association.

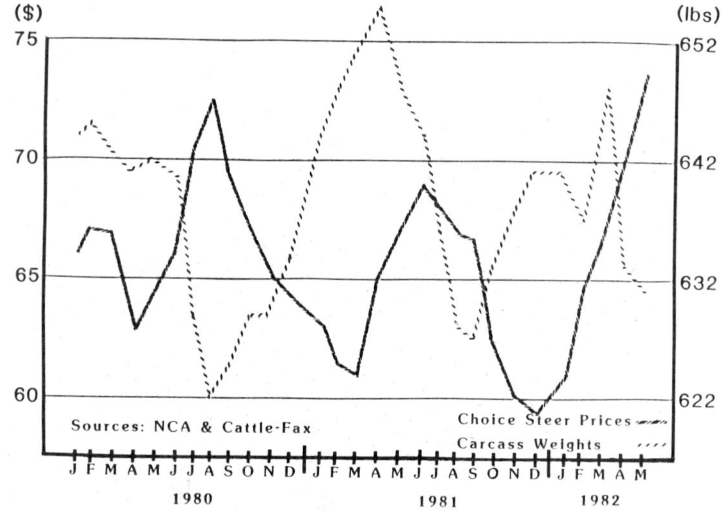

Figure 18.11. Monthly averages of choice steer prices and carcass weights, 1980 to 1982. Courtesy of NCA and Cattle-Fax.

lower prices. When prices begin to drop, feeders typically feed their cattle longer, anticipating a price increase. This usually has an antagonistic effect on prices by increasing the tonnage of beef and more yield grade 4–5 slaughter cattle. As cattle prices increase, feeders will usually market their cattle at lighter weights, desiring to market as many cattle as possible before a price break occurs.

Quality Grades and Yield Grades

Even though not all carcass beef is quality graded or yield graded, these grades are a significant part of the pricing structure of beef. Figure 18.12 shows that poorer yield grades are heavily discounted and that the premium paid for Choice over Good fluctuates widely, both between and within years.

FORECASTING BEEF PRICES

One of the largest variables affecting profitable beef production is projecting the market price at some future date. For example, the cattle feeder currently purchasing a

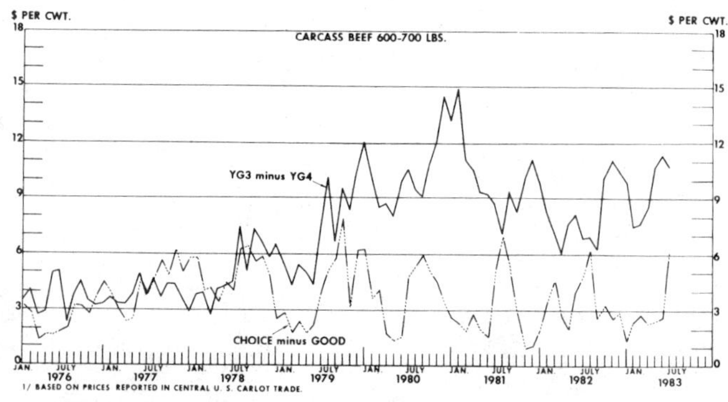

Figure 18.12. Price spreads for carcass beef (monthly average, 1976–1983). Courtesy of the Western Livestock Round-Up.

pen of feeder cattle will be selling them at an unknown price some 100 to 150 days in the future.

With all the data and technology available today, the evidence is that cattle prices cannot be predicted very accurately a few months in advance. Even though hedging can be used to lock in profits or reduce losses, the futures market is not an effective marketing tool to forecast prices. If the long-term cattle cycles are functioning as in the past, then prices between years could be forecast reasonably well.

Seasonal variations in prices can be evaluated for possible use in forecasting prices. Seasonal marketings are those that follow a more or less uniform pattern within the period of a year, and tend to repeat that pattern from one year to the next. Seasonal price changes are a direct reflection of seasonal marketing and to a lesser degree, seasonality in demand. Seasonality in marketings results from biological factors and management practices. Most calves are born during the spring and weaned and sold in the fall. Cows are typically culled and sold in the fall.

Figure 18.13 shows the seasonal variation in slaughter steer prices. The solid line is average price while the shaded band represents the price range. The seasonal trend line would be expected to stay within the shaded area during most years. The seasonal price patterns are shown with index numbers where an average month is given an index of 100, high months are greater than 100 and months in which seasonal price cows occur have indices of less than 100.

Even though seasonal prices for slaughter steers show some strength during spring and summer months, the shaded area is quite large, indicating a high degree of instability. This means that little confidence can be placed in this seasonal trend, and it is of limited use as a marketing tool in price forecasting.

Seasonal price trends for yearling feeder steers show a trend similar to slaughter steers (Figure 18.14). This weak seasonal pattern is of limited value in forecasting prices.

The seasonal price pattern for 400 to 500 lb steer calves is nearly an inverse of the marketing pattern during the year (Figure 18.15). Highest prices generally occur during the March, April, and May period as few calves are marketed and demand by grass operators is quite high. Cattle producers could use this information as a tool to market

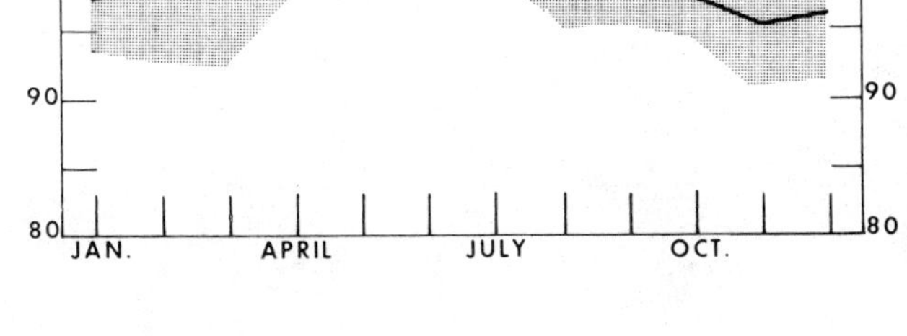

Figure 18.13. Seasonal price trends in 900 to 1100 pound slaughter steers (Omaha market, 1971 to 1980). Courtesy of Western Livestock Round-Up.

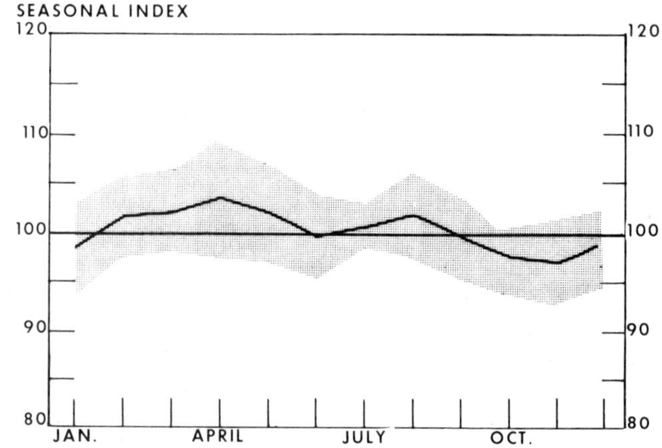

Figure 18.14. Seasonal price trends for 600 to 700 pound yearling steers (Oklahoma City market, 1971 to 1980). Courtesy of Western Livestock Round-Up.

their calves if production costs favor it. Even though the shaded area is quite large, it did not drop much below 100 during the spring months.

Prices for cull slaughter cows tend to exhibit the strongest seasonal tendencies of any class of cattle (Figure 18.6). This is to be expected because of marketing patterns. Cattle producers can use this information to forecast slaughter cow prices since the shaded area did not go below 100 during the spring months and was not above 100 during the late fall months.

Jacobs (1983) reported some interesting information on within year variations in fed cattle prices. His comments follow:

"Forecasting within year variations in fed cattle prices has been demonstrated to be a very hazardous activity. Study of the 'track record' of major forecast sources quickly discloses our 'disabilities.' In a study by this writer of forecast errors of three nationally respected forecast sources, the average forecast errors for quarterly forecasts only one quarter ahead varied from $5.26/cwt. (most accurate) to $6.69/cwt. (least accurate) for

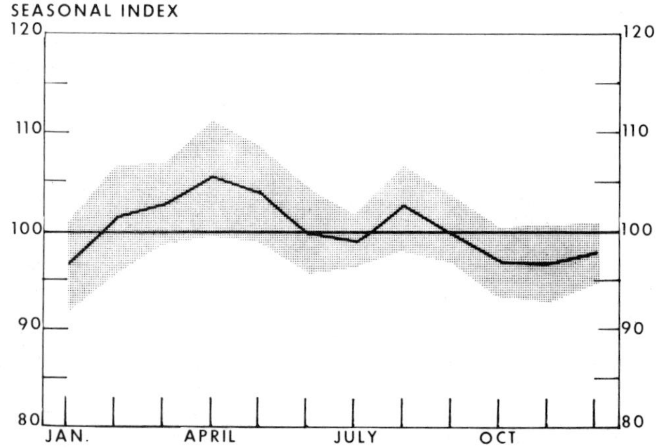

Figure 18.15. Seasonal price trends in 400 to 500 pound feeder steer calves (Oklahoma City market, 1971 to 1980). Courtesy of Western Livestock Round-Up.

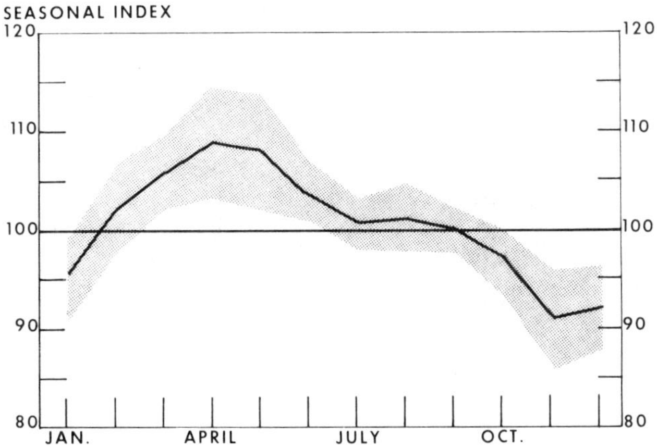

Figure 18.16. Seasonal price trends in utility, slaughter cows (Omaha market, 1971 to 1980). Courtesy of Western Livestock Round-Up.

the 25 quarters ending 2nd Q 1982. The relative magnitude of these errors can be appraised by comparing them with the average error of $3.91/cwt. that would have occurred had they simply forecast each quarter to be the same as the prior quarter.

"Sources of within year price variations (and of forecast errors) are seemingly unpredictable relationships between live and retail prices and marketing margins. In brief, if retail prices of beef could be forecast with 100 percent accuracy, errors in derived live price forecasts would still be quite large—unless fluctuations in marketing margins could be accurately forecast. The simple, if discouraging, fact is that within year live price variations are far more related to fluctuations in marketing margins than in retail prices. In 1982, for instance, only a third of the month-to-month variations in farm price was associated with variations in retail prices while two-thirds were directly associated with shifts in retail marketing margins. In 1981, the contrast was even more vivid, as 76 percent of the farm price variance was explainable by shifts in margins while only a half percent was explainable by retail price movements. Such a relationship is particularly 'unhandy' because published data on retail prices and margins are typically two months behind.

"Within year variations are where the action is! When the numbers cycle is relatively quiet (as in the past four years), within year price movements are much more active than between year shifts (which are more predictable from conventional supply–demand analysis). The maximum between year difference in the past four years has only been $3.83 (1979's $67.67 vs. 1981's $63.84)—while the within year range in monthly fed cattle prices has averaged a more hefty $11.52 per cwt.

"Improved understanding of retail pricing behavior would seem to be a precondition to more accurate forecasts. Observations and analysis of this behavior in the past suggests the following should help.

"Three recent price cycles are shown diagramatically in Figure 18.17. Starting with a cyclical low in Apr. 1980, three cycles were completed in the 30 mos. ending Oct. 1982—thus averaging 10 mos. per cycle. Variations in live and carcass price were associated mostly with expansion and contraction of marketing margins. In the average run-up period (averaging 4 mos.), fed cattle prices increased by $9.65/cwt. Retail prices

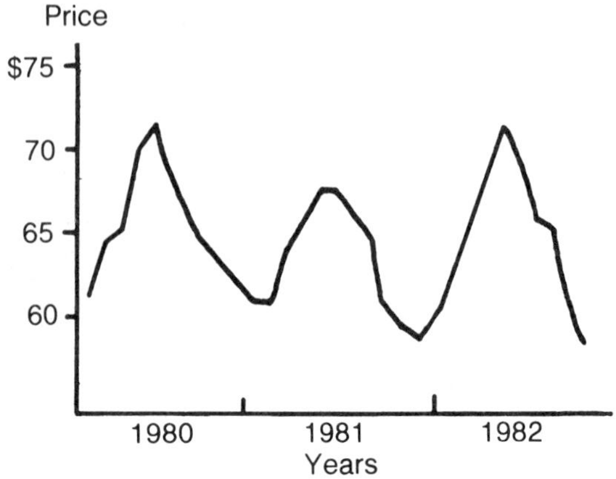

Figure 18.17. Three recent 'within year' price cycles of fed cattle (900 to 1100 pound Choice steers, Omaha). Courtesy of Vic Jacobs.

increased an average of only 6.2¢ per lb. while 72% of the $9.65 run-up in farm price was traceable to a 16¢ decline in retail margins. During the three 'run-downs' (averaging six months each), fed cattle prices dropped by an average of $11.08 per cwt. Meanwhile, retail prices dropped only 4.4¢ with 83% of live and carcass price drop traceable to a 21¢ inflation of marketing margins.

"Price lags at retail tend, however, to be somewhat obscured, in comparisons such as the above, because of the belated 'catchup' that occurs in the last month or two just prior to the cyclical tops and bottoms. If only the first three months of the past 3 run-ups (averaging four months) are examined, we find no increase in retail price, rather a 0.9¢ decrease while live cattle prices increased nearly $7/cwt!. Similarly, if we look only at the first four months of the six months run-down periods, we find no decrease in retail price—rather a 1.8¢ increase during the same months that live prices declined $8.56 per cwt.!

"Thus, while the absolute tops and bottoms in retail prices may lag live and carcass prices by only around a month, they lag much further than that during most of each boom or bust move—but belatedly make a fast catchup move just prior to the turning point. In fact, this belated but rapid move may cause the turning point because of its effect on retail purchase volume.

"Observations of these within year price cycles suggest certain behavioral characteristics:

"(1) Cause–effect reversal. While live and carcass price is derived from retail price over the long run (years), the reverse appears to occur over short periods (months). Retail prices (the administered price) appear to be adjusted upward and downward after the fact of changed carcass prices—and in response to shrunken or inflated marketing margins.

"(2) Demand for live animals and carcasses thus overshoots to nonsustainably high and low levels as retail marketing margins are allowed to shrink or inflate (temporarily) to nonsustainable levels. As a result, retail purchase volume is higher (or lower) than it will be after retail prices are belatedly adjusted to 'correct' margins. This lag in retail adjustment then causes—

"(3) Price overruns on both ends of the cycle. Fast run-ups occur while retail prices lag—to be followed by fast run-downs after retail prices belatedly slow consumer purchases."

If producers could accurately forecast prices, they likely would change the supply pattern.

SHRINK

Shrink is loss of weight which can apply to live cattle, carcasses, or retail cuts. It is most widely used in market terminology in marketing feeder cattle and slaughter cattle. Shrink is usually expressed as the percentage of liveweight loss.

Major factors affecting shrink are (1) time in transit, (2) distance, (3) weather, and (4) handling methods and type of feed fed.

It is not uncommon for feeder cattle to shrink from 2% to 8% and slaughter steers to shrink 2% to 6%.

Buyers and sellers should know how to account for shrink differences in arriving at fair market prices (Tables 18.3 and 18.4). Buyers of feeder cattle should also know the costs involved in the cattle regaining lost weight.

FUTURES MARKETING

Futures markets are a highly standardized, regulated method of exchanging forward contracts.

In the cash cattle market, it is not uncommon to forward contract a prescribed number of cattle at a mutually agreed upon price for delivery at a predetermined date in the future. The language of a futures market contract specifies the number and kind of cattle, and the month when a delivery may be made. Price is the only variable which is determined when a contract is bought or sold.

Trading of futures contracts is regulated by the exchange trading the contract and by the Commodity Futures Trading Commission—the government agency overseeing all futures activities.

The Chicago Mercantile Exchange (C.M.E.), located at 444 W. Jackson Blvd., Chicago, IL 60006, is the exchange involving both slaughter cattle and feeder cattle contracts (Figure 18.18). The C.M.E. suggests these advantages of the futures markets in helping meet the needs of cattle producers: (1) provide protection from adverse price fluctuation, (2) another tool in marketing alternatives, (3) more working capital is kept free, and (4) adds to the borrowing capacity.

People who use the C.M.E. fall into two general groups—hedgers and speculators. Hedgers are typically beef producers who produce, feed, and market cattle. They seek to transfer the price risk. Speculators may not own or produce cattle. They buy or sell contracts in the hope of making a profit on price changes. They plan to use their knowledge to make profits. Commodity prices can be very volatile, and the potential for profit and loss is greater in commodity trading than in many other kinds of investments.

Who are the speculators? They come from all walks of life—doctors, lawyers, professors, brokers, housewives—and a strong percentage are cattle producers.

Contracts traded on the C.M.E. are shown in Table 18.5. Specifications for these contracts change periodically, therefore the C.M.E. or a broker should be consulted.

Table 18.3. Buying Cattle. Change in Price to Compensate for Shrink

Asking Price	Percent Shrink 2	3	4	6	8
			Dollars per cwt		
100.00	102.04	103.09	104.16	106.38	108.69
98.00	100.00	101.03	102.08	104.26	106.52
96.00	97.95	98.96	100.00	102.12	104.34
94.00	95.92	96.90	97.91	100.00	102.17
92.00	93.87	94.84	95.83	97.87	100.00
90.00	91.84	92.78	93.75	95.75	97.83
88.00	89.80	90.72	91.67	93.62	95.69
86.00	87.76	88.66	89.58	91.94	93.48
84.00	85.71	86.60	87.50	89.36	91.30
82.00	83.67	84.54	85.42	87.23	89.13
80.00	81.63	82.47	83.33	85.11	86.96
78.00	79.59	80.41	81.25	82.98	84.78
76.00	77.55	78.35	79.17	80.85	82.61
74.00	75.51	76.29	77.08	78.72	80.43
72.00	73.47	74.23	75.00	76.60	78.26
70.00	71.43	72.16	72.92	74.47	76.09
68.00	69.39	70.10	70.83	72.34	73.91
66.00	67.35	68.04	68.75	70.21	71.74
64.00	65.31	65.98	66.67	68.09	69.57
62.00	63.27	63.92	64.58	65.96	67.39
60.00	61.22	61.86	62.50	63.83	65.22
58.00	59.18	59.79	60.42	61.70	63.04
56.00	57.14	57.73	58.33	59.57	60.87
54.00	55.10	55.67	56.25	57.45	58.70
52.00	53.06	53.61	54.17	55.32	56.52
50.00	51.02	51.55	52.08	53.19	54.35
48.00	48.98	49.49	49.99	51.06	52.17
46.00	46.94	47.43	47.90	48.93	49.99
44.00	44.90	45.37	45.81	46.80	47.81
42.00	42.86	43.31	43.72	44.67	45.63
40.00	40.82	41.25	41.63	42.54	43.45

Deciding to Hedge

Hedging permits producers to establish a buying or selling price for their livestock months before they are ready for sale or purchase. In hedging, a producer is taking a position opposite the cash position or what the producer owns. The cattle feeder, for example, who owns cattle, sells a future contract at a favorable price to cover the animals on feed. Packers wishing to cover future purchases would buy futures contracts, as would cattle feeders planning to buy yearlings.

This contract runs to some specified maturity date, usually months in advance. At that point, the producer has two options: (1) buy back futures contracts to offset the

Table 18.4. Selling Cattle. Net price After Shrink Allowance

Asking Price	Percent shrink 2	3	4	6	8
			Dollars per cwt		
100.00	98.00	97.00	96.00	94.00	92.00
98.00	96.04	95.06	94.08	92.12	90.16
96.00	94.08	93.12	92.16	90.24	88.32
94.00	92.12	91.18	90.24	88.36	86.48
92.00	90.16	89.24	88.32	86.48	84.64
90.00	88.20	87.30	86.40	84.60	82.80
88.00	86.24	85.36	84.48	82.72	80.98
86.00	84.28	83.42	82.56	80.84	79.12
84.00	82.32	81.48	80.64	78.96	77.28
82.00	80.36	79.54	78.72	77.08	75.44
80.00	78.40	77.60	76.80	75.20	73.60
78.00	76.44	75.66	74.88	73.32	71.76
76.00	74.48	73.72	72.96	71.44	69.92
74.00	72.52	71.78	71.04	69.56	68.08
72.00	70.56	69.84	69.12	67.68	66.24
70.00	68.60	67.90	67.20	65.80	64.40
68.00	66.64	65.96	65.28	63.92	62.56
66.00	64.68	64.02	63.36	62.04	60.72
64.00	62.72	62.08	61.44	60.16	58.88
62.00	60.76	60.14	59.52	58.28	57.04
60.00	58.80	58.20	57.60	56.40	55.20
58.00	56.84	56.26	55.68	54.52	53.36
56.00	54.88	54.32	53.76	52.64	51.52
54.00	52.92	52.38	51.84	50.76	49.68
52.00	50.96	50.44	49.92	48.88	47.84
50.00	49.00	48.50	48.00	47.00	46.00
48.00	47.04	46.56	46.08	45.12	44.16
46.00	45.08	44.62	44.16	43.24	42.32
44.00	43.12	42.68	42.24	41.36	40.48
42.00	41.16	40.74	40.32	39.48	38.64
40.00	39.20	38.80	38.40	37.60	36.80

contracts which were originally sold; or (2) deliver the cattle to one of the specified delivery points (Table 18.5).

Hedging guarantees the selling price; however, it does not necessarily guarantee a profit. The profit depends on placing the hedge at a price that exceeds all costs. It is extremely essential that costs are known before becoming involved in hedging.

The two types of costs which must be fully understood are: (1) production costs and (2) *basis*. *Basis* is the difference between the futures market price and the producer's local price. The spread between these prices includes transportation, shrink and other marketing costs as well as differences in supply and demand. Producers should know their *basis* and watch future quotations daily even if they do not have a hedged position, because this tells them what they could lock in through the use of a hedge.

Figure 18.18. Cattle futures trading on the Chicago Mercantile Exchange. Courtesy of the Chicago Mercantile Exchange.

Hedging is a tool to be used in managing risk because the selling price is known. There are times when futures are favorable for hedging—at other times they are not. The management rule is this: If a producer cannot figure a satisfactory profit (Table 18.6), do not hedge. In this case the producer should take a chance on the cash market.

The decision to hedge depends on the market situation and the degree of risk a producer is willing to assume. Each producer has a different degree of risk-carrying ability. In Table 18.6, if a producer does not hedge in this situation the \$2/cwt profit is by-passed in hopes of a higher profit on the cash market. Some cattle producers do not like hedging because they would rather gamble for higher profits—they are also gambling for higher losses.

Choosing the day to place a hedge is an extremely tough management decision. Before placing the hedge, a producer needs to consider all the market facts on supply and demand and how prices might be affected. After all the facts have been examined, the producer can come to two possible conclusions:

1. Higher prices are coming. In this case, the producer may want to wait to place the hedge at a higher price.
2. Lower prices are likely. This means "locking in" a price on the futures market now.

Table 18.5. Specifications for Slaughter Cattle and Feeder Cattle Contracts

	Contract	
	Slaughter Cattle	**Feeder Cattle**
Size	40,000 lbs	44,000 lbs
Delivery months	Jan., Feb., Apr., June., Aug., Oct., Dec.	Jan., Mar., Apr., May, Aug., Sep., Oct., Nov.
Delivery sites	Par delivery: Peoria and Joliet, IL; Omaha, NB; Sioux City, IA; Guymon, OK and Greeley, CO.	Par delivery: Omaha, NB; Sioux City, IA; Oklahoma City, OK; Others: St. Paul, MN; Greeley, CO; Kansas City, MO; St. Joseph, MO; Dodge City, KS; Amarillo, TX; Billings, MT, and Montgomery, AL.
Last trading day	20th of delivery month	20th of delivery month

[a] Contact the Chicago Mercantile Exchange for more detail and changes in contracts.

Placing the Hedge

All orders of C.M.E. contracts are executed through its members. Some members specialize in beef cattle commodities while others are large, diversified brokerage houses that are also members of the New York Stock Exchange.

Producers, placing a hedge, must contact one of the member firms, establish an account and work directly through a broker. The broker also helps producers understand the beef market and the workings of the C.M.E.

Margin money is required by the C.M.E. to bind the broker to the transaction. In turn, the broker must ask the customer for this "good faith" money. The margin money paid is only a small percentage of the value of the contract. A minimum margin is

Table 18.6. Example of Computing the Profit Potential

Item	Value
Futures market price (percent)	$62.00
Less "basis"	−2.00
Projected cash price (if hedging is used)	$60.00
Less production costs	−58.00
Profit (per cwt)	$ 2.00

established by the C.M.S. (example $1200 on a $28,000 valued live cattle contract) although certain brokerage houses may require more than the minimum.

In addition to initial margin money, traders must maintain a margin reserve, i.e., if the market movement results in a loss of more than a certain amount, the broker will ask for additional margin money. Most hedgers arrange with their banks to provide more money if needed. In this case, the broker goes directly to the bank. When the producer profits in the futures market, the excess margin money can be taken out of the account at any time.

The following examples show how the futures market can be used to "lock in" solid profits. Other figures can be substituted for a specific producer's situation.

Slaughter Cattle Hedge Example

This example is based on a typical feeding situation, assuming 100 calves bought in November for finishing by the following August.

Step 1: Check the futures potential.

August live cattle futures	$65.00/cwt
Subtract "basis"	−1.75/cwt
Expected price	$63.25/cwt

Step 2: Determine profit potential.

Expected price	$63.25/cwt
Costs (projected)	−61.00/cwt
Profit potential	$ 2.25/cwt

Step 3: Decision making.

In this case, a $2.25 per cwt profit is available through selling the August futures.

When Prices Fall

Step 4: Sell cattle in August.

Hedge is placed to set a price regardless of what cattle prices may do. Here is what happens if prices decline as the 100 head come to market:

Price in August for 100 head	$61.25/cwt
Costs	−61.00/cwt
Feeding profit	$ 0.25/cwt

Step 5: Buy back your futures contract. Futures contracts tend to move with cash prices.

Futures sold at	$65.00/cwt
Buy back futures at	−63.00/cwt
Futures gain	$ 2.00/cwt

Here are the results:

	With hedge	No hedge
Feeding profit	$0.25	$0.25
Futures gain	+2.00	0.00
Realized profit	$2.25/cwt	$0.25/cwt

In this case, the feeder made an extra $2.00 per cwt, or $20.00 per head.

When Prices Rise

If selling price is set with a futures contract, extra profits in a rising market may be missed.

Step 6: Buy back futures contract.

Cash price next August	$66.00/cwt
Production costs	−61.00/cwt
Feeding profit	$ 5.00/cwt
Futures sold at	$65.00/cwt
Buy back futures at	67.75/cwt
Loss on futures	$ 2.75/cwt

Here are the results:

	With hedge	No hedge
Feeding profit	$5.00	$5.00
Loss from hedge	−2.75	0.00
Realized profit	$2.25/cwt	$5.00/cwt

The profit with the hedge was the same when prices rose as when they fell (Steps 5 and 6). Hedging was done to assure price protection. Windfall profits from a price rise were sacrificed in order to be insured against price declines. If futures price moves against producers, they can get out at any time.

Feeder Cattle Hedge Example

Step 1: Check the futures potential.

November feeder cattle futures	$85.00/cwt
"Basis"	−5.00/cwt
Expected purchase price	$80.00/cwt

Step 2: Decision making.

An $80 per cwt purchase price for fall delivery appears reasonable. Anticipating a substantial demand for feeders in the fall and a narrow profit margin for spring delivery of slaughter cattle, the cattleman executes a long hedge.

Step 3: Buy yearlings in November.

Regardless of hedging, it is important that the feeder buy the best quality cattle at the best possible price. Yearlings have risen to a cost (laid in the lot) of $90.00/cwt.

Step 4: Lift hedge.

Future prices, following cash, also have risen. After buying the feeders, have your broker liquidate the long position, thus lifting your hedge.

Futures bought at	$85.00/cwt
Sell back futures at	95.00/cwt
Futures gain	$10.00/cwt

Here are the results:

	With hedge	No hedge
Yearling purchase price	$90.00	$90.00
Futures gain	−10.00	0.00
Realized cost of feeders	$80.00/cwt	$90.00/cwt

Through futures a purchase price of $80.00/cwt was established. In this example, insulation from a "bull" market in feeder cattle was provided. Had yearling prices gone down, the hedge also would have assured a $80.00 price, although there is always the opportunity to lift hedges early in an adverse market.

Stocker Hedge Example

Ranchers pasturing calves typically have been hurt the least in heavy price swings of the cattle cycle. Feeder futures are available for them to price calves, judge the merits of forward contracts offered by order buyers, and to hedge.

Running calves on grass from 350 lbs to 650 lbs, March through October, provides ranchers with ample hedging opportunities. Let us consider one situation:

Step 1: Check the futures potential.

October feeder cattle futures	$90.00/cwt
"Basis"	−5.00/cwt
Expected selling price	$85.00/cwt

Step 2: Determine the profit potential.

Expected price	$85.00/cwt
Costs (projected)	78.00 /cwt
Profit potential	$ 7.00/cwt

Costs include those for calves, in and out transportation, gain and all other anticipated costs.

Step 3: Decision making.

The futures are offering a $45.50 profit per head ($7.00/cwt × 650 lbs). One-third of your pasturing calves are hedged.

Step 4: Sell yearlings and buy back futures.

October local auction price	$81.00/cwt
Costs	78.00/cwt
Stocker profit	$ 3.00/cwt

Futures sold at	$90.00/cwt
Buy back futures at	86.00/cwt
Futures gain	$ 4.00/cwt

Here are the results:

	With hedge	No hedge
Stocker profit	$3.00	$3.00/cwt
Futures gain	4.00	0.00
Realized profit	$7.00/cwt	$3.00/cwt

Anticipated profit was $7.00/cwt. If at any time during the summer a better cash contract was offered, it would have been accepted and the hedge lifted at the same time.

WORLD BEEF MARKETING INVOLVING THE UNITED STATES

International beef trade and beef as a world food item were given a brief overview in Chapter 1. Marketing of beef between countries is an intricate interaction between production efficiency and relative international political, social, cultural, and economic institutions—all of which vary from country to country.

The United States continues to be a major importing country of lean, fresh or frozen, boneless beef used in its processed beef products. The 1.0 to 1.5 billion lbs of imported beef have been supplied primarily by Australia, New Zealand, Canada, and Argentina.

The United States, which imports approximately 10% of its domestic consumption, has been a prime market for other beef exporting countries. This is because the United States is a net beef importer and a relatively high income country.

Beef imports do have an effect on beef prices in the United States. This effect has usually been most serious when domestic production of beef is exceptionally high and thus prices are usually low. The counter-cyclical meat import law was passed in 1979 which was designed to place a quota on beef imports when the beef supply in the United States is also very large. The formula which determines the import quota is given as:

Base Figure		Production Adjustment Factor		Countercyclical Adjuster		
1,204.6 million lbs. (ave. imports 1968–1977)	×	3-yr. moving ave. of production / ave. prod 1968–1977	×	5-yr. moving ave. per capita cow beef / 2-yr moving ave. per capita cow beef	=	adjusted base quantity

Many countries have experienced higher beef production costs from rising energy, feed, and financing cost. These increased costs, along with world-wide recession which has dampened beef demand, has resulted in growing competition for slower growing world beef markets.

Very few beef products were exported during the 1960s and 1970s. The value of live beef cattle and beef products exported were approximately 5 times higher in 1981 ($2 billion) than in 1971. The Meat Export Federation (see Chapter 23) has been taking an active, coordinating role in expanding the U.S. beef export markets.

Export markets for U.S. beef by-products are well established in the European Economic Community (EEC). Heart, liver, tongue, kidneys, and other edible offal along with tallow and greases are the primary products exported to the EEC. Hides are also exported to the EEC, the Orient, and Mexico. The EEC market, which has a complicated system of duties and tariffs designed to protect their producers and bolster their incomes, appears to be an excellent market in the future for U.S. beef by-products.

Fresh beef and veal export markets for the United States are in the Bahamas, Jamaica, and Canada. The newly emerging Japanese beef market grew from basically zero to $158 million in 1981. The Japanese market has received considerable attention during recent years as a major potential market for U.S. fed beef. This potential has not materialized to the high expectation of several U.S. beef exporters, and it is meaningful to the U.S. beef industry to understand the reasons why.

Japan, highly dependent on imports, has traditionally followed two import policy goals. The first goal is to maintain and protect its domestic industry for reasons of national security. The second goal is to assure adequate supplies of product at the lowest possible cost by diversifying its sources of raw materials and points of production, as well as by importing finished product from low cost sources, consistent with the quality desired. These goals are accomplished with a complex system of quotas, tariffs, duties, licenses, and health regulations.

These restrictions have been the subject of repeated negotiations with the Japanese government. Despite these negotiations, Japan remains a highly protected market, particularly for the high-quality, grain-fed beef most U.S. consumers prefer.

Despite high hopes by U.S. producers, the Japanese market is likely to remain protected, primarily for two reasons. First, the market for high-quality beef remains a very small proportion of domestic Japanese beef consumption, with the major source consumed coming from Australian and New Zealand grass-fed beef. These countries continue to have the greatest production efficiency for this product.

A second reason is that Japan continues to be one of the largest buyers of feed grains and wheat in the United States, and we compete with other countries for this market as well. Any concession in beef access would be part of a total agricultural trade negotiating package. Since feed grains and wheat exports have made significant contributions to our balance of payments, it is highly unlikely major concessions could be gained for both beef and grain access; and, in the event of tradeoffs, an increased share of the Japanese beef market is an inferior bargaining position to those in the grain sector. On the bright side, the large Japanese positive balance of payments points to the need to negotiate larger market shares for those agricultural products for which we can effectively compete, and the Japanese recognize the duality of encouraging increased market access or accepting the possibility of rising world protectionism.

There are those in the U.S. beef industry who feel strongly that any future expansion of beef production in the United States will come only with expanded beef export markets.

SELECTED REFERENCES

Publications

Beef Exports and Imports—Current Situation and Outlook. Beef Business Bulletin (NCA), 23 September 1983.

Cothern, J. H. *World Beef Trade.* Western Livestock Round-Up. Oct 1982.

Futures Trading in Feeder Cattle, 1981. Chicago Mercantile Exchange, 444 W. Jackson Blvd., Chicago IL 60606.

Futures Trading in Live Beef Cattle, 1981. Chicago Mercantile Exchange, 444 W. Jackson Blvd., Chicago IL 60606.

Jacobs, V. *Within Year Variation in Fed Cattle Prices . . . and Retail Pricing Behavior.* Missouri Agric. Outlook Letter, No. 110. June 14, 1983.

Livestock Market News. USDA. 13 January 1981.

McCoy, J. H., 1979. *Livestock and Meat Marketing.* Westport, Conn.: AVI Publishing Co., Inc.

Packers and Stockyards Resume. USDA. 28 December 1979.

Petritz, D. C., Erickson, S. P., and Armstrong, J. H., 1982. *The Cattle and Beef Industry in the United States: Buying, Selling, Pricing.* Purdue Univ. Coop. Ext. Serv. CES Paper 93.

Seasonal Variation in Cattle Prices. Western Livestock Round-Up. Jan 1983.

Ward, C. E., 1982. *Cattle Pricing Fact and Fiction.* Oaklahoma State Univ. Coop. Ext. Serv. No. 459.

Visuals

"Your Profit in Futures." 16 mm film (15 min). Available on loan from the Farm Film Foundation, 1425 H. St., N.W., Washington, D.C. 20005.

"Hedging Effectively." 35 mm. slide–script presentation (modules for cattle and hogs). 30 min, $35 per set. Chicago Mercantile Exchange, Agric. Marketing Dept., 444 W. Jackson Blvd., Chicago, IL 60606.

"Improve Cattle Handling for Greater Profits." Slide/cassette version, new edition. Livestock Conservation Institute, 1100 Jourie Blvd., Suite 143, Oak Brook, Il 60521.

19

Herd Health Programs for Beef Cattle

Effective herd health programs involve both prevention and treatment of health problems affecting beef cattle. Greater emphasis should be on preventative measures since these enhance the profit picture, whereas treatment programs typically attempt to minimize economic losses.

Beef producers should be knowledgeable of stresses, diseases, parasites, and other health related conditions that may be unique to their area and especially to their specific operations. The importance of strict sanitation practices cannot be overemphasized. Herd health management programs which are most successful involve cooperative planning by producers and their veterinarians. Veterinarians need to understand the economic factors influencing beef cattle operations and also the specifics of a total management program for individual operations. Producers should be aware of their limitations in evaluating, treating, and caring for sick animals and the proper timing in seeking a veterinarian's assistance. Local veterinarians need to understand their limitations and how to work effectively with good pathologists in well-equipped diagnostic laboratories.

This chapter will help producers understand some major health problems affecting beef cattle which should be helpful in planning economically effective prevention and treatment programs.

THE STRESSES

Cattle have naturally occurring and acquired defense mechanisms which can counteract most disease organisms. Exposure to severe stress conditions, however, can overcome defense mechanisms and cause sickness or death in animals.

Stress factors include fatigue, hunger, thirst, dust, anxiety, ammonia buildup, adverse weather, parasites, weaning, castration, dehorning, shipping, poor sanitation, mixing with other cattle, and unnecessary or abusive handling.

Preconditioning can be utilized to prepare calves to better withstand the stress of movement from their production site into and through marketing channels. *Pre* means before some event. *Condition* means to process, to prepare. Preconditioning is intended to be a complete health management program to reduce sickness and death rate and to improve weight gains.

A complete preconditioning program for calves has been defined as follows:

Weaning—All calves should be weaned at least 30 days prior to entry into marketing channels.

Nutrition—Calves should be adjusted to trough and bunk and accustomed to a beginning feedlot ration. Medicated feed is recommended (chlortetracycline 350 mg, sulfamethazine 350 mg).

Vaccination—Calves should be vaccinated for Clostridial group, Infectious Bovine Rhinotracheitis, Para-Influenza$_3$, and Pasteurella at least two weeks prior to movement.

Worming—All calves to be dewormed with injectable, drench, or feed additive dewormer if worms are a problem in the area.

Grub, Lice Control—All calves treated with approved grubicide at least two weeks prior to shipment.

Castration and Dehorning—Calves should be castrated and dehorned at least 30 days prior to movement, preferably when one month of age.

Implanting—Calves should be properly implanted with approved implant.

Identification and Certification—All preconditioned calves should be identified with an ear tag. A properly signed certificate should accompany the calves stating how and when they were processed.

This concept of preconditioning has been poorly interpreted or badly abused by producer, buyer, and veterinarian alike. Furthermore, some feeders want to buy replacement cattle at the cheapest price and in as thin condition as possible. They frequently overlook the immediate health status and prior immunization of animals they are purchasing, hoping to compensate for losses that might occur through compensatory gains. Thus the economic success of preconditioning varies considerably. The concept of preconditioning is valid, and it has economic merit if it is implemented correctly.

DISEASES AND HEALTH PROBLEMS

Table 19.1 identifies some common disease and health problems encountered in beef cattle. A specific diagnosis by a veterinarian or a diagnostic laboratory may be essential to identify specific diseases and causes. For certain diseases, a veterinarian should prescribe specific treatment. Required withdrawal periods should be observed for specific products used in treatment.

Disease is defined as any abnormal structural or functional change in tissues of the body. It is important to understand vital body signs since diseases typically affect vital signs to some degree.

Temperature, respiration rate, and pulse are vital signs as they are fundamental for maintenance of life. Without them, all other body functions, while very important, are secondary.

Cattle producers need to recognize abnormal changes in condition and appearance of animals but also to what extent abnormalities may affect vital signs. Usually the first signs of disease are slight and may go unnoticed. An observant individual who

Table 19.1. Common diseases and other health problems

Diseases/Causes	Estimated annual loss	Signs	Prevention	Treatment
Acidosis, Rumenitis (Liver Abscess Complex) Overloading after "starvation" period and using a high concentrate diet. S. necrophorus found in hepatic lesions.		Initially, animals off feed. Symptoms of "feed intoxication."	Reduction of concentrates (less than 75%), more roughage Chlortetracycline 70 mg/day. Bacitracin methylene disalicylate 70 mg/head for 5 days, wait 25 days and repeat.	Streptomycin I.M.; Antiacids; Antifermentatives; Chlortetracycline 70 mg/head/day; Tylosin 60–90 mg/head/day.
Alkali: disease (see Selenium, Chapter 20)				
Anaplasmosis Anaplasma marginale (protozoa)		Anemia, fever icterus, weakness and emaciation. Use only laboratory diagnosis.	Control of insects. Care in spreading disease by dehorning. Vaccine is available. Chlortetracycline Cattle up to 700 lb—350 mg/head/day. Cattle 700–1000 lb—500 mg/head/day. Cattle 1500 lb and over—750 mg/head/day. Chlortetracycline (for carrier stage of anaplasmosis) 5.0 mg/lb body weight/day for 60 days.	Acute cases—blood transfusions; Chlortetracycline; Oxytetracycline.
Anthrax Anthrax bacillus		Sudden death (1 to 2 hours after infection). "Sawhorse on its side" appearance; failure of blood to clot; delayed rigor mortis.	Vaccination (recommended only in areas where disease occurs).	Antibiotics and antiserums (success relatively poor).
Asthma (cow) occurs in the fall when cows are moved from dry, sparse forage to lush, green pasture. Believed to		Symptoms vary from slight to severe. Respiratory distress, with abnormal increase in depth and rate of respiration.	Moderate the feed change. Have cows filled with dry feed before moving on lush, green pasture.	None very effective.

Table 19.1. (*Continued*)

Diseases/Causes	Estimated Annual Loss	Signs	Prevention	Treatment
be due to improper amino acid (tryptophan) metabolism.		A grunt usually accompanies respiration.		
Blackleg Clostridium chauvoei (bacteria).		Muscular depression, gaseous swelling in muscles, lameness.	Vaccination after calf is three months of age.	Penicillin I.M.
Bloat Variety of causes.		Excessive accumulation of gas in the rumen and reticulum causes distension of left side. Right side is also distended in more severe cases. Labored breathing may occur.	Oxytetracycline (75 mg/head/day) Penicillin (75 mg/head/day); Poloxalene (1–2 g/100 lb body weight/day) Proloxalene drench. Feed stemmy roughage if possible.	Release of gas by hose or trocar.
Bluetongue Caused by a virus spread by bloodsucking insects (primarily the sandfly or punky gnat).		Varies. High fever; depressed; profuse slobbering; crusty muzzle; ulcers of dental pad and lips; poor reproduction.	Maintenance of closed herd and control of bloodsucking insects. Vaccine is available but recommended only in infected herd. Vaccination of pregnant animals may cause brain damage to fetus.	None. Disease runs its course in several weeks. Death may occur due to secondary infection of pneumonia.
Bovine Viral Diarrhea (BVD) Virus. Laboratory diagnosis imperative for accuracy.		(Feeder Cattle) Ulcerations throughout digestive tract. Diarrhea (often containing mucus or blood.	Vaccination prior to exposure. Avoid contact with infected animals.	Symptomatic treatment; antibiotics; sulfonamids; "Hyper" serums; force feed.
BVD (Bovine Viral Diarrhea) Virus.		(Breeding Cattle) Abortions	Vaccination only when indicated.	None Symptomatic treatment.
Brisket Disease or "High Altitude Sickness" Congestive right heart failure due to stress of high altitude. Greater	0.2% to 2% of cattle raised at high altitudes. Incidence in "introduced" cattle has been 40%.	Edema in brisket. (see Figure 19.1)	Keep cattle below 7,000 ft. altitude. Practice genetic selection (primarily bulls) using pulmonary arterial pressure test.	Take affected cattle to elevations below 7,000 ft.

Table 19.1. (*Continued*)

Diseases/Causes	Estimated Annual Loss	Signs	Prevention	Treatment
predisposition if cattle graze on locoweed at high altitudes.			(must be done at high altitude).	
Brucellosis: Brucella abortus (bacteria)		Abortions.	Calfhood vaccination (in some states) between the ages of 2–4 months. Test and slaughter reactors.	None.
Cancer Eye Genetic predisposition; environmental factors (dust, wind and ultraviolet light) enhance the development.	20 mil	Cancerous cell growth on the eyeball, eyelid or third eyelid.	Genetic selection; reduction of other predisposing factors.	Surgery; immunotherapy; electrothermal.
Coccidiosis: Protozoan. Eimeria zumii; Eimeria bovis		Fluid feces, bloody feces and straining.	Avoid crowding, filth and remove affected animals from lots in early stages. Amprolium, 5 mg/kg/day for 21 days. Decoquinate 22.7 mg per 100 lb per head per day for 28 days.	Sulfonamide in drinking water; 1 lb. drug/100 gal. of water for 2–3 days; 10 mg/kg for 5 days.
Crooked Calf Disease (See Chapter 20).				
Diarrhea (See BVD, Coccidiosis, Enterotoxemia, Johne's Disease, Scours and Salmonellosis).				
Encephalomyelitis, Sporadic Bovine (Buss Disease) Virus. (Psittacosislymphogranuloma group).		Fever, weakness, uncoordination, depression.	Not known (remove infected animals from herd).	Chlortetracycline; Oxytetracycline.
Enterotoxemia "Overeating" Toxins of Clostridium perfringens type D (bacteria)		Sudden death "downers." Animals usually fed on high concentrate diets. Diarrhea, although many animals die before this occurs.	Increase concentrate slowly. Vaccination. Chlortetracycline 0.1 mg/lb of body weight/day.	Reduce concentrate. Short course of the disease precludes treatment.

Table 19.1. *(Continued)*

Diseases/Causes	Estimated Annual Loss	Signs	Prevention	Treatment
Fescue toxicity (See Chapter 20).				
Footrot (infectious Pododermatitis): Spherophorus necrophorus		Lameness, foot swelling.	Cattle up to 700 lb Chlortetracycline 70 mg/head day. Ethylene Diamine Dihydroiodide 400–500 mg/head/day for three weeks or 500 mg/head/day continuously in salt or feed. Pens and lots free of objects which will injure feet.	Sulfonamides; Oxytetracycline; Chlortetracycline; Erythromycin. Tylosin (Injectable).
Founder (Laminitis): Similar to an allergic reaction. Due to an overconsumption of grain.		Lameness, inflammation between bony part of the foot and the hoof wall. Toes may grow long and turn up. Painful straightlegged gait in extreme cases.	Keep cattle from overconsuming grain. High level of management on high concentration rations.	Trim long toes. Sell animal for slaughter. Antihistamines and steroid therapy may be useful but not economical.
Grass Tetany (See Chapter 20, Range, Pasture and Grazed Forage management).				
Haemophilus somnus (Infectious Embolic Memingo-Encephalitis): Haemophilus somnus. Infections difficult to diagnose. Needs laboratory confirmation.		Multiple infective emboli.	Vaccine (2 injections). Chlortetracycline, 350 mg/head/day for 28 days. 350 mg sulfamethazine/head/day for 28 days. Oxytetracycline 0.05—2 mg/head daily for three to five days.	Antibiotics; Sulfonamides.
Hardware Disease: Sharp objects perforate the reticulum—can cause irritation of heart sac, lungs or abdominal cavity.		Loss of appetite, reduced milk production, abdominal pain and sometimes labored breathing. Similar to signs of pneumonia.	Keep wire, nails and other sharp objects from being eaten by cattle.	Mild cases may heal if animal is immobilized and feed intake is reduced. Magnets sometimes used in treatment. Surgery may be performed in more severe cases.

Table 19.1. (*Continued*)

Diseases/Causes	Estimated Annual Loss	Signs	Prevention	Treatment
Infectious Bovine Rhinotracheitis (IBR): Virus. Laboratory diagnosis imperative for accuracy.		Respiratory involvement. Fever, vaginitis, abortion, eye and nose discharges. Abortions or stillborn calves. Pneumonia in calves.	Vaccination prior to exposure.	Chlortetracycline 350 g/day; Sulfamethazine 350 mg to keep down secondary infections. "Hyper" serums.
Johne's Disease, Paratuberculosis: Mycobacterium, paratuberculosis. Laboratory confirmation necessary.		Chronic diarrhea.	No vaccine available. Test: remove reactors. Hygenic program for calf raising, separate from adult herd.	Temporary treatment from symptomatic treatment but no complete cure.
Leptospirosis: Leptospira ssp L. pomona L. grippotyphosa L. sejore L. canicola L. Hardjo (bacteria)		(Breeding Cattle) Fever, off feed, abortions, icterus, discolored urine.	Vaccination at least annually (in high risk areas, more frequently). Control rodents. Avoid contact with wildlife and other infected animals.	Streptomycin; Penicillin I.M. Oxytetracycline 1 mg/lb. body weight/7 days; Chlortetracycline 1 mg/lb./body weight per 7 days.
Leptospirosis:		(Feeder Cattle)	Vaccination on arrival.	Dihydrostreptomycin
Leukosis: Not known; virus suspected. Laboratory diagnosis necessary.		Enlarged lymph nodes.	Prohibiting sale of animals for breeding purposes. Prevention of infected animals going into non-infected herds.	None.
Listeriosis: Listeria monocytogenes.		Fever circling. One eye, or one ear paralyzed.	No vaccine. Organism frequently found growing in silage.	Sulfamethazine; Antibiotics.
Liver abscesses		Cattle off feed. Acidiosis. Observed on slaughter.	Bacitracin 70 mg/head/day 250 mg/head/day for five days—repeat in 25 days; Chlortetracycline 70 mg/head/day; Oxytetracycline 75 mg/head/day; Tylosin 8–10 g/head/day.	None specific. Correct faulty feeding practices.

Table 19.1. (*Continued*)

Diseases/Causes	Estimated Annual Loss	Signs	Prevention	Treatment
Lump Jaw (Actinomycosis) Actinomyces bovis.		Lumps on bony tissues of head.	Removal of objects causing head punctures (nails, splinters). Ethylene Diamine Dihydroiodide, 50 mg/head/day in feed or salt.	Ethylene Diamine Dihydroiodide 300–500 mg/head/day for 2–3 weeks; Streptomycin I.M.; Sodium idodide.
Mastitis Usually a bacterium. Creates the disease when injury, chilling or any type of trauma occurs to the udder.		Inflammation of mammary gland. Udder is hard, hot, painful to touch and commonly discolored. Milk is yellow, thick and stringy.	Reduce injury from rough handling and other types of trauma.	Milk and "strip" clean infected quarter(s). Inject systemic and broad spectrum antibiotics directly into teat canal of infected quarter(s).
Malignant Catarrhal Fever: Virus. Frequently confused with other viral infections.		High fever—Inflammation of respiratory, digestive and urinary systems.	No vaccine. Requires positive diagnosis to differentiate from other diseases.	Non-specific nursing. Usually treatment ineffective.
Malignant Edema: Clostridium septicum (Bacteria).		History of wounds, fever and swelling around wound.	Vaccination	Penicillin I.M.; Streptomycin I.M.
Navel ill Neonatal septicemia and bacteremia (bacteria)		Umbilical abcesses and hernias. Swollen joints in young calves.	Iodine on umbilical cord immediately after birth.	Penicillin-Streptomycin combination (1 to 2 gm streptomycin every 12 hours for 3 days).
Pine Needle Abortion Cows eat needles or buds from Ponderosa pine.		Abortion	Prevent consumption of needles.	None.
Pinkeye (Bovine Infectious Keratoconjunctivitis) Moraxella bovis (bacteria, carried by insects, primarily the face fly).		Watery eyes, swelling, corneal ulceration. 5 lb to 40 lb reduction in weaning weight.	Control of flies. Isolate infected animals. Select breeding cattle with eyelid pigmentation.	Sulfonamides; Antibiotics; Powders in the eyes. Patch over eye.
Pneumonia, Bacterial		Pneumonia	350 mg/head/day of H. somnus bacterin.	Specific serums.

Table 19.1. (*Continued*)

Diseases/Causes	Estimated Annual Loss	Signs	Prevention	Treatment
Pasteurella ssp. Haemophilus somnum				
Poisoning (Plant) (See Chapter 20, Range, Pasture and Grazed Forage management).				
Polioencephalomaiacia (Forage poisoning): Etiology not known. Complete autopsy essential.		Sudden deaths, blindness, uncoordination, "downers."	None known	Not known Vitamin B complex (thiamine) may be helpful.
Pulmonary emphysema (See Asthma)				
Rabies: Virus. Usually bites from infected wild fur-bearing animals.		Symptoms vary from furious to dumb type. (Can be transmitted to humans). *Never* insert hand in suspect animal's mouth.	Vaccination of animals, dogs and cats that act as intermediary between man and wild fur-bearing animals when necessary (epidemic).	None Human health hazard.
Red Water Disease, Bacillary Hemoglobinuria Clostridium hemolyticium.		Sudden deaths, red urine, bloody diarrhea. (Not to be confused with leptospirosis).	Vaccination Clostridium hemolyticium bacterin	Penicillin I.M.
Ringworm Fungus.		Round, crusty and thickened circles, with hair denuded inside circle. More common in winter months.	Avoid contact with infected animals and areas.	Local application of fungicide material (equal parts of tincture of iodine and glycerin). Avoid contact with hand as it is contagious to humans.
Rumenitis (see acidosis)				
Salmonellosis: Salmonella typhimurium and S. dublin (bacteria)		Diarrhea (bloody diarrhea. High temperature often confused with coccidia infections). Highly fatal.	Contaminated lots and feed must be eliminated. Neomycin 35 g/ton of complete feed. Avoid contact with infected animals.	Neomycin, 140 mg per ton of complete feed. For calves, chlortetracycline 0.5 mg/lb. body.

Table 19.1. (*Continued*)

Diseases/Causes	Estimated Annual Loss	Signs	Prevention	Treatment
Scours, Calf Colibacillosis "Septicemia" E. coli. plus stress factors.		Diarrhea, weakness and dehydration. Rough hair and coat.	See last part of Chapter 19.	
Scours (Viral) Virus (reovirus and corona virus)		Acute diarrhea, high mortality. Affects calves shortly after birth.	Vaccination with specific viral vaccines.	Symptomatic. See last part of Chapter 19.
"Shipping Fever" Para-influenza Virus + bacteria + stress. May be accompanied by other viral diseases.		Respiratory involvement. Differential diagnosis.	Vaccination; Chlortetracycline 350 mg/head/day; Sulfamethazone, 350 mg/head/day; Oxytetracycline, 0.5—2 mg/head /day for three days.	Sulfonamides; Antibiotics; Antiserums; Serums; "Hyper" serums; Tylan 200.
Sudden Death Syndrome		Sudden death without the animal showing signs of sickness.	Not known. This disease usually affects cattle which have been on a high-energy grain ration for at least 100 days. Proposed that death is due to anaphylatic shock caused by rumen bacterial endotoxin.	
Tetanus Clostridium tenani (bacteria)		Spasms, contractions of voluntary muscles, high mortality rate.	Avoid contamination of open wounds.	Three phases—(1) Remove organism with antibiotics; (2) Tranquilizers or chloral hydrates to relax muscles; (3) High doses of tetanus antitoxin (up to 300,000 units every 12 hrs.). Keep cattle in quiet, dark area.
Trichomoniasis Trichomonas fe-		Abortion (2 to 4 months), usually		Broad spectrum antibiotics.

Table 19.1. (*Continued*)

Diseases/Causes	Estimated Annual Loss	Signs	Prevention	Treatment
tus (protozoa)		followed by an infected uterus. Usually rare, bulls spread the disease.		Isolate carriers.
Tuberculosis		Usually none.	Periodic testing slaughter reactors.	None.
Ulcers, Gastric Sudden change of feed. Ration too high in concentrates.		Not observed clinically, "tarry" stools.	Change roughage concentrate ratio. Reduce stresses.	None.
Urinary calculi "Water-Belly": Change in ration, mineral imbalance.		Straining to urinate, dribbling of urine.	Salt (NaCl) 1–2% of ration; Ammonium chloride 0.74–1.25 oz. per head per day. Keep calcium levels higher than phosphorus in diet.	Surgery.
Vesicular stomatitus (virus)		Inflammation or blistering of tongue, lips, dental pad, nose, teats and feet. Froth around mouth, "smacking" of lips and tongue hangs out.	Control flies and contact with infected animals.	None.
Vibriosis Vibrio fetus (bacteria)		Repeat services Abortions (1–2%)	Vaccination of females and bulls prior to breeding. Use of artificial insemination. Untried bulls on virgin heifers. Avoid sexual contact with infected animals.	None.
Warts, Viral Papillomatosis (virus)		Warts on all parts of body.	Sanitation in vaccinating, tattooing to prevent transmission of virus from animal to animal.	Vaccine—repeated doses; Autogenous vaccines when needed.

Figure 19.1. Fluid accumulation (edema) in the dewlap, brisket and underline are characteristic signs of brisket disease. Congestive heart failure at high altitudes (over 7,000 feet), rather than an organism, cause the physiological problem. Courtesy of J. L. Shupe, Utah State University.

detects early symptoms and initiates appropriate action can be most effective in controlling diseases.

An abnormal temperature of cattle can be one of the first noticeable signs that health problems are occurring. When an animal's temperature is above normal, it is considered a fever; if it is below normal, it is called hypothermia. Fever is a more common occurrence than hypothermia. Cattle have a range in normal temperature since time of day, amount of physical activity, pregnancy, and other factors cause the temperature to fluctuate. The normal temperature range for cattle is 100.4 °F to 103.1 °F. When body temperature goes one degree above the normal upper limit, the animal is considered to have a fever. Even with an extremely high fever, the temperature seldom goes above 107 °F unless the animal experiences heat stroke when the temperature may exceed 110 °F.

Pulse is the rhythmic, periodic thrust felt over an artery in time with the heartbeat. The pulse can be obtained by putting the fingers over the superficial arteries and pressing the arteries against a hard or boney structure. The common location in cattle is the lower part of the jaw. The pulse varies depending on age, size, sex, atmospheric condition, time of day, excitement, and other factors. The range for a normal pulse rate (heart beats per minute) is 40 to 70.

Respiration is the act of breathing by taking in oxygen and expiring carbon dioxide. The respiratory system is affected by many primary and secondary diseases. Respiration rate is the number of inspirations (expansion of chest or thorax area) per minute. The normal range is 10 to 30 per minute. Some of the same factors identified for temperature and pulse can cause variations in the rate of respiration.

Vaccination Methods

Since vaccination is a vital part of disease prevention and treatment, methods of injection are important because they affect the speed with which the vaccine enters the animal's system.

Vaccines should always be measured accurately and administered according to the manufacturer's recommendations. Sterile conditions should be utilized during vaccination, since other diseases and infections can be easily transmitted.

Following are several methods of administering vaccines and therapeutic agents:

Intramuscular (I.M.)—probably most common—vaccine is injected in a well-muscled area of the flank, hip, or shoulder without danger of striking a bone. Absorption is rapid due to good blood supply. A three-fourth-inch needle is commonly used, allowing complete penetration of skin and partial penetration of muscle. Volume of injection should be moderate to prevent an abscess forming.

Intranasal—relatively rare—used to create local resistance to disease in the respiratory tract. Most of the antibody protection will be in an animal's general system and must be in small amounts to permit absorption.

Intravenous (I.V.)—used commonly on small calves—seldom used for vaccines due to amount of time required to locate injection site. Drug is rapidly available to animal's system in larger volume and tissue irritation is avoided. Best site is jugular vein, located between the neck muscles and throat on the side of the neck (ideal site is approximately one-third of the distance between jaw and chest). Site should be wet with alcohol, 1½- to 2-inch needle should be inserted at a 30 degree angle pointing toward the body. Slight suction on the syringe should produce blood in the fluid and injection should be slow and steady.

Oral—common in small animals—used to administer therapeutic drugs to cattle. It encourages local production of antibodies in digestive tract at site of attack. Administration is simple in smaller calves. The greatest danger is failure to get the drug far enough into the throat to ensure swallowing; aspiration of the drug can lead to pneumonia or, in large enough quantities, drowning.

Subcutaneous (Sub.Q.)—injection under the skin—slow but sustained absorption due to small blood supply. Injection site is where the skin is loose, usually in the neck. Loose skin is gathered and needle inserted through gathered area. A ½-inch needle can be used, permitting administration with one hand.

THE PARASITES

Parasites cause millions of dollars of loss to the cattle industry annually. These losses take the form of reduced weight gains, increased amount of feed per pound of gain, lower milk production, reduced hide value, additional trim on carcasses, and even death loss. The losses are not always apparent and may be excessive before cattle producers recognize there is a serious problem.

There are two basic types of parasites: external and internal.

External Parasites

There are two primary groups of external parasites: (1) insects (which include flies, lice, and mosquitos) and (2) arachnids (includes ticks and mites). Figures 19.2, 19.3, 19.4, and 19.5 identify some common flies, lice, mites, and ticks affecting cattle. External parasites live off flesh or blood of cattle. These parasites can also mechanically transmit pinkeye, mastitis, anaplasmosis, blue tongue, and other infectious diseases. Buzzing and biting of flies annoys cattle which decreases gains and increases feed costs. Lice and mites cause cattle to rub which occasionally results in large areas of bare skin. Fence maintenance costs increase when posts and wire are used to satisfy the itching caused by lice and mites.

Lice are flat, wingless insects with three pairs of legs. The body of the louse is divided into three segments with the body of most lice being less than 3⁄16 inch long. The blood-

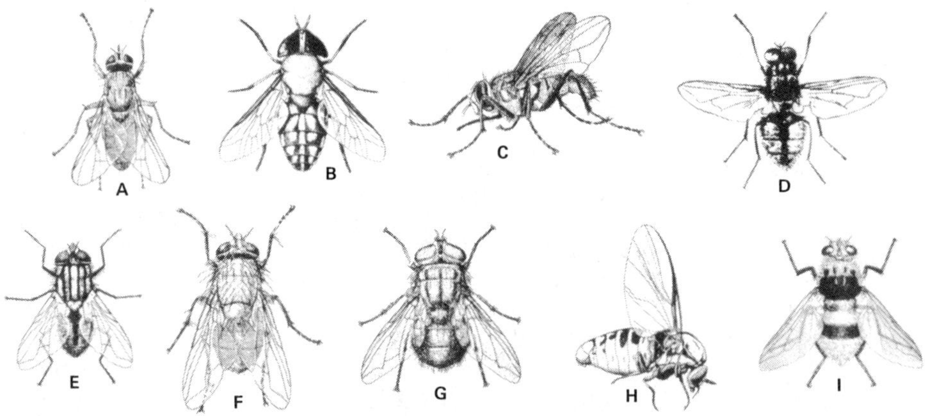

Figure 19.2. Flies. A. Horn fly is most common fly around cattle. It is about half as big as the housefly and similar in appearance. It spends most of the life cycle on cattle, typically resting around the base of the horn, except to lay eggs. B. Horsefly (also called deer fly, ear fly, or gadfly) ranges up to an inch in size. Specimen shown is Greenhead Horsefly, one of the many species. C. Stablefly, similar to common housefly, has a stinging bite. It doesn't remain on cattle, usually resting in the shade. D. Face fly (male). It is slightly larger than common housefly and breeds in fresh manure. Pinkeye is carried by this fly when it feeds on the mucus membranes around the eyes. E. Housefly has four black stripes on its back and feeds on secretion around eyes, nose and other orifices. It can carry disease and eggs of other smaller parasites. F. Blowfly is usually metallic or yellow color and approximately ½ inch in length. The blowfly will lay eggs in an open wound where the blowfly maggot or larva feeds off the flesh, which increases the size of the wound and prevents healing. G. Screwworm fly (adult) is probably the deadliest fly. It is bluish-green, with a yellowish-red face and three blue stripes on its back between the wings. It resides only in the southern parts of the United States and Central America where the screwworm (maggots or larvae of the screwworm fly) feeds on the flesh of open wounds. H. Buffalo gnat (adult female) is a small black fly less than ⅙ inch long which sucks blood from the ears, eyes and nose of cattle. I. Heel fly (adult female) has an ½ inch long black and yellow-striped, hairy body, with orange hairs at the posterior end of the abdomen. The eggs are laid on legs and lower parts of bodies of cattle during the spring. The eggs hatch, burrowing through the skin and migrating to the back of the animal forming bumps under the hide. The larvae (called grubs) winter in the back of cattle, emerging through the hide, then hatching and becoming an adult fly in the spring. Courtesy of USDA and Bull-O-Gram, Rainbow's End Ranch, Rt. 1, Box 43, Elfrida, AZ 85610.

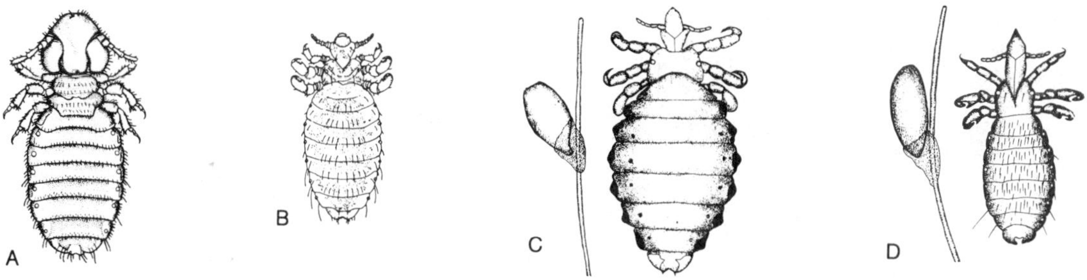

Figure 19.3. Lice. A. Chewing cattle louse (a biting louse) has a yellow body with eight dark crossbars. B. Hairy cattle louse (blood-sucking). C. Short-nosed cattle louse (blood-sucking), with a louse egg attached to an animal hair. D. Long-nosed cattle louse (blood-sucking), with a louse egg attached to an animal hair. Courtesy of USDA and Bull-O-Gram, Rainbow's End Ranch, Rt. 1, Box 43, Elfrida, AZ 85610.

sucking lice are bluish slate in color. Lice spend their entire lives on the host and are especially abundant during the winter when the host's hair is longer and thicker.

Mites have four pairs of legs and undivided bodies. They are microscopic in size and are whitish in color. They live off the skin tissue of the host and cause cattle scab, mange, itch, or scabies. Mites are spread by contact with infected animals or from objects which have been in contact with infected animals. The psoroptic mite (not shown) causes common scab. The mites feed on the skin of the host, which becomes covered with scabs that form over the mites and the hair comes out in patches. Scabs usually appear first on the neck and around the base of the horns.

Controlling External Parasites

Table 19.2 lists several commonly used insecticides for controlling external parasites. The most popular method of application is the use of systemic insecticides. In this situation, the material is applied to only one area of the body (usually the back) and the insecticide spreads throughout the entire system (Figure 19.6). Insecticides can also be applied by spraying, fogging, using back rubbers, ear tags or dust bags. They can also be applied by preparing a large quantity of solution and placing it in an underground vat. The cattle are then either herded into the vat one by one or lowered into the vat in a hydraulic cage. This is the most thorough method of applying insecticides; however, it is usually the most expensive and least practical for most cattle operations.

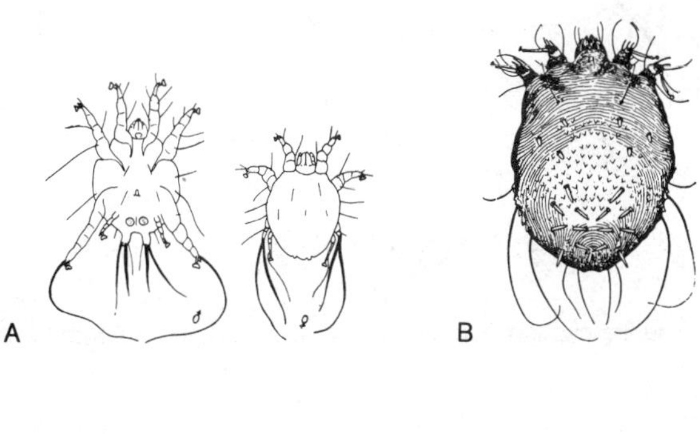

Figure 19.4. Mites. A. Chorioptic mite is responsible for tail mange which is the most common type of mange in cattle. It develops chiefly in the tail region and spreads to other areas of the body. B. Sarcoptic mite causes scabies which is the most severe form of mange. Lesions, characterized by a crusted appearance, and actual thickening of folds of skin, appear around the head and neck and spread to other parts of the body. Courtesy of USDA and Bull-O-Gram, Rainbow's End Ranch, Rt. 1, Box 43, Elfrida, AZ 85610.

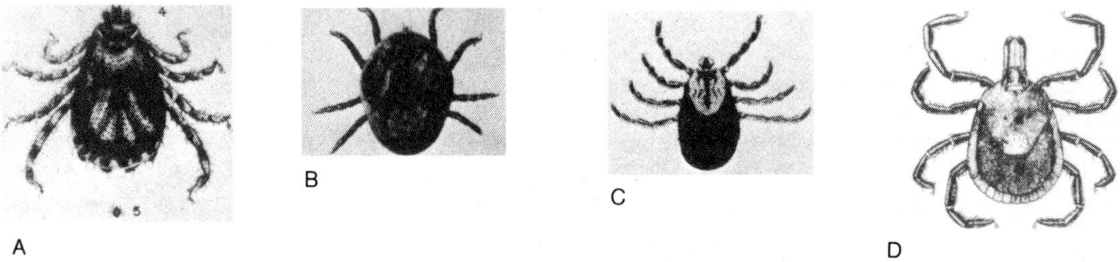

Figure 19.5. Ticks. A. Cattle fever tick (mature male) was previously a serious danger as a vector for the protozoa which caused Babesiasis. It has been eradicated in the United States and is no longer such a serious threat. B. Spinose ear tick nymph (fully engorged) is a soft tick whose larvae infest the ears of livestock. They are common in the arid region of the southwest, especially in cattle kept in confinement. C. Winter tick is different from other hard ticks in that it is a single-host tick and attacks cattle in the winter and early spring. D. Lone star tick (female) is a hard tick with silvery markings on its shield. The male has a pattern of markings while the female has only one dot. The tick is a vector for Rocky Mountain spotted fever. Courtesy of USDA and Bull-O-Gram, Rainbow's End Ranch, Rt. 1, Box 43, Elfrida, AZ 85610.

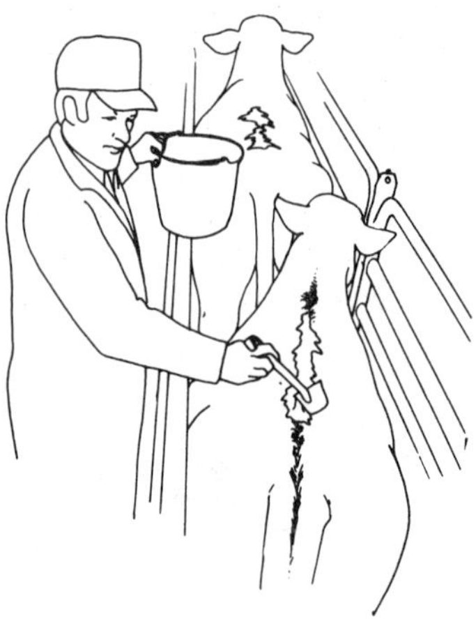

Figure 19.6. A systemic insecticide being poured along the animal's back. Courtesy of Battaglia, R. A. and Mayrose, V. B., 1981. *Handbook of Livestock Management Techniques.* Minneapolis: Burgess Publishing Co., copyright ©.

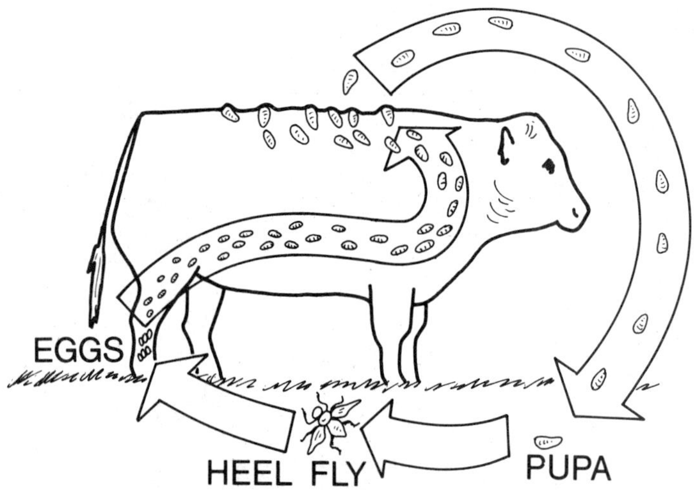

Figure 19.7. The life cycle of the cattle grub. Drawn by CSU Graphics.

Table 19.2. Selected Common Insecticides (Application and Control) for External Parasites

Insecticide	Form	Method of Application	Parasites Controlled
Coumaphos (Co-Ral)[a]	4% Ready mixed	Pour on	Grubs (heel fly larvae) Lice Horn flies (some effectiveness)
	5% Dust	Dust bags or shake on	Lice Horn flies Screwworms Face flies
	11.6% Emulsifiable concentrate or 25% wettable powder	Spray (.375–.5%)	Lice Face flies Horn flies Ticks Mites
Crufomate (Rulene)[b]	12% Ready mixed	Pour on	Grubs Horn flies Lice
	25% Emulsifiable concentrate	Spray (.375%) or dip (.25%)	Grubs Hornflies Lice Face flies
	5% in oil	Back rubber	Lice Horn flies Face flies Ticks (some effectiveness)
Toxaphene	40% Wettable powder or 60% emulsifiable concentrate	Spray or dip	Lice Mites Horn flies Ticks Face flies
Lindane	3% Smear	Direct to wound	Screwworms
Permethrin (Atroban)[g] (Ectiban)[f] (Gardstar)[e] Fenralerate (Ectrin)[c] (Starbar)[d] Flucythrinate (Guardian)[h]	Impregnated into ear tag	Ear tag	Horn fly, face fly, Gulf coast tick and spinose ear tick. Aids in control of house fly and stable fly.

[a] Registered trademark of Chemagro Corp.
[b] Registered trademark of Dow Chemical Co.
[c] Registered trademark of Diamond Shamrock Corp.
[d] Registered trademark of Zoecon Corp.
[e] Registered trademark of Y-Tex Corp.
[f] Registered trademark of ICI Amercas.
[g] Registered trademark of Burroughs Wellcome Co.
[h] Registered trademark of American Cyanamid Co.

Insecticides are used to break the life cycle of the parasite. Figures 19.7 and 19.8 show the life cycle of the cattle grub and when treatment should be given to be effective in breaking the life cycle.

Several methods may have to be used to control external parasites in any one operation. Continuous inspection for parasite infestation is the best indicator of effective control.

Internal Parasites

While the presence of external parasites is quite obvious to cattle producers, internal parasites can easily escape notice. Internal parasites are not only inside the animal, but these parasites and their eggs are microscopic in size. The loss to cattle producers is often a slow, draining, continuous process that goes unnoticed because of the difficulty in measuring it.

The most common class of internal parasites is the roundworm (Nematodes) with the other three basic classes being tapeworms (Cestodes). Some of the more common internal parasites are identified in Table 19.3.

Controlling Internal Parasites

In most areas, internal parasites cannot be totally eliminated. Cattle producers, however, should keep the level of infestation below the level of economic loss.

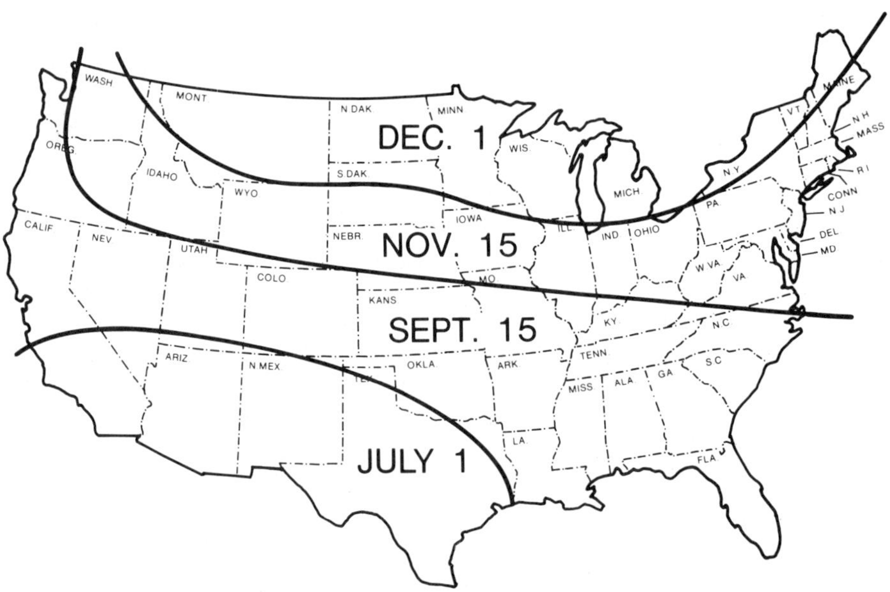

Figure 19.8. Cattle located in the areas outlined should be treated prior to the date specified for maximum grub control during heel fly activity. Dates given are subject to seasonal variation and apply to the common grub. The northern species appear later. Drawn by CSU Graphics.

Table 19.3. Identification of the Common Internal Parasites

Common Name	Scientific Name	Length and Shape	Location	Effect
		Roundworms (Smooth, Long and with a Cyclindrical Shape)		
Large stomach worm (barberpole or wire worm	Haemonchus placei	15mm to 40 mm	Abomasum	Sucks blood; anemia.
Medium stomach worm (brown stomach worm)	Ostertagia ostertagia	6mm to 9mm	Abomasum	Reduction of nutrient absorption; profuse, watery diarrhea.
Small stomach worm (bankrupt worm	Trichostongylus axei	5mm	Abomasum	Reduction of nutrient absorption; profuse, watery diarrhea.
Cooperids	Cooperia (various species)	5mm to 8mm	Small intestine (first 10 to 20 feet)	Congest mucosa; profuse diarrhea; weight loss.
Hookworm	Bunostomum phlebotomum	9mm to 18mm	Small intestine (first few feet)	Sucks blood; anemia; weight loss.
Intestinal threadworm	Stongyloides papillosus	3.5mm to 6mm	Small intestine	Intestinal hemorrhages.
Thread-necked intestinal worm	Nematodirus (various species)	12mm to 25mm	Small intestine	Diarrhea; loss of appetite; weakness.
Nodular worm	Oesophagostomum radiatum	12mm to 15mm	Colon	Loss of appetite; diarrhea; loss of weight; weakness.
Large-mouth bowel worm	Chabertia ovina	12mm (body is bent)	Colon	Sucks blood; intestinal edema; diarrhea.
Lungworm	Dictyocaulus viviparus	80mm	Trachea and bronchioles	Coughing; rapid, shallow breathing; secondary bacterial infections.
		Tapeworms (Long Flat Worms Whose Bodies Are Divided Into Segments)		
Tapeworm	Most common is Moniezia expansa and Moniezia beuedeni	Several feet	Small intestine	Reduce nutrient absorption; occasional diarrhea.
		Flukes (Trematodes—They Have a Flat Body with One or More Suckers)		
Common liver fluke	Fasciola hepatica	30mm (leaflike)	Liver (bile ducts)	Hemorrhage of liver; cystic and enlarged bile ducts.
Giant liver fluke	Fasciola gigantica	75mm (leaflike)	Liver	Similar to common liver fluke.
Giant liver fluke	Fasciladides magna	75mm (oval)	Liver	Similar to common liver fluke.
Lancet fluke (lesser liver fluke)	Dicrocoelium dendriticum	12mm (long and slender)	Liver	Similar to common liver fluke.
Stomach flukes (conical flukes or amphistomes)	Paramphistomes (several species)	15mm	Rumen and reticulum	Diarrhea; loss of appetite; loss of weight.

Control of internal parasites involves interrupting their life cycle. This can be accomplished in several ways among which are: (1) the presence of unfavorable climatic conditions; (2) development of resistance in the cattle; (3) management of the cattle to prevent their ingestion of infective organisms; (4) destruction of intermediate hosts and environmental features hospitable to the parasite outside the cattle's bodies; and (5) therapeutic chemical treatment of the cattle.

Wet and warm weather favors proliferation of certain internal parasites. Areas which are colder and more arid (western states) do not seem to have the same internal parasite problems compared to the southern states. While cattle producers cannot control the climate, they can use weather to assist in controlling parasites.

Some cattle have more resistance to worm infestations than do other cattle. This resistance may be due to a tolerance developed because of earlier infestations. Some cattle, particularly the Brahman, show evidence that this natural resistance is inherited. Cattle stressed with poor nutrition and diseased conditions are less resistant to internal parasites.

Close confinement and overgrazing are both conducive to spreading parasitic infections. Under these conditions, cattle are more likely to eat infected manure or graze the lower parts of plants where parasites are more numerous. Pasture rotation can help reduce the number of parasites by causing infective organisms to die before they are ingested.

In some cases, the intermediate host can be destroyed which will reduce the number of parasites. An example is the snail which is the intermediate host for the fluke. Scattering manure, renovating pastures, plowing, and draining marshes and swamps all upset the hatching process and reduce the concentration of infective organisms.

Anthelmintics are drugs which are available to kill parasites inside cattle. These chemical wormers have varying degrees of effectiveness on different parasites and different forms of the same parasite (Table 19.4). Wormers come in drench, feed additive, or bolus form.

The best time to worm depends on the type of operation and its location. In general, the best time to worm a cow herd is just before calving. Calves may be wormed when they are worked prior to weaning. In areas where worms have not been a serious problem, a periodic fecal check for parasite eggs and comparing the performance of some treated and nontreated animals is useful in determining if a total treatment program is necessary.

Since feedlot cattle originate from many sources (usually unknown), a routine worming of incoming cattle will usually be economical.

Ivermectin, an antibiotic (kills parasites rather than bacteria) produced by the fermentation of Streptomyces avermitilis, has been shown the most potent, broad-spectrum anthelmintic yet discovered. Research has given evidence of its effectiveness against stomach worms, lung worms, blood-sucking lice, and scabies. Ivermectin is still an investigational drug since it has not yet been cleared by the FDA for use in cattle.

COW–CALF HEALTH MANAGEMENT PROGRAMS

Respiratory diseases continue to be the major cause of disease loss in beef cattle. The costs to cattlemen for treatment, weight loss, death loss, and culling in weaning calves are estimated to be about a third of a billion dollars annually.

Table 19.4. Selected Anthelmintic Drugs Used to Control Internal Parasites

Chemical	Trade Name	Form	Effective Against	
Thiabendazole	Omnizole	Bolus	Hemonchus	
	TBZ	Paste	Ostertagia	
	Thiabenzole	Drench	Trichostryongylus	Adult and immature forms
		Pellet	Cooperia	
			Nematodirus	
			Oesophagostomum	
			Bunostomum	
			Chabertia	
			Stongyloides	Adult only
			Neoascaris	
Levamisole	Levasole	Injectable	Hemonchus	
	Tramisole	Bolus	Ostertagia	
		Drench	Trichostryongylus	Adult and immature forms
		Soluble powder	Cooperia	
		Feed additive	Nematodirus	
		Oral gel	Oesophagostomum	
			Bunostomum	
Phenothiazine		Bolus	Hemonchus	Adult immature forms
		Drench	Ostertagia	
		Feed additive	Trichostryongylus	
			Cooperia	
			Nematodirus	Adult only
			Oesophagostomum	
			Bunostomum	
			Chabertia	
Crufomate	Rulene	Drench	Ostertagia	
			Trichostryongylus	
			Cooperia	Adult only
			Oesophagostomum	
			Bunostomum	
			Trichuris	
Coumaphos	Co-Ral	Feed additive	Hemonchus	
	Baymix		Ostertagia	
			Trichostryongylus	Adult only
			Cooperia	
			Nematodirus	
Morantel Tartrate	Rumatel	Bolus	All major cattle worms	
		Feed additive		
Haloxon	Loxon	Bolus	Hemonchus	
		Drench	Ostertagia	Adult only
			Trichostryongylus	
			Cooperia	
			Nematodirus	Some effectiveness
			Bunostomum	
Trichlorfon	Neguvon	Drench	Hemonchus	
			Ostertagia	Adult only
			Cooperia	
			Oesophagostum	

Bovine Respiratory Disease (BRD) is seldom the result of a single factor. BRD usually is caused by a combination of stress, virus infection, and invasion of the lungs by pathogenic bacteria such as Pasteurella and Hemophilus. Stress undermines the natural defenses built in the lining of the trachea and bronchi, and respiratory viruses (such as IBR, PI_3, BVD, and others) further damage these natural defenses. Ultimately, pathogenic bacteria find a wide open road into the lungs where they localize, multiply, and cause severe damage called BRD, pneumonia, or shipping fever.

Immunity against IBR, BVD, and PI_3 can be achieved by the administration of virus vaccines to cattle. Modified live and inactivated virus vaccines are available in single or combination forms. The routes of administration of these vaccines are intramuscular (IBR, PI_3, BVD) or intranasal (IBR, PI_3 only). Both intramuscular and intranasal vaccines provide adequate immunity.

Bacterial infections can be controlled with bacterins (vaccines) which are composed of killed organisms. Two injections 14 or more days apart are needed for adequate protection whenever bacterins are used.

Currently there are no immunizing agents available for Chlamydia, Mycoplasma, and Rickettsia which are other microorganisms sometimes associated with BRD.

Recommended Practices for the Control of Bovine Respiratory Disease (BRD) in the Cow–Calf Herd

The following management guidelines are a joint effort involving the animal health committee (NCA) and the American Association of Bovine Practitioners. The management outline attempts to present the ideal immunization program. There are regional and area differences in disease incidence and the general health of cow herds.

Management needs vary from operation to operation; therefore, an individual operation may need more or fewer treatments. Consult with a veterinarian and follow label directions on any product used. Several different management options are presented for each individual cattleman's consideration.

Program "A"—for young calves (1–3 months of age)

1. Respiratory diseases
   a. IBR–PI_3 (use killed vaccine intramuscularly or modified live vaccine intranasally)
   b. Pasteurella
   c. Hemophilus somnus
2. Other diseases
   a. Leptospirosis (single or multiple strains available)
   b. Clostridial diseases (includes Cl. chauvei, Cl. septicum, Cl. novyi, Cl. sordellii, Cl. perfringens C and D, and Cl. haemolyticum)
3. Other procedures which should be done at this time:
   a. Implant with approved growth stimulants (except animals to be kept for breeding)
   b. Castrate and dehorn
   c. External and internal parasite control. (Some products are not recommended to be used in very young calves)

Program "B"—for older calves (3–4 weeks prior to weaning)

1. Respiratory disease
   a. IBR–PI_3: If killed vaccine was administered at 1 to 3 months of age, it should be readministered as a booster, or a modified live vaccine intranasally or intramuscularly may be used instead of the killed product. If this is the initial vaccination and not a booster, however, use the modified live vaccine intranasally or intramuscularly.
   b. BVD: BVD vaccine is available singly or in combination with modified live IBR virus, PI_3 virus, or pasteurella. Since calves are not apt to be heavily stressed when vaccinated 3 to 4 weeks before weaning, a "combination" BVD vaccine could be used at this time.
      Some cattlemen and veterinarians prefer not to administer BVD simultaneously with IBR–PI_3 modified live vaccine intramuscularly because of the added stress on the animal. This is true in very young or heavily stressed animals which can be vaccinated against BVD 2 to 3 weeks later; in this case, at weaning time. It must be emphasized that calves stressed in any form (e.g., castrated, dehorned, branded, shipped, and so forth) should not be vaccinated against BVD; otherwise adverse reactions could occur.
   c. Pasteurella and Hemophilus somnus: Two injections, 2 to 4 weeks apart, are needed whenever Pasteurella and Hemophilus bacterins are used. If the first injection is given at this time, follow with the booster injection at weaning time.
2. Other diseases (not generally part of BRD complex) which may be considered at this time:
   a. Clostridial diseases booster
   b. Leptospirosis
   c. Vibriosis (heifer and bull calves kept for breeding)
   d. Brucellosis (for heifer calves): Requirements vary among different states. Consult veterinarian regarding specific state regulations. It is a one- time vaccination and generally must be administered by a licensed veterinarian or other designated animal health official. All brucellosis vaccinations must be officially reported to the appropriate state agency.
3. Other procedures
   a. Vitamin A (injectable preferred)
   b. Implant with approved growth stimulants (except animals to be kept for breeding).
      Note: If immunization program is begun early enough, both initial and booster shots may be administered prior to weaning. This will reduce stress at weaning and reduce shrink and amount of time required to regain weaning weight. This regime is subject to local practices, individual herd health needs and various combinations of vaccines. The above recommendations, early immunization, processing of cattle, and other management practices are designed to reduce stress at weaning.
   c. Treat for internal and external parasites.

Calves—weaning time

1. If original and booster shots have both been given, no additional immunization is needed at this time. Treat for internal and external parasites, if not done 3 to 4 weeks prior to weaning.

2. If the first vaccination series as listed in Program "A" (Young Calves, 1 to 3 months of age), above, has been given, give boosters as listed under Program "B" (3 to 4 weeks prior to weaning) and treat for internal and external parasites.
3. For calves that have not previously been vaccinated, several alternatives are feasible, depending upon type of confinement, feed, equipment, availability of labor, and final disposition or destination of the calves. Calves must be handled twice for optimum results: two programs ("C" and "D") are outlined for your selection:

Program "C"—at weaning

1. Respiratory disease
   a. IBR–PI_3 intranasally or I.M. (killed vaccine)
   b. Pasteurella bacterin
   c. Hemophilus bacterin
2. Other diseases
   a. Clostridial bacterin
3. Other procedures
   a. Internal parasites
   b. Implant
   c. Vitamin A

14 to 21 days later (providing calves are consuming 2% to 3% of their body weight of feed, or are on pasture and eating well).

1. Respiratory disease
   a. IBR–PI_3 booster if killed vaccine was used at weaning.
   b. BVD vaccine
   c. Pasteurella booster
   d. Hemophilus booster
2. Other diseases (not generally considered part of BRD) which may be prevented at this time.
   a. Clostridial booster
   b. Leptospirosis bacterin
   c. Vibriosis bacterin
   d. Brucellosis (heifers).
3. Other procedures
   a. Treat for external parasites

Program "D"—at weaning

1. Respiratory disease
   a. Pasteurella bacterin
   b. Hemophilus bacterin
   c. BVD vaccine
2. Other diseases
   a. Clostridial bacterin
3. Other procedures
   a. Treat for internal parasites
   b. Implant
   c. Vitamin A

14 to 21 days later (providing calves are consuming feed equivalent to 2% to 3% of their body weight or are on pasture and eating well).

1. Respiratory disease
   a. IBR–PI_3 intranasally or I.M.
   b. Pasteurella booster
   c. Hemophilus booster
2. Other diseases (not generally associated with BRD) which may be considered at this time.
   a. Clostridial booster
   b. Leptospirosis bacterin
   c. Vibrio vaccine (heifers and bulls)
   d. Brucellosis (heifers).
3. Other procedures
   a. Treat for external parasites

A summary of programs "A," "B," "C," and "D" is given in Table 19.5.

Additional Management Tips

High levels of antibiotics may be fed during the weaning period on the advice of a veterinarian. Do not attempt to feed antibiotics unless calves are consuming 2 to 3 percent of body weight of feed. The need and the level of antibiotics should be determined by a veterinarian. Prior to working cattle, consult with a veterinarian concerning the use of epinephrine to treat shock or sensitivity reactions in cattle. It is a potent drug. You should know the clinical signs of shock and sensitivity, indications for epinephrine and its dosage. Always take time to observe cattle that have been treated.

Management of Calves at Weaning

1. Calves should be eating some dry feed 2 to 4 weeks prior to weaning.
2. Vaccination procedures should be reviewed and changed as necessary, depending upon health conditions of specific lots of cattle, environmental conditions, and prevalence of various diseases in the immediate area.
3. An adequate fresh water supply is essential, preferably from a source which cattle can see or hear running.
4. Vitamin A prior to or at weaning is generally recommended.
5. Check feed and water consumption—both should increase during the weaning period.
6. Provide good high-quality hay. Calves should be consuming 2% to 3% of body weight of feed before either feed or water is medicated.
7. Check calves 2 or 3 times daily.
8. Seek professional help from veterinarian when needed and before a major problem arises.

Marketing and Transportation

A. Minimize stress factors as follows:
   1. It is crucial to get calves moving through marketing channels quickly.
   2. Avoid crowding and bruising.
   3. Avoid conditions of extreme temperature variations, dust or wetness.
   4. Feed and water calves before shipping.

B. Other factors to improve health:
   1. Upon arrival at the feedlot, cattle should first be fed hay prior to having access to water. Be sure to have adequate water facilities available. Begin a limited feeding of grain and protein supplement.
   2. Segregate sick animals.
   3. Tractor exhaust stacks must be tall enough for gases to clear trailer well.
   4. Avoid ammonia buildup in trucks, yards, barns or sheds from excess urine, manure and moisture. Ammonia contributes to respiratory disease.
   5. Start adequate treatment promptly. Identify sick cattle and treat as recommended by your veterinarian.

Cow Herd, Replacement Heifers and Bulls

It must be emphasized that vaccination and adequate handling of calves are part of, but not a substitute for, a total herd health management program. It is essential that an adequate breeding herd vaccination program be implemented if maximum benefits are expected from vaccinating or preconditioning calves or both.

The following recommendations are intended to ensure immunization of the breeding herd against diseases of recognized significance.

A. Replacement heifers and bulls (generally 10–15 months of age). The immunization outlined below should be boostered annually, no later than 30 days prior to breeding.
   1. Replacement cattle with unknown history status. Heifers not pregnant.
      a. First working
         1. Immunization: (a) IBR–PI_3, (b) Vibriosis, (c) Leptospirosis, (d) Clostridial diseases, (e) Haemophilus, and (f) Pasteurella.
         2. Other treatment: (a) Treat for internal parasites and (b) Vitamin A.
      b. Second working
         1. Immunization: (a) BVD, (b) Leptospirosis booster, (c) Vibriosis booster, (d) Clostridial booster, (e) Haemophilus booster, and (f) Pasteurella booster.
         2. Other treatments: (a) Treat for external parasites, depending upon grub development and season. (b) If heifers are of eligible age, vaccinate for brucellosis. If heifers are older than of eligible vaccination age, they should be tested, possibly twice, for brucellosis.
   2. Replacement cattle sufficiently immunized by calfhood and weaning programs.

This is recommended as optimum management, starting after calves are weaned and assuming they have had at least minimal recommendations suggested for calves.
   a. Booster vaccinations: (1) IBR, (2) BVD, (3) Leptospirosis, (4) Vibriosis, and (5) Clostridial.
   b. Other treatments: (1) Internal and external parasites. (2) Vitamin A.

Mature Cows and Bulls

Assuming that the breeding herd of mature cows and bulls has been previously immunized either as calves or as herd replacements, the following booster immunizations are recommended:

Booster immunizations	*Not less than 30 days prior to breeding*		*Last trimester of pregnancy*
IBR	*X*	*or*	*X–Killed or intranasal vaccine*
BVD	*X*		**?*
Clostridial	*X*	*or*	*X*
Leptospirosis	*X*		
Vibriosis	*X*		

Frequently veterinarians recommend that IBR, BVD, and Clostridial immunizations be administered during the last trimester of pregnancy because it conveys a greater passive immunity to the calf. One should be cautioned, however, that if IBR is administered during the last trimester of pregnancy, that it be a killed product or an intranasal vaccine.

BVD vaccines are modified live virus vaccines and ordinarily should not be administered to pregnant animals. There are experimental studies and reports of practitioners having administered BVD vaccines during the last trimester of pregnancy to increase passive immunity in the calf without adverse effects. We are cautioned again, however, that all biological products should be administered in accordance with recommendations of the manufacturers.

Annual boosters are frequently recommended. It is important, however, that cattlemen consult with their veterinarian as to the appropriate schedule of immunization for their herd.

Special Notice: Brucellosis Status

Cow–Calf producers should pay special attention to brucellosis since the infection still exists in many herds in some states. Further regulations regarding vaccination and testing are due to change. If new cattle are to be brought into a herd, they should originate from a negative herd, be isolated from other animals in the herd of destination and be retested before being commingled with the new herd. Vaccination should be used as recommended by a veterinarian. *Vaccinate all eligible heifers for brucellosis. Purchase only vaccinated heifers.*

Calf Scours (Causes)

Since diarrhea in the calf is second only to reproductive diseases as a cause of losses to the cattle industry, it will be discussed in more detail than some other diseases. The four major causes of diarrhea are scours, enterotoxemia, salmonellosis and coccidiosis. A summary of each of the latter three diseases is presented in Table 19.1.

Scours is by far the most important cause of diarrhea in calves. Although scours can affect calves up to a month of age, it is most commonly observed in calves from birth up to 2 weeks in age. Scours may be caused by any of seven different viruses and numerous different bacteria. In addition, there are several different subtypes of each of these; for example, three corona viruses, eight enteroviruses, and 300 serotypes

Table 19.5. A Summary of Practices Recommended for the Control of Respiratory Disease in a Beef Cow–Calf Herd

Calves Age	Immunizations Against BRD	Other Immunizations Which May Be Considered	Additional Procedures
PROGRAM "A"			
Young calves 1 to 3 mo. old	IBR/PI_3: killed vaccine intramuscularly or intranasal modified live vaccine Pasteurella Hemophilus somnus	Leptospirosis Clostridial diseases: the "7 way" bacterin is preferred	Castrate and dehorn Implant (except animals to be kept for breeding) with approved product
PROGRAM "B"			
Older calves 3 or 4 wk. before weaning	IBR/PI_3: booster vaccination given at 1–3 mo. of age (may use killed vaccine, or intranasal vaccine, or live intramuscular vaccine) –Use only modified live vaccine is this is the first vaccination BVD: Do not give to stressed or very young calves. Not always recommended in conjunction with live IBR/PI_3 vaccine. Often given at weaning time. Pasteurella and Hemophilus somnus: booster shots if given earlier (1–3 mo. old) –if given now for the first time, booster, is needed 3–4 wk. later	Leptospirosis Clostridial diseases: a booster is recommended at this time Vibriosis: for heifers and bull calves kept for breeding Brucellosis:–only heifers –State regulations vary; consult with your veterinarian	Vitamin A: Injectable vitamin is preferred Implant:–use approved product –do not implant animals to be kept for breeding

At weaning time–If initial and booster shots had been administered as outlined in PROGRAMS "A" and "B," no additional immunizations are needed at this time.

–If the first vaccination series as listed in Program "A" was given but was not followed by Program

Table 19.5. (*Continued*)

"B" (3–4 wk. before weaning) give boosters listed in Program "B" at weaning time.
–If calves have not been vaccinated at all before weaning, the following programs ("C" and "D") are offered. It must be emphasized that PROGRAM "C" and "D" are not always an adequate substitute for PROGRAMS "A" and "B"; there may be a degree of risk when calves receive various vaccinations for the first time at weaning, or worse, after weaning.

PROGRAM "C"			
Weaning time	IBR/PI_3:–give intranasal vaccine or –killed vaccine (in muscle Pasteurella and Hemophilus somus: give first shot	Clostridial bacterin: the "7 way" is recommended	Parasites: treat for internal parasites Vitamin A: the injectable form is preferred
14–21 days after weaning	IBR/PI_3: if killed vaccine was given at weaning, 1 2nd shot of killed vaccine is needed now. BVD vaccine Pasteurella and Hemophilus: give booster	Chlostridial diseases: a booster is needed now Leptospirosis bacterin Vibriosis: for replacements and	Treat for external parasites
PROGRAM "D"			
At weaning	Pasteurella and Hemphilus: first shot BVD vaccine	Clostridial bacterin: 1st shot	Parasites: treat for internal Vitamin A: Injectable
14–21 days after weaning	IBR/PI_3: use modified live vaccine, either intranasal or intramuscular Pasteurella and Hemphilus: booster	Clostridial diseases: booster Leptospirosis bacterin Vibriosis bacterin Brucellosis: heifers only	Parasites: treat for external parasites

of the bacteria E. coli. Vaccination or immunity against one type has little affect against the others, and many different stress factors may predispose the calf to these organisms. Hence the complicated nature of the calf scour disease in being able to prevent or treat it.

There are two clinical forms of calf scours. One is sudden death, often without signs of sickness and before diarrhea has had a chance to develop. It occurs because the organisms, or toxins produced by them, gain access to the blood, and death occurs within 4 to 24 hours.

The second and most common clinical form of scours is diarrhea. Stools vary in consistency from watery to semisolid and are lighter in color than normal. This is in contrast to coccidiosis and enterotoxemia in which the feces are darker than normal and may contain visible blood. In the early stages, the calf continues to eat but later becomes dehydrated, weak, and stops eating. The course of the disease generally is 2 to 3 days, but may vary from 1 to 7 days.

Calf Scours (Prevention)

Scours occur if a calf's exposure to the disease exceeds the calf's resistance to the disease. Prevention of calf scours occurs when exposure is decreased and resistance is increased.

Exposure to calf scours can be decreased by calving and raising calves in dry, clean areas. If cows have to calve in a confined area, provide a dry, clean individual stall for the cow. The cow and calf should be moved to clean pasture as soon as possible after calving. Calving in corrals and keeping cows and calves in corrals, increases the risk of exposing the animals to scour-causing organisms.

Bringing new cows or calves into the herd just prior to calving or during calving increases the risk of exposure to calf scours. Calves needed for grafting would best be selected within the herd rather than purchasing calves from outside herds.

The major factor in increasing a calf's resistance to scours is to give the calf the first-milking colostrum, preferably 2 to 3 hours after the calf is born. Colostrum contains antibodies needed by the calf since the calf is born with little natural immunity against disease organisms. Because of rapid changes in the digestive system, the calf is not able to absorb a sufficient amount of antibodies after the calf is 12 hours old.

Both cows and calves can be vaccinated (orally and intramuscularly) for a few types of scours. The antibodies produced are for specific organisms only, and scours could still occur if caused by other organisms. Immunity, even for the specific organisms, is effective only when the colostrum is consumed by the calf and the antibodies are absorbed from the digestive tract.

Calf Scours (treatment)

Treatment of calf scours is similar regardless of cause. Diarrhea occurs because of alterations in intestinal function resulting from increased secretion from the body or decreased absorption of water and electrolytes (sodium, chloride, potassium, and bicarbonate) from the intestinal lumen. The electrolytes are lost in the feces.

Loss of bicarbonate causes acidosis. Loss of potassium causes lethargy and muscle weakness. Loss of water and sodium chloride results in dehydration. Treatment should be directed toward correction of the dehydration, acidosis, and electrolyte loss.

During diarrhea, cells that line the intestine are sloughed off, and it takes two or three days for new cells to develop. The major aspect of treatment, therefore, is to keep the calf alive while the affected intestinal cells are replaced and prevent secondary diseases, such as pneumonia.

Most dehydrated calves suffer from hypothermia (body temperature lower than normal). Therefore, the calf could be provided with supplemental heat in a dry location.

Clinical signs of dehydration occur when fluid loss reaches approximately 6% of the body weight. Fluid losses of 8% result in depression, sunken eyes, dry skin, and the calf will have difficulty standing. A 10% fluid loss will have additional clinical signs of the legs being colder than the rest of the body. A 12% fluid loss usually results in death.

There are several commercial electrolyte powders available for administering fluids to the calf. If they are not available, producers can economically prepare some mixtures which have proven very satisfactory.

Formula 1

1 tablespoon baking soda
1 teaspoon salt
8 ounces (250 cc) dextrose (do not use table sugar)
Add warm water to make a gallon
Administer up to 1 quart

Formula 2

1 package (1 ounce) fruit pectin
1 teaspoon of Lite® salt
2 teaspoons baking soda
1 can beef consomme soup
Add warm water to make 2 quarts
Give 1 quart orally every 4 to 6 hours.

Either of these home-made products will provide an adequate nutrient supply for a period of 24 to 48 hours. Do not give milk or milk replacers to the calf during treatment as E. coli grows rapidly in milk. Return the calf to the cow as soon as it is able to follow its mother.

Antibiotics (penicillin plus streptomycin or tetracyclines) are given during fluid therapy to prevent secondary infections. They should be given at proper treatment levels each day for at least 3 days or until the calf has recovered.

SELECTED REFERENCES

Publications

Bailey, J. H. *Parasites: The Profit Robbers.* Proc., The Range Beef Cow Symposium VII. 7–9 December 1981.

Blakely, J. and King, D. D., 1978. *The Brass Tacks of Animal Health.* St. Louis: Doane Agric. Service, Inc.

Control of Bovine Respiratory Disease in the Cow–Calf Herd. (Management guidelines prepared by Animal Health Committee (NCA) and the American Association of Bovine Practioners. Beef Business Bul. 17 July 1981.

Dip Vat Management. Feedlot Management. 1981 Planner Issue.

Foster, L. Electro-Thermal Treatment of Cancer Eye. Proc., The Range Beef Cow Symposium 3–5 December 1979.

Jensen, R. and Mackey, D. R., 1979. *Diseases of Feedlot Cattle.* Philadelphia: Lea and Febiger.

Kimberling, C. V. *Brucellosis.* Proc., The Range Beef Cow Symp. VII, 7–9 December 1981.

Koehler, R. G. and Sand, R. S. *Knock Out Insect Pests and Improve Performance. BEEF,* Feb. 1981.

Pierson, R. E., Kainer, R. A., and Teegarden, R. M. *Classifications of Pneumonias in Cattle.* Proc., The Range Beef Cow Symp. VII, 7–9 December 1981.

Schoenfelder, D. *Herd Health. External and Internal Parasites.* Charolais Bull-O-Gram, August/ September 1979.

Teegarden, R. M., 1982. *Calf Scour Seminars.* Colo. State Univ. Ext. Public.

Ward, J. K. and Nielson, M. K., 1979. *Pinkeye (Bovine Infectious Keratoconjunctivitis) in Beef Cattle.* J. of Anim. Sci. 38:1179.

White, G. and Wohlgemuth, K., 1983. *Calf Scours: Causes, Prevention, Treatment.* Great Plains Beef Cattle Handbook, GPE-3156.

Williams, R. E., Westby, E. J., Hendrix, K. S., and Lemenager, R. P., 1981. *Use of Insecticide-Impregnated Ear Tags for the Control of Face Flies and Horn Flies on Pastured Cattle.* J. of Anim. Sci. 53:1159.

Visuals

"Beef Cattle Castration." Sound filmstrip. Voc. Educ. Productions, Calif. Polytechnic State Univ., San Luis Obispo, CA 93407.

"Dehorning Beef Cattle." Sound filmstrip. Voc. Educ. Productions, Calif. Polytechnic State Univ., San Luis Obispo, CA 93407.

"Bovine Dehorning Procedure." 13 minute videotype. Instructional Media Center, Michigan State Univ., East Lansing, MI 48824.

20

Range, Pasture, and Other Grazed Forage Management

The total land area in the United States is approximately 2.2 billion acres. Nearly 1 billion acres are grazed by livestock, primarily cattle (Table 20.1).

Range and pasture areas in the United States are envisioned somewhat differently by different individuals. Typically the range area is conceived as the arid and semiarid lands of the 17 Western States (further broken down by some into the 11 Western States and the major 6 states in the Great Plains area). Pasture areas can be considered as irrigated pastures of the West in addition to grazed areas of the Eastern United States.

Rangeland has been defined by the Society for Range Management as "land on which the native vegetation (climax of natural potential) is predominantly grasses, grass-like plants, forbs, or shrubs suitable for grazing or browsing and present in sufficient quantity to justify grazing or browsing use. Rangelands include natural grasslands, savannahs, shrublands, most deserts, tundra, alpine communities, coastal marshes, and wet meadows." While this latter definition is inclusive for all grazed forage, the discussion in this chapter will use the former definition which separates the range and pasture areas.

Forage which is available for grazing, in addition to that which is harvested, is the backbone of the cattle industry. Over 80% of the feed consumed by cattle is forage crops, most of which is grazed. This forage can be converted by cattle from an unusable product for human consumption to a highly preferred consumer product with minimal cultural energy expended. Beef producers are interested in having large amounts of highly palatable, nutritious forage available for grazing. They are also interested in systems of grazing which will optimize pounds of beef produced without sacrificing the productivity of the range and pastures over an extended number of years.

GRAZED FORAGE RESOURCES

The available range, pasture, and forage resources in the United States and how they are utilized are very diverse. Their diversity is caused primarily by different climatic and soil conditions.

Table 20.1. Agricultural and Nonagricultural Uses of Land, United States, 1977

Major Land Use	Acres (Millions)	% of Total
Agricultural:		
Cropland	413	18.2
Cultivated cropland	368	16.2
Cropland used for hay	33	1.5
Other cropland	12	0.5
Pastureland and range	987	43.6
Forest land grazed	61	2.7
Farmsteads	11	0.5
Total agricultural land	1,472	65.00
Non-agricultural:		
Forest land not grazed	601	26.5
Urban, transportation, and other built-up areas	90	4.0
Other land (includes 9 million acres of small water areas)	102	4.5
Total non-agricultural land	793	35.0
Total land area	2,265	100.0

Source: USDA.

Annual precipitation, sometimes associated with other climatic influences, can have marked effects on forage supply (Figure 20.1). Figure 20.2 compares pasture and feed conditions between two consecutive years. Drought, many times associated with overgrazing, can reduce the desirable plant species which affects the future forage supply even in years of adequate precipitation. Extended drought periods for more than a year usually result in herd reductions, shipping in high-priced feed or moving cattle to areas having a more abundant feed supply.

PLANT TYPES AND THEIR DISTRIBUTION

The general classification of plants used primarily for grazed forage are grasses, forbs, and browse. Some grazing areas will contain all of these different classifications; however, more typically there will be mixed species of one or two classes. Since there are hundreds of different species of each plant class, only some of the more important species will be discussed.

Figure 20.3 shows the distribution of some major grasses in the United States. Cool-season grasses are more productive in spring and early fall, while warm-season grasses are more productive in late summer. The most important native grasses are found in the western two-thirds of the United States, while tame pasture predominates from the eastern Great Plains eastward.

The wheatgrasses, bluestems, gramma grasses, buffalo, and switchgrass are the more

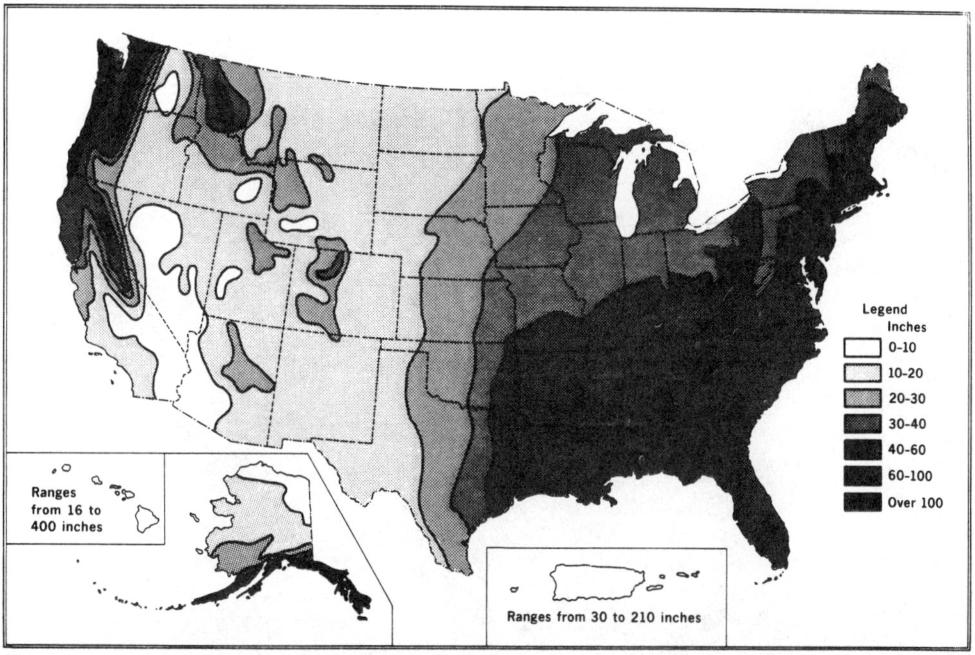

Figure 20.1. Average annual precipitation in the United States. (CAST Report No. 95).

common grasses on western ranges. Orchard grass, smooth bromegrass, reed canary grass, the fescues, and bermudagrass (primarily Coastal) account for most tame grass production in the United States.

Forbs are herbaceous annual and perennial dicots and monocots other than grasses. Legumes are the most important forbs with many of the other forbs being called weeds by cattle producers. Many forbs are fair to good range forage plants. Some highly palatable range forbs are sensitive to overgrazing and are used as indicators of range condition.

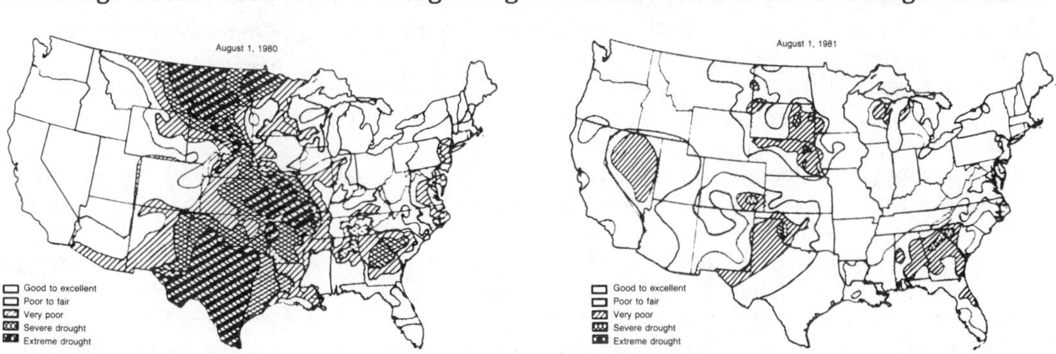

Figure 20.2. Range and nonirrigated pasture conditions, 1 Aug. 1980 and 1 Aug. 1981. Courtesy of the Western Livestock Round-Up.

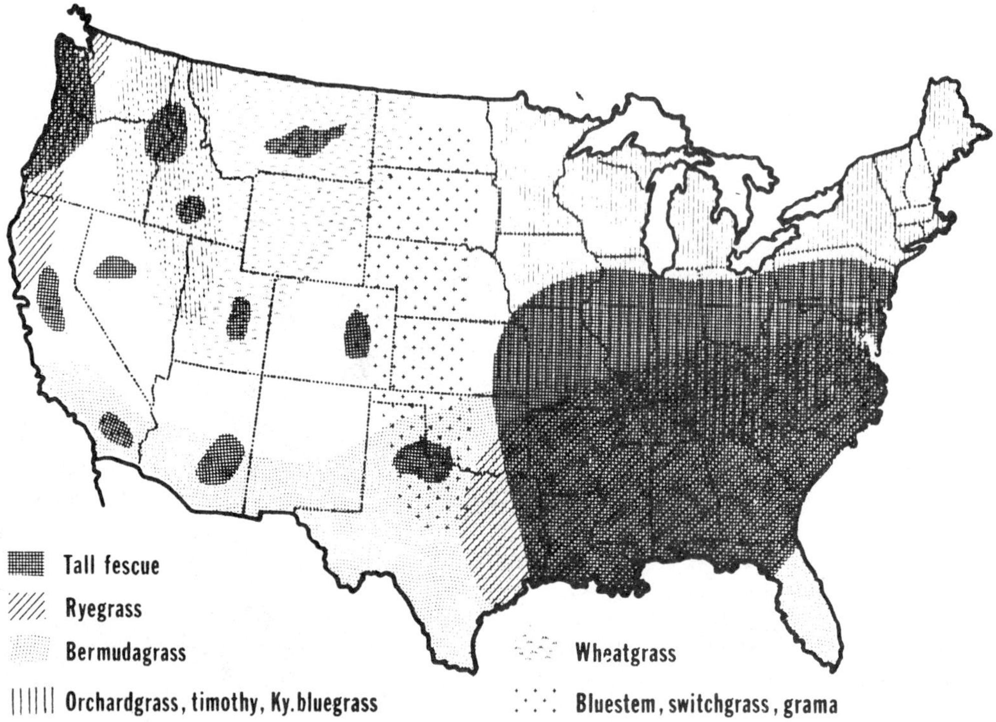

Figure 20.3. Distribution of some of the more important cool season and warm season grasses in the United States. (Courtesy of the Amer. Soc. of Anim. Sci., *J. Anim. Sci.* 57:546, Suppl. 2).

More common perennial range forbs are halfshrub sundrop, pitchers sage, compass plant, roundhead, and slender lespedeza.

Legumes are identified by the characteristic butterfly-like flowers. Legumes are usually higher in protein than most other grazed plants and are also important for the nitrogen fixation process which enhances soil fertility.

The most important forage legumes are alfalfa, the trefoils, the lupines, sweet clover, kudzu, and true clovers. Alfalfa is widely distributed since it is adapted to nearly all states.

Browse consists of perennial woody shrubs and vines used as forage by cattle. Some examples of range browse are sagebrush, shadscale, winterfat, mountain mahogany, and four-winged saltbrush.

Cattle grazing on range which includes grass, forbs, and browse typically change their consumption pattern of each as the grazing season progresses. Grass comprises a higher percentage of the diet in early summer. In late summer there is an increase in the consumption of forbs and browse. Possibly cattle perceive that in late summer the protein content of the grasses has significantly decreased, while protein content of forbs and browse is at least twice that of grass.

GRAZING AREAS IN THE UNITED STATES

Western

The western half of the United States is highly dependent on snow accumulations in the mountains to supply needed water for irrigation and to have adequate soil moisture to enhance spring plant growth. Thus many cattle producers in the West watch for mountain snow pack reports (Figure 20.4).

Pacific West

Most cattle grazing operations in the Pacific Northwest are from cow–calf operations. Nearly three-fourths of the brood cow's diet is from on-site grazed forage with supplemental hay fed in the winter time.

Northern California with its mixture of valley lands and timbered mountain ranges has the typical cow–calf operations of the Pacific Northwest states. In the California annual grass type, there is no one predominant type of cattle operation.

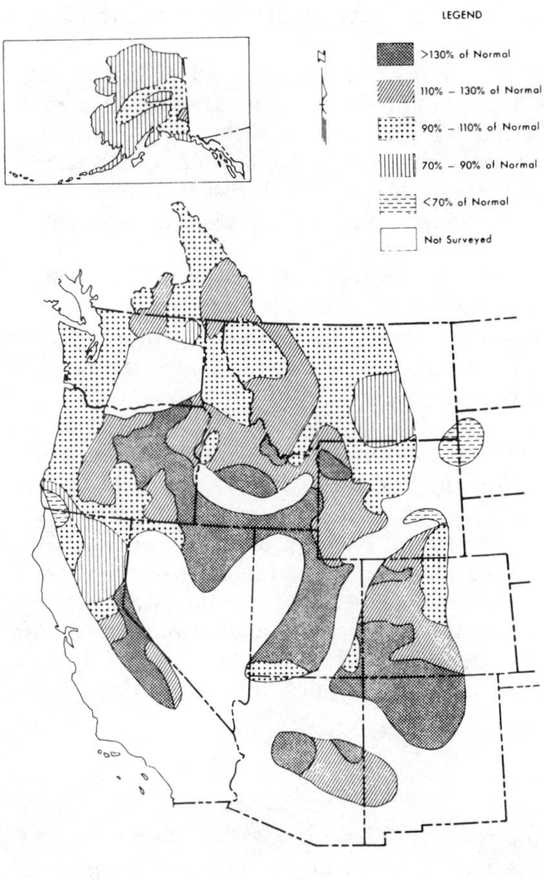

Figure 20.4. Mountain Snowpack Report, April 1, 1982. Courtesy of Western Livestock Round-Up.

The mountain grasslands which lie between the valleys and timbered mountains typically are grazed for two months each in the spring and fall.

Productivity of mountain bunchgrass is high, yet nutritional value drops below adequate levels in late summer and fall as forage dries. During about one-third of the fall period, seasonal rains followed by several days of adequate growing temperatures permit fall regrowth which has good protein content.

The California annual grasslands, prominent in much of the California foothills, are used during fall, winter, and spring periods. They are a very good source of forage in connection with yearling or feeder operations. Coupled with the use of annual grasslands and rations for feeders or stockers are improved pastures and grazing of crop aftermath. Some cow–calf operators make use of the annual grass type until it dries, then move their herds to higher elevation timber types or to renovated or suitable mountain shrub areas for summer grazing.

Chaparral Mountain Shrub occupies mountain areas over 3000 feet elevation, yet grows to sea level in some situations. The shrub species vary from deciduous, to semi-deciduous, to evergreen. Some brush types are so dense that understory vegetation is almost absent. In other cases, the shrub overstory permits a highly productive understory. Shrub control efforts and grass reseeding have improved the forage supply in areas where a mixture of forage types, natural low density of shrubs, or shrub control exists.

Ranchers living within a reasonable distance of mountain-timbered summer range often combine grazing of sagebrush range with mountain summer grazing to achieve forage supplies for all but the winter feed period. Ranch operations in the heart of sagebrush country try to make this range serve their needs for spring, summer, and fall. Ideally, however, sagebrush range should be used as spring–fall range. Many producers have small tracts of irrigated meadows interspersed with sagebrush, or they provide supplemental feeds in addition to their sagebrush range.

Throughout the Pacific West, forests provide considerable livestock grazing on a range basis. The most valuable of these for range grazing is ponderosa pine. In prime condition, ponderosa pine is open forest, with an abundance of green feed of good protein content during the hot season. This characteristic gives the ponderosa pine type a premium or extra incremental value over lower elevation range.

Coastal Douglas fir and interior mixed conifer forests occupy western Oregon, western Washington, and the northerly slopes of all interior mountains. The grazing productivity of the forest floor is poor (0 to 50 pounds of air-dry herbage per acre) prior to logging but may increase greatly following logging. The openness of a particular logging or harvest system regulates the amount of usable ground cover produced. Because these tree species are so shade tolerant, they tend to naturally provide a closed canopy in about 20 years. After tree seedlings are re-established, the forage produced during this 20-year period is often abundant, but the nutritive quality is variable and uncertain. With better utilization of forest residues, treatment of slash, and reseeding of improved ground cover, possibilities exist for summer season herbage production of 1200 to 1800 pounds per acre.

Intermountain

Livestock operators of the Intermountain region which consists of Idaho, Nevada, Utah, western Colorado, and Wyoming make use of seasonally-productive range lands by moving animals from one geographical range to another. The desert ranges are used during the winter (1 November to 5 April), and foothill or intermediate elevation ranges are used during the spring (5 April to 1 July), and some are used in the fall (1 October to 1 November). Mountain ranges are used during the summer and fall (1 July until about 15 October).

Of great importance is the comparative nutrient value of different forage plants during various seasons and the ability or inability of these forage species to meet requirements for optimum livestock production. Animals in the intermountain area do not need a supplement during the spring and summer range grazing seasons if the plants are growing. During fall and winter, supplements often are necessary because the forage nutrient content is marginal for animal needs, and inclement weather may seriously reduce daily intake (Figure 20.5).

Intermountain (Spring Range)

A scarcity of suitable spring range in the intermountain area generally is a limiting factor for successful cattle production. During recent years, it has become common practice to seed depleted foothill range to provide more suitable forage for spring grazing.

Grass species such as crested wheatgrass, intermediate wheatgrass, or Russian wildrye often are planted to help provide needed nutrients when native foothill ranges are mature and deficient in nutrients (Figure 20.6). An alternative, when foothill ranges become dry and dormant, is to move animals to higher elevations where feed is still green and growing. This is sometimes difficult because it puts an additional grazing load on ranges that must also be grazed in the summer.

Intermountain (Winter Range)

Vegetation of desert ranges of the Intermountain-Great Basin region is composed primarily of browse species with various quantities of grass. Generally, desert browse

Figure 20.5. Cattle may need a protein supplement, shown here in block form, when forage matures and decreases in protein content. Courtesy of Elanco Products Company.

Figure 20.6. Cattle grazing a cleared area, reseeded to crested wheatgrass, in western U.S. Courtesy of the Colorado Rancher and Farmer.

plants meet the protein requirements for livestock during gestation and are exceptionally high in carotene. They may be slightly deficient in phosphorus, however, and decidedly low in energy furnishing constituents. Grasses, during the winter, are markedly deficient in protein, phosphorus, and carotene but are good energy sources. Forbs are generally sparse on desert ranges and are unimportant in the diet during winter grazing. If the diet is largely grass, phosphorus, and digestible protein may be markedly deficient, but if the diet is largely browse, energy may be decidedly deficient.

Animals on many winter ranges may require a particular supplement to meet requirements when properly grazed, but with increased grazing intensity, the quantity and even the type of supplement needed may change. Overgrazing may result in a need for a greater quantity or even a more expensive supplement over a longer period of time.

Southwest

Most cattle operations consist of cow–calf and cow–calf–yearling. Calves are usually sold as stockers and feeders. In desert areas, small base herds are more common, with steers purchased during wet periods to use abundantly-produced annual grasses and forbs.

Desert ranges can be grazed yearlong. Grasses like black grama cure well and maintain some nutritive value yearlong. Desert grasslands are quite valuable for grazing despite aridity. Most of the grasses are highly preferred. During the growing season, grasses have a crude protein content of 8% or higher, generally at or above minimum levels for livestock. Protein drops off to 4% to 5% in winter, requiring some supplementation. Desert shrub areas have a more erratic type of forage production. Most of the forage is produced following rains (usually late winter and spring). It consists of annuals such as alfilaria, six weeks grama, and other annual forbs and grasses. Some shrubs, principally four-wing saltbush, provide nutritious, palatable feed. Four-wing saltbush retains a protein content in excess of 11 percent yearlong.

The foothills type rangelands including pinyon–juniper and chaparral–mountain shrub ecosystems provide mainly spring–fall grazing for cow herds. The pinyon–juniper is an abundant ecosystem but relatively unproductive of forage. It has historically been heavily grazed because of its proximity to ranch headquarters and the better parts have been plowed for crop production. The chaparral–mountain shrub ecosystems generally provide limited grazing because the dense brush and tree cover make grazing difficult. Their primary use is spring–fall, or spring, summer and fall. Oak savanna areas of eastern Arizona and western New Mexico are productive of foliage and are often used yearlong.

The mountains of the Southwest commonly support a forest cover of ponderosa pine, with occasional open, mountain grassland areas. Part of the precipitation falls in the winter and spring from Pacific storms, and part during the summer from storms originating in the Gulf of Mexico. Grazing by cow herds occurs during the summer and fall.

Throughout the Southwest, brush has increased since domestic livestock grazing began several hundred years ago. Palatable, perennial grasses have decreased, and in some areas, like the Desert Shrub, they have been largely replaced by annual grasses and forbs.

Great Plains

The Great Plains occupies approximately one-third of the United States and includes parts of ten states (Montana, North Dakota, South Dakota, Wyoming, Nebraska, Kansas, Colorado, Oklahoma, Texas, and New Mexico). This vast area of semiarid, high-plateaued grasslands, once known as the Great American Desert, varies in topography from extremely flat to quite rolling. In general, it is covered with short grasses and mid-grasses, but in the Black Hills of South Dakota, it is covered with trees. On the eastern edge and scattered throughout on deep, sandy soils are large areas of highly productive tall grass prairie.

Soils in the Great Plains range from very deep, black soil, high in organic matter in the more humid areas to very thin—almost colorless soils that are low in organic matter in drier regions. Generally, the Great Plains is characterized by extremely cold winters, especially in the North, and warm summers with relatively low precipitation and humidity. Wind is common and often accompanied by sudden variations in temperature. Precipitation varies from 15 to 25 inches annually in the southern portion to 12 to 15 inches in the Northern Great Plains. It is higher in the eastern edges of the Great Plains, and for example, may exceed 30 inches in eastern Kansas and Oklahoma. Growing seasons range from 110 to 120 days in the Northern Great Plains to over 240 days in Texas and New Mexico.

Northern Great Plains

The Northern Great Plains covers an estimated 300,000 square miles, which is approximately one-tenth of the total land area of the United States. The range in the Northern Great Plains occurs in the western three-fourths of North Dakota, South Dakota, and Nebraska, and the eastern two-thirds of Montana and the eastern one-half and the eastern one-fourth, respectively, of Wyoming and Colorado. Approximately one-fourth of the Northern Great Plains area is under cultivation. Production of both cultivated crops and range forage varies greatly from year to year and from one area to another within the Northern Great Plains. Droughts are frequent and records indicate that precipitation in certain parts of this area has dropped below 75% of the normal on the average of once in every 5 to 8 years. Native range can best withstand these successive drought periods.

While cultivated crops may be of importance, especially in local areas, grasslands are the principal resource base for the agricultural economy prevalent in the Northern Great Plains.

Livestock in the Northern Great Plains graze the native range species for about 8 to 9 months of the year from April to December. During cold and stormy winter months, cattle are generally fed in holding pastures near farmsteads or on range areas where shelter is provided by breaks in the terrain or creek bottoms where trees are generally present. Many range areas in the Northern Great Plains furnish perhaps one-half of the nutrients needed by cattle. An additional 30% comes from grassland hay, which is useful during the longer winter periods in the North. The rest of the annual feed requirements comes from supplemental feeds and cultivated cropland, including crop aftermath.

Some livestock operations in the Northern Great Plains are a cow–calf enterprise or a steer operation, but in most cases the livestock producers run an all age group where they sell weaner calves, yearlings, and in a few isolated instances, even two-year-old steers. In many cases where topography is highly variable and vegetation composition is complex, both cattle and sheep can be managed to more fully utilize the range forage resource.

The major range types in the northern Great Plains consist of short grass and mixed grass prairies. The Sandhill ranges are also important vegetation types and occur mainly in west-central Nebraska with smaller areas in adjacent states. Precipitation varies from 15 to as high as 20 to 22 inches annually. Since moisture is rapidly absorbed into the soil, there is little runoff; therefore, effective precipitation is greater in the sand areas.

The sagebrush-saltbush grassland that occurs in the Northern Great Plains is called the Red Desert and covers some 43,000 square miles. It is generally the most arid part of the Northern Great Plains with a precipitation ranging from 7 to 12 inches annually and soils are often alkaline or saline. This area is grazed mainly by sheep in the fall and winter when snow is available.

In the Black Hills of South Dakota and the Bighorn Mountains of Wyoming and other local areas, grasses are open forest type that includes ponderosa pine, Douglas fir, and some spruce at higher elevations. This covers almost 20,000 square miles in the Northern Great Plains and produces considerable amounts of forage used primarily by cattle during the summer.

Southern Great Plains

The Southern Great Plains is a leading cattle-producing and livestock farming area that includes over 130 million acres of southeastern Colorado, western Kansas, western Oklahoma and Texas, and parts of northeastern New Mexico. Like the plains area to the north, the climate of the Southern Great Plains is extremely variable from year to year. Rainfall is generally limited, humidity is low, and high winds cause high evaporation rates.

Year-long grazing of range is common practice in the Southern Great Plains, especially with breeding herds. Cultivated crops are often used to supplement winter range, and wheat pasture is an important winter and early spring forage. In the extreme northern part of this area, ranges may be used from April or May to October or November as is the case farther north in the Northern Great Plains.

The vegetation and topography are excellent for cattle. The environment is generally more suitable for production of young animals. Calves are sold primarily as stockers and feeders.

Some supplemental feeding is practiced in winter for increasing protein. The use of crop residues in parts of the Southern Great Plains and wheat pasture in fall and spring is common. Sorghum pastures for late summer grazing is rapidly increasing throughout most of the Southern Great Plains.

The tall grass prairie in the eastern portion is among the world's most productive grassland areas. The vegetation is comprised primarily of tall grasses, some of which exceed 6 feet in height. Precipitation ranges from 25 to 38 inches annually, most of which occurs during the summer growing season. The forage of this area can support many cattle. Compared to more xeric, or drier areas, the vegetation does not cure well. Protein supplement is needed throughout the winter months. Areas which once largely grazed steers in the spring and summer are now being converted to cow–calf operations.

The principal range forage types in the western portion of the Southern Great Plains consist of short and mid-grass vegetation that occurs on the heavier textured soils. On more sandy soils along the streams in this area, a mid-grass and a tall-grass vegetation type occurs. The shortgrass vegetation type is dominated by blue grama and buffalograss (Figures 20.7 and 20.8).

Wheat Pasture

Cattle graze several types of small grain pastures throughout the United States. The most noted of these are winter wheat pasture areas in several southern Great Plains states where millions of stocker cattle are grazed each year (Figure 20.9). Wheat pasture is also important in some adjoining states but to a much lesser extent than in Kansas, Oklahoma, and Texas.

Figure 20.7. A brush and grass range forage combination in the Great Plains area. Note the water tank, filled by a windmill, which is common to several, separately fenced pastures. Courtesy of the *Colorado Rancher and Farmer.*

Figure 20.8. Cattle grazing native range in the Flint Hills, a very productive grazing area in Kansas. Courtesy of Kansas Agric. Expt. Sta.

Wheat is planted in late summer, then depending on moisture and other conditions, the fall and winter growth is available for cattle grazing. Typically the cattle will be removed from wheat pastures in early spring, and a crop of wheat will be produced. The usual grazing period is from November to the middle of March. Grazing can continue to the end of May if a decision is made not to harvest a crop of grain.

Daily gains on wheat pasture are usually 1.5 to 2.0 lbs per day. There can be serious problems with bloat with average annual death losses of 2% to 3% expected. Bloat is most frequent during a 2- to 4-week period in the early spring when the wheat breaks dormancy. As the wheat matures, bloat is reduced. Some producers feed low-quality roughages (wheat straw or sorgum–sudan hay) feeling that it will improve the utilization

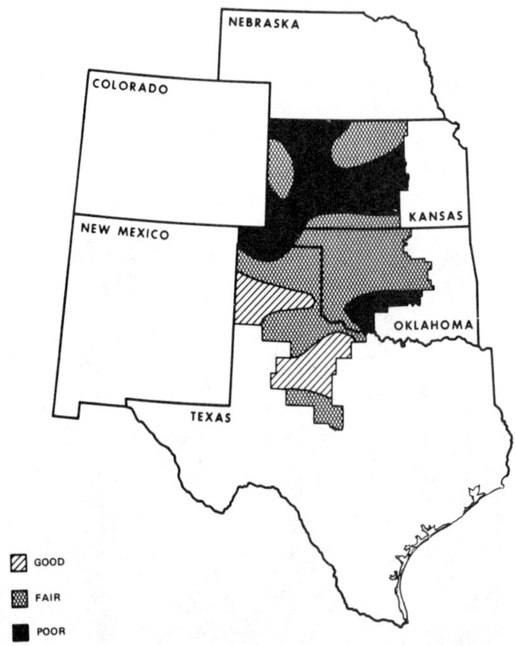

Figure 20.9. Major winter wheat pasture area in the United States. Amount of top growth available for grazing, 1 Dec. 1978 is depicted. Courtesy of Western Livestock Round-Up.

of wheat pasture and reduce bloat. Research shows that ad libitum consumption of these roughages is low with little effect on the incidence of bloat (*J. Anim. Sci.* 56:1021).

Stocking rates vary with the amount of top growth available, and more numbers can typically be put on irrigated wheat than dryland wheat. Stocking rates are more critical after the jointing of the wheat as it starts to mature. If stocking rates are not high enough at this time, the wheat will mature ahead of the cattle's ability to graze it. Thus the cattle will spot graze and avoid the more mature wheat plants.

Cattle may have to be moved from wheat pasture during periods of heavy rainfall or irrigation to prevent trampling damage and a significant reduction to the wheat crop.

Grazing Fees

Rangelands in the 17 Western States (includes Great Plains) have both private and public ownership. Many ranches will have a combination of privately deeded land and grazing permits for use of public lands. Grazing associations also exist where several ranchers cooperatively graze cattle on land privately owned by a grazing association, public grazing land, or a combination of the two.

Public lands are both state and federally owned. State lands primarily involve millions of acres given to states by the Federal Government. These state lands are usually referred to as "school sections" since revenues from two sections, from each township (see Appendix Figures A.1 and A.2), were to help support the public schools. Cattle producers have 10-year lease agreements of this land with extended leases given to them if terms and conditions are met. Lessees own all improvements on the land. If the land is sold or leased to another individual, the original lessee is reimbursed for an appraised value of improvements.

Federally owned lands are managed by the U.S. Forest Service or the Bureau of Land Management (BLM). Permits to graze Forest Service land are issued to qualifying permittees who have a privately owned livestock operation. Only cattle owned by permittees are allowed to graze the lands managed by the Forest Service. Forest supervisors control the pattern of range use. They may adjust number of cattle to graze a certain area and length of time the cattle can be on the range. Any changes made are difficult and usually cause disagreement between cattle producers and Forest Service personnel.

Permits can be terminated whenever the Forest Service decides to use the land for other purposes. In recent years, there has been a considerable amount of discussion as to whether these federally owned lands should have only a single use (recreation for the general public) or have a multiple use which would include grazing of cattle.

The BLM manages the other public land in a similar manner as the Forest Service. Grazing permits are issued, and BLM personnel determine the manner in which the land will be grazed.

Annual grazing fees are assessed to those beef producers who have grazing permits. Fees are based on animal unit month (AUM). An animal unit includes cattle over 8 months of age or a cow with a calf less than 6 months of age. A more precise definition used for an AUM is the TDN required to maintain a 1000 lb cow for 30 days or 300 lbs of TDN.

Table 20.2. Average Grazing Fees for Federal Range

Year	Forest Service	BLM
	Dollars per AUM	
1970	$0.60	$0.44
1975	1.11	1.00
1980	2.36	2.36
1983	1.40	1.40

Total forage costs on national forests (BLM forage costs are similar) may have one or more of the following components: (1) the annual grazing fee, (2) investment and maintenance of improvements, and (3) investment in the grazing permit. Grazing fees which are governed by Congress have been increasing over the years (Table 20.2).

Table 20.3 shows a comparison of 1978 forage costs from alternate feed costs in a western state using a cow–calf pair as the basis. These costs are for range forage only and do not include costs associated with the livestock such as transportation, salt, or herding.

Grazing on federal land is the cheapest forage source. This is particularly true if

Table 20.3. Cost of Range Forage from Alternative Sources, Colorado, 1978

Forage Source	Cost per AUM[a]
	Dollars
Deeded range:	
1950 purchase	9.33
1960 purchase	11.88
1970 purchase	16.20
1977 purchase	40.17
National forest grazing permit:	
Grazing fee only	1.29
Grazing fee and investment in improvements	1.85
Grazing fee and investment in improvements and grazing permit	6.77
Bureau of Land Management grazing permit:	
Grazing fee only	1.19
Grazing fee and investment in improvements	1.85
Grazing fee and investment in improvements and grazing permit	4.65
Colorado state school grazing land	9.90
Grazing Association (Farmers Home Administration financed)	9.27
Rental rate deeded range	7.47

[a] Assumes a cow and a calf as an animal unit based on Forest Service and BLM definition which are equal to 1.3 animal units for calculating AUMs of forage consumed (yearling would be 0.7 animal unit).

purchase of grazing permits is not required. Forage cost rates on state land and grazing associations are comparable with rental of deeded range lower than the other two.

Deeded range is the most expensive source of forage—interest on invested capital being the largest component of costs. Only deeded rangeland purchased in the 1950s is comparable with costs of other forage sources. Land values in more recent years have caused forage cost from this source to far exceed costs which livestock grazed can reasonably be expected to support.

Midwest

A large concentration of highly fertile soil and favorable growing conditions are located in the Corn Belt States (Iowa, Missouri, Illinois, Indiana, Ohio, and parts of some of the adjoining states). In most cases, pasture production cannot compete economically on the most productive land where corn and soybeans can be grown. There are thousands of acres of productive land in the Midwest, however, where row crop farming is not feasible because of the slope of the land. This land has major erosion problems if it is plowed annually. Pasture production on this type of land is excellent, and beef cow numbers are high in several of the states in the Midwest. For example, Iowa, with its 1.7 million head of cows, has more beef cows than any one of the 11 Western States and ranks 6th in the United States in beef cow numbers. Beef cows are concentrated in the rolling hill area of Southern Iowa where it is estimated that more than 650,000 cows are located within a 75-mile radius of Ottumwa, Iowa.

Bromegrass, orchardgrass, and bluegrass are among the most common grasses in the Midwest. The grasses are grown in purestand pastures or interseeded with some of the more common legumes, such as, alfalfa, clovers, birdsfoot trefoil, and lespedeza. These cool-season pastures are highly productive during spring and early summer. Their productivity, however, is low in July and August. Thus cattle performance declines during the midsummer months. If adequate amounts of rainfall are received during late summer and early fall, there will be a regrowth of highly nutritious cool-season forage which can be utilized. Freezing temperatures which occur in the fall terminate new growth, and beef producers rely on the grazing of matured forages and crop aftermath. Harvested forage is then provided when available forage for grazing is depleted, or when snow prevents an adequate amount of dry matter to be grazed.

There is increased utilization of warm-season grasses to complement cool-season forage. For example, in past years the tall-grass prairie extended into the eastern Corn Belt. Switchgrass, Indiangrass, and big blue stem are being utilized in Iowa and Missouri to provide available green forage during July and August. Preference is given to seeding these as a single plant species because they are easier to manage. Also, warm-season grasses do not perform well when mixed with cool-season grasses or legumes because of differences in growth patterns and competition.

Warm-season grasses have to be managed differently than the same grasses grown further west, because higher rainfall in the Midwest and competition from weeds is a greater problem. Switchgrass is often the first choice among farmers trying a warm-season grass for the first time because it is easier to establish. Switch, however, is considered a lower quality grass compared to big bluestem or Indiangrass.

Grazing management of cool-season species is similar to warm-season grasses with one noticeable exception. Warm-season grasses store their energy reserve for regrowth above ground in the lower 8 to 10 inches of the plant, whereas cool-season grasses and legumes place their food reserves in the root systems and crowns. This is the main reason that warm-season grasses will not tolerate close grazing.

Southeast

Until the 1940s, cattle production was incidental to the principal uses of the cleared eastern forest land for raising cotton and tobacco. Prior to this time, the relatively infertile soils were becoming depleted and cash returns were dropping. Ways were sought to rehabilitate seriously eroded cotton and tobacco lands and to provide alternate sources of cash income.

Many eroded croplands were planted in trees; however, that did not solve the cash income problem since 15 to 20 years were required for the planted trees to reach pulpwood size. Interest in cattle raising increased as an income source. This interest quickened when it became known that forest soils could produce high yields of forage when soil fertility needs were met. In addition, high-yielding species or strains of grasses and legumes were developed, among them the pioneer Coastal bermudagrass.

As a result of improved pastures in the Southeast, beef cattle numbers increased dramatically. In the 12 Southeastern States (Florida, Mississippi, Louisiana, Alabama, Georgia, Arkansas, Kentucky, Tennessee, North Carolina, South Carolina, Virginia, and West Virginia), beef cow numbers increased from 2.8 million in 1950 to more than 9.5 million in 1980. This is approximately 25% of the nation's cowherd and would be even higher if the coastal area of Texas was included which would be a similar forage environment to some of the other states.

There is a great diversity of pasture operations in the Southeast (Figure 20.10). There are many highly developed and intensively managed pasture situations for purebred, commercial cow–calf or yearling operations. Other pastures include "hill" pastures, generally Kentucky bluegrass pastures with mixtures of other grass and legume species—augmented by bottomland pastures grazed seasonally.

In some areas of the Southeast, cattle may graze entirely on farm woodlands and

Figure 20.10. Brahman cattle grazing Bahiagrass pasture in Louisiana. Courtesy of Dr. L. I. Smart, Louisiana State University Agricultural Center.

Figure 20.11. Cattle grazing native forage in a pine stand on the coastal plain in Alabama. Courtesy of Winrock International.

pastures (Figure 20.11). Many depend largely on industrial forest lands and national forests. It is not unusual for cattle to graze yearlong in the forest with no provision for rotation, except that associated with range burning. Systems for grazing pine forest ecosystems have been developed that are compatible with tree growing. Cattle also graze hardwood forests where generally it is often highly damaging to tree growth.

GRAZING SYSTEMS

Grazing systems are extremely diverse as there are hundreds of combinations of climatic differences, plant species, terrains, types of cattle, and preferences of cattle producers. There is no one best system, so producers must select the systems which fit their needs and goals.

Range and pasture grazing systems are primarily a function of three factors: (a) amount of forage produced, (b) forage quality, and (c) the efficiency by which the forage is harvested (Heitschmidt, et al. 1983). All systems of grazing management are centered around the basic principle of grazing management which is the control and frequency and severity of defoliation of individual plants (Heitschmidt et al. 1983). This principle needs to be carefully assessed over several grazing seasons as well as a single grazing season. Effective grazing systems are determined by stocking rates, intensity and frequency of grazing, grazing of single or mixed stands of forage, and other factors which will optimize efficient beef production.

Some grazing intensities and frequencies are reflected in the following grazing terms commonly used:

Continuous—Allowing grazing on a specific area throughout the entire grazing season (does not refer to level of use or imply year-long grazing).

Rotational—Cattle moved from pasture to pasture on a particular schedule based on calendar, level of use, or stage of growth.

Deferred—Not grazed until the primary forage plants have matured, usually the setting of seed.

Rested—No grazing for at least a one-year period.

Rested-rotational—Scheme incorporating at least one rested pasture on a rotational basis.

Season—Grazed for the duration of a particular season of the year.

Short duration (high intensity)—Heavy stocking rates for short periods. Each unit may be grazed several times during a grazing season.

The objective of most grazing systems is to optimize animal performance and maintain or improve the productive potential of the vegetation. Kinds of available animals and the type of vegetation, along with the seasons of grazing, help determine the stocking rate as part of the overall grazing plan. Matching dietary preferences and nutrient requirements of cattle to kinds and quality of vegetation is the first important step in determining the grazing management system.

The season of grazing is determined primarily by recognizing two important relationships. First, animal performance is greatest when the forage is grazed soon after it is produced and before it matures. Second, to obtain range improvement, the desirable forage species must be allowed to reproduce. Rangelands generally provide flexibility where they can be combined with tame pastures.

The optimum stocking rate is not a constant but will vary from season to season and year to year. Grazing pressures which are too high for long periods of time will reduce the cattle's nutrient intake and thus reduce their performance.

The correct stocking rate for a particular pasture will depend on the desired level of animal performance, ability of vegetation to withstand grazing and range or pasture improvement goals. Grasses such as bermudagrass, bahiagrass, K. R. bluestem, buffalograss, and common curly mesquite are well suited for intensive grazing. Most taller grasses and bunch grasses do not withstand intensive grazing. Where mixed stands of these grasses exist, the species which eventually predominate will depend on the intensity and length of grazing.

Cattle are rotated to different pastures to control grazing distribution problems. If grazing distribution is not a problem, there will be little benefit from rotational grazing.

Cattle are selective in their grazing for two reasons: (1) they respond to topography, distance to water and other factors. Salt, water distribution, and forced movement of cattle are three practices which are utilized to prevent wide differences in overgrazing of certain areas; (2) some species of plants in mixed stands are more palatable than others. Some grazing systems will prevent overgrazing of the more palatable species.

Currently there is considerable interest in high-intensity low frequency (HILF) grazing systems. Some of these systems are excellent in optimizing animal and forage performance; however, they can be expensive because additional labor and fencing are required. Range improvements will occur under HILF if moderate stocking rates, which will give acceptable levels of cattle performance, are utilized. Under this system, it is possible to harvest 70% of the forage and maintain an excellent range instead of the "take half and leave half" grazing recommendation.

In recent years, there has been an intense interest and acceptance of HILF grazing system in the arid rangelands. Strong proponents of certain HILF systems have generated considerable controversy. Much of the attention has focused on the Savory Grazing Method (Figure 20.12). Excellent discussions on the Savory Grazing Method and the basic principles of grazing systems are discussed by Heitschmidt, et al. 1983 and Savory 1983.

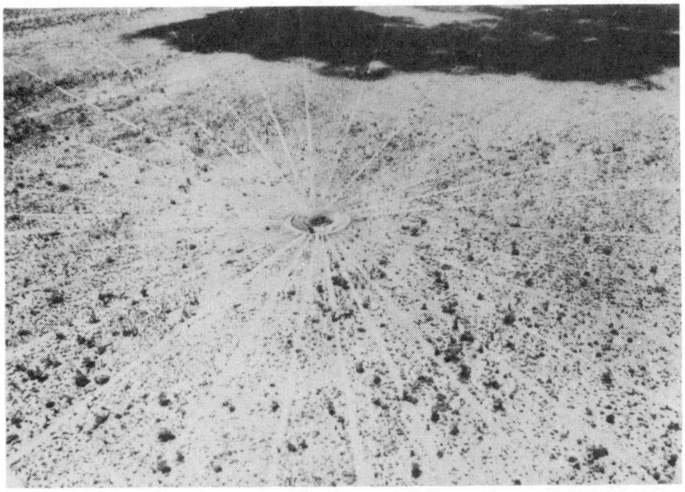

Figure 20.12. Aerial photograph demonstrating the Savory Grazing Method. The cleared lanes are fence lines which divide several thousand acres into 30 pastures. The circular area in the center is the water tank (common to all pastures) and working area. Cattle will graze each pasture for only a short period of time. Courtesy of Western Livestock Journal and Allan Savory.

Range and pasture areas which permit year-around grazing, with satisfactory animal performance, are usually the most economical for cattle producers. This is true if input costs can be moderated. The primary objective is to have green forage (which is typically higher in nutrient content than dry, mature forage) available to cattle on a year-round basis. In the intermountain area, such a grazing system might be as follows: early spring (crested wheatgrass), late spring–early summer (grazing cool season grasses in foothills area), late summer–fall (grazing warm season grasses or cool season grasses in the high mountain pastures as shown in Figure 20.13), and winter (grazing on the desert shrub areas at low elevations).

Figures 20.14 and 20.15 demonstrate an excellent planning system to provide a yearly

Figure 20.13. Cattle graze foothill areas in early spring and the high mountain pastures (far background) during the summer. In the fall and early winter, the forage aftermath from the hay meadows along the river bottom will be grazed. This is a common grazing system in the intermountain area of the western United States. Courtesy of *Colorado Rancher and Farmer.*

Guide for Year-Around Forage Supply

Use the following 3 tables to help plan a year-around forage program to fit your land and livestock needs. Table 1 is used to evaluate your present forage production and to plan additional forage if needed. In Table 2, you can list present or planned livestock and determine their forage needs. Table 3 provides data for use in completing Table 1.

Table 1. WORKSHEET—YOUR FORAGE SUPPLY

Kind of Forage	Total Available		Animal Unit Months per Acre[2]												Total
	Acres	AUM[1]	Jan.	Feb.	Mar.	Apr.	May	June	July	Aug.	Sept.	Oct.	Nov.	Dec.	
Total AUM's Available															

[1] AUM = Acres in column 2 x total AUM's for each kind of pasture as shown in last column in Table 3.
[2] AUM per acre = Acres in column 2 x AUM per acre for each month given in Table 1.

Table 2. WORKSHEET—YOUR LIVESTOCK NEEDS

Kind of Livestock	Number[1]	AU Factor	Animal Units[2]	Animal Units to Be Carried Each Month[3]											
				Jan.	Feb.	Mar.	Apr.	May	June	July	Aug.	Sept.	Oct.	Nov.	Dec.
Mature Cattle		1.0													
Yearlings		.7													
Calves (over 3 mos.)		.3													
Ewes		.2													
Horses		1.0													
Total AUM's of Forage Needed															

[1] Enter number of each kind of livestock.
[2] Animal units = Number in column 2 x AU factor in column 3.
[3] Enter the total animal units as shown in column 4 which will be on hand each month.

Table 3. GUIDE FOR PLANNING A YEAR-AROUND FORAGE SUPPLY

Forage	Animal Unit Months Per Acre												Total
	Jan.	Feb.	Mar.	Apr.	May	June	July	Aug.	Sept.	Oct.	Nov.	Dec.	
Pasture[1]													
Continuous grazing													
Unimproved bluegrass					.6	.6	.3	.2	.2	.1			2.0
Improved bluegrass				.2	1.1	1.1	.6	.3	.4	.3			4.0
Trefoil-grass					.8	1.1	1.1	1.1	.5	.4			5.0
Deferred grazing													
Trefoil-grass					.5	.5	1.2	1.5	.4	.6			4.7
Orchard or bromegrass				.5	1.5	1.5	.8			1.2	1.0		6.5
Reed canary grass				.5	1.5	1.5	.8			1.4	1.0		6.7
Tall fescue				.5	1.5	1.4	.6			1.2	1.6		6.8
Warm season grass						.2	1.0	1.3	1.0	.5			4.0
Sudan or sudan-sorgham							1.5	2.5	1.0		.3		5.3
Alfalfa-grass					1.4	1.4	1.0	1.0	.6	.5	.1		6.0
Alfalfa-grass, after 1 cut							1.2	1.2	.5	.5	.1		3.5
Alfalfa-grass, after 2 cuts								1.0	.4	.4			1.8
Cornstalks										.4	1.0	.3	1.7
Sorghum-set aside acres									2.0	2.0	1.0	1.0	6.0
Hay and Silage													
Legume hay													10
Grass hay													6
Oat silage													4
Corn silage													18
Other													

[1] **Production for these forages assumes nitrogen fertilizer per acre at rates of 80 lbs. for bluegrass and warm season grasses; 150 to 200 lbs. for tallgrasses and 120 lbs. for sorghum. Phosphorus and potassium would be supplied for all forages as needed. Alfalfa should be rested from Sept. 15 to Oct. 20.**

Cooperative Extension Service, Iowa State University of Science and Technology and the United States Department of Agriculture cooperating. Robert L. Crom, director, Ames, Iowa. Distributed in furtherance of the Acts of Congress of May 8 and June 30, 1914.

Figure 20.14. Worksheets for planning a year-around forage supply under midwest conditions. Courtesy of Frank Schaller, Ames, Iowa.

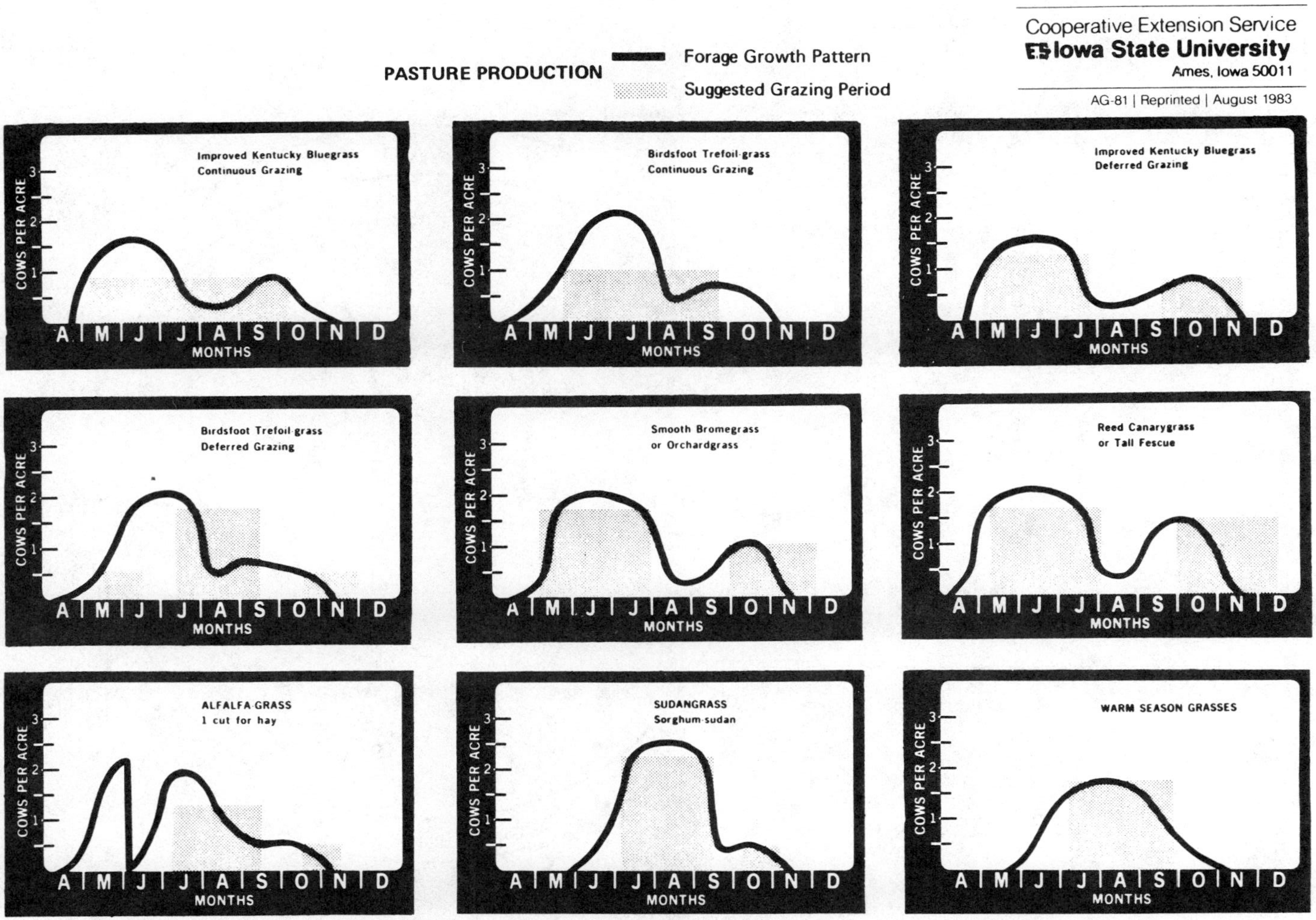

GUIDE FOR YEAR-ROUND FORAGE SUPPLY

Prepared by Frank Schaller, extension agronomist.

Figure 20.15. Some forage choices for planning a year-around forage supply under midwest conditions. Courtesy of Frank Schaller, Ames, Iowa.

Figure 20.16. Forage management fields for year around forage supply in Kentucky (adapted from Absher, 1979).

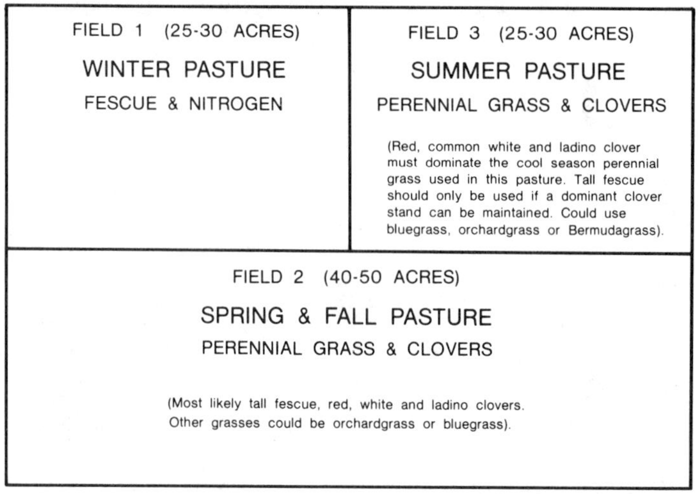

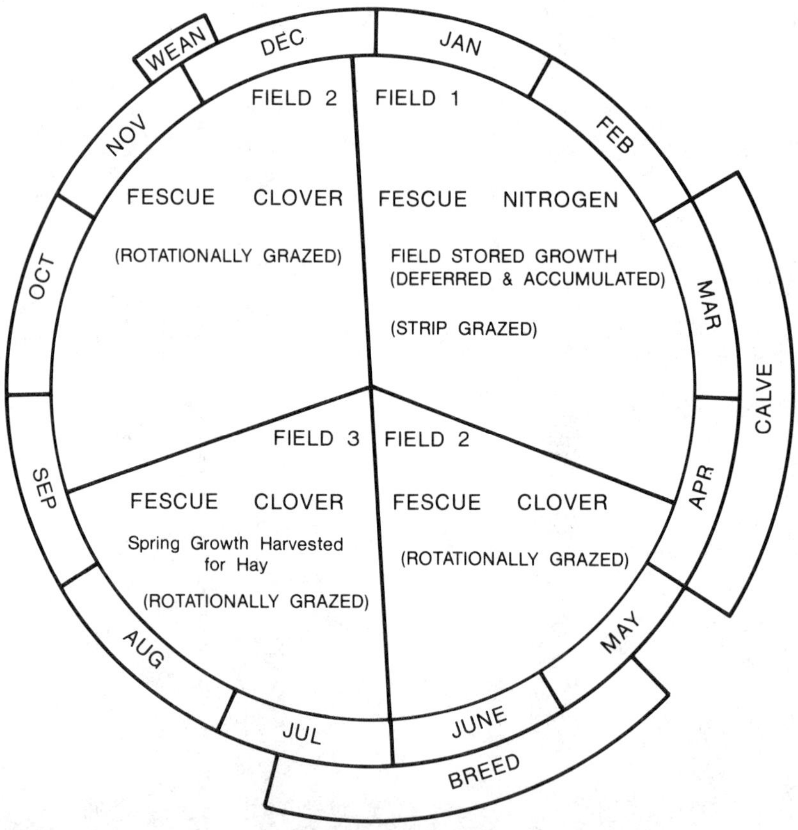

Figure 20.17. Cow forage management calendar-clock in Kentucky (adapted from Absher, 1979).

forage supply under midwest conditions. Figure 20.15 shows seasonal forage choices for grazing and hay. Figure 20.14 gives a worksheet where forage supply can be matched with animal requirements.

An example of year-around grazing in the Southeast is shown in Figures 20.16 and 20.17. Cattle that can be grazed on green forage most of the year will have excellent performance. Areas of the southern Great Plains and the Southeast where winter pasture from small grains can be produced will yield very economical cattle performance through the winter months (Figure 20.18).

SLAUGHTER CATTLE PRODUCTION ON GRAZED FORAGE

There was much discussion and considerable research during the 1970s involving the fattening of cattle on grass. Most of the grass fattening emphasis occurs when feed grain prices are very high, and large losses are experienced in commercial feedlots. Feed grain prices are variable depending on yearly production, exports and government programs (Figure 20.19 and 20.20).

A few individuals continue the rhetoric that feeding grain to cattle rather than humans is a moral issue. Feed grain prices will be the primary determining factor in how much beef is produced from grain feeding. Numerous beef industry personnel feel that grass fattening of beef may increase in future years due to increasing costs of cultural energy and a reduction in underground water. The latter would significantly reduce the amount of grain grown in several important cattle feeding states in the Great Plains area.

The large renewable forage base for cattle grazing assures people in the United States

Figure 20.18. Steer grazing on rye and rye grass pasture in northern Georgia. Courtesy of M. K. Cook, Univ. of Georgia.

Figure 20.19. Yearling heifers on a grain-on-grass system in Kentucky. Cattle graze pastures, then fed grain for 90 to 150 days to the high good-low Choice grade. Courtesy of Kentucky Agric. Expt. Station.

of a supply of beef even under adverse economic conditions. At present, about one-third of the nation's beef supply is accounted for by the commercial cattle feeding industry. Without a feeding industry, the cow herd would be reduced in size to accommodate the cattle being grown and finished on grass. Total annual tonnage of beef would be reduced and be more seasonal in supply. Palatability characteristics of grass-finished beef would be less uniform than the current preferences expressed towards grain-fed beef.

MAJOR HEALTH PROBLEMS ASSOCIATED WITH GRAZING PLANTS

Poisonous Plants

Poisonous plants are a major economical loss to the cattle industry since 3% to 5% of cattle are affected each year. These losses result from death, reduced gains, cattle

Figure 20.20. Calves can be self-fed grain while obtaining part of their nutrient supply from pasture. Courtesy of Elanco Products Company.

treatment costs, abortions, photosensitizations, and birth defects. Additional economic losses occur when management costs such as spraying, fencing, altered grazing programs, and forage loss occur in attempting to reduce cattle exposure to poisonous plants.

Hundreds of plants are poisonous to cattle. Some are toxic at all times, while others are toxic only under certain conditions. Some plants are quite palatable to cattle, but others will be consumed if they are the only plants available in any quantity. When cattle are not watered regularly or are allowed to become hungry, they are more likely to eat poisonous plants in lethal amounts.

There are very few known treatments for cattle poisoned by plants. Many times cattle are poisoned in remote range or pasture areas, and the effect of poison is so rapid that the animal is seriously affected before treatment can be applied.

Prevention of loss from poisonous plants is generally a matter of range, pasture, and cattle management. Producers should be aware of poisonous plants which appear in their areas and conditions under which they could be toxic. Also, producers should be aware of the animal symptoms associated with plant poisoning. Symptoms will vary depending on amount of poisonous plants ingested over a given period of time and variation in individual animal response.

Other preventive measures include developing grazing plans to prevent poisonous plant ingestion, not overgrazing, keeping cattle from undue stress including hunger and thirst, providing adequate salt and other necessary minerals, and controlling poisonous plants where physically and economically feasible.

If cattle show signs of being poisoned, a local veterinarian should be consulted for proper diagnosis, treatment, and prevention.

Table 20.4 lists some of the more common poisonous plants with brief information on distribution, control, poisoning symptoms, and treatment of poisoned cattle. Producers will need more extensive information when plant poisoning problems are encountered. An additional informational source is the USDA Poison Plant Research Laboratory, Logan, UT 84321.

Grass Tetany

Grass tetany occurs throughout the United States and may also be known as winter tetany, grass staggers, wheat poisoning, magnesium tetany, and hypomagnesemia. It is observed primarily in the spring, but it can occur in the fall and winter. Cows nursing calves under two months of age are more frequently affected.

Many times clinical symptoms are not observed, with dead animals the only history of the problem. Affected animals may become excitable, expressing a wild stare with erect ears, and appear to be blind. They are uncoordinated and tend to lean backward or stumble. An affected animal often has trembling muscles and grinding teeth followed by violent convulsions, deep coma, and death.

Positive diagnosis is difficult because the symptoms can easily be confused with other diseases. A blood test can be helpful as the serum of magnesium is typically low (normal is 2.25 mg per 100 ml of serum, whereas affected animals are usually below 1.0 mg).

Prevention of grass tetany may involve a combination of several factors. Magnesium mineral (magnesium oxide) should be provided. Legume or legume-grass pasture should be utilized since tetany seldom occurs where legumes are grazed. Less susceptible animals (heifers, stockers, dry cows, or cows with calves over 4 months) can be grazed on high-

Table 20.4. Selected Poisonous Plants[a] Which Have the Largest Economic Loss on Cattle

Plant	Distribution	Cause of Poisoning	Prevention or Control[b]	Sign of Poisoning	Effective Treatment
Arrowgrass	Widely distributed in marshy, alkaline pastures and native hay areas throughout the U.S.	Hydrocyanic acid	Prevent consumption—especially with plants retarded by drought or frost	Distress, cyanosis, rapid respiration, salivation, convulsions, rapid death	Limited; immediate treatment of solution (10 cc of 10% sodium thiosulfate and 10 cc of 10% sodium nitrate)
Astragulus (nitro-containing	Western Canada, Western United States and Northern Mexico	3-nitropropionic acid or 3-nitro-1-propanol	Use grazing programs which favor the more desirable forages; application of herbicides; prevent grazing	*Acute* Respiratory distress and muscular weakness, primarily in the pelvic limbs *Chronic* Labored and rapid respiration, wheezing or roaring sound, weakness in pelvic limbs, knuckling at fetlocks, goose stepping, and knocking together of the hind feet when walking	None
Bracken fern	Throughout United States		Prevent overgrazing	High fever; loss of appetite, depression, labored breathing, salivation, nasal, and rectal bleeding.	None
Larkspur (two groups—tall and low, based on height at maturity)	Primarily in the 11 Western States	Alkaloids	Grazing of infested areas later in grazing season. Apply herbicides (2, 4-D, 2, 4, 5-T, silvex, picloram, dicamba) if economical	Nervousness, staggering gait, salivation, bloat, sudden death	Limited; immediate injections of physostigmine salicylate can be helpful.

Table 20.4. (*Continued*)

Plant	Distribution	Cause of Poisoning	Prevention of Control[b]	Sign of Poisoning	Effective Treatment
Locoweed (several species)	Foothills, plains and semi-arid desert areas of the 17 Western States		Keep off infested areas until good feed is available. Apply herbicides if economical (2, 4-D, dicamba or picloram)	Abortion, birth defects, depression, nervousness, loss of sense of direction, poor feed consumption, and poor gains	None
Lupines	Foothills and mountain ranges in sagebrush and aspen areas (Most Western States)	Alkaloids	Birth defects prevented if consumption not permitted between 40 and 70 days of gestation. Apply herbicides if economical (2, 4-D or 2, 4, 5-T)	Crooked-legged calves and other birth defects. (Fig. 20.21). Nervousness, loss of muscular control, convulsions, coma, frothing at mouth	None
Milkvetch (Several species)	Throughout most of United States	Miserotoxin or nitro compounds	Prevent grazing as plants are palatable. Use herbicides on some species if economical (Silvex or 2, 4, 5-T)	*Acute* Tendency to stand on toes, weakness, irregular gait, posterior paralysis *Chronic* Nervousness, muscular weakness, respiratory distress	None
Milkweed (Several species)	Most of 17 Western States	Glucosidic substances called cardenolides	Adequate forage as plant is unpalatable. Herbicides available but usually not economical	Difficult breathing, (grunting), loss of muscular control, bloat, spasms	None
Nitrate-accumulating plants (examples—oat hay, sorghum, kochia, pig-	Throughout United States	Nitrate (lethal if plants contain more than 1.%; Sublethal at 0.5% to 1.5% plant levels). Ni-	Be cautious during drought as plants increase in nitrate content. Check nitrate content of	*Acute* Blue coloration of mucous membranes, staggering gait, shortness of breath,	Dilute high nitrate feeds with low nitrate feeds. None in acute cases

Table 20.4. (*Continued*)

Plant	Distribution	Cause of Poisoning	Prevention of Control[b]	Sign of Poisoning	Effective Treatment
weed, Russian thistle, and nightshade)		trates are converted to nitrites which produce methemoglobin, a type of hemoglobin which can't carry oxygen	plants if this type of poisoning is suspected.	death *Chronic* Watering eyes, unthriftiness, reduced milk flow and reduced gains	
Ponderosa pine needles	Most of 11 Western States		Needles are usually grazed only when cattle are stressed (weather, feed changes and hunger)	Abortion during the last trimester, retained placenta, calves may be weak if near term	None
Selenium-accumulating plants (sometimes called alkali disease)	Primarily in North and South Dakota, Montana, Wyoming, Colorado, and Utah	More than 5 parts per million of selenium in plants *Acute* Selenium accumulation greater than about 100 ppm *Chronic* Grasses and grains containing from 5 to 40 ppm of selenium	Prevent consumption	*Acute* Loss of appetite, depression, coma, death *Chronic* Dullness, rough hair coat, emaciation, lameness, possible hoof deformities	Remove from range containing plants with high selenium concentrations
Threadleaf snakeweed (Broomweed)	Drier range areas, primarily in the southwestern United States		Avoid grazing. Plants susceptible to 2, 4-D	Abortion, sometimes death. Weight loss and diarrhea	None
Waterhemlock	Throughout most of U.S.	Cicutoxin (a highly poisonous unsaturated alcohol that has a strong carrot-like odor)	Provide more palatable forage; prevent cattle from consuming roots brought to the surface by plowing or cleaning ditches; keep	Cattle die suddenly with few observed signs of poisoning	None

Table 20.4. (*Continued*)

Plant	Distribution	Cause of Poisoning	Prevention of Control[b]	Sign of Poisoning	Effective Treatment
			animals away (3 weeks) from plants sprayed with herbicide as palatability is increased; Apply herbicide (2, 4-D or 2, 4, 5-T)		

[a] Other plants which occasionally prove poisonous to cattle are bitter rubberweed, chokecherry, cocklebur, copperweed, desert baileya, death camas, drymary (inkweed), greasewood, groundsel (riddel and threadleaf) halogeton, hemp dogbane, horsetails, oak, pinque, poison hemlock, rayless goldenrod, St. Johnswort, sneezeweed, spring parsely, tansy mustard, tansy ragwort and, western false hellebore.

[b] The use of some herbicides listed may be banned in certain states. Directions should be followed carefully.

risk pastures. Some producers have fed a corn silage supplement while cows and calves were on wheat pasture which has greatly reduced the incidence of grass tetany. If soil magnesium is low, a program can be implemented to increase the level in the soil.

Early treatment of affected animals is important as cattle which have been down more than 12 to 24 hours will seldom recover. Animals should be handled gently to prevent excitement. Two hundred cubic centimeters (cc) of a sterile solution of magne-

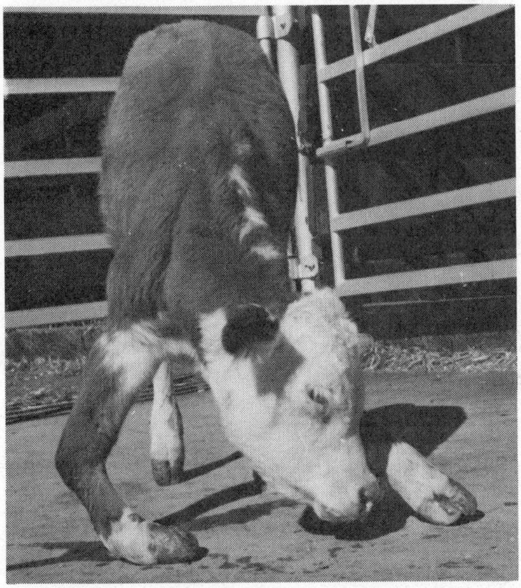

Figure 20.21. Crooked legs and other congenital deformities occur in newborn calves if cows graze toxic species of the lupine plant between 40 and 70 days after breeding. Cleft palate may also occur in this "crooked calf disease or syndrome." Courtesy of USDA Poisonous Plant Research Laboratory, Logan, Utah.

sium sulfate (epsom salts) injected under the skin, in at least four different sites, will give a high level of magnesium in the blood in 15 minutes.

Fescue Toxicity

Fescue toxicity is a serious problem with cattle grazing tall fescue throughout the tall fescue growing areas in the eastern United States when the grass is the principle diet of the animal. It is often incorrectly referred to as "summer slump" because of extreme symptoms during warm weather. Poor animal gains, reduced conception rates, intolerance to heat, failure to shed winter hair coat, elevated body temperature, and nervousness characterize this syndrome.

Fescue toxicity is associated with a fungus. The fungus lives within the leaves, stems, and seed of the tall fescue plant and is not visible externally. Tall fescue pasture, hay, and seed contain the fungus, although the fungus dies in about a year in stored seed. It is likely that the fungus causes the grass to produce some toxic compound, as yet unidentified, which is very harmful to animals eating the forage. Surveys made in several southeastern states show over 95% of tall fescue pastures are heavily infected with the fungus.

Average daily gain of steers on tall fescue pastures has been poor, usually about 1 lb per day or less for the grazing season.

The introduction of a fungus-free fescue has increased gains dramatically (Figure 20.22). A four-year grazing study with steers has shown significantly higher gains (1.82 lbs per day versus 1.09 lbs per day) where steers grazed fungus-free fescue. Another study with cows and calves showed average daily calf gains of 2.65 lb for non-infected fescue and 1.92 lb gains for calves on fungus-infected grass. Average daily weight changes of cows were 1.0 lb on non-infected fescue and −0.40 lb on fungus-infected fescue. The average daily gain of steers fed fungus-free hay was 1.45 lb but only 0.63 lb when fed fungus-infected hay.

Figure 20.22. Yearling heifers grazing on low endophyte tall fescue in Kentucky. The fungus-free fescue shows great promise of reducing the problem of poor animal performance in hot weather. Courtesy of Kentucky Agric. Expt. Sta.

Fescue toxicity effects can be reduced by one or more of the following methods:

1. Growing a legume with tall fescue. Planting ladino or red clover in fescue sod is cheap and practical.
2. Feeding hay other than tall fescue in winter when cattle are grazing toxic fescue pasture. Bermudagrass or sericea lespedeza hay are good choices.
3. Avoid heavy application of broiler litter or nitrogen fertilizer on tall fescue when it is the sole source of pasture and hay.
4. Replanting infected fescue pastures with fungus-free seed. Old seed, where the fungus has died, can be planted, but the seed rate must be increased to make up for reduced germination. Fescue pastures should not be allowed to make seed during the same year it is to be destroyed and replanted with fungus-free seed.
5. Plant seed of the new improved winter-productive Triumph variety.

Sweet Clover Poisoning

Sweet clover toxicosis occurs when spoiled or moldy sweet clover is fed. It occurs primarily with hay and silage and less frequently on pasture. Coumarin (substance responsible for the characteristic odor of sweet clover) is converted to dicoumarin which interferes with the synthesis and metabolism of vitamin K. This toxic compound prevents normal blood clotting resulting in hemorrhages and associated symptoms.

Preliminary symptoms include stiffness, lameness, and swellings (blood clots) beneath the skin and over most of the body. Fetal reabsorption, stillbirths, and neonatal deaths may also occur.

Prevention alternatives include feeding moldy sweet clover with caution. Not all moldy sweet clover is toxic, and the absence of observable mold is not absolute assurance that dicoumarin is not present in the forage. Suspect sweet clover can be fed for 7 to 11 days, then replaced with alfalfa for an equal period of time. Do not feed sweet clover for 2 weeks prior to calving or after calving.

Other preventive measures include stacking and baling of sweet clover only when it is well cured and dry and avoiding the use of large, tightly bound bales.

Direct blood transfusions, using blood from cattle not consuming sweet clover, can be helpful to cattle suspected of poisoning. Intramuscular injections of vitamin K will counteract the effects of dicoumarin. Poisoned animals should be removed from sweet clover and fed alfalfa which is high in vitamin K and calcium.

Veterinary assistance should be obtained since the symptoms of sweet clover poisoning and anaplasmosis are quite similar.

SELECTED REFERENCES

Publications

Absher, C. W., Murdock, L. W., and Kaiser, C. J., 1979. *A Beef Forage System.* Kentucky Agric. Ext. Service ID-5.

Cook, C. W., et al, 1983. *Alternate Grass and Grain Feeding Systems for Beef Production.* Colo. State Univ. Expt. Sta. Bul. 579S (Revised).

Heitschmidt, R. and Walker, J., 1983. *Short duration grazing and the Savory Grazing Method in perspective.* Rangelands 5:147.

James, L. F., Keeler, R. F., Johnson, A. E., Williams, M. C., Cronin, E. H., and Olsen, J. D., 1980. *Plants Poisonous to Livestock in the Western States.* USDA Agric. Info. Bull. 415.

Launchbaugh, J. L. and Owensby, C. E., 1978. *Kansas Rangelands. Their Management Based on a Half Century of Research.* Kansas Agric. Expt. Sta. Bull. 622.

Madsen, A. G. and Gee, C. K., 1978. *Sources and Costs of Range Forage in Colorado.* Colorado Agric. Expt. Sta. Gen. Series 976.

Opportunities to Increase Red Meat Production from Ranges of the United States. Phase I—Non-Research. 1974. USDA.

Reid, R. L. and Klopfenstein, T. J., 1983. *Forages and Crop Residues: Quality Evaluation and Systems of Utilization.* J. Anim. Sci. 57:534 (Suppl. 2).

Savory, A., 1983. *The Savory Grazing Method or holistic resource management.* Rangelands 5:155.

Stuedemann, J. A. (Ed.), 1977. *Forage Fed Beef: Production and Marketing Alternatives in the South.* Southern Coop. Series Bull. 220.

Triplett, C. M., 1982. *Forage Systems for Backgrounding and Finishing Cattle in Georgia.* Georgia Coop. Ext. Serv. Bull. 753.

Visuals

"Grass tetany—clinical signs in a natural environment." 16 mm film or videotape. Southern Piedmont Conservation Research Center, USDA, ARS, P.O. Box 555, Watkinsville, GA 30677.

21 Facilities and Equipment for the Cattle Producer

Adequate facilities and equipment are necessary to handle and feed cattle for optimum performance. Stresses can be minimized for both cattle and producers. Understanding cattle behavior (Chapter 16) is most useful in designing facilities and equipment.

Tables 21.1 and 21.2 give space requirements and general facilities guidelines needed by beef cattle.

CATTLE RESTRAINT

Restraining and controlling cattle are necessary for many production and management practices such as health care, identification, showing, marketing, weighing, breeding, pregnancy checking, and implanting growth stimulants.

Cattle restraining facilities should be constructed to provide safety for those working the cattle and to prevent unnecessary excitement or stress on the cattle. Cattle behavior should be understood (see Chapter 16) so that working facilities and equipment conform to natural instincts of cattle. Forcing cattle or attempting to out-muscle them usually results in less productive cattle which develop poor dispositions.

Working Chutes

Working chutes can be used effectively as the only restraint method for several management practices such as vaccinations, spraying, and applying pour-on insecticides while the animals are crowded head-to-tail in the chute. For more rigid confinement, a squeeze chute can be placed at the end of the working chute.

Many commercial and homemade designs are available, but the best of them share the following characteristics: (1) they have V-shaped sides or an adjustable side (18 to 30 inches) so that large or small cattle can be handled, (2) the "V" measures 14 to 17 inches wide at the bottom and remains so for the first 2 feet from the bottom, at which

Table 21.1. Cattle and Facility Space Requirements

Corral and Pen Dimensions			
	To 600 Lbs	600 Lbs to 1200 Lbs	1200 Lbs or Cow–Calf
Holding area (sq ft per head)	14	17	20
Crowding pen (sq ft per head)	6	10	12
Working chute with vertical sides			
Width (in.)	18	22	26
Desirable min. length (ft)	20	20	20
Working chute with sloping sides			
Width at bottom inside clear (in.)	15	15	16
Width at top inside clear (in.)	20	24	26
Desirable min. length (ft)	18	18	18
Working chute fence			
Height solid wall (in.)	45	50	50
Depth of posts in ground (in.)	36	36	36
Overall height (7 ft minimum clearance below cross ties to walk under)			
Top rail, medium-sized, docile cattle (in.)	55	60	60
Top rail, large-sized or wild cattle (in.)	60	66	72
Corral fence			
Recommended height (in.)	60	60	60
Depth of posts in ground (in.)	30	30	30
Loading chute			
Width (in.)	26	26	26–30
Length (in.)	12	12	12
Rise (in. per ft)	3	3	3
Ramp height for gooseneck trailer (in.)	15		
Pickup truck (in.)	28		
Van type truck (in.)	40		
Tractor trailer (in.)	48		
Double deck (in.)	100		

point it flares to approximately 24–30 inches at hip and head height. Straight-sided permanent chutes measure 26 to 28 inches in width, (3) the sides should be solid so the cattle cannot see through and become distracted, (4) the crowding pen and working chute should be curved so that cattle do not sense and resist the tunnel effect, and (5) the working chute exit gate-squeeze chute entry gate is usually made of bars (not solid) so as to give the illusion to the cattle of being able to escape and thereby cause them to enter.

When constructing the working chute, build an access gate into it, adjacent to the

Table 21.2. Space Requirements and General Facility Guidelines for Feedlot Cattle

	FEEDLOT, sq ft per head
20 in barn and 30 in lot	Lot surfaced, cattle have free access to shelter
50	Lot surfaced, no shelter
150–800	Lot unsurfaced except around waterers, along bunks and open-front buildings and a connecting strip between them
20–25	Sunshade
	BUILDINGS WITH FEEDLOTS, sq ft per head
20–25	600 lb to market
15–20	Calves to 600 lb
½ ton per head	Bedding
	COLD CONFINEMENT BUILDINGS, sq ft per head
30	Solid floor, bedded
17–18	Solid floor, flushing flume
17–18	Totally or partly slotted
100	Calving pen
1 pen 12 cows	Calving space
	FEEDERS, in. per head along feeder
All animals eat at once:	
18–22	Calves to 600 lb
22–26	600 lb to market
26–30	Mature cows
14–18	Calves
Feed always available:	
4–6	Hay or silage
3–4	Grain or supplement
6	Grain or silage
1 space per 5 calves	Creep or supplement
	WATER REQUIREMENTS
Cattle per 1000 lbs	29 gallons per day @ 50 °F
	18 gallons per day @ 90 °F
Automatic waterers for cattle:	
20–40 head	½ gpm minimum
	2 gpm preferred
	WATERERS
40 head per available water space in drylot	
	ISOLATION AND SICK PENS
40–50 sq ft per head	
Pens for 2%–5% of herd	
	MOUNDS
25 sq ft per head	Minimum
50 sq ft per head	If windbreak on top of mound; 25 sq ft per head each side
	SLOPES
Floors, Pavements	
½–¼ in. per ft	Bunk aprons with step
1 in. per ft	Bunk aprons without step

Table 21.2. (*Continued*)

¼–1 in. per ft	Solid floors toward slats, flumes
½% or more	Longitudinal bottoms of gutters, flumes
1%	Gravity pipe to lagoon
Earth	
4:1 to 5:1	Mound sideslopes
5% max	Mound longitudinal
4%–6%	Lots
DAILY MANURE PRODUCTION PER 100 LB LIVEWEIGHT	
60 lb, 1 cu ft	Feces & urine
6.91 lb, 0.12 cu ft	Solids

squeeze chute end, so that the working chute can be entered to perform palpations and castrations.

The step-by-step procedure for moving cattle into and through the working chutes is as follows:

1. Move the cattle to the working alley and crowding pen that funnels into the working chute.
2. Start the cattle into the working chute. It will be more efficient to get one animal started down the chute instead of trying at random to drive all of them at once. If the cattle will simply not cooperate, it may be necessary to enter the crowding pen and force the cattle with the sparing use of whip, cane, or tail-twist.

   With several large animals in a relatively small area, the producer has risk of being squeezed, stepped on, or kicked. Avoid using the electric prod or "hot shot." It will make cattle nervous and less cooperative, increasing the risk of injury not only while cattle are being worked this time but also when they are reworked later.
3. Keep the cattle moving down the working chute by walking along the outside and talking to them, slapping them on the rump, and tail-twisting as necessary.
4. Place a bar completely across the working chute behind the last animal and also behind the first animal in the chute. This should be placed just above the hock and will prevent the animal from backing up.
5. Open the gate to the squeeze chute and allow the first animal in the working chute to enter (Battaglia and Mayrose 1981).

Squeeze Chute and Headgate

The combination of squeeze chute and headgate can be used to advantage on any type of farm or ranch for such management techniques as dehorning, castrating, branding, implanting, ear tagging, stomach tubing, artificial insemination, and blood testing. It is made even more valuable when it is positioned at the terminal end of a working chute.

There are many commercial and homemade designs available. A workable combination should consist of the following: a squeeze mechanism, a headgate with head and nose bars, a tail gate, removable solid side panels measuring approximately 24 inches

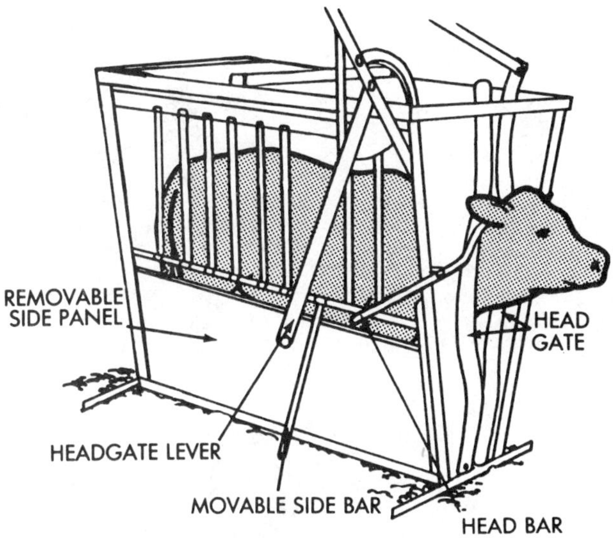

Figure 21.1. The component parts of a typical squeeze chute and headgate. Courtesy of Battaglia, R. A. and Mayrose, V. B., 1981. *Handbook of Livestock Management Techniques.* Minneapolis: Burgess Publishing Co., copyright ©.

from the ground, and removable side bars for easy access to the animal's side (Figure 21.1).

There are three basic headgate designs used in beef operations: straight-bar headgate, positive-type headgate, and curved-bar headgate. The straight-bar headgate generally is designed to automatically catch an animal as it walks through the chute. The greatest advantage of the straight-bar headgate is its protection against choking the animal. The main disadvantage is that the animal can move its head up and down easily, which can create some problems in certain techniques that require head immobilization.

The positive-type headgate operates something like a guillotine. The main advantage of this headgate is almost complete head control, both sideways and up and down. The headgate almost completely immobilizes the head without the necessity of a head and nose bar. This type of headgate is popular among feedlot operators.

It is absolutely necessary that the headgate be adjusted to the correct height of the animal being worked to prevent choking. In addition, it is mandatory that the animal's head be released from the headgate before the gate is opened for the animal to exit. If this is not done, the animal can lunge forward and break its neck or choke to death.

Regardless of the headgate design used, it is important that the headgate be adjusted for the size of animals being worked to prevent choking or escape.

The curved-bar headgate combines most of the safety of the straight-bar headgate and some of the up and down head restraint of the positive-type headgate. This headgate is an excellent compromise for the cow–calf feedlot operator.

Rope Halters

Rope halters are considered a necessity on many farms and ranches where cattle are kept. Halters are used to teach cattle to lead and are also used where the head

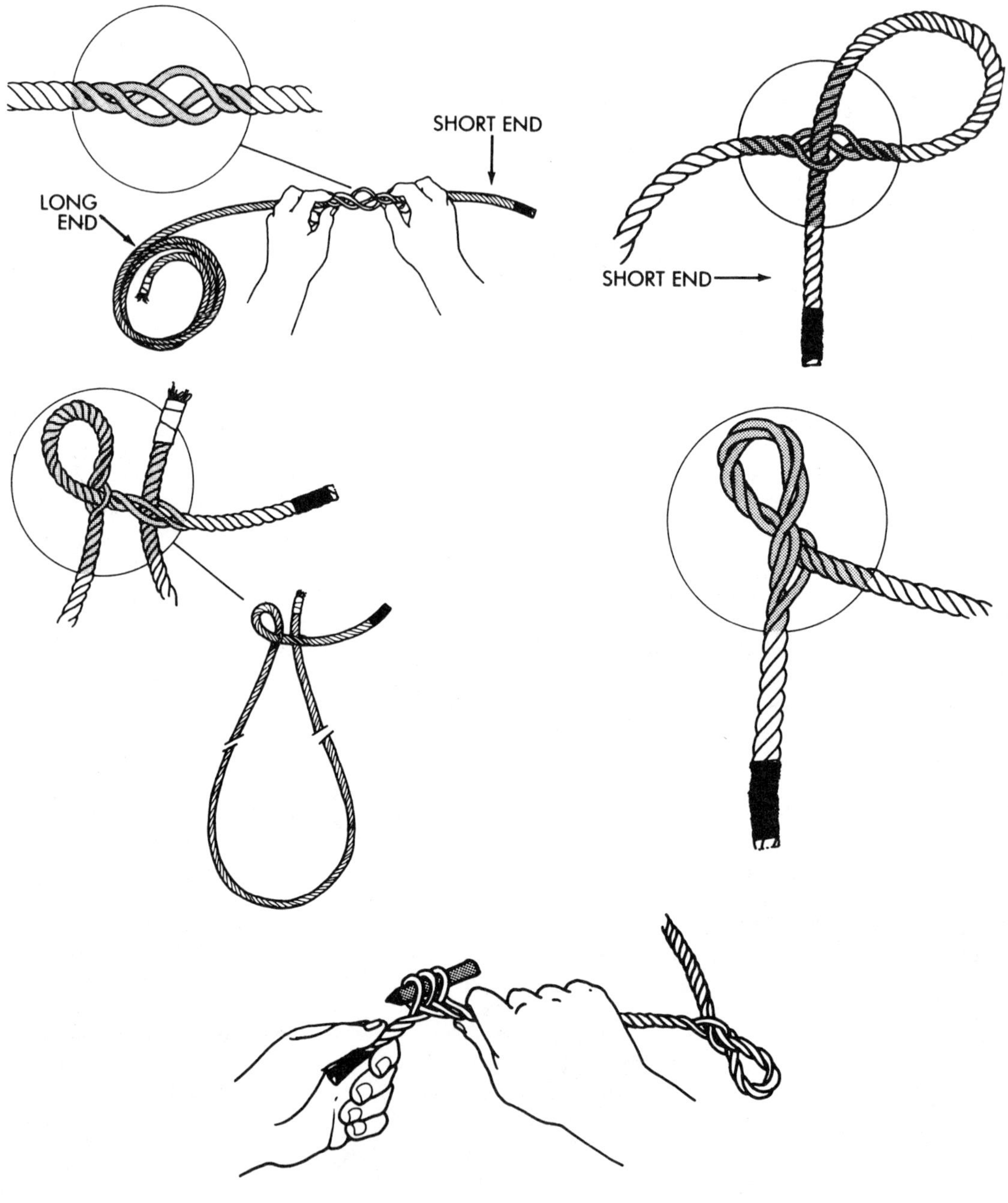

Figure 21.2. Diagrams and description for making an adjustable rope halter. Description and diagrams courtesy of Battaglia, R. A. and Mayrose, V. B., 1981. *Handbook of Livestock Management Techniques.* Minneapolis: Burgess Publishing Co., copyright ©.

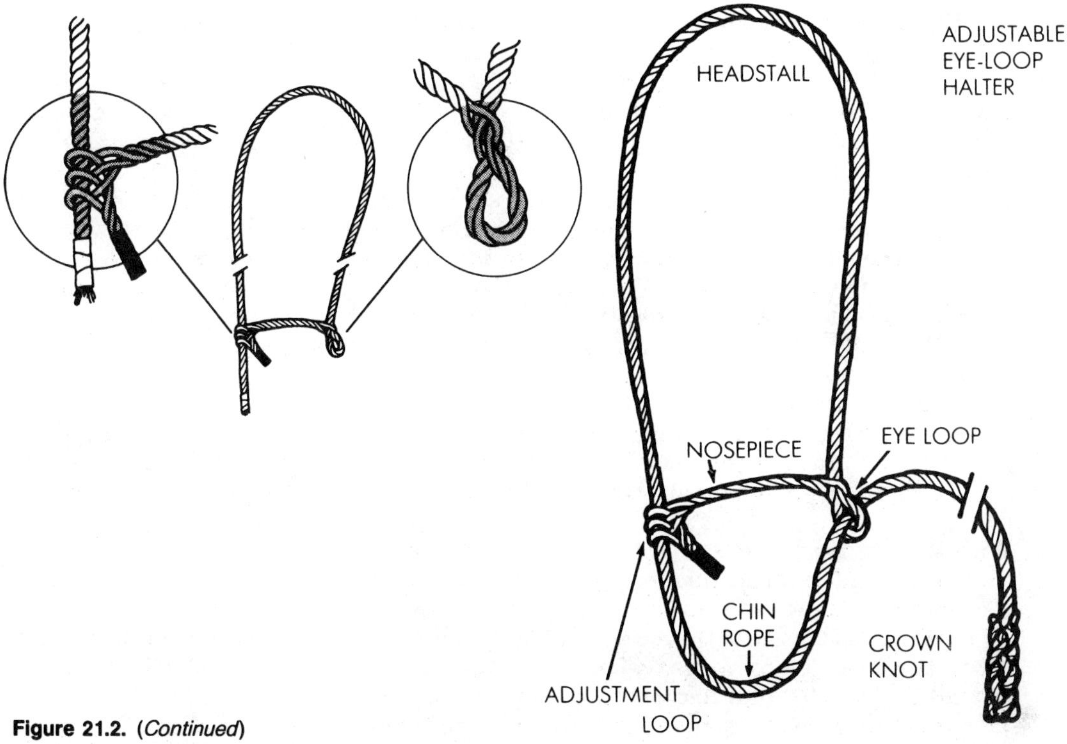

Figure 21.2. (*Continued*)

needs to be tightly restrained. Figure 21.2 shows the construction of a low-cost, adjustable rope halter.

Description

1. Select a 12′ to 15′ length of ½″ three-strand rope. Any of the rope types from cotton through nylon will work. The choice will depend upon the strength needed, durability required and cost.
2. Finish one end of the rope by whipping, clamping, dipping or heat-treating. (The method selected will depend upon your wishes and the type of rope selected.) If you do not finish the rope ends, they will unlay and fray. Temporarily finish the other end of the rope with tape or string. A crown knot may be formed into this end after the halter is constructed.
3. Mark a point with your hand about 12″ to 15″ from the whipped end of the rope. Refer to this 12″ to 15″ length as the *short end* of the rope—the remaining length is the *long end.*
4. Place the short end to your right, the long end to your left. Grasp the rope at the 12″ to 15″ mark between the thumb and first two fingers of both hands. Separate your right and left hands by about 2″. Rotate the rope clockwise with your right

hand and counterclockwise with your left. This will open the strands of the rope between your hands.

5. Isolate any one of the opened strands with the thumb and index fingers of your left hand. Use your right hand to insert the whipped end of the short end of the rope under this strand opening until the loop formed has an eye opening of about double the rope diameter.
6. Now position the rope so that the eye loop is in your left hand with the short end pointing toward 3 o'clock and long end exiting toward 6 o'clock. Grasp the eye loop and the single strand running across the short end of the rope between your left thumb and index finger. With your right thumb and index finger, grasp the short end of the rope at a point near the eye loop. Twist the eye loop and short end of the rope with your hands until you have isolated two strands between your right thumb and index finger.
7. Use your left hand to insert the long end of the rope from bottom to top under and through these two strands. Pull it completely through until all the slack is gone. If done properly, one side of the loop will show three strands lying smoothly side by side. This is important because they will be positioned against the animal's face.
8. With the eye loop to your right, grasp the short end of the rope between your left thumb and index finger about 2″ from the whipped end. Two inches farther from the whipped end, grasp it in the same manner in your right hand. Open the strands by twisting clockwise with your right and counterclockwise with your left hand. When the strands are opened wide, push your hands together. This will cause the strands to "buckle" and fold over, forming three loops.
9. Line these three loops up in order and work into them a sharpened stick of diameter equal to that of the rope. Use your right hand to feed the long end of the rope into the loops, starting at the one closest to the eye loop. Remove the stick from one loop at a time as you run the long end of the halter through them.
10. Run the long end of the rope into and through the eye of the loop. This completes the halter.
11. Permanently finish the long end of the rope in the chosen manner. Consider crowning the end because a crown knot and back splice creates a convenient handle.
12. Always place the halter on the animal so that the eye loop is on the left side. Lead from the left.

Vertical Tilt Tables

In many cow–calf operations, calf tables are used for restraining the calf during the branding, vaccination, castration, and dehorning process (Figure 21.3). The calves are relatively small being only a few weeks to a few months of age. Some cow–calf operations continue to restrain the calves by roping the hindlegs of the calf from a horse. Another person "tails" or "flanks" the calf to its side and sits on the calf while others dehorn, castrate, vaccinate, and brand the calf.

Another type of vertical tilt equipment is used for hoof trimming (Figure 21.4). Hooves can also be trimmed in squeeze chutes and in upright non-tilting chutes especially constructed for hoof trimming.

Figure 21.3. Dehorning a calf in a calf table. Courtesy of Powder River Enterprises.

Grooming Chute

The grooming chute is used only when cattle are gentle and trained to lead. It is not constructed to replace the squeeze chute or foot-trimming chute for restraint. A grooming chute is used to clip and prepare an animal for show or sale. Many commercial and homemade designs are available for use. The best of them have a metal frame made of pipe or square tubing, rough wood floor, simple headgate mechanism, and swingable or removable side bars that allow easy access to the side of the animal (Figure 21.5).

Casting

Casting refers to causing the animal to lie down on its side due to pressure exerted on its muscles and nerves by a series of carefully placed and tightened ropes. This

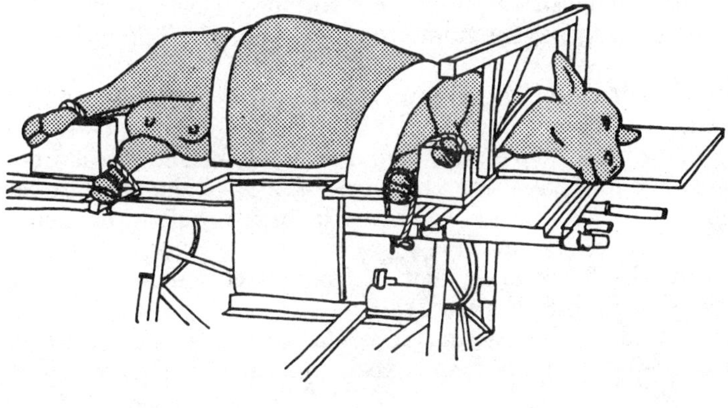

Figure 21.4. Animal restrained on tilting hoof-trimming table. Note feet are tied to minimize movement during the trimming process. Courtesy of Battaglia, R. A. and Mayrose, V. B., 1981. *Handbook of Livestock Management Techniques.* Minneapolis: Burgess Publishing Co., copyright ©.

Figure 21.5. Diagram of a grooming chute. Courtesy of Battaglia, R. A. and Mayrose, V. B., 1981. *Handbook of Livestock Management Techniques.* Minneapolis: Burgess Publishing Co., copyright ©.

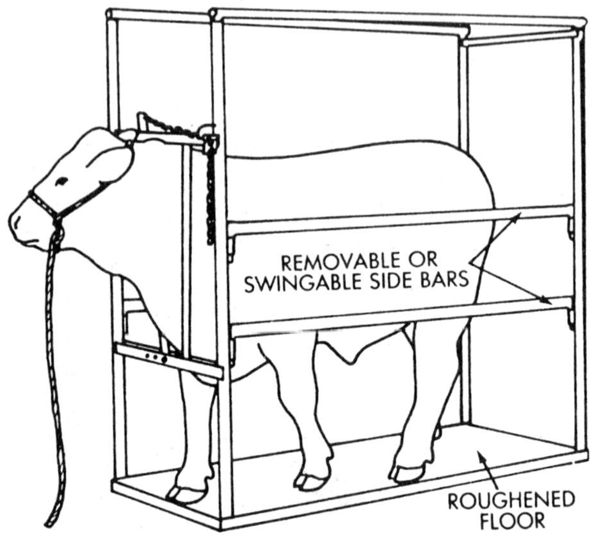

restraint should be used sparingly and only when near-total immobilization is required and no chutes or tables are available. It can cause pregnancy complications and displaced abomasums from the twisting, or bloat and pneumonia from having the animal lie on its side too long.

The casting method to follow is only one of several, but it is the standard method for animals of any size. To minimize the risk of injuring the penis of a bull, wrap the rope with a soft material such as a cotton bag. Before any animal is cast, the lay-down area should be checked for rocks, dirt clods, and any other debris that could cause cuts or bruises.

The following is a step by step procedure for the half-hitch casting method.

1. Halter the animal and lead it to the lay-down area. Secure the rope to a post or ring approximately 18″ from the ground.
2. Tie the end of a 35- to 40-foot, ⅝- to ¾-inch, 3-strand, loose-weave cotton rope loosely around the animal's neck, using a bowline knot to form the loop. Position the knot approximately 6 inches below the ridge of the spine to one side or the other (Figure 21.6).
3. Place a wrap of rope immediately behind the animal's forelegs. Form this wrap by dropping the whole coil of the rope over the animal's back to the side opposite the bowline knot. Reach under the animal (carefully, to avoid being kicked), grasp the rope, and bring it out from under its body. Run the working end of the rope under the standing part connected to the bowline knot. This will form a half hitch about 12 inches behind the bowline knot (Figure 21.7).
4. Place a second wrap of rope in the flank area, just in front of the udder if it is a cow, by repeating the processes in Step 3 (Figure 21.8).
5. Begin pulling on the working end of the casting rope. As pressure is exerted on the nerves and muscles, the animal will lie down. Keep pressure on the rope and

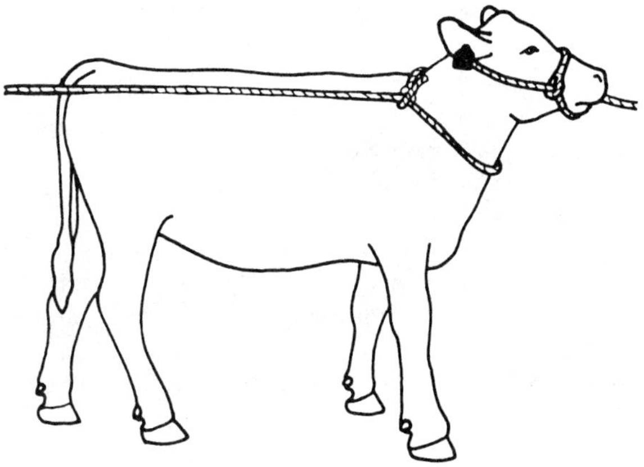

Figure 21.6. Courtesy of Battaglia, R. A. and Mayrose, V. B., 1981. *Handbook of Livestock Management Techniques.* Minneapolis: Burgess Publishing Co., copyright ©.

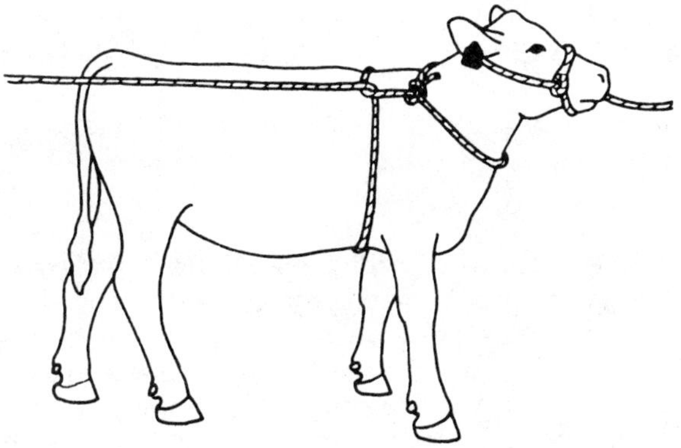

Figure 21.7. Courtesy of Battaglia, R. A. and Mayrose, V. B., 1981. *Handbook of Livestock Management Techniques.* Minneapolis: Burgess Publishing Co., copyright ©.

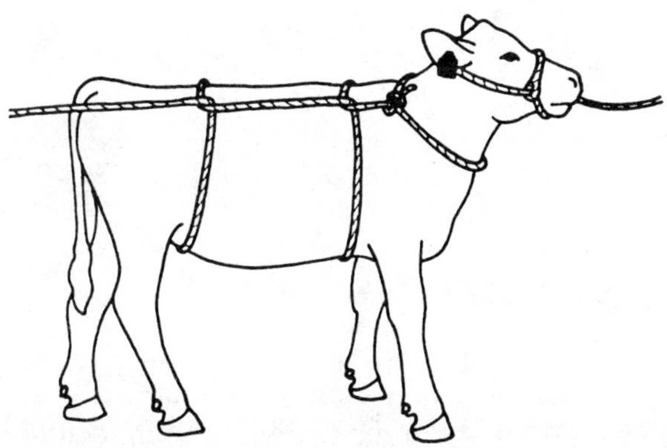

Figure 21.8. Courtesy of Battaglia, R. A. and Mayrose, V. B., 1981. *Handbook of Livestock Management Techniques.* Minneapolis: Burgess Publishing Co., copyright ©.

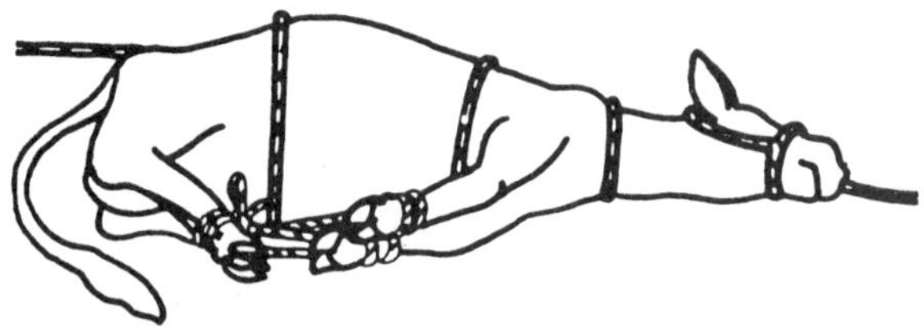

Figure 21.9. Courtesy of Battaglia, R. A. and Mayrose, V. B., 1981. *Handbook of Livestock Management Techniques.* Minneapolis: Burgess Publishing Co., copyright ©.

have an assistant tie the animal's legs before beginning the management practice. A constant, steady pull should be maintained on the working end of the casting rope (Figure 21.9).

6. After the management technique has been completed, release the legs, loosen the casting rope, and allow the animal to regain its feet (Battaglia and Mayrose 1981).

Cow–Calf Facilities

Figure 21.10 shows some simple handling facilities for small cow herds where cattle need to be handled for doctoring, marketing, weighing, and other management activities. Corral designs for small and large cow herds are shown in Figure 21.11. Cattle will move easier through facilities where there is a crowding area and the chutes are curved with solid sides.

A combination calving barn and working facility is shown in Figure 12.12.

Feedlot Facilities

Figure 21.13 demonstrates a combination of several items which should be considered in the facilities for a commercial feedlot operation. Mounds to keep cattle dry and out of the mud, location of feed bunks and waterers, sick pens, feed alleys, and working alleys for cattle are all important considerations. Well-drained pens with waste control are necessary for feedlots to meet government pollution regulations. Feed storage and processing, cattle loading and unloading facilities, and cattle processing facilities are necessary ingredients for a cattle feeding operation. It is important that many of these processes operate simultaneously in using labor and equipment efficiently and where one activity does not interfere with another. A selected working area design is shown in Figure 21.14.

Helpful Hints for Handling Cattle

Cattle can be handled most effectively by understanding cattle behavior, adapting human behavior to cattle behavior and constructing handling facilities which utilize cattle psychology in a positive manner.

Some of the most important items are:

1. Cattle have panoramic vision. They can see all around themselves without turning their heads.
2. Cattle tend to move toward light. Loading cattle at night is enhanced with a non-glaring light near truck or trailer gate. Cattle will move easier into a dark area by single file rather than trying to force a group into a dark area. Loading chutes

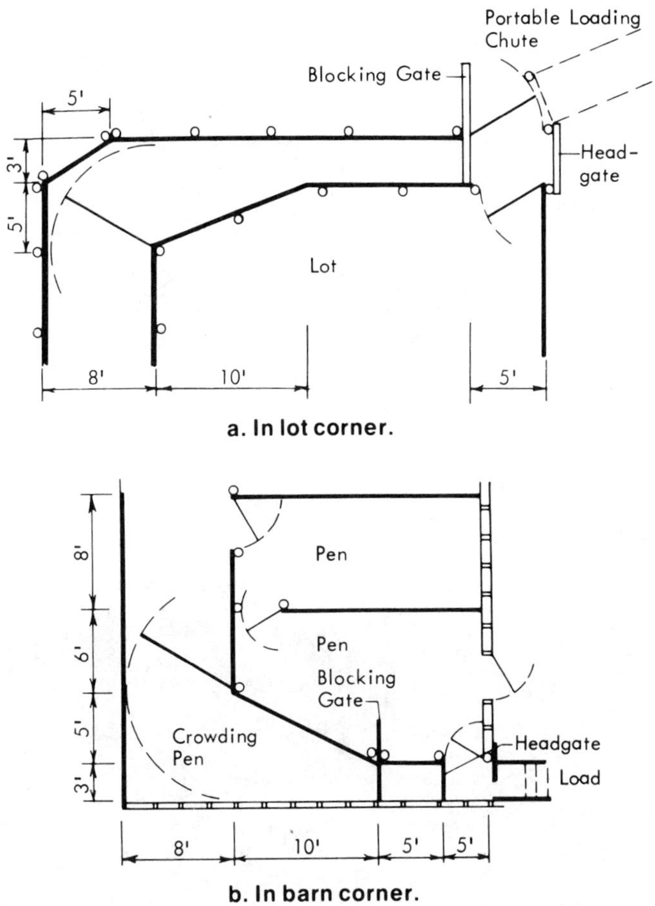

Figure 21.10. Simple layouts of handling facilities for small herds. Courtesy of Beef Housing and Equipment Handbook, MWPS-6, 3rd Ed., 1975 ©. Midwest Plan Service, Ames, Iowa 50011.

and squeeze chutes should face north and south as cattle do not move easily into direct sunlight.

3. Shadows across chutes and scales should be prevented as cattle are fearful of shadows.
4. Hot-shots, whips, and prods should be used sparingly, especially in cow–calf operations. Poor dispositions are developed by cows and calves handled in an abusive manner.
5. Cattle usually respond negatively to abuse, loud noises, and other confusion which might excite them. Noise-producing equipment should be kept as far from cattle as possible. Metal chutes and alleys should be constructed to prevent excessive clanging and banging noises.
6. Cattle are creatures of habit, thus by using similar repetitive patterns, they can be handled quite easily.
7. A single animal does not handle easily. It will resist going into a chute or pen by itself. The animal will usually become excited and injure itself, the handler, or damage the facilities.

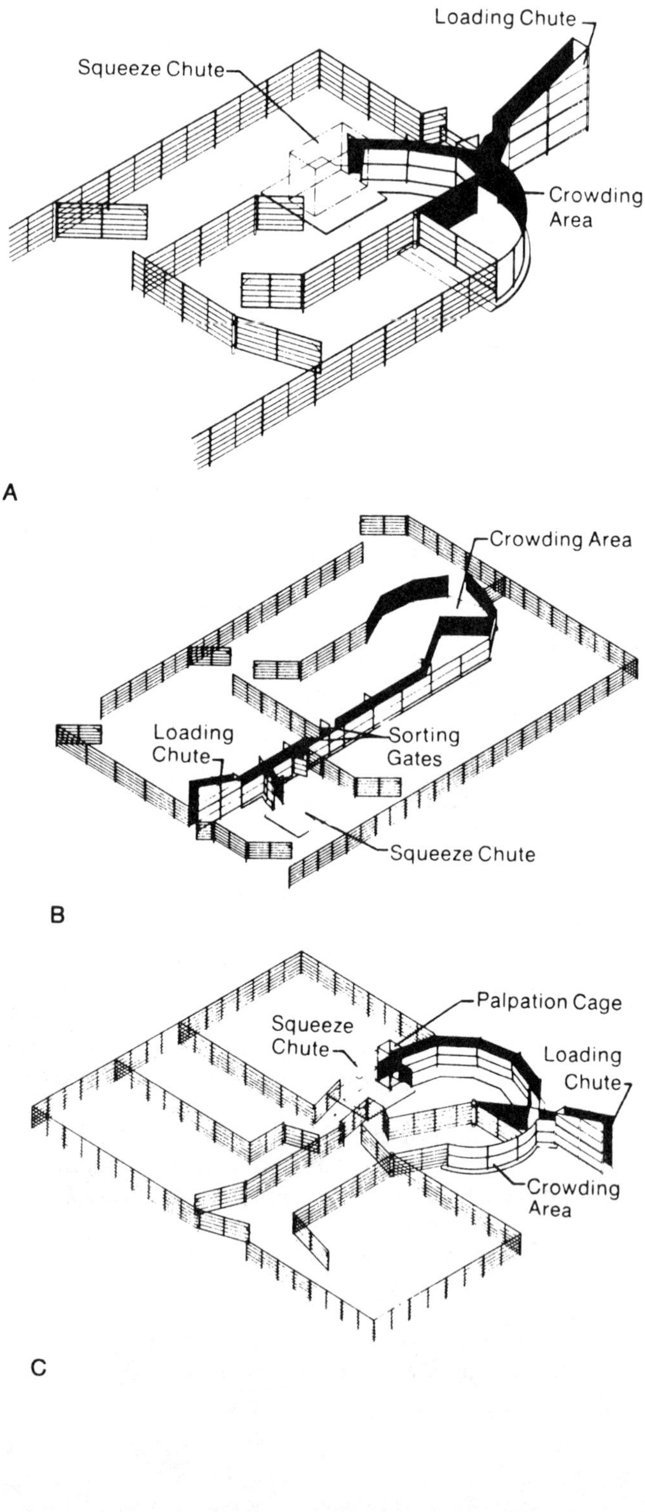

Figure 21.11. Several corral designs. A. 20 to 75 head capacity. B. Up to 400 head capacity. C. 70 to 400 head or more capacity. *Source:* USDA.

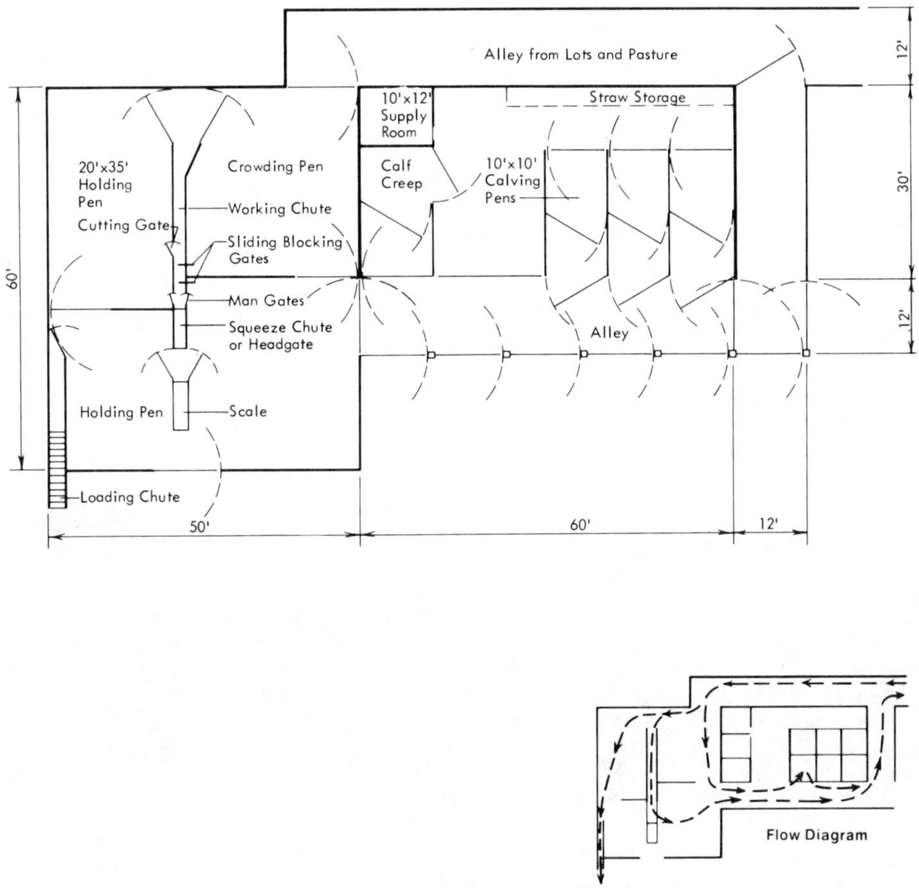

Figure 21.12. Calving barn, corral and working facility for cow–calf producer. Courtesy of *Beef Housing and Equipment Handbook, MWPS-6,* 3rd Ed., 1975 ©. Midwest Plan Service, Ames, Iowa 50011.

8. Sides of crowding pen, loading chute, and single-file chute should be solid. This prevents outside distractions to the cattle, and they will move much easier through the facilities.
9. Curved chutes are preferred over straight chutes because the cattle cannot see what is at the end of the chute until they are almost there. Also, a curved chute utilizes the natural tendency of cattle to circle around the handler. The catwalk should be inside the curved chute which puts the handler at the best working angle with the cattle.
10. The cattle handler's movements should be slow and deliberate because sudden movement will frighten the cattle and make them difficult to handle.

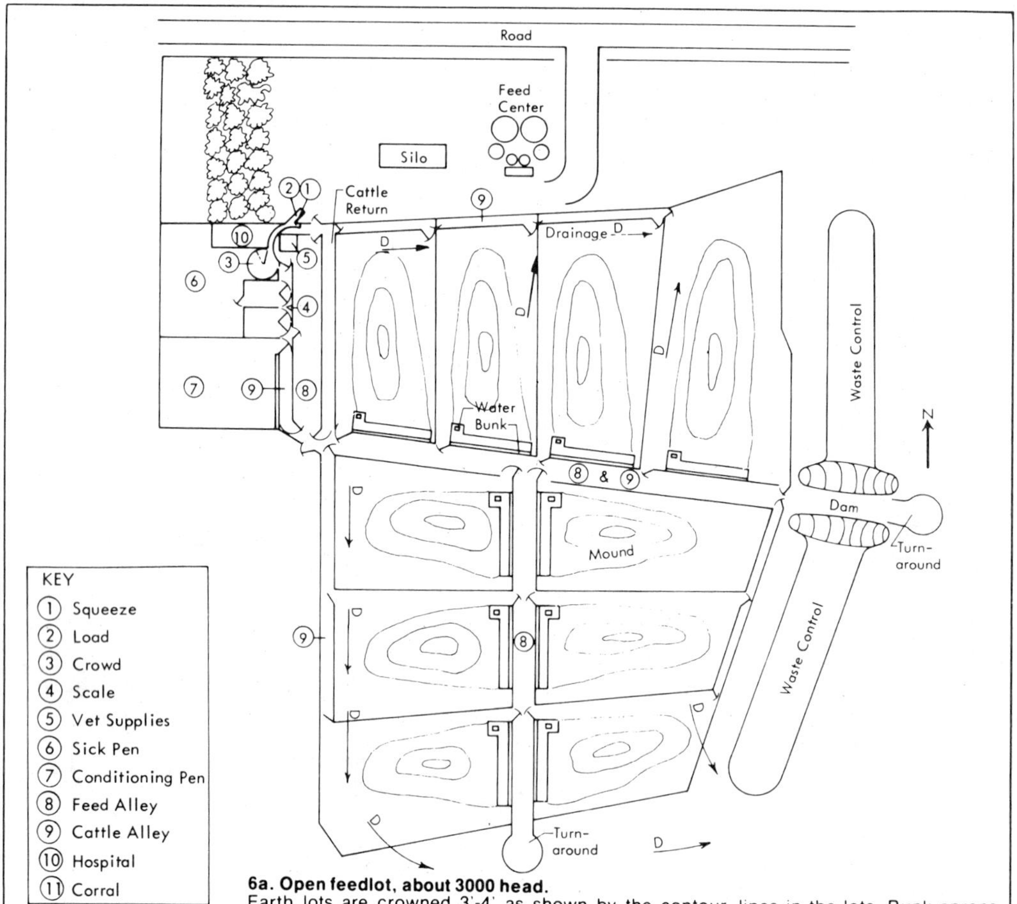

6a. Open feedlot, about 3000 head.
Earth lots are crowned 3'-4' as shown by the contour lines in the lots. Bunk aprons are paved. Handling and hospital facilities are combined.

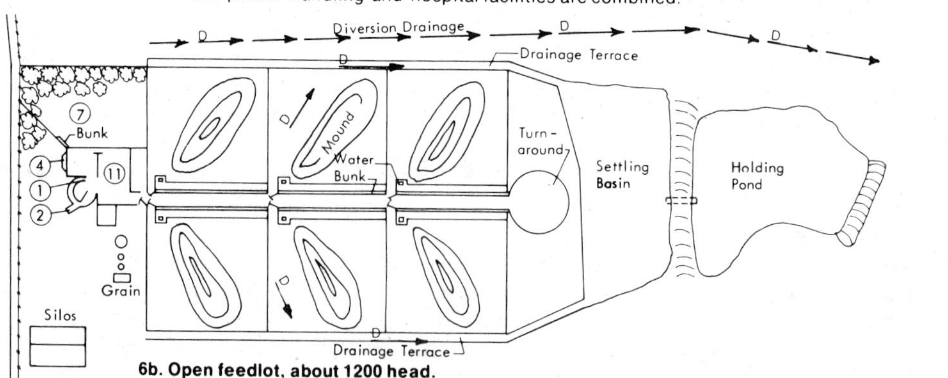

6b. Open feedlot, about 1200 head.
This lot is on a very gentle slope; mounds are only about 3' high. Because drainage is poor at the back of the lots, cattle are handled in the feed alleys. Each lot is 250' square.

Figure 21.13. Open feedlot layout for a 3000 head capacity feedlot. Courtesy of *Beef Housing and Equipment Handbook, MWPS-6,* 3rd Ed., 1975 ©. Midwest Plan Service, Ames, Iowa 50011.

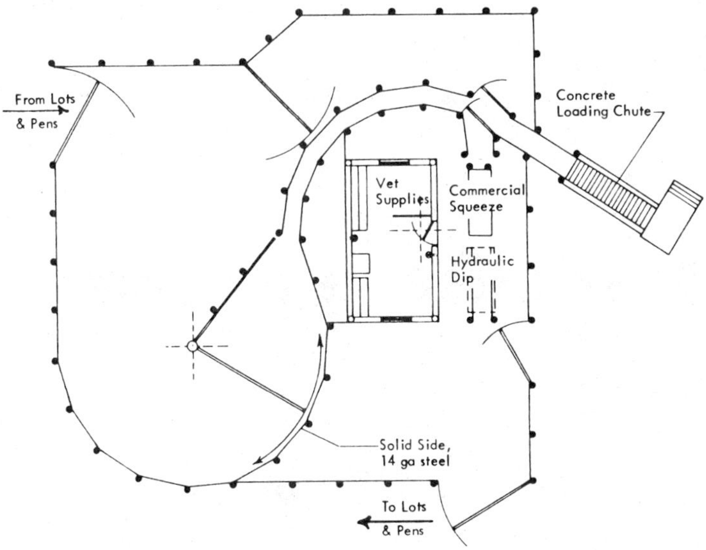

Figure 21.14. Working area for feedlot. Area could be covered by a roof if so desired. Courtesy of *Beef Housing and Equipment Handbook, MWPS-6,* 3rd Ed., 1975 ©. Midwest Plan Service, Ames, Iowa 50011.

SELECTED REFERENCES

Publications

Battaglia, R. A. and Mayrose, V. B., 1981. *Handbook of Livestock Management Techniques.* Minneapolis: Burgess Publishing Co.

Beef Housing and Equipment Handbook, 1975. Midwest Plan Service. Iowa State University, Ames, IA 50011.

Elam, C. J., 1971. *Problems Related to Intensive Indoor and Outdoor Beef Production Systems.* J. Anim. Sci. 32:554.

Grandin, T., *Livestock Handling Made Easy.* Proceedings of the 1983 International Ranchers Roundup. Texas Agric. Ext. Serv., San Angelo, TX.

Grandin, T. *Efficient Curved Corrals.* Angus Journal, October 1980.

Grandin, T. *Renovating Processing Facilities.* Feedlot Management, January 1983.

Grandin, T. *Your Processing Chute May Be Squeezing Down Your Profits.* BEEF, August 1979.

Grandin, T. *Select Equipment with Care.* Angus Journal, August 1979.

Rains, D. C. *Handling Facilities for Beef Cattle.* Angus Journal, August 1983.

Rider, A. R. and Harp, S. *Corrals and Handling Facilities for Beef Cattle.* Hereford Journal. July 1978.

Visuals

"Handling Equipment and Facilities." Sound filmstrip. Vocational Education Productions, Calif. Polytechnic State Univ., San Luis Obispo, CA 93407.

"Bovine Restraint." 13 min. videotape. Instructional Media Center, Michigan State Univ., East Lansing, MI 48824.

22 Keeping Up-To-Date in the Beef Industry

Someone has said that the only thing constant in this life is change. It has also been stated that cattle producers may be engulfed with change or engulfed because they do not change. Change is not without frustration, or as one old cowboy puts it, "I'm all for progress; it's just those changes I don't like."

Not all change is good. Undirected or mismanaged change may be more disastrous than no change at all. The balance needed is summarized in the statement, "Tradition for tradition's sake is no worse than change for change's sake."

One needs to reflect occasionally on the purpose and source of change. Man is motivated to seek and understand the unknown. The statement, "Ye shall know the truth, and the truth shall make you free," continues to cause human beings to strive and reach for a fuller understanding of the world around them. It is obvious that the application of the biological truths of nutrition, health, and genetics have resulted in more productive cattle with less death loss. Fluctuating economic conditions dictate a considerable amount of change. Economics can change slowly over an extended period of time or change rapidly, almost overnight. The latter may create the need for immediate change.

Intelligent producers recognize that life is a continual learning process and truths flow to those who seek them. How refreshing it is to see individuals who did not respond to the retirement pressure at 60 to 70 years of age but continued to learn for several more decades. Individuals who handle change best seem to be involved with directing change rather than having change direct them. Great intellectual individuals are continually seeking and learning; individuals with small minds seem to know everything.

How much more knowledge (true relationships) is waiting to be discovered? No doubt Sir Isaac Newton, who discovered many great truths, had vision and understanding when he said, "I am but a child playing on the sands by the sea. I have found a few pretty pebbles washed by the spray, but beyond me, unfathomed and unexplored, lies the great ocean of whose depth and mysteries I know nothing."

A considerable amount of man's effort is spent identifying, teaching, and learning true relationships.

Claims have been made that there have been more scientific achievements and advancement in technology in the past 150 years than in all previous recorded world history. This explosion of knowledge continues at a rapid pace in a society that is increasing in specialization. The beef producer and others in the beef industry therefore find themselves surrounded with voluminous amounts of information. Some of this information can or will have a direct effect upon their specific operations by changing levels of productivity or changing the economic environment in the area, country, or world.

The extensive amount of available information and limitations of time force an individual to set priorities as to how much time can be spent with the many sources of information. The management plan, which has been developed and is continually assessed for a specific operation, should help set these priorities. The importance of having useful information is best summarized by the following statement made by a beef producer, "No decision is any better than the information on which it is based."

It is important to recognize that not all written or verbalized information is valid or useful. Some research reports contain gross errors while some popular publications attempt to capitalize on the emotional and sensational. Occasionally information is highly selected and there are obvious attempts to avoid telling the entire story. The decision maker must move from the "opinion type" or "selected type" of information to the "verified type" of information. Different information sources must be critically evaluated for their validity and reliability in providing useful information in helping make effective decisions.

Some primary sources of information for the decision maker are shown in Figure 22.1.

In the sciences, the scientific method is used to move the unknown to the known (Figure 22.2). Those concepts and ideas which are determined to have cause and effect

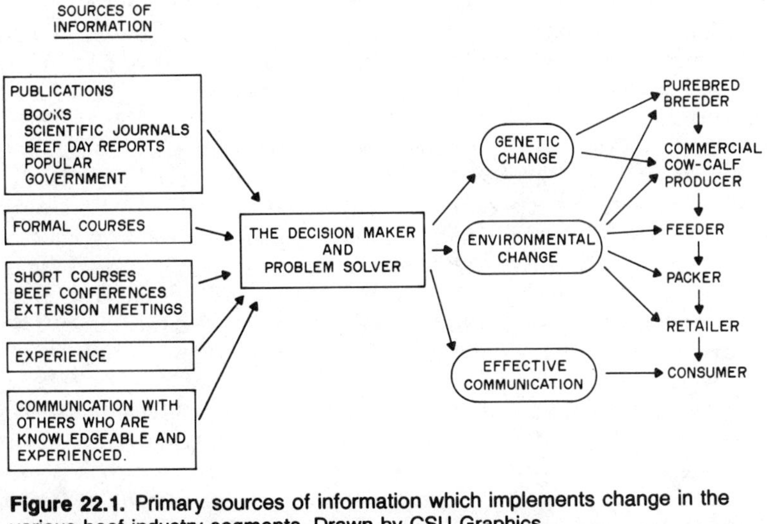

Figure 22.1. Primary sources of information which implements change in the various beef industry segments. Drawn by CSU Graphics.

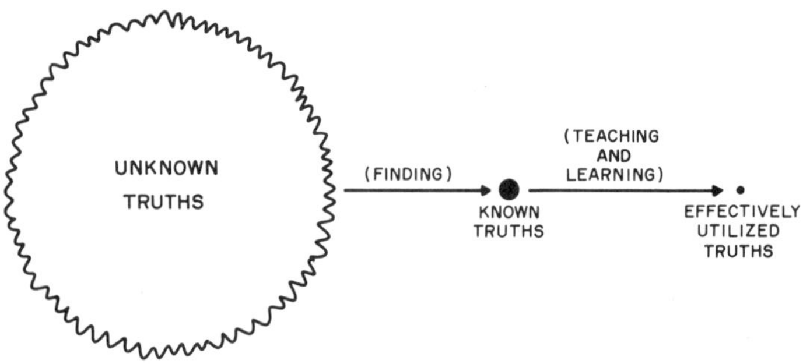

Figure 22.2. Man's process of finding and utilizing true relationships. Size of each area approximated for demonstrational purposes. The magnitude of the unknown truths cannot be quantified at this time. Drawn by CSU Graphics.

relationships become additions to the pool of known information. The educational process attempts to move the known truths, understood by relatively few people, to where many more people can understand and effectively apply these truths.

The purpose of the scientific method is to analyze well-designed experiments statistically to understand biological relationships more clearly in cattle. Producers need to know that if cattle are treated a certain way, what would be the predictability of the expected response. An excellent presentation on confidence in research results is given in the Okla. Agric. Expt. Sta. MP–114, April 1983 Animal Research Report. It is quoted here for reference:

"The inherent variability among animals used in an experiment can lead to problems in interpreting the results. When the animals on one treatment gain more rapidly than those on another treatment, it may be that, by chance the animals on the first treatment were faster gaining animals than those on the second, or it may be due to the fact that the first treatment caused the animals to gain faster. Scientists use statistical analysis procedures to calculate the probability that such differences are due to chance rather than treatment.

"In some of the articles in the Animal Science Research Report, the notation $P < 0.05$ is used. This means that the probability that the differences referred to in the article resulted from chance is less than 5 in 100. In other words, it is highly unlikely that results obtained were due to chance. Sometimes the writers state that two averages are 'significantly different.' This means the same thing, that is, that the probability that the difference was due to chance is less than 0.05. In such cases, it is very likely that the treatment caused the differences that were observed and that they did not result from random factors such as allotment of the animals to the treatment groups.

"In some articles there are tables of values given in such a way that by each value in the table there is a $\pm$ and then another value, such as: 3.6 ± 0.5. The 3.6 is called the mean, or average, of the sample of animals that was studied and is an estimate of the mean of the larger group of animals from which the sample came. The 0.5 is called the standard error and is a measure of the accuracy of the estimation procedure. If the standard error is quite small in relation to the size of the average with which it is associated, this suggests that the estimation procedure was accurate. The interpretation is that the probability is 0.68 that the value being estimated by the sample mean is within one standard error of the estimate. *Example 1:* If we took a random sample of calves from a very large herd and calculated the mean weight of those calves to be 425 pounds and calculated the standard error to be 10 pounds, then we would say that the probability is 0.68 that the mean weight of the total group of calves is between

415 and 435 pounds. *Example 2:* Assume that we compared the average daily gain of calves on two different rations and found that the difference in gain between the two groups was 0.15 pounds per day and that the standard error of the difference was 0.10 pounds per day. Since the 0.10 is nearly as large as the 0.15, we do not have great confidence that the 0.15 is a highly accurate estimate of the true difference between the two treatments involved. "Some papers contain 'correlation' coefficients. These are measures of the relationship between traits or variables. These relationships may be positive or negative. Positive relationships mean that when one variable is higher than average, more often than not the other variable is also higher than average. These positive values may be anywhere between zero and 1.0. Values near 1 indicate a strong tendency for larger values for one trait to be associated with larger values for the other. Negative correlations merely mean that larger than average values in one trait are more often than not associated with smaller than average values of the other trait, and the nearer the values are to −1, the stronger this negative relationship. When it is stated that a correlation efficient is statistically significant, it means that there is very strong evidence that there is in fact a correlation in the direction indicated and that the relationship found probably was not due to chance."

SOURCES OF WRITTEN INFORMATION

Selected Books

Baker, F. H. (editor), 1984. *Beef Cattle Science Handbooks,* Vol. 19 and 20. Winrock International. Rt. 3. Morrilton, AR 72110. Volumes 1–18 from Agriservices Foundation, 3699 East Sierra Ave., Clovis, CA.

Ensminger, M. E., 1976. *Beef Cattle Science.* Danville, IL: The Interstate Printers and Publishers, Inc.

Lasley, J. F., 1981. *Beef Cattle Production.* Englewood Cliffs: Prentice-Hall, Inc.

Minish, G. L. and Fox, D. G., 1982. *Beef Production and Management.* Reston, VA: Reston Publishing Co., Inc.

Neumann, A. L., 1977. *Beef Cattle.* New York: John Wiley and Sons.

O'Mary, C. C. and Dyer, I. A., 1978. *Commercial Beef Cattle Production.* Philadelphia, PA.: Lea and Febiger.

Preston, T. R. and Willis, M. B., 1974. *Intensive Beef Production.* New York: Pergamon Press.

Simpson, J. R. and Farris, D. E., 1982. *The World's Beef Business.* Ames: Iowa State University Press.

Thompson, G. B. and O'Mary, C. C., 1983. *The Feedlot.* Philadelphia, PA: Lea and Febiger.

Books give a broad coverage of beef production and the beef industry. Books, covering the principle areas (nutrition, breeding, reproduction, meats, and others) and some of the specific beef industry segments, are identified at the end of each chapter in this book under Selected References.

Books are excellent sources of material for overview and basic principles. Most of these books, however, are outdated in several areas at the time they are printed. This is particularly true where the current direction of the beef industry needs to be assessed. Preston's book, *Intensive Beef Production,* is excellent from the standpoint of review of research studies up to 1974. Extensive references are given if one desires to refer to the original research.

SELECTED SCIENTIFIC PUBLICATIONS (United States and Canada)

American Journal of Veterinary Research of the American Veterinary Medical Association. 930 N. Meacham Road, Schaumburg, IL 60196. Monthly.

Figure 22.3. The Journal of Animal Science is one of the primary scientific publications in which beef cattle research results are published. Courtesy of CSU Photo Lab.

Canadian Journal of Animal Science. Published by the Agricultural Institute of Canada, Suite 907, 151 Slater Street, Ottawa, Canada KIP 5H4.

Journal of Animal Science. Published by American Society of Animal Science, 425 Illinois Bldg., 113 North Neil St., Champaign, IL 61820 (Figure 22.3).

Journal of Food Science. Institute of Food Technologists, Suite 2120, 221 North LaSalle Street, Chicago, IL 60601.

Journal of Range Management. Society of Range Management, 2760 West Fifth Ave., Denver, CO 80204. Published bimonthly.

SELECTED SCIENTIFIC PUBLICATIONS (International)

Animal Production. Journal of the British Society of Animal Production. Longman Group Limited, Fourth Avenue, Harlow, Essex, England CM 19 5AA.

Journal of Agricultural Science. Cambridge University Press, Bentley House, 200 Euston Road, London NWI 20B.

Australian Journal of Agricultural Research. CSIRO, 314 Albert Street, East Melbourne, Victoria, Australia 3002.

Theriogenology. An International Journal of Animal Reproduction. Geron-x Inc., Box 1108, Los Altos, CA 94022.

Tropical Animal Health and Production. Longman Group Limited, Fourth Avenue, Harlow, Essex, England CM 19 5AA.

The Indian Journal of Animal Sciences. Krishi Bhavan New Delhi, India 110001.

World Animal Review. Published by the Food and Agriculture Organization of the United Nations, via delle Terme di Caracalla, 00100 Rome, Italy. Quarterly.

World Review of Animal Production. c/o International Publishing Enterprises, Via di Tor Vergata, 85/87 Rome, Italy.

SELECTED POPULAR PUBLICATIONS

The Advanced Animal Breeder. The official publication of National Association of Animal Breeders, Box 1033. Columbia, MO 65205.

Animal Nutrition and Health. Watt Publishing Co., Mount Morris, IL 61054.

BEEF. 1999 Shepard Road, St. Paul, MN 55116 (Figure 22.4).

Beef Business Bulletin. A weekly newsletter giving an update of items affecting the beef industry. Published by the National Cattlemen's Assoc., P.O. Box 3469, Englewood, CO 80155 (Figure 22.6).

Beef Digest. Shield Publishing Co., P.O. Box 511, 446 Harrison, Lindsborg, KS 67456.

Better Beef Business. P.O. Box 12740, 1520 Taney, North Kansas City, MO 64116 (Figure 22.4).

CALF News. Suite 303, 18345 Ventura Blvd., Tarzana, CA 91356.

Cattle-Fax. A marketing report and analyst service associated with NCA. Marketing information provided by telephone and a weekly UPDATE publication. Contact Cattle-Fax, P.O. Box 3947, Englewood, CO 80155.

Drovers Journal. 300 W. Adams St., Chicago, IL 60606. Weekly (Figure 22.5).

Farm Journal. 230 West Washington Square, Philadelphia, PA 19105.

Feedlot Management. P.O. Box 67, Minneapolis, MN 55440. Monthly (Figure 22.4).

Feedstuffs. Miller Publishing Co., 2501 Wayzata Blvd., Minneapolis, MN 55405.

High Plains Journal. P.O. Box 760, Dodge City, KS 67801.

Ideal Beef Memo. Rt. 1, Box 79, Huxley, IA 50124 (Figure 22.4).

Livestock Breeder Journal. Official publication for the beef associations in Georgia, Virginia, West Virginia, Kentucky, North Carolina, South Carolina, Maryland, New York, and Vermont.

Meat Industry. 90 Throckmorton Ave., P.O. Box 72, Mill Valley, CA 94942. Monthly publication for meat packers and processors.

The National Provisioner. 15 W. Huron St., Chicago, IL 60610. Weekly publication for the meat industry.

Figure 22.4. Selected monthly publications which emphasize cow–calf and feedlot production. Courtesy of CSU Photo Lab.

Figure 22.5. Selected weekly publications which emphasize current articles, sales, and market reports. Courtesy of CSU Photo Lab.

Record Stockman. 105 Livestock Exchange Bldg., Denver, CO 80216. Weekly (Figure 22.5).
Southern Beef Producer. 226 Largo Dr., Nashville, TN 37211.
Successful Farming. 1716 Locust Street, Des Moines, IA 50336.
Rangelands. Society for Range Management, 2760 West Fifth Ave., Denver, CO 80204.
The Cattleman. 410 E. Weatherford St. Fort Worth, TX 76102.
Western Livestock Journal. P.O. Drawer 17F, Denver, CO 80217. Weekly. Also publishes the monthly publication, Livestock (Figure 22.5).

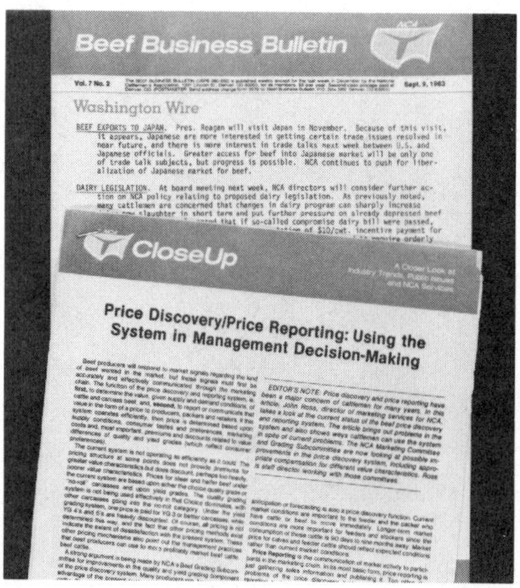

Figure 22.6. The Beef Business Bulletin, published by the National Cattlemen's Association, is an excellent source of current material affecting the beef industry. It also contains informative articles of importance to beef producers. Courtesy of the National Cattlemen's Association.

Western Livestock Round-up. Market information publication from cooperative project of USDA and 17 western states. Contact the Extension Economist in any of the land grant universities located in the western states (Figure 22.8).

Breed Journals and Publications

Most breeds have a breed publication (Figure 22.7). The following represent the most numerous breeds in the United States. Other breed publications can be identified by writing to the respective breed association. (See Chapter 23).

American Hereford Journal. 715 Hereford Drive, P.O. Box 4059, Kansas City, MO 64105. Monthly.
Angus Journal. Frederick and Brookside, St. Joseph, MO 64501. Monthly.
Brahman Journal. P.O. Box 220, Eddy, TX 76524.
Limousin World. 6408 So. College Ave., P.O. Box 2205, Fort Collins, CO 80522. Monthly.
Polled Hereford World. 4700 East 63rd St., Kansas City, MO 64130. Monthly.
The Santa Gertrudis Journal. P.O. Box 2386, Fort Worth, TX 76101.
Simmental Shield. P.O. Box 511, Lindsborg, KS 67456. Monthly.

Government Publications (USDA Publications).

Write to: USDA, ERS Information Division, Publications Unit, Washington, DC for the following publications:

Agricultural Statistics
Cattle
Cattle on Feed
Handbook of Agricultural Charts
Livestock and Meat Situation. For meat animals; gives production, prices, outlook, consumption, imports, exports, and usually a feature article.

Figure 22.7. Selected breed publications containing information primarily for purebred breeders. Courtesy of CSU Photo Lab.

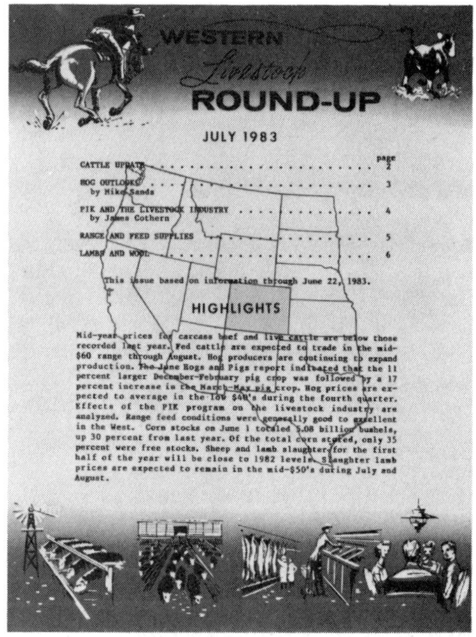

WESTERN Livestock ROUND-UP

JULY 1983

	page
CATTLE UPDATE	2
HOG OUTLOOK by Mike Sands	3
PIK AND THE LIVESTOCK INDUSTRY by James Cothern	4
RANGE AND FEED SUPPLIES	5
LAMBS AND WOOL	6

This issue based on information through June 22, 1983.

HIGHLIGHTS

Mid-year prices for carcass beef and live cattle are below those recorded last year. Fed cattle are expected to trade in the mid-$60 range through August. Hog producers are continuing to expand production. The June Hogs and Pigs report indicated that the 11 percent larger December-February pig crop was followed by a 17 percent increase in the March-May pig crop. Hog prices are expected to average in the low $40's during the fourth quarter. Effects of the PIK program on the livestock industry are analyzed. Range feed conditions were generally good to excellent in the West. Corn stocks on June 1 totaled 5.08 billion bushels, up 30 percent from last year. Of the total corn stored, only 35 percent were free stocks. Sheep and lamb slaughter for the first half of the year will be close to 1982 levels. Slaughter lamb prices are expected to remain in the mid-$50's during July and August.

Figure 22.8. The Western Livestock Roundup is an excellent source of marketing information. The supplement in the February issue provides a useful overview of beef industry numbers. Courtesy of CSU Photo Lab.

Livestock Meat Wool Market News. Weekly summary of numbers marketed and prices for live animals, carcass, and retail cuts.

Livestock Slaughter. Shows number and tonnage of meat animals slaughtered on a monthly basis.

Meat Animals. Production, Disposition and Income.

National Food Situation. Per capita consumption of foods, food prices, outlook and special articles usually related to the consumer and food.

State Publications

Most states have a Crop and Livestock Reporting Service, in their State Department of Agriculture which cooperates with the USDA. An annual agricultural statistics book which lists cattle numbers, prices, etc. is usually published. Addresses of Departments of Agriculture for specific states can be obtained by contacting: National Association of State Departments of Agriculture, 1616 H Street, N.W., Washington, DC 20006.

SELECTED LIBRARY REFERENCE BOOKS

Agrindex. A monthly bibliography providing references to current agricultural literature (including cattle) collected from worldwide sources. Input is made from cooperating centers, located in several countries to Food and Agriculture Organization of the United Nations.

Animal Breeding Abstracts. Contains abstracts under headings such as cattle and breeds, growth and production, meat production, milk production, progeny and performance tests, genetics, reproduction and artificial insemination, and miscellaneous.

Biological Abstracts. Contains abstracts from numerous research publications throughout the world. Some of the subject headings are animal production, genetics, nutrition, and veterinary science.

Food Science and Technology Abstracts. Published by the International Food Information Service (IFIS) whose purpose is to collect, abstract, collate and distribute information on all matters concerned with research and new developments in Food Science and Technology. To collect this information, over 1200 journals from over 50 countries, patents from 20 countries, and books are systematically searched by cooperating organizations or individuals in several countries.

Nutrition Abstracts and Reviews. Contains abstracts from numerous research publications relating to nutrition. Some of the most pertinent headings would be Cattle (growth and fattening), Cattle (milk production), and Cattle (reproduction).

Readers Guide to Periodical Literature. Contains indexed references from many popular publications. The publications most frequently referred to in the cattle section are *Farm Journal* and *Successful Farming.*

University Publications

Experiment Station Bulletins, Extension Service (Bulletins, Newsletters, Handbooks, and Fact Sheets), Field Day Reports—(Figures 22.9 and 22.10). Many of these are semi-technical scientific reports and publications. They usually have not met the same criteria for those listed in the scientific publication category; such as, statistical analysis and peer review for publication. Most universities have a publications catalog that lists research and extension bulletins and pamphlets which are available.

Examples: *Beef Cattle Research in Texas* (1982), Consolidated PR–3916-3963. Selected articles: Effect of 48-hour Calf Removal Regimes, Prior to Controlled Breeding, on Reproductive Performance; Analysis and Synthesis of Optimal Beef Cattle Systems; Forage Utilization Systems for Wintering and Breeding Replacement Heifers; Cattle Feedlot Design Strategies Under Alternative Price Relationships; Cholesterol Content of Raw and Cooled Beef Muscles with Different Amounts of Marbling.

Figure 22.9. Selected beef research reports, publish annually by Land Grant Universities (Departments of Animal Science), in states where beef cattle are of primary importance. Courtesy of CSU Photo Lab.

Figure 22.10. Beef cattle handbooks are available from Beef Cattle Extension Specialists in several states. Courtesy of CSU Photo Lab.

Okla. Agric. Expt. Sta.—1982 Animal Science Research Report. Selected articles: Growth Rates and Relationships Among Frame Size, Performance Traits, and Scrotal Circumference in Young Beef Bulls; Protein versus Energy in Receiving Diets for Stocker Cattle; Limestone and Potassium for Feedlot Steers.

Cow–Calf Management Guide. Cattleman's Library. Published by the Cooperative Extension Services of Colorado, Idaho, Montana, Oregon, Utah, Washington, and Wyoming. Contact: Dept. of Agricultural Information, College of Agriculture, University of Idaho, Moscow, ID 83843.

Great Plains Beef Cow–Calf Handbook. Cooperative Extension Service. Great Plains States (Montana, Wyoming, Colorado, New Mexico, North Dakota, South Dakota, Nebraska, Kansas, Oklahoma, and Texas. Contact the Beef Cattle Extension Specialist in one of these states).

Great Plains Beef Cattle Feeding Handbook. Cooperative Extension Service. Great Plains States (Montana, Wyoming, Colorado, New Mexico, North Dakota, South Dakota, Nebraska, Kansas, Oklahoma, and Texas. Contact the Beef Cattle Extension Specialist in one of these states).

Kearle, W. G. 1972. Economic comparisons of the cow–calf and cow–yearling systems for northern plains cattle ranching. Univ. of Wyo. Experiment Station Research Journal 67, December, 1972.

Nebrasks 1983 Beef Cattle Report, MP44. Selected articles: Bull Exposure Influences Early Return of Cows to Estrus; Lasalocid for Growing—Finishing Cattle; Restructured Beef—Impact on Future Industry; Lice Affect Yearling Gains.

Indiana 1982 Beef Cattle Day. Selected articles: Supplying Monensin to Cattle on Pasture; Computers—An Aid to Beef Production; Use of Insecticide Ear Tags for Face Fly and Horn Fly Control.

Seidel, G. E., *et al.,* 1982. *Embryo Transfer Procedures for Cattle.* Colo. Experiment station Gen. Series 1011.

Southern Regional Beef Cow-Calf Handbook. Cooperative Extension Service, Southern States. Contact the Beef Cattle Extension Specialist at one of the land grant universities in the southern states.

CONFERENCES, SHORTCOURSES, BEEF CATTLE DAYS AND SYMPOSIA

Examples: Beef Improvement Federation holds an annual convention at different locations in the United States. Proceedings are published. (A. L. Eller, Animal Science Department, VPI, Blacksburg, VA 24061.)

Breed Associations typically hold annual educational conferences. (Contact specific breed association.)

Range Beef Cow Symposium is held on odd-numbered years. Location alternates between Nebraska, Wyoming, South Dakota, and Colorado. (Contact Beef Extension Specialist in one of these states.) Proceedings are published (Figure 22.11).

California Livestock Symposium (held annually).

Cornbelt Cow–Calf Conference, Ottumwa, IA. Contact: Clifford I. Iverson, Terminal Bldg., Industrial Airport, Ottumwa, IA 52501 (Figure 22.11).

International Stockman's School—held annually and sponsored by Winrock International, Route Three, Morrilton, AR 72110. A Beef Cattle Science Handbook covering the proceedings is published yearly (Figure 22.11).

National Cattlemen's Association—Conferences are held periodically to keep members informed in the areas of cattle production and business management, i.e., National Beef Profit Conference, Denver, CO, June 28–29, 1982.

Figure 22.11. Proceedings from conferences and shortcourses provide useful reference material for many beef producers. Courtesy of CSU Photo Lab.

Most state universities located in important cattle producing states will hold an annual livestock, research, feeders, cattlemen's, or beef day. Usually research progress reports are given and other pertinent topics are covered. Written reports covering the research and other proceedings are typically available. (Contact Beef Extension Specialist.)

Producer oriented extension meetings and workshops are held on a county, area, or state basis within several important beef states. (Contact County Extension Office or Beef Extension Specialist at the land grant university.)

LAND GRANT UNIVERSITIES

Most land grant universities have teaching, research and extension activities involving beef cattle. For more background information see year 1862 in Chapter 24.

Reference has been made to available publications and beef cattle extension specialists located at land grant universities. Addresses of land grant universities are included to obtain publications and personally contact beef extension specialists.

Write to Animal Science Departments, located at the following universities, for information on beef cattle.

Auburn University
Auburn, AL 36830

The University of Arizona
Tucson, AZ 85721

University of Arkansas
Fayetteville, AR 72701

University of California
Davis, CA 95616

Colorado State University
Fort Collins, CO 80523

The University of Connecticut
Storrs, CT 06268

The University of Delaware
Newark, DE 19711

University of Florida
Gainesville, FL 32611

The University of Georgia
Athens, GA 30602

University of Hawaii
Honolulu, HI 96822

University of Illinois
Urbana, IL 61801

Purdue University
West Lafayette, IN 47907

Iowa State University
Ames, IA 50011

Kansas State University
Manhattan, KS 66506

University of Kentucky
Lexington, KY 40506

The Louisiana State University
Baton Rouge, LA 70893

University of Maine
Orono, ME 04473

University of Maryland
College Park, MD 20742

University of Massachusetts
Amherst, MA 01002

Michigan State University
East Lansing, MI 48824

University of Minnesota
St. Paul, MN 55108

Mississippi State University
Mississippi State, MS 39762

University of Missouri
Columbia, MO 65201

Montana State University
Bozeman, MT 59717

University of Nebraska
Lincoln, NE 68503

University of Nevada
Reno, NV 89507

University of New Hampshire
Durham, NH 03824

The State University of New Jersey
New Brunswick, NJ 08903

New Mexico State University
Las Cruces, NM 88003

Cornell University
Ithaca, NY 14853

North Carolina State University
Raleigh, NC 27607

North Dakota State University
Fargo, ND 58102

The Ohio State University
Columbus, OH 43210

Oklahoma State University
Stillwater, OK 74074

Oregon State University
Corvallis, OR 97331

The Pennsylvania State University
University Park, PA 16802

University of Rhode Island
Kingston, RI 02881

Clemson University
Clemson, SC 29631

South Dakota State University
Brookings, SD 57007

The University of Tennessee
Knoxville, TN 37901

Tennessee State University
Nashville, TN 37203

Texas A&M University
College Station, TX 77843

Utah State University
Logan, UT 84322

University of Vermont
Burlington, VT 05401

Virginia Polytechnic Institute
Blacksburg, VA 24061

Washington State University
Pullman, WA 99164

West Virginia University
Morgantown, WV 26506

University of Wisconsin
Madison, WI 53706

University of Wyoming
Laramie, WY 82071

SELECTED REFERENCES

Publications

Hubbard, D. D. 1980. *Beef Cattle Publications and Visual Materials: 1980 Listing.* USDA Misc. Public. 1398. Washington, DC.

List of available publications of the USDA. 1973. List No. 11. USDA. Washington, DC.

Confidence in research results. Okla. Agric. Expt. Sta. MP–114, April 1983.

23
Beef Industry Organizations

The beef industry covers the production, processing, and distribution of beef and the by-products of cattle. Each segment, operating quite independently of each other, has different economic criteria and management problems. Each segment markets a different type of beef product (breeding animal, feeder animal, slaughter animal, carcass, fresh or cooked retail product).

Beef organizations have originated to inform, promote, regulate, support research, more effectively compete, and have political influence in areas directly affecting their businesses.

The major organizations representing individuals or companies within each beef industry segment and the major organization having an indirect effect on each segment are shown in Table 23.1. The three national organizations that have a significant impact on profitability to individuals in the beef industry are: (1) U.S. Meat Export Federation, (2) the Beef Industry Council of the National Live Stock and Meat Board, and (3) the National Cattlemen's Association (NCA). Also the major organizations involved with the marketing process as animal or product moves from one segment to another are shown in Table 23.1.

An organizational challenge exists with the approximate 1.5 million cattlemen, many of whom desire to operate independent of any organizational structure and other cattle producers who have major sources of income other than from their cattle operations. There are many organizations who represent or affect the interests of cattlemen. Even though many cattlemen dislike it, the era of total independence of cattlemen with complete individual freedom to determine their individual destiny has passed. The traditional independent philosophy of cattlemen and their organizations requires reevaluation and adjustment to optimize their success in the planning of change. The future of beef will be charted by or from the results of group action, mainly that of significant organizations. It appears that the influence of organizations on the beef industry will be determined primarily by, (1) the industry's capacity for linking and coordinating action of

Table 23.1. Major Organizations Representing or Affecting the Beef Industry

Segments	Major Organizations Representing Each Segment	Other Major Organizations Directly Affecting each Segment
Seedstock producers	Amer. Natl. CowBelles Beef Improvement Federation Breed Associations Natl. Cattlemen's Assn. Natl. Live Stock and Meat Board State Beef Boards State Cattlemen's Assn. U.S. Beef Breeds Council	(1) State Depart. of Agriculture (2) USDA Animal and Plant Health Inspection Service (3) USDA Packers and Stockyards Administration (4) Amer. Assn. of Bovine Practitioners (5) Occupational Safety and Health Administration
Commercial cow–calf producers	Amer. Farm Bureau Federation Amer. Natl. CowBelles Beef Improvement Federation Breed Associations Cattle-Fax County Livestock Growers Natl. Cattlemen's Assn. Natl. Grange Natl. Live Stock and Meat Board State Beef Boards State Cattlemen's Assn.	(6) Environmental Protection Agency (7) Amer. Society of Animal Science (8) USDA-Forest Service (9) USDA-Soil Conservation Service (10) USDA-Agric. Research Service (11) USDA-Extension Service (12) USDA-Statist. Reporting Service (13) Society for Range Mgt. (14) Amer. Regist. of Certif. Animal Scientists (15) Livestock Public. Council (16) Internatl. Embryo Transfer Society (17) Beef AI organizations
Feeders	American Farm Bureau Federation Amer. Natl. CowBelles Cattle-Fax Natl. Cattlemen's Assn. Natl. Live Stock and Meat Board State Beef Boards State Cattle Feeders Assn.	(1),(2),(3),(4),(5),(6),(10),(11), (12),(14),(15) (18) Amer. Feed Manuf. Assn. (19) Natl. Feed Ingredient Assn. (20) Food and Drug Admin. (21) USDA Office of Transportation (22) Animal Health Institute
Packers and processors	Amer. Assn. of Meat Proc. Amer. Meat Institute Amer. Meat Science Assn. Institute of Food Technologists Natl. Cattlemen's Assn. Natl. Food Processors Assn. Natl. Live Stock and Meat Board Natl. Meat Assn. Natl. Meat Canners Assn. State Beef Boards State Meat Dealers	(1),(2),(3),(4),(5),(6),(12),(14) (15),(20),(21) (23) Federal Trade Commission (24) Labor Unions (25) Natl. Perishable Transport. Assn. (26) USDA Food Safety and Quality Service (27) USDA Agric. Marketing Service
Meat retailers/ food service operations	Amer. Assn. of Meat Proc. Amer. Institute of Food Dist. Natl. Assn. of Meat Purv.	(1),(2),(3),(5),(6),(20),(21),(23) (24),(25),(26),(27) (28) Food and Drug Law Institute

Table 23.1 (*Continued*)

Segments	Major Organizations Representing Each Segment	Other Major Organizations Directly Affecting each Segment
	Natl. Assn. of Retail Grocers of the United States	(29) USDA Food and Nutrition Service
	Natl. Frozen Food Assn.	(30) Joint Labor Mgt. Comm. of the Retail Food Industry
	Natl. Live Stock and Meat Board	(31) Natl. Restaurant Association
	Natl. Restaurant Assn.	(32) Food Service and Lodging Inst.
	State Meat Dealers	
	State Restaurant Assn.	
Consumer	Community Nutrition Institute	(1),(2),(3),(5),(6),(20),(23),(26)
	Congress Watch	(28),(29)
	Consumer Alert Council	(33) American Public Health Assn.
	Consumer Federation of Amer.	(34) Food and Energy Council
	Natl. Consumers League	(35) National Academy of Sciences
	Vegetarian Inform. Service	
Marketing Points Between Segments	Amer. Stockyards Assn.	(12),(20),(21),(23)
	Cattle-Fax	(36) Agriculture Trade Council
	Natl. Auctioneers Assn.	(37) USDA Economic Management Staff
	Natl. Cattlemen's Assn.	(38) USDA Marketing and Inspec. Mgt.
	Natl. Livestock Grading and Marketing Assn.	(39) USDA World Agric. Outlook Board
	Natl. Live Stock and Meat Board	(40) USDA Animal Air Transp. Assn.
	Natl. Livestock Producers Assn.	
	U.S. Meat Export Federation	

[a] Repeated numbers refer to the same organizations identified in a previous segment.

its organizations and forming alliances with other groups having common interests, and (2) effective organizational leaders who can represent members and who have character and commitment to their organizations.

An NCA report, *The Future for Beef,* identified the organizational status for beef and several other competing meat industries as follows:

Chicken Broilers. The broiler industry is served at the national level by one organization. This organization handles government affairs, information and public relations. It is not involved in product advertising. Individual integrated companies do their own advertising of branded products, if there is advertising at all.

Pork. The pork producing industry's work at the national level is handled largely by one organization, funded mostly through check-off funds. This organization takes care of government affairs, research and education, public information, and most of the product promotion.

Beef. Just as the cattle and beef businesses are segmented and fragmented to a greater degree than are other parts of agriculture, the industry has a multiplicity of organizations at both the state and national levels. As a result, it is more difficult compared to competing industries to achieve the kind of industry support and organization that can maximize the results of industry-funded programs. More unity, however, can and will be achieved.

For several years, the producing and feeding segments of the industry have worked together. In more recent years, the purebred segment also has become more united. In the years ahead, breed registry associations will become even more involved in the general cattle industry. They are now becoming more active in the National Cattlemen's Association (NCA), and today 20 breed organizations are affiliated members of NCA. As a result, breed associations have direct input—through their directors on the board and through their representatives on each of the standing committees, and on the Purebred Advisory Council. The seedstock industry is in a position to bring additional talents and skills to the solution of over-all industry problems.

Meanwhile, it is obvious, all cattlemen must work jointly in the following areas:

1. The impact of government on cattlemen will grow at a slower rate than in the past, but the impact will remain great. Government affairs will continue as a major function of the industry collectively.
2. In order to help make beef more competitive, the industry will have to guide, encourage, and support research and development efforts in production technology and also in product development and marketing.
3. Cattlemen will need more economic research and information to help them collectively and individually make sounder decisions.
4. The industry needs an expanded beef marketing program—including consumer and market research, product development, product information, promotion, and merchandising. In addition to improving beef production and distribution efficiency, the industry will have to support programs to improve beef acceptance.
5. Public information efforts should be expanded to improve public and government acceptance and understanding of beef economics and beef production and marketing methods.

Eventually, there may be single, unified beef industry organizations at both state and national levels. Meanwhile, there will have to be continued efforts to coordinate beef industry programs in order to maximize results of funding now provided by the individual cattle producer to a multiplicity of farm organizations and promotion organizations.

Major Beef Industry Organizations or Organizations Affecting the Beef Industry

The names, addresses, and functions of the major organizations representing and affecting the beef industry are listed for reference.

AGRICULTURAL COUNCIL OF AMERICA
Suite 708, 1625 Eye Street, N.W.
Washington, DC 20006

Farm and commodity organizations, agribusiness firms, ranchers, and individual farmers constitute this council. Purpose is to provide communication between the farmer, the urban consumer, and decision-makers in Washington; find a more unified approach to problems. Developed "Its Your Heartbeat America Program" to educate the public on agriculture and its role in America.

AGRICULTURAL RESEARCH INSTITUTE (ARI)
Number 835
2100 Pennsylvania Avenue, N.W.
Washington, DC 20037

ARI was organized by industrial scientists from both the public and private sectors to identify and promote the kinds of research and policies needed to ensure the best long-term utilization and management of agricultural resources for the national welfare.

AGRICULTURAL TRADE COUNCIL (ATC)
Suite 675, 1800 M Street, N.W.
Washington, DC 20006

Comprised of companies and organizations which import and export agricultural commodities, livestock, and food. Maintains legislative, research, and educational services to develop agricultural world trade.

AMERICAN ASSOCIATION OF BOVINE PRACTITIONERS
P.O. Box 2319
West Lafayette, IN 47906

AMERICAN ASSOCIATION OF MEAT PROCESSORS (AAMP)
224 East High St.
Elizabethtown, PA 17022

Retail operators of meat processing plants, locker plants, frozen food centers, and freezer food suppliers form this association.

AMERICAN FARM BUREAU FEDERATION
225 Touhy Ave.
Park Ridge, IL 60068

Farm Bureau is a general farm organization with a voluntary membership. Producers of every agricultural commodity produced in this country are represented by this membership. Farm Bureau provides a structure and a philosophy which enables producers of agricultural commodities to reconcile their inevitable commodity differences and develop the unity of purpose and action necessary to solve their problems.

AMERICAN FEED MANUFACTURERS ASSOCIATION
1701 North Ft. Meyer Dr.
Arlington, VA 22209

AMERICAN FORAGE AND GRASSLAND COUNCIL (AFGC)
121 Dantzler Ct.
Lexington, KY 40503

The AFGC is the central organization of and spokesman for North America's forage-base agriculture. Members are representatives of agencies providing research, education, and public services. Additional members are manufacturers and marketers of forage-farming inputs and producers, marketers, and users of forage crops. Primary objective is that Americans can and should make more efficient use of forages in the profitable production of meat, milk, and fiber.

AMERICAN HEALTH INSTITUTE
(Address unknown)

The National trade association for major animal health product manufacturers who produce biologicals, pharmaceuticals, and feed additives.

AMERICAN INSTITUTE OF FOOD DISTRIBUTION (AIFD)
28–06 Broadway, P.O. Box 523
Fair Lawn, NJ 07410

Canners, packers, freezers, manufacturers, brokers, wholesalers, retailers-cooperatives; chains, independent retailers, growers, trade associations, banks, advertising agencies, government agencies, supply houses, and others working for or with the food trades. Serves as central information service for the food trades. Issues statistical and analytical digests on food distribution and related topics. Maintains library of information on the food industry.

AMERICAN MEAT INSTITUTE (AMI)
P.O. Box 3556
Washington, DC 20007

Meat packers, processors, sausage manufacturers, meat suppliers, canners, and others related to the industry. Activities include marketing, research, improved operating methods and products, conservation, spoilage prevention, and industrial education.

AMERICAN MEAT SCIENCE ASSOCIATION
National Live Stock and Meat Board
444 North Michigan Ave.
Chicago, IL 60611

Persons engaged in meat research, extension and education in universities, industry, government, and other organizations. To afford an opportunity for exchange of ideas and information; to foster education, research, and development in the field of meat science.

AMERICAN NATIONAL COWBELLE ASSOCIATION
5420 South Quebec
P.O. Box 3881
Englewood, CO 80155

Supports the livestock industry with labor and finances; promotes the livestock industry through information and publicity; encourages its producers with understanding and love; instills hope and faith in the future of cattlemen and their industry.

AMERICAN PUBLIC HEALTH ASSOCIATION
1015 15th St., S.W.
Washington, DC 20005

Professional organization of physicians, nurses, educators, sanitary engineers, environmentalists, social workers, optometrists, podiatrists, community health specialists, pharmacists, dentists, industrial hygienists, and interested consumers. Seeks to protect and promote personal and environmental health. Services include: promulgation of standards; establishment of uniform practices and procedures; development of the etiology of communicable diseases; research in many areas of public health.

AMERICAN REGISTRY OF CERTIFIED ANIMAL SCIENTISTS (ARCAS)
309 West Clark St.
Champaign, IL 61820

Organization certifies qualified animal scientists as credible experts in their field based on intense scholarly preparation and practical experience.

AMERICAN SOCIETY OF ANIMAL SCIENCE (ASAS)
309 West Clark St.
Champaign, IL 61820

The basic purpose of the ASAS is to increase knowledge and understanding of animals, especially farm animals, and to improve the care and productivity of animals both in commercial production and in research.

AMERICAN STOCKYARDS ASSOCIATION (ASYA)
805 Livestock Exchange Building
Omaha, NE 68107

Stockyard owners united to maintain and preserve a system of sound, competitive, economical, and public livestock markets; seek just and equitable regulatory conditions affecting livestock marketing; promote the welfare of the livestock industry.

ANIMAL AIR TRANSPORTATION ASSOCIATION
6180 South 56 Ct.
Davie, FL 33314

Organized in 1976 to help improve animal transportation, particularly by air.

BASIC AND TRADITIONAL FOOD ASSOCIATION
1707 N St., N.W.
Washington, DC 20036

Coalition of consumers, industries, public interest groups and trade associations. Purpose is to educate the general public about food and nutrition, and to focus attention on those foods which are "inherently delightful, nutritious and safe." Encourages increased production and use of traditional foods. Increases visibility of basic and traditional foods where foods are sold; develops educational materials for all age groups; advances the area of preparing basic and traditional foods.

BEEF AI ORGANIZATIONS

American Breeders Service
Deforest, WI 53532

Carnation Genetics
P.O. Box 938
Hughson, CA 95326

Midwest Breeders Coop.
Shawano, WI 54166

New Breeds Industries Inc.
Box 236A
Manhattan, KS 66502

NOBA
P.O. Box 607
Tiffin, OH 44883

Select Sires Inc.
11740 U.S. 42
Plain City, OH 43064

Tri-State Breeders Coop.
Westby, WI 54667

The AI organizations are the primary sources of commercial semen used in artificial insemination programs. The sire directories list bulls on whom semen is available (Figure 23.1).

BEEF IMPROVEMENT FEDERATION (BIF)

c/o A. L. Eller
Animal Science Dept.
Virginia Polytechnic Institute
Blacksburg, VA

BIF coordinates the performance testing programs of approximately 50 organizations from several states and including the national breed associations. One of the primary objectives is to establish accurate and uniform procedures for measuring and recording beef cattle performance data.

BREED ASSOCIATIONS

Members are breeders or owners of each specific breed of cattle. Purposes are to promote the breed, record performance of the cattle, issue registrations and keep

Figure 23.1. Sire Directories are available from beef AI organizations who commercially sell semen. Courtesy of CSU Photo Lab.

the herd book. Dairy breeds are included as they are used because of crossbreeding programs, and all cattle eventually produce beef products. Major breed associations are listed as follows:

AMERIFAX
Amerifax Cattle Association
P.O. Box 149
Hastings, NE 68901

ANGUS
American Angus Association
3201 Frederick Boulevard
St. Joseph, MO 64501

ANKINA
Ankina Breeders Inc.
5803 Oaks Rd.
Clayton, OH 45315

ANKOLE WATUSI
Ankole Watusi International Registry
Star Rt., Box 45
Hebron, ND 58638

BARZONA
Barzona Breeders Association of America
Box 631
Prescott, AZ 83602

BEEFALO
American Beefalo Breeders
116 Executive Park
Louisville, KY 40207

BEEF FRIESIAN
Beef Friesan Society
25377 Weld County Road 17
Johnstown, CO 80534

BEEF-MASTER
Beef-Master Breeders Universal
Suite 350, GPM South Tower
800 Northwest Loop 410
San Antonio, TX 78216

BEEFMASTER
Foundation Beefmaster's Association
200 Livestock Exchange Building
4701 Marion St.
Denver, CO 80216

BEEFMASTER
National Beefmaster Association
Ellis, KS 67637

BELTED GALLOWAY
The Belted Galloway Society Inc.
P.O. Box 5
Summitville, OH 43962

BLACK MAINE-ANJOU
American Black Maine-Anjou Association
4310 North Central Expressway
Box 64841
Dallas, TX 75206

BLONDE D'AQUITAINE
American Blonde D'Acquitaine Association
Rt. 3, Box 230
Grandview, ID 83624

BLONDE D'AQUITAINE
National Blonde D'Acquitaine Foundation
Rt. 1, Box 132
Payson, UT 84651

BRAFORD
International Braford Association
Box 1030
Ft. Pierce, FL 33454

BRAHMAN
American Brahman Breeders Association
1313 LaConcha Lane
Houston, TX 77054

BRANGUS
International Brangus Breeders Association
9500 Tioga Dr.
San Antonio, TX 78230

BROWN SWISS
Brown Swiss Cattle Breeders Association
Box 1038
Beloit, WI 53511

BUFFALO
American Buffalo Association
Wallace, KS 67761

BUFFALO
National Buffalo Association
Box 706
Custer, SD 57730

CHAROLAIS
American-International Charolais Association
1610 Old Spanish Trail
Houston, TX 77054

CHAR-SWISS
Char-Swiss Breeders Association
407 Chambers St.
Marlin, TX 76661

CHIANINA
American Chianina Association
Box 890
Platte City, MO 64079

CORRIENTE
North American Corriente Association
Pukwana, SD 57370

DEVON
Devon Cattle Association Inc.
P.O. Box 68
Uvalde, TX 78801

DEXTER
American Dexter Cattle Association
P.O. Box 56
Decorah, IA 52101

DUTCH BELTED
Dutch Belted Cattle Association of America
Box 358-Highland County
Venus, FL 33960

GALLOWAY
American Galloway Breeders
Suite 311, 1204 W. University Dr.
Denton, Texas 76201

GALLOWAY
Galloway Cattle Society of America
(Address Unknown)

GELBVIEH
American Gelbvieh Association
5001 National Western Drive
Denver, CO 80216

HEREFORD
American Hereford Association
P.O. Box 4059
Kansas City, MO 64101

HOLSTEIN-FRIESIAN
Holstein-Friesian Association of America
P.O. Box 808
Battleboro, VT 05301

ILLAWARRA
International Illawarra Association
1722-JJ South Glenstone Ave.
Springfield, MO 65803

LIMOUSIN
North American Limousin Foundation
100 Livestock Exchange Building
Denver, CO 80216

LONGHORN
Texas Longhorn Breeders Association of America
3701 Airport Freeway
Fort Worth, TX 76111

MAINE-ANJOU
American Maine-Anjou Association
567 Livestock Exchange Building
Kansas City, MO 64102

MARCHIGIANA
American International Marchigiana Society
P.O. Box 342
Atlanta, TX 75551

MARCHIGIANA
Marky Cattle Assn.
111 Livestock Exchange Building
Denver, CO 80216

MILKING SHORTHORN
American Milking Shorthorn Society
1722-JJ South Glenstone Ave.
Springfield, MO 65804

MURRAY GREY
American Murray Grey Association
Box 30085
Billings, MT 59107

NORMANDE
American Normande Association
Rt. 1
Verdon, NE 68457

NORMANDE
North American Normande Association
123 Airport Rd.
Ames, IA 50010

NORWEGIAN RED
North American Norwegian Red Association
(Address Unknown)

PINZGAUER
American Pinzgauer Association
123 Airport Rd.
Ames, IA 50010

POLLED HEREFORD
American Polled Hereford Association
4700 East 63 St.
Kansas City, MO 64130

POLLED SHORTHORN
American Polled Shorthorn Society
Box 156
Virginia, IL 62691

RED ANGUS
Red Angus Association of America
P.O. Box 776
Denton, TX 76201

RED BRANGUS
American Red Brangus Association
Department H, Box 1326
Austin, TX 78767

RED POLL
American Red Poll Association
P.O. Box 35519
Louisville, KY 40232

ROMAGNOLA
American Romagnola Association
(Address unknown)

SALERS
American Salers Association
101 Livestock Exchange Bldg.
Denver, CO 80216

SANTA GERTRUDIS
Santa Gertrudis Breeders Interntl.
P.O. Box 1257
Kingsville, TX 78363

SCOTCH HIGHLAND
American Scotch Highland Breeders Association
P.O. Box 81
Remer, MN 56672

SHORTHORN
American Shorthorn Association
8288 Hascall St.
Omaha, NE 68124

SIMMENTAL
American Simmental Association
1 Simmental Way
Bozeman, MT 59715

SOUTH DEVON
North American South Devon Association
Box 68
Lynnville, IA 50153

TARENTAISE
American Tarentaise Association
123 Airport Rd.
Ames, IA 50010

WELSH BLACK
Welsh Black Cattle Association
Route 1, Box 76B
Shelburn, IN 47879

WHITE PARK
White Park Cattle Association of America
419 North Water St.
Madrid, IA 50156

CATTLE-FAX
Cattle Marketing Information Service Inc.
P.O. Box 3947
Englewood, CO 80155

A non-profit corporation governed by cattlemen, under the NCA. Cattle-Fax provides a staff of skilled, knowledgeable market analysts. Local, state, regional, national, and international market information and analysis. Analysis of factors affecting the market.

COMMUNITY NUTRITION INSTITUTE (CNI)
1146 19th Street, N.W.
Washington, DC 20036

A non-profit citizen's organization specializing in food and nutrition. CNI's Hunger Action Division provides training and technical assistance to community groups engaged in anti-hunger activities. CNI's Consumer Division (CD), a nationally recognized advocate of the consumer interest, contributes significantly to federal policies on food safety, inflation and other economic issues, nutrition labeling, food advertising, and public participation in regulatory processes. A Food Policy Resource Center assists in monitoring and analyzing these issues and in preparing testimony which CNI/CD regularly presents to the Congress.

CONGRESS WATCH
215 Pennsylvania Ave., S.E.
Second Floor
Washington, DC 20003

CONSUMER ALERT COUNCIL
1024 J. St.
Modesto, CA 95354

CONSUMER FEDERATION OF AMERICA
1012 14th Street, N.W.
Washington, DC 20036

COUNCIL FOR AGRICULTURAL SCIENCE AND TECHNOLOGY (CAST)
250 Memorial Union
Ames, IA 50011

The purposes of CAST are: (a) to serve as a resource group from which the public and government may seek information on the science and technology of food and agricultural issues of current concern, and independently to identify developing issues of broad public concern on which food and agricultural scientists and technologists from the relevant disciplines can provide information, (b) to organize task forces of food and agricultural scientists and technologists from the relevant disciplines to assemble and interpret factual information related to these issues, and (c) to disseminate the information in a usable and effective form to the public, the news media, and the government, as appropriate.

FEDERAL TRADE COMMISSION
Pennsylvania Ave. at Sixth St., N.W.
Washington, DC 20580

The basic objective of the FTC is the maintenance of strongly competitive enterprises as the keystone of the American economic system. Although the duties of the Commission are many and varied under law, the foundation of public policy underlying all these duties is essentially the same: to prevent the free enterprise system from being stifled, substantially lessened or fettered by monopoly or restraints on trade, or corrupted by unfair or deceptive trade practices.

FEDERATION OF SCIENTIFIC AGRICULTURAL SOCIETIES (FSAS)

Principal objective is to represent the scientific societies, particularly in identifying national goals and priorities related to education and research in food, agriculture and natural resources, and in urging executive and legislative support of the priority areas. Objectives of FSAS and CAST are largely complementary.

FOOD AND DRUG ADMINISTRATION (FDA)
Department of Health, Education and Welfare
Rockville, MD 20857

The FDA is a regulatory agency responsible for protecting the public against impure and unsafe foods (with the exception of products containing red meat and poultry), drugs, cosmetics, medical devices, radiation emitting products, biologics, and veterinary products.

FOOD AND DRUG LAW INSTITUTE
Suite 280
1200 New Hampshire Avenue, N.W.
Washington, DC 20036

Manufacturers and distributors of food, drugs, and cosmetics are industrial members. To promote development of essential knowledge about the laws of food, drugs, and related laws.

FOOD AND ENERGY COUNCIL INC.
Building 4, Suite 202
Vandiver West Business Park
Columbia, MO 65201

Membership: rural electric cooperatives, public power districts, investor-owned utility companies, equipment manufacturers and a few individuals.

Objectives: (1) To cause the general public to be more aware of the critical interdependence of food supplies and of energy supplies as well as the impact each has upon the American system, economy and lifestyles; (2) Encourage the full development and use of all available and practical energy resources, but especially the use of coal and uranium supplies during the next few decades; (3) Foster better energy management, energy conservation and the substitution of energy resources whenever economically feasible.

FOOD MARKETING INSTITUTE
Suite 700
1750 K St., N.W.
Washington, DC 20006

Independent grocers, chain stores and wholesalers. Maintains liason with government and consumers. Conducts educational conferences and seminars on store security systems and personnel divisions.

FOODSERVICE EQUIPMENT DISTRIBUTORS ASSOCIATION
332 South Michigan Ave.
Chicago, IL 60604

Distributors of food service equipment (for example, ovens, ranges, dishwashers) for hotels, restaurants, and institutions.

INSTITUTE OF FOOD TECHNOLOGISTS (IFT)
221 North LaSalle
Chicago, IL 60601

IFT is a world-wide society of professional food technologists, scientists, engineers, executives, and educators in the field of food technology and other individuals interested in food technology because of their work in closely related fields.

Objectives have the ultimate goal of promoting the application of science and engineering to the improved production, processing, packaging, distribution, preparation evaluation, and utilization of foods in order to provide better and more adequate foods for mankind.

INTERAMERICAN CONFEDERATION OF CATTLEMEN
Avenue Justo Arosemena Y Calle 32
Segundo Piso, Edif. Vallarino
Panama City, Panama

Federation of associations of cattlemen in the Americas. To promote the welfare of the cattle industry, exchange information, organize forums.

INTERNATIONAL EMBRYO TRANSFER SOCIETY
3101 Arrowhead Road
Laporte, CO 80535

An organization of professional persons with particular interest in the field of animal embryo transfer.

"The society shall further the sciences of animal embryo transfer by: (1) promoting more effective research, (2) disseminating scientific and education information, (3) fostering high standards of education; (4) maintaining high standards of ethics, and (5) cooperating with other organizations having similar objectives."

INTERNATIONAL LIVESTOCK BRAND AND THEFT CONFERENCE
Tuffy Cooper, Pres.
Box 2183
Hobbs, NM 88240

An association of offical livestock inspection and theft investigation agencies. An organization concerned with animal identification (for example, brands, ear marks, wattles) and the recording of them by states.

Goals: Develop uniformity in member states brand inspection certificates, recording of brands and marks and research into new identification methods, that is, electronic identification and freeze branding. Closely involved in rustling and anti-theft prevention.

LIVESTOCK CONSERVATION INSTITUTE
Suite 143
1100 Jorie Blvd.
Oak Brook, IL 60521

LCI serves as a clearinghouse for all sectors of the livestock and meat industry in sponsoring research and educational programs designed to eradicate diseases among livestock and to reduce the tremendous losses that result from parasites, bruises and other injuries incurred in handling or shipping.

Objectives: (1) to help the livestock and meat industry make money by improving operations and to save money by reducing man-made losses; (2) to work with all segments of the livestock and meat industry in assuring the public an adequate supply of wholesome meats at affordable prices; (3) to encourage safe and humane handling of all livestock from conception to slaughter; (4) to initiate and implement methods and procedures in the production, marketing, transportation, processing, and handling of livestock and meat products that will improve these operations and thus make them more profitable for those legitimately involved in the various steps; and (5) to maintain strong communications and good working relationships between its subscribing members and educational groups, governmental agencies at all levels, and the general public in respect to health standards among livestock.

LIVESTOCK MARKETING ASSOCIATION (LMA)
301 East Armour Blvd.
Kansas City, MO 64111

The livestock marketing sector's Trade Group is composed of five interrelated organizations which function together to advance the interests of the livestock industry through the maintenance of a free, competitive marketing sector. The Trade Group's other major activity is to develop and provide commercial services for its subscribers.

The Trade Group's interrelated organizations are: (1) LMA; (2) Livestock Marketing Services Corporation; (3) Livestock Merchandising Institute; (4) Livestock Laws Reform Commission; and (5) Livestock Marketing Council.

LIVESTOCK MERCHANDISING INSTITUTE (LMI)
4900 Oak St.
Kansas City, MO 64112

LMI is an independent organization made up of trustees who are leaders in the various business activities relating to the livestock economy.

Objective: To help maintain equitable and stable profits for the livestock industry.

Purpose: To fully explore the potential for such profits offered by any and all business concepts and methods relating to the merchandising of livestock; and to report its findings to the livestock industry.

LMI is not a trade association. Its programs and activities are designed to complement and encourage the marketing interest of the livestock industry's trade associations rather than to duplicate their efforts or compete with them in any way. The Institute is a non-profit and non-political organization. It is supported through the financial investments of organizations, companies, and individuals.

LIVESTOCK PUBLICATIONS COUNCIL
927 Elmview
Eminitas, CA 92024

Numbered among the several objectives are: (1) to promote understanding and cooperation among publications serving the livestock industry; (2) to encourage and support research and activities designed to further the livestock industry; (3) through cooperative effort to foster relations between publishers and legislators, administrators, and regulators at all levels of government, as well as between publishers and people in all segments of the livestock industry and allied enterprises.

MEAT IMPORTER'S COUNCIL OF AMERICA (MICA)
1901 North Fort Myer Dr.
Arlington, VA 22209

Importers of fresh chilled and frozen meats and others indirectly associated with such importation. Promulgates industry guidelines and standards; circulates information to the industry.

NATIONAL ACADEMY OF SCIENCES (NAS)
2101 Constitution Ave.
Washington, DC 20418

NAS is a private honorary organization which serves as an adviser to the federal government on scientific questions. It is an independent source of counsel to the government on matters of science and technology. In each case, such counsel is provided as a formal report of the deliberations of a study project and its conclusions and recommendations; all reports are publicly available. The Academy created the National Research Council (NRC) as part of its structure. The NRC, as it is called, is the operating arm of the Academy and currently comprises some 900 committees and panels. Each study made by the Academy is conducted by one or more of these NRC committees or panels, the costs being met out of contracts between the Academy and the government and private foundations.

Also under the same address are: National Academy of Engineering, National Research Council and Institute of Medicine.

NATIONAL AGRICULTURAL CHEMICALS ASSOCIATION
115 Fifth St., N.W.
Washington, DC 20005

Firms engaged in producing or formulating agricultural chemical products, or producing ingredients for products used to protect against agricultural fungus diseases, insect pests, rodents, and weeds.

NATIONAL AGRICULTURE LEGAL FUND, INC.
P.O. Box 841
Ben Franklin Station
Washington, DC 20004

A national organization established to fund litigation on behalf of agriculture. Their goal is to restore an environment in which agriculture can profitably produce and maintain adequate supplies of food and fiber to meet the needs of the public. Membership is comprised of individuals and organizations representing the broad spectrum of agriculture.

NATIONAL AGRI-MARKETING ASSOCIATION
8340 Mission Rd.
Prairie Village, KS 66206

Persons engaged in agricultural advertising and marketing for manufacturers, advertising agencies and media. To promote the highest standards of agri-marketing, provide for exchange of ideas, encourage the study and better understanding of agricultural advertising, selling, and marketing.

NATIONAL AGRICULTURAL MARKETING OFFICIALS
Marketing Services Section, Department of Agriculture
Parole Plaza Bldg.
Anapolis, MD 24101

State officials responsible for administration of state agricultural marketing programs. To improve the marketing, handling, storage, processing, transportation, and distribution of agricultural products.

NATIONAL AGRICULTURAL TRANSPORTATION LEAGUE
Gale Research Company
Book Tower
Detroit, MI 48225

Truckers, truck brokers and others concerned with longhaul agricultural trucking.

NATIONAL AMERICAN INDIAN CATTLEMEN'S ASSOCIATION
Suite 918
1660 South Albion St.
Denver, CO 80222

Indian cattle producers. "To carry on all activities necessary for the betterment of the Indian cattle industry and to act as a clearinghouse for the accumulation and dissemination of information concerning the Indian cattle industry." Compiles statistics and conducts research.

NATIONAL ASSOCIATION OF ANIMAL BREEDERS (NAAB)
401 Bernadette Dr., P.O. Box 1033
Columbia, MO 65202

Farmer cooperative and private businesses interested in the improvement of farm livestock. Stimulates and encourages research in artificial insemination and reproduction. Establishes guidelines for AI, Certified Semen Services (CSS), Disease Prevention, CSS Health Program, State and Foreign Health Requirements.

NATIONAL ASSOCIATION OF MEAT PURVEYORS
252 West Ina Rd.
Tucson, AZ 85704

A non-profit organization of hotel, restaurant, and institution (HRI) supply houses who purvey meats and other food items to food service establishments and who have joined together to cooperatively enhance the interests of the industry. The association brings to its members the combined experience of the leading members of the industry throughout the United States and Canada.

Major objectives are: (1) to disseminate among the members in the industry, information that may be desired by the members or may be beneficial to the members and the industry; (2) to aid in the standardization of the products processed and distributed by the industry; (3) to act as the industry's representative in all matters pertaining to government and its agencies; and (4) to encourage the growth of product sales and to assist in the development of new products.

NATIONAL ASSOCIATION OF RETAIL GROCERS OF THE UNITED STATES
P.O. Box 17208
Washington, DC 20041

Independent food retailers. Provides the industry with services in store planning and engineering, operational methods, and procedures.

NATIONAL AUCTIONEERS ASSOCIATION
135 Lakewood Dr.
Lincoln, NE 68510-2487

An association of professionals who have organized to promote and advance the auction profession . . . promote the mutual interest of its members; formulate and maintain the high ethical standards of its members; encourages the enactment of just and reasonable laws, ordinances, and regulations affecting auction selling and the auction profession; making the public more aware of the advantage of auction selling; and improve the business conditions affecting the auction profession.

NATIONAL CATTLEMEN'S ASSOCIATION
5420 S. Quebec St., P.O. Box 3469
Englewood, CO 80155

The National Cattlemen's Association (NCA) is a national spokesman for several segments of the nation's beef cattle industry—including cattle breeders, producers, and feeders (Figure 23.2). The association represents approximately 245,000 professional cattlemen throughout the country. Membership includes individual members, 51 affiliated state cattle organizations, and 20 affiliated national breed organizations. Businesses in allied industries may be associate members. Current membership numbers approximately 30,000 commercial producers, 3,000 purebred breeders and 3,000 associate members.

The association's principal services are government affairs, public information, management education, and economic information for members. Associated with NCA

Figure 23.2. Headquarter's building for the National Cattlemen's Association and Cattle-Fax. Courtesy of the National Cattlemen's Association.

is Cattle-Fax, a market analysis service for cattlemen. The association also has offices in Washington, D.C.

NCA is an organization of commercial and purebred cattle growers, stocker operators and feeders. Every active cattleman member has an opportunity, and is encouraged to participate in policy formulation and implementation.

To assure representation of cattlemen from all areas, NCA is organized into seven regions (Figure 23.3), with a vice president representing each region on the executive committee.

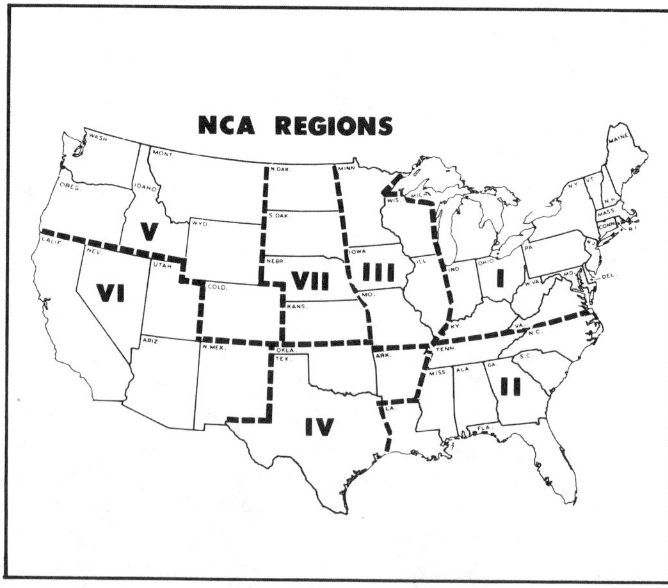

Figure 23.3. The seven designated regions of the National Cattlemen's Association. Courtesy of the National Cattlemen's Assocation.

Figure 23.4 shows the basic organizational structure of NCA.

The committees, appointed by the president, develop policy resolutions which are presented for review and disposition to the membership body during the annual business meeting of the NCA. The board of directors is charged with the responsibilities of carrying out the dictates of the membership (Figure 23.4). The active committees in NCA are (1) Animal Health and Identification, (2) Beef Promotion and Consumer Relations, (3) Environmental Management, (4) Foreign Trade, (5) Grading and Inspection, (6) Labor, (7) Marketing, (8) Membership and Revenue Development, (9) Private Lands and Water, (10) Public Lands, (11) Research and Education, (12) Tax, and (13) Transportation.

NATIONAL FARM-CITY COUNCIL INC.
Kiwanis International Building
101 East Erie St.
Chicago, IL 60611

Purpose is to bring about better understanding between the rural and urban segments of society. Members of the National Farm-City Council are organizations and individuals prominent in the agri-business complex.

Kiwanis International is the "coordinating agency." Kiwanis furnishes a headquarters address, basic office services, and the cooperation of many of its nearly 8,000 clubs.

Annual focal point of Farm-City Council effort is National Farm-City Week around Thanksgiving Day. Activities during the week include tours of farms by urban businessmen and tours of cities, factories and other businesses, by rural residents.

NATIONAL FEED INGREDIENTS ASSOCIATION (NFIA)
One Corporate Place
West Des Moines, IA 50265

The NFIA is the national spokesman for the feed ingredient industry.

Purpose: The general purpose of NFIA is to bring together ingredient suppliers, users, marketers and other related people to discuss and solve common problems.

Objectives: Research support; legislation; government regulations; public relations; mutual problem solving, sponsor continuing education, and learning among the members and others in the industry.

NATIONAL FOOD BROKERS ASSOCIATION (NFBA)
1916 M St., N.W.
Washington, DC 20036

NFBA's membership only qualified firms and persons whose business is to sell, at the wholesale buying level, food and non-food products for a commission or brokerage fee compensation. NFBA has always directed its efforts toward improvement of business conditions, not only as they affect brokers but also for the continuing progress of the food industry. The association has consistently fought for elimination of unfair business practices and championed the true competitive drive among large and small businesses.

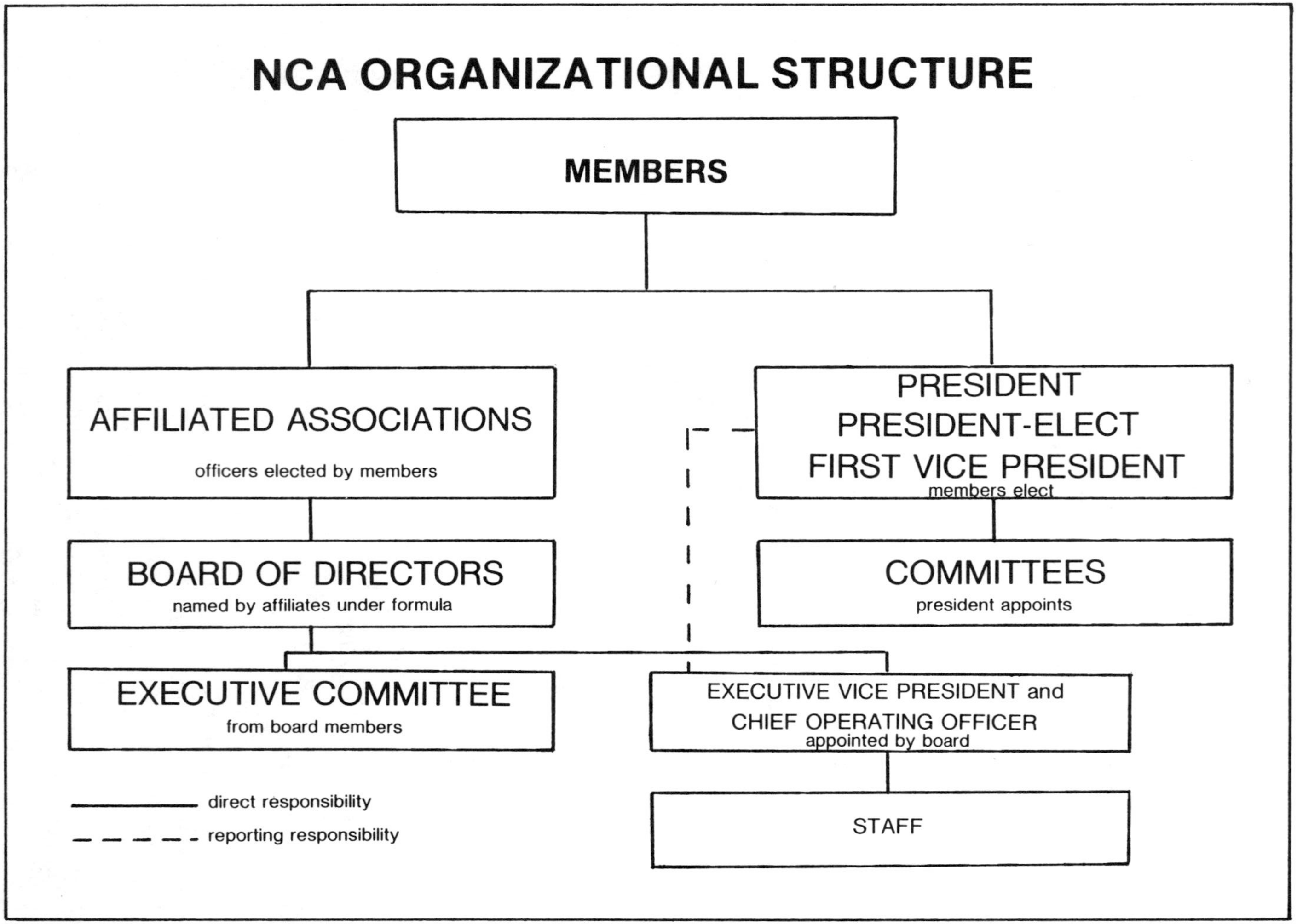

Figure 23.4. The organizational structure of the National Cattlemen's Association. Courtesy of the National Cattlemen's Association.

NATIONAL FOOD PROCESSORS ASSOCIATION (NFPA)
(Formerly National Canners Assn.)
1133 20th St., N.W.
Washington, DC 20036

NFPA is a non-profit trade association supported by food processing companies and suppliers to the industry. The association's board of directors and executive committee, chosen from the food processing industry, establish policy and provide overall direction to the association.

NATIONAL FOOD SERVICE ASSOCIATION
P.O. Box 1932
Columbus, OH 43216

Food service firms; factories, schools, hospitals, clubs, restaurants, and institutions.

NATIONAL GRANGE
1616 H St., N.W.
Washington, DC 20006

Fraternal organization of rural families. Promotes general welfare and agriculture through legislative, social, education community service, home economics, youth, aid for handicapped, cooperatives, insurance, and credit union programs.

NATIONAL HOT DOG AND SAUSAGE COUNCIL
Two North Riverside Plaza
Chicago, IL 60606

Conducts public relation campaigns to the consumer on behalf of the processed meat industry.

NATIONAL INSTITUTE FOR THE FOOD SERVICE INDUSTRY
20 North Wacker Drive
Chicago, IL 60606

Non-membership, educational foundation supported by segments of the food service industry such as restaurateurs, food service companies, food and equipment manufacturers, distributors, and trade associations. Dedicated to the advancement of professional standards in the food service industry through education and research. Offers food service management courses, career information, scholarships and teacher training grants, and research.

NATIONAL LAND FOR PEOPLE
2348 North Carnelia
Fresno, CA 93711

A non-profit, non-government funded organization which is opposed to "big business" involvement in beef production. The organization is oriented towards small-farm groups with issues centered around reclamation. No direct ties to the beef industry. Beef feedlots are considered deplorable practices.

NATIONAL LIVE STOCK AND MEAT BOARD
44 North Michigan Ave.
Chicago, IL 60611
(312) 467-5520

Funded by state checkoff systems of several cents per head on livestock marketed and slaughtered (see Chapter 2). The checkoffs vary as some are voluntary and some are legislated. All checkoff money is identified by species and goes to one of the three divisions within the organization— Beef Industry Council, Pork Group, or Lamb Committee. These three divisions then collectively fund the Meat Board's red meat programs (nutrition, research, information, education) as well as promotional activities for each species. The Meat Board's promotional work is directed toward the retailer by developing promotional services for retailers' use and ways to merchandise meat. Involved in USDA projects for the nutrient labeling of beef, school lunch programs, educational kits, filmstrips and other materials. The nutrition department establishes meat's credibility. Its concern is proving the worth of meat in the human diet through mailing out "Food and Nutrition News." Diet-disease question—an administrative arm of the American Meat Science Association. Symposia held on meat in nutrition and health-credentials.

NATIONAL LIVESTOCK PRODUCERS ASSOCIATION
307 Livestock Exchange Building
Denver, CO 80216

Largest livestock cooperative marketing group in the nation. Federated headquarters for 12 livestock marketing cooperatives, 6 affiliated livestock credit corporations and a subsidiary commodity company. They are owned and controlled by member marketing associations which are in turn owned and controlled by their member patrons. Livestock producers make up the membership in areas from the Eastern Corn Belt to the West Coast. They do hedging and forward contracting for members through the commodity company.

NATIONAL MEAT ASSOCIATION (NMA)
734 15th Street, N.W.
Washington, DC 20005

Independent meat packers and meat processors; suppliers to the industry. Divisions: Beef; Pork; Processed Meats.

NATIONAL LIVESTOCK GRADING AND MARKETING ASSOCIATION
Livestock Standardization Branch
USDA
Washington, DC 20250

An association of states whose objectives and purposes are to: Standardize grades between states through the use of USDA grades for livestock; encourage and assist states in the correlation of live and carcass grades; and promote and encourage more livestock buying power in states through uniform livestock grading and improved techniques in marketing.

NATIONAL MEAT CANNERS ASSOCIATION
P.O. Box 3556
Washington, DC 20007

Members are 35 companies whose primary or secondary business is sterile processed meat products.

Purpose: Promote the interests of packers of commercially sterile prepared meats and meat food products packed in containers; encourage scientific and practical research; collect and disseminate research information; inform the trade and public of the advantages of processed meat product usage.

NATIONAL RESTAURANT ASSOCIATION
311 First St., N.W.
Washington, DC 20001

Restaurants, cafeterias, clubs, contract feeders, drive-ins, caterers, institutional food services, and other members of the food service industry; also represents establishments belonging to non-affiliated state and local restaurant associations in governmental affairs. Supports food service education and research in several educational institutions; conducts traveling management courses and seminars for restaurant personnel; distributes educational pamphlets, books, and films about the industry.

NATIONAL RENDERERS ASSOCIATION
3130 Des Plaines Ave.
Des Plaines, IL 60018

NATIONAL SOCIETY OF AUCTIONEERS
135 Lakewood St.
Lincoln, NE 68510

Professional society of auctioneers. Conducts educational seminars and sponsors Certified Auctioneer Institute Program, Indiana University.

NATIONAL SOCIETY OF LIVESTOCK RECORD ASSOCIATIONS
210 Utah Ave.
West Plains, MO 65775

The society provides an opportunity for qualifying livestock registries to cooperate and work together in solving problems of common interest; develop programs and procedures on matters of common interest; study and develop improved procedures for processing and preserving records and pedigrees; develop testing improvement and promotional ideas; aid in establishing useful and workable health regulations and ecology programs; develop effective, workable and protective regulations and procedures for artificial insemination and embryo transplants; aid in formulation of practical tax legislation; aid in developing adequate facilities and workable procedures for blood typing; help develop improved and more meaningful shows and fairs; aid in developing workable export and import procedure and regulations and in the development of programs for service and consultation to commercial livestock producers.

OCCUPATIONAL SAFETY AND HEALTH ADMINISTRATION (OSHA)
U.S. Department of Labor
200 Constitution Ave., N.W.
Washington, DC 20210

OSHA was established pursuant to the Occupational Safety and Health Act of 1970. It develops and promulgates occupational safety and health standards; develops and issues regulations; conducts investigations and inspections to determine the status of compliance with safety and health standards and regulations. OSHA has regional offices established in ten areas throughout the United States.

PUBLIC LANDS COUNCIL
425—13th St., N.W. #1020
Washington, DC 20004

Established in 1969 to represent the interests of ranchers who hold leases and permits to graze livestock on the public lands in the western states. The Council is dedicated to the principle of sound management of the Federal lands for grazing and all other multiple use purposes. Membership of the Council includes both individual livestock men and sheep and cattle organizations in the 13 western states: Arizona, California, Colorado, Idaho, Montana, Nevada, New Mexico, North Dakota, Utah, Washington, and Wyoming.

SOCIETY FOR RANGE MANAGEMENT
2760 West Fifth Ave.
Denver, CO 80204

Objectives are: (1) develop an understanding of range ecosystems and principles applicable to management of range resources; (2) assist all who work with range resources to keep abreast of new findings and techniques in the science and art of range management; (3) improve the effectiveness of range management to obtain from range resources products and values necessary for man's welfare; (4) create a public appreciation of economic and social benefits to be obtained from the range environment; and (5) promote professional development of its members.

STATE BEEF COUNCIL

Essentially each state has a check-off system obtained from the beef industry in that state. They are involved with the school system; hotel, restaurant and institutional industry; retailers and all other segments of the industry. Their objective is to facilitate better knowledge by consumers of beef's proper role in the diet and educate those consumers in selection, preparation and proper cookery methods to best utilize the beef product.

STATE CATTLE AND FEEDERS ASSOCIATION

Several states have a state cattlemen's association which combines both cow–calf producers and feeders. Other states have separate organizations representing each group. Members discuss and take action on legislation, research, marketing, and other current issues influencing beef producers in their respective states and in the nation.

Most state cattle organizations are affiliated with the National Cattlemen's Association (NCA).

NATIONAL ASSOCIATION OF STATE DEPARTMENTS OF AGRICULTURE
1616 H St., N.W.
Washington, DC 20006

TANNERS COUNCIL OF AMERICA
411 Fifth Ave.
New York, NY 10016

UNITED STATES BEEF BREEDS COUNCIL
(no established office—correspondence is addressed to current president's office)
President Greg Martin
Room 100, Livestock Exchange Building
Denver, CO 80216

Their function is to create a more unified effort by purebred breeders to promote beef and the industry.

UNITED STATES DEPARTMENT OF AGRICULTURE (USDA)
Washington, DC 20250

It works to improve and maintain farm income and develop and expand markets abroad for agricultural products. The Department helps curb and cure poverty, hunger and malnutrition. It works to enhance the environment and maintain production capacity by helping landowners protect soil, water, forests, and other natural resources. Rural development, credit, and conservation programs are key resources for carrying out national growth policies. The USDA, through inspection and grading services, safeguards and assures standards of quality in the daily food supply.

USDA—AGRICULTURAL MARKETING SERVICES (AMS)

AMS administers standardization, grading, inspection, market news, marketing orders, regulatory, and related programs.

AMS provides a Market News service which gives current, unbiased information to producers, processors, distributors, and others to assist them in orderly marketing and distribution of farm commodities.

Standardization and grading of live cattle, swine, sheep, wool and mohair are also carried out by AMS.

USDA—AGRICULTURAL RESEARCH SERVICE (ARS)

ARS staff administers a basic and applied research program in animal and plant production; use and improvement of soil, water and air; processing, storage, and distribution of farm products; and food safety and consumer services.

USDA—ANIMAL AND PLANT HEALTH INSPECTION SERVICE (APHIS)

APHIS was established to conduct regulatory and control programs to protect and improve animal and plant health for the benefit of man and the environment. In cooperation with state governments, APHIS administers Federal laws and regulations pertaining to animal and plant health and quarantine, humane treatment of animals, and control and eradication of pests and diseases.

USDA—ECONOMIC RESEARCH SERVICE (ERS)

ERS carries out a national program of research relating to production and marketing of farm commodities. It includes evaluations of the organization and performance of major commodity subsectors; costs and returns to farmers and marketers; situation and outlook; commodity projections; price spreads; and analysis of the United States farm commodity programs.

USDA—ECONOMICS AND STATISTICS SERVICE (ESS)

ESS formulates, develops and administers a program of economics, statistical, and other social science research, analysis, and information related to food, agriculture, and rural resources and communities; a program to collect and publish statistics related to food, agriculture, and rural resources and communities. Results of these activities are made available to users through research and statistical reports and through outlook and situation reports on major commodities, rural areas, and national and international economies.

USDA—EXTENSION SERVICE

The Extension Service serves as partners with state governments, through their land-grant universities and county governments to form the Cooperative Extension Service. These levels of government share in financing and conducting educational programs to help the public learn about and apply to every day activities the latest technology and management knowledge developed through research by landgrant universities, the USDA and other sources. Major areas of assistance are agricultural production, marketing, natural resources, home economics, food and nutrition, 4-H youth development, community, and rural development.

USDA—FARMERS HOME ADMINISTRATION

The FHA provides credit for those in rural America who are unable to get credit from other sources at reasonable rates and terms.

Applications for loans are made at the agency's 2,200 local county and district offices, generally located in county seats. A county or area committee of three individuals, at least two of whom are farmers, certifies or recommends as to eligibility of individual farm loan applicants and reviews borrower's progress.

USDA—FOOD AND NUTRITION SERVICE (FNS)

FNS administers programs to make food assistance available to people who need it. These programs are operated in cooperation with state and with local governments.

Programs include: family nutrition program, food distribution, special nutrition programs, nutrition education, and training.

USDA—FOOD SAFETY AND QUALITY SERVICE (FSQS)

FSQS, in cooperation with state agencies, offers official grading or inspection for quality of manufactured dairy products, poultry, eggs, meat, and fresh and processed fruits and vegetables. Grading is based on grade standards developed by the USDA for these products.

The FSQS examines the product at the manufacturing, processing, or packing plant of the supplier's warehouse. If the product meets contract specifications, it is stamped with a USDA grademark.

USDA—FOREST SERVICE

The Forest Service has the federal responsibility for national leadership in forestry. Some of their goals are: generation of forestry opportunity to accelerate rural community growth; encouragement of optimum forest land ownership patterns.

USDA—OFFICE OF TRANSPORTATION (OT)

The OT is responsible for development of USDA transportation policies for agriculture and rural development, including coordination of USDA programs in education, information, loans, grants, and research to meet policy goals, and representation of agricultural and rural transportation interests before regulatory bodies.

USDA—PACKERS AND STOCKYARDS ADMINISTRATION

Administers the Packers and Stockyards Act, helps maintain free and open competition in marketing of the nation's livestock, poultry, and meat production. The P&SA Act prohibits unfair, deceptive, discriminatory, and monopolistic marketing practices which would deprive markets of a reasonable profit for their services, and would unduly affect consumer prices for meat and poultry.

USDA—SCIENCE AND EDUCATION ADMINISTRATION

Livestock and veterinary sciences program. The objectives are: (1) improved efficiency of production of safe, high quality animal protein and fiber through development of new knowledge to improve genetic characteristics; (2) improved reproductive efficiency; (3) prevention and control of diseases, insects, and pests and hazards; (4) improved feeding and management practices and machinery, buildings, equipment, and energy use; (5) reduction of rural housing construction and operating costs and improvement of water and waste systems; and (6) reduction of problems associated with insects affecting man.

USDA—SOIL CONSERVATION SERVICE (SCS)

The SCS has the responsibility for developing and carrying out a national soil and water conservation program in cooperation with landowners and operators.

USDA—WORLD FOOD AND AGRIC. OUTLOOK SITUATION BOARD (WFAOSB)

The WFAOSB coordinates and reviews all commodity and aggregate agricultural and food data and analyses used to develop outlook and situation material prepared within the USDA. The aim is to improve consistency, objectivity, and reliability of material being disseminated to the public.

USDA—UNITED STATES MEAT ANIMAL RESEARCH CENTER (MARC)
Clay Center, NE

MARC develops, through research, new technology for meat animal production. The goals are to improve carcass merit and reduce production costs of cattle, sheep, and hogs. The center is being developed by the federal government and is administered by the Agricultural Research Service, USDA.

U.S. ENVIRONMENTAL PROTECTION AGENCY (EPA)
401 M St., S.W.
Washington, DC 20460

The EPA was created to permit coordinated and effective governmental action on behalf of the environment. EPA endeavors to abate and control pollution systematically, by proper integration of a variety of research, monitoring, standard setting, and enforcement activities. EPA is designed to serve as the public's advocate for a livable environment.

U.S. MEAT EXPORT FEDERATION (USMEF)
7200 Stapleton Plaza
3333 Quebec St.
Denver, CO 80207

The purpose of USMEF is to coordinate and unify efforts in areas of foreign market development, inter-governmental trade policies, and fair and reciprocity in world meat trade. Every private U.S. business whose activity is in any way associated with successful profitable livestock and meat production, marketing, and consumption may be involved.

VEGETARIAN INFORMATION SERVICE

During their 1982 conference, task forces were established to work on the following objectives: promotion of meatless diets, establishment of treatment standards for farm animals, raising public awareness of animal rights, introduction of animal issues to the political process, and planning more training and mobilization sessions.

WESTERN STATES MEAT PACKERS ASSOCIATION
88 First St.
San Francisco, CA 94105

Meat packers, processors and jobbers; to promote interest of independent meat packers in 14 western states.

SELECTED REFERENCES

Publications

Baker, F. H. *The potential of beef organizations to contribute to beef directions of the 80's and 90's.* Beef Directions for the 80's and 90's. International Symp. Queenstown, MD., April 1981.

Berry, W. T., Jr. *The Beef Industry—and Organizations Which Serve Cattlemen.* NCA Beef Business Bull., 21, January 1983.

The Future for Beef. National Cattlemen's Association Beef Business Bulletin. 5 March 1982.

Robbe, D., 1982. *The Organizational Structure of the Beef Industry.* Master of Science Project. Colo. State Univ.

24

The Past and Future of the Beef Industry

The beef cattle industry has had an exciting past. Many books and films have captured this excitement, allowing others to relive it. The future of the beef industry appears to be equally as exciting and challenging as the past.

There are those who say, "Don't worry about history and the past because you can't do anything about it." There is much truth to that statement, but considerable insight into the cattle industry can be gained by studying the past. The quotations of two noted individuals appear to be worthy of serious consideration

"The farther back you look, the farther forward you are likely to see."
Winston Churchill

"Those who don't remember the past are compelled to relive it."
George Santayana

The latter quote is most meaningful to those individuals who have observed or studied the repeated beef cattle cycles (see Chapter 18) or to those who understand the several marked changes in beef cattle type (see Chapter 17).

Scientific and Historical Evidences of Early Cattle

3 TO 4 MILLION YEARS AGO Fossil remains of cattle have been found in northwestern India. One species was believed to be a direct ancestor of the ox. The head was relatively long and narrow. Bulls were horned and females were hornless. The horn spread was between 6 to 7 feet. They were approximately 5½ feet high at the shoulders.

3 TO 4 MILLION YEARS AGO Fossil remains found in West Central Italy (after those found in India). These had more slender limbs than those found in Asia. They were given the name *Leptobos etruscus*. Other specimens were found in Southern France.

1 MILLION YEARS AGO Fossils were identified that provided recognition of three different forms of ox in Europe, also a bison and a water buffalo. The most noted of these was the Auroch (the great wild ox), classified as *Bos primigenius.* They were extremely large, standing some six feet high at the shoulder. Measurements of 10 to 12 feet between the tips of the horns were common.

ONE-HALF MILLION YEARS AGO Competition for feed from other forage-loving animals occurred. Over the next several hundred thousands of years, four glaciation periods posed climatic challenges to cattle. It is believed that cattle were protected with heavy layers of thick, shaggy hair, similar to that of Scotch Highland cattle today. During this time, cattle had an additional enemy—man, who came in search of food and skins.

40,000 to 50,000 BC Fossilized bones reveal that prehistoric man cooked his meat and cracked open the bones for the marrow inside.

30,000 to 15,000 BC Paintings of bison found in caves in Northern Spain in 1875. More spectacular paintings (varied from 3 inches to 17 feet in length) of Aurochs and *Bos longifrons* were found in the Lascaux Caves in France in 1940. Evidence that some of these paintings, in the same caves, occurred thousands of years apart.

10,000 to 8000 BC Neanderthal man, prevalent in Europe, chased the Auroch for food. Man was likely assisted in the hunt with his first domestic animal—the dog. Cro-Magnon man improved his hunting skill by driving the cattle over steep cliffs, a later practice of American Plains Indians when hunting bison.

8000 to 4000 BC Domestication of cattle occurred during this time. There is insufficient information to identify the exact time of domestication. Fossil remains give evidence that Denmark, Switzerland, Iraq, Palestine, Egypt, and Iran were early locations for domestic cattle. In these locations, Aurochs (*Bos primigenius*), Celtic Shorthorns (*Bos longifrons*), and cattle of Zebu descent have been identified.

6500 BC Some evidence indicates that the earliest domestication date for cattle was in Greece at this approximate date.

4500 BC The oldest archaeological record for Zebu cattle was from a figurine found in Mesopotamia.

3000 BC Fossil remains of humped cattle and art showing Auroch type cattle in Crete. Pre-Roman colonizers, migrating from Greece, had cattle. Humped cattle also appeared in Egypt about this time.

2500 BC The tombs in Egypt show the following pictures as evidence of early cattle production:

—Tomb of Akhethetep shows boys playing with calves whose forelegs are hobbled.

—Tombs of Deir show a calf being delivered. Also shown are cattle being thrown for slaughter or branding.

—There is a group of six oxen pulling a plow.

—Tomb of Huy depicts cattle being branded.

—Tomb of Auta contains a picture of a bull branded with the number 113.

—Cattle being fattened.

—Cattle being stunned for slaughter, skinned, and meat being boned. There are also models of slaughter houses, killing floors, smoke houses, and meat shops.

2000 to 1500 BC The year 2000 BC marks the estimated beginning of the patriarchs of the Old Testament (Abraham, Jacob, and others). Many references are made to cattle. The treasured Ark and Covenant were carried on a cart pulled by a select yoke of oxen (I Samuel 6:7).

500 BC Romans were probably the first people to discover the principles of good farming and stock raising.

65 BC Julius Caesar provided the first historical record of wild cattle (Uri or Aurochs) when he wrote:

> "There is a third kind of these animals which are called Uri. In size these are but little inferior to elephants, although in appearance, color, and form they are bulls. Their strength and speed are great. They spare neither men or beasts when they see them. . . . Those who kill most receive great praise when they exhibit their horns as trophies of their success. These Uri, however, even when young cannot be tamed. In the expanse of their horns, as well as in form and appearance, they differ much from our domesticated oxen."

Auroch cattle existed in the wild state for several centuries. The last known survivor died in 1627 in a Polish Park.

30 BC Roman cattle producers got advice from Virgil (farmer's son, poet, and small-time husbandman) in his four poems called Georgics. Following is an excerpt from one of the poems:

> "Next, when calving is o'er, men's whole thought goes to the offspring; and they stamp them anon with brands distinguishing each one as preference dictates: these of maintaining a true breed, these of sacred office, and these as laboring oxen upturning the rugged loamclods and straining across them, while others at grass go forth as an army to pasture."

100 AD Trajan built his forum in Rome. Some of the decorative bull sculptures showed an advanced stage of breeding for edible meat production.

300 AD Age of the Barbarians. They killed cattle, scattered the herds, and killed many of the knowledgeable cattle producers. This was a noted setback to sound cattle breeding and management.

493 AD There was a revival of fairs in Italy. They were held to stimulate trade with other countries. There were impressive sales of work oxen and cows at these fairs. Other significant cattle fairs were held in France and Russia.

HISTORICAL LANDMARKS IN THE DEVELOPMENT OF THE CATTLE INDUSTRY IN THE AMERICAS

981 Erik the Red, the famous Norseman, brought cattle to Greenland. The cattle were supposedly of Spanish (Andalusian) origin. During the fifteenth century, several thousand people succumbed to illness and all died along with their livestock.

1131 Oldest fair for the exhibition of cattle established at Smithfield, near London. The fair, considered as the granddaddy of all fat stock shows in England and America, had a very slow start. It was not until 1798 that the Smithfield Show emerged as being an important influence on the cattle industry.

1200 The Castile area in North Central Spain became one of the first known areas of running domesticated cattle on the open range. This practice spread southward into Andalusia (extreme Southern Spain). Cattle ranching was practiced here, and its influence was carried into North and South America when the cattle of Andalusia were transported to the West Indies. When Columbis left Spain for the New World, cattle ranching in Andalusia was two and a half centuries old.

1493 Columbus, on his second voyage, brought cattle to Santo Domigo (West Indies). This is considered by many to be the first significant introduction of cattle to the New World. These first cattle, in addition to later shipments of cattle from Spain to the West Indies, soon developed into a thriving cattle industry during the next 20 years. During that time period, the number of cattle reached surplus levels and some ran wild. These several hundred head of cattle, shipped from Spain to the West Indies, would be the progenitors of millions of head of cattle that would range from Canada to Argentina. No other cattle types would mix with these Spanish cattle until after the beginning of the Nineteenth Century.

1513 Ponce de Leon brought cattle into Florida on his first expedition and also on his second trip some eight years later. There is no evidence that these cattle or the settlers survived. Several other cattle importations into Florida during the next 40 years failed to be successful because of resistance by Indians.

1519 Cortez was the first individual to bring livestock (there is some question if cattle were included in all of his several expeditions) directly from Spain to the American mainland. Cortez not only conquered Mexico but also started the cattle ranching business in Mexico. He branded some of his Aztec prisoners with the letter "G" on their cheek. It is possible that the first cowboys on the American continent were branded before the first cows.

1521 About the same time as Cortez, de Villalobos shipped cattle from Santo Domingo into Mexico. These cattle were likely descendants of the first cattle introduced by Columbus in 1493.

1528 de Vaca was the first Spaniard to see the American bison while crossing Texas. The French who were among the first white men to explore North America called the large, hump-backed animals they encountered on the plains "les boeufs," meaning oxen or beeves. Through a sequence of word changes, "Boeuf" became the English "Buff," then "Buffle" or "Buffler," and finally "Buffalo." Scientists use the Latin terminology of Bison bison (genus and species) for this animal pointing out that its correct name is "Bison." The true buffalo has no hump, thus the name buffalo belongs only to the water buffalo in Asia and the African Buffalo (these two buffalo belong to the genus *Bubalus*).

1529 A brand book was established in Mexico City, where all cattle owners were required to register their brands.

1540 Coronado was the first of Spanish origin to enter the southwest part of the United States with cattle on foot. He entered into what is now known as Arizona with 500 cattle which were part of 6500 head of livestock associated with Coronado's traveling army. Some exhausted cattle left behind became ancestors of thousands of wild cattle in the territory.

1565 Cattle were taken from the West Indies into Florida. These Spanish cattle were the first to persist in a breeding herd in the territory, later to become part of the United States. St. Augustine has been considered the first permanent settlement of Europeans in the United States.

1598 Juan de Onate founded the first self-supporting colony of San Juan at the junction of the Chama and Rio Grande Rivers in New Mexico. Stock breeding began in this area where 7000 head of livestock were originally involved.

1607 Cattle were brought to Virginia from England when the first English Colony was founded at Jamestown. By 1620 Virginia had 500 cattle, and by 1833, the numbers were estimated from two to five thousand.

1624 Captain Edward Winslow brought the first cattle, three heifers and a bull, from England to the Plymouth Colony in Massachusetts.

1625 The Dutch, who settled in New York in 1614, brought from Holland the coarse, Flemish-type bulls and cows.

1630 The Massachusetts Bay Colony imported 30 cows. During the next three years, cattle arrived on almost every ship. No doubt many choice cattle of various breeds found their way from England to the Colonies. Among these were noticeably the "Rubies" or the bright Red Devon cattle. During this same time period, John Mason, founder of New Hampshire, started a Danish Colony and imported several groups of the large, yellow Danish cattle for draft purposes. Few cattle were imported from England after 1650. The early pioneers, carving out new homes in a wilderness, had little time for selective cattle breeding. Crosses of cattle from the Red Devons, the yellow Danish cattle, and the red-and-white and black-and-white Dutch cattle were evident.

1636 William Pynchon and his son, John, established the first meat packing plant in Springfield, Massachusetts. Beef and pork were slaughtered and packed in barrels; thus the terms, beef packer and packing plant.

1660s The cattle industry began to flourish in the Carolinas. At that time some of the European-bred cattle began to encounter numbers of wild Spanish cattle. These latter cattle were descendants of those brought by the Spaniards from the West Indies to Florida.

1691 While it was the Spanish conquistadors that trailed the first cattle into the Southwest, it was the Spanish padres who insured the survival of cattle ranching in this area. One pioneer missionary who probably had the greatest influence on development of the Southwest was Eusebio Francisco Kino. He founded and managed a livestock breeding establishment at his first mission, Delores, in Old Mexico. From that location he sent hundreds of cattle into nineteen other missions in Arizona, Senora, and Southern California. He personally supervised many of these livestock operations. Kino never profited from these cattle operations but gave cattle freely to the Indians. During this same time period, the first cattle entered Texas, near the Louisiana line, a result again of the extension of Spanish missions.

1756 Brighton Market, near Boston, became the first recorded public auction market. It became the slaughter center for the Northeast, even though most slaughtering was still done on the farm.

LATE 1700s The farmer and cattle raiser began to push westward from the original colonies, following earlier hunters, trappers, and traders. The Allegheny Mountains presented a physical barrier. It was not long, however, until the emigrants had floated hundreds of cattle into the Eastern Mississippi Valley by way of the Allegheny and Ohio Rivers. Daniel Boone's wilderness road through the Cumberland Gap, a southern trail around the Appalachians into Tennessee, and more importantly, trails across Pennsylvania, provided access routes for the westward movement of surplus cattle from the Eastern United States.

1769 The first cattle (of Spanish origin) arrived in California and were part of the mission founded at San Diego.

1783 Shorthorn cattle exported from England to Virginia. The Shorthorn had the first herd book record of any breed established by George Coates in 1822.

1788 Cattle, probably of Longhorn origin, shipped by boat from California to Oregon Territory.

1817 Grain feeding of cattle occurred in Ohio in the early 1800s. New York City received its first shipment of Ohio grain-fed steers in June of 1817. They were driven for nearly one thousand miles at approximately seven miles per day. Cattle starting on these drives to the Eastern United States weighed from 900 to 1000 lbs and usually lost from 100 to 150 lbs during the drive period. During the next few decades, Illinois became the center of surplus cattle production. Many cattle produced in Illinois were driven to Ohio where they were fattened and sold as Ohio cattle.

The westward movement of cattle was based primarily on the economics of fed-beef production. Ohio enjoyed a near monopoly of beef production until a few years later when Central Indiana took over. By the mid 1850s, Illinois assumed the leadership, but in the late 1880s, Iowa became a more efficient producer of beef. In the early 1900s the westward shift of efficient beef production had moved into Eastern Nebraska.

1817 The famous importation of Shorthorn cattle (ancestors of the American Shorthorns or Durhams) was made by Lewis Sanders of Kentucky when he brought them from England. Henry Clay also imported Herefords from England to Kentucky that same year.

1834 During this year, the 21 Spanish missions reported total cattle numbers of 396,400. By 1842 cattle numbers in these missions had been reduced to approximately 29,000. This reduction in numbers was due in part to the drought in 1840, but primarily due to Mexican Republic authorities enforcing secularization. This policy was to emancipate the Indians and divide mission properties among them. The padres retaliated by the wholesale slaughtering of cattle, primarily in 1834. Since the primary value of cattle was in their hides and tallow, they were slaughtered for that purpose. Most of these products were shipped by ox-carts and mules to San Pedro (now Los Angeles). Boats then took these products, primarily to the Boston area. In 1834 the year's shipment from San Pedro was 100,000 hides and 2500 centals (100 pounds to the cental) of tallow and several cargoes of locally made soap.

1836 First public cattle auction sale held in Ohio.

1830–1870 The development of the Western United States had its stimulus, primarily from missionaries coming to the Indians, the gold seekers, freedom from religious persecution (Mormons), and those who hungered for the acquisition of land. Cattle (oxen) were primarily responsible for the movement of thousands of emigrants and their many tons of supplies. Most covered wagons were drawn by six- and eight-yoke teams of oxen. Their hoofprints dug deep into the major trails (Oregon, California-Overland, Santa Fe and their many cut-off trails). An estimated 250,000 emigrants traveled the California-Overland Trail between 1859 and 1869. On 14 August, during the 1850 Gold Rush, the register at Fort Laramie recorded the presence of over 39,000 men, more than 3000 women and children, 36,000 oxen, 7300 cows, 9000 wagons and other livestock.

1849 First importation of Brahman cattle to America occurred when James Bolton Davis brought two cows and a bull to his South Carolina plantation. It is believed that the Civil War caused all purebred Brahmans to become extinct. In 1854 two Brahman bulls were brought to Louisiana and used in crossbreeding. In 1885 Texas breeders imported two Nellore-type Brahmans from Calcutta. These early importations, along with later importations and the grading-up breeding process, increased Brahman numbers, particularly in Southeastern United States. The largest importations of Zebu cattle into the United States came from Brazil, through Mexico, in the 1920s.

1842 Cattle from Texas moved northward into Missouri. A few years later, herds of 1000 plus head moved from Texas into Ohio and Illinois.

1850s Cattle from Texas followed the Gold Rush into California. Cattle prices in Texas of $14.00 per head while cattle in California brought $100 per head greatly stimulated the movement of thousands of cattle.

1853 Joseph Mallory of Piatt County, Illinois, created a cattle sensation when he moved a herd from Indian territory to his home farm where he fattened them, then later shipped them into New York. Soon after this, Illinois became an active feeding center for Texas and Indian territory cattle. Independence, Westport, and Kansas City became the big cattle markets.

1860s The Civil War changed the cattle picture in Texas and eventually in the North. The Union Army blocked the gulf ports, and the Confederate Army demanded the services of all ablebodied men. The latter situation caused the cattle to run wild and multiply at a very rapid rate. By 1865 cattle numbers were at six million head in Texas. When the war ended, cattle were worth $5.00 per head in Texas and $50.00 in the North. Inequality in numbers and prices set the scene for historic cattle drives with their well-known trails. During the next thirty years, some ten million head of cattle would be driven primarily over four famous trails—the Shawnee, the Chisolm, the Western, and the Pecos River. These trails ended at the Kansas railheads, Dakota-Wyoming-Montana pastures for fattening, or connected with feeder trails into Colorado, Kansas, and Missouri.

1862 Land grant colleges were endowed under the Morrill or Land Grant Act of 1862. Congress granted every state 30,000 acres of land for each senator and representative it had in Congress. The lands were to be sold, proceeds invested, and income used to create and maintain a college for agriculture and mechanical arts.

United States Department of Agriculture (USDA) was established as a separate government agency.

The Homestead Act encouraged land settlement and family farms, which pushed the cattle grazing lands farther west.

1865 The Union Stock Yards and Transit Company of Chicago opened for business.

1869 First transcontinental railroad was completed.

1870 to 1890s Railroads developed new centers of meat packing as stockyards with affiliated meat plants were built in Midwestern cities. Kansas City yards founded in 1871, St. Louis in 1873, Omaha in 1884, Denver and St. Paul in 1886, Sioux City and St. Joseph in 1887, and Oklahoma City in 1910.

1871 Peak of cattle drives. Most of these cattle were Longhorns. Some historians have stated that the Texas Longhorn made more history than any breed of cattle the civilized world has known.

1873 Barbed wire was invented.

Angus were imported (Figure 24.1).

1875 Silos and refrigerator cars came into existence.

1884 Federal disease eradication programs began.

1880 Many cattle died because of the severe winter. Cattle producers realized the need to store and preserve winter feed.

Figure 24.1. During the late 1800s and early 1900s, the English breeds (primarily Angus, Hereford, and Shorthorn) increased in numbers which resulted in the Longhorn range cattle losing their identity. The American Angus Association used this oil painting in 1973 to help commemorate the 100th anniversary of importation of Angus cattle into the United States. The painting represents the first Angus bulls brought to Victoria, Kansas by George Grant to cross on his Longhorn cows. Courtesy of Frank C. Murphy for the American Angus Association.

1881 The Breeders' Gazette originated.

American Hereford Association established.

1882 American Shorthorn Association organized.

1883 American Angus Association established.

1886 Severe blizzards, drought, and overgrazing plagued cattle producers in the Northern Great Plains.

1887 Hatch Act passed by Congress. This act created agricultural research funds for college experiment stations.

1888 Crested wheatgrass introduced from Russia.

1895 Commercial feed industry began in Chicago.

1896 Simmental introduced. They had very little influence on the cattle industry at that time.

1897 National Livestock Association established with headquarters in Denver. This was the forerunner of The American National Cattlemen's Association.

Amalgamated Meat Cutters and Butcher Workmen of North America (AFL-CIO) founded.

1900 International Livestock Exposition at Chicago opened. It was held each year shortly after Thanksgiving so that it could be a culminating show of the year after all other state and regional shows had been held. Also, the show at this time would permit a great source of Christmas beef which was a traditional dish in the Chicago and surrounding area.

Polled Hereford Association organized.

1903 Emergence of the five major meat packing companies (Swift, Armour, Cudahy, Wilson, Morris).

1904 Value of protein supplementation to grains was shown.

1906 Model T Ford.

American Society of Animal Science was organized.

Beefmaster breed established.

Meat Inspection Act passed which provided for inspection (for wholesomeness, freedom from disease, and adulteration) of all meat intended for interstate and foreign commerce.

1909 Airplane.

1912 The term "vitamin" was coined.

First successful AI of cattle (Russia).

1914 The Smith-Lever Act was passed by Congress which brought the Extension Service into existence.

1916 Stock Raisers Homestead Act passed which permitted the homesteading of 640 acres for stock-raising purposes.

1918 Santa Gertrudis breed founded.

1920 Tractors. This had a tremendous effect on the cattle industry by increasing forage and grain production. Mechanization increased the feed supply. But also the reduction of 24 million head of horses and mules (1920–1960) made additional feed available to the cattle industry.

The first self-service meat counters open in retail stores.

1921 Packer and Stockyards Act passed by Congress which provided marketing regulations.

1922 Founding of National Live Stock and Meat Board. First industry agency to enhance the public image of meat and build consumer dollar demand.

1924 The American Brahman Breeders Association was organized.

1926 USDA sets up Federal Meat Grading Service. Beef carcass grades officially accepted but used only on a voluntary basis.

1930 U.S. Range Livestock Research Station established at Miles City, Montana.

1931 Drought and depression plagued the cattle industry.

1932 Brangus breed established.

1934 Taylor Grazing Act passed by Congress.

1936 Charolais imported from Mexico.

Artificial insemination began in dairy cattle.

1941 First performance bull testing station (Texas).

1942 Large commercial feedlots emerge. Large numbers of semi-trailer trucks hauled both grain and cattle.

1945 National Association of Animal Breeders organized.

1946 Beneficial results obtained from feeding antibiotics.

1950 Beef carcass grade standards amended. Previous grades of Prime and Choice were designated Prime: previous Good grade called Choice.

1951 "Snorter" dwarfism reported.

First successful bovine embryo transfer.

1952 First successful breeding using frozen semen.

American National Cowbelles, women's auxillary of the American National Cattleman's Association, was created for women desiring to promote beef and the beef industry.

1953 Stilbestrol discovered.

1954 Red Angus Association of America formed.

1955 First Beef Cattle Improvement Association formed in Virginia.

1956 Performance Registry International (PRI) established in Texas. This organization was created to record performance records of beef cattle.

USDA carcass grades amended. Commercial grade divided into two grades, Standard and Commercial.

1958 Humane Slaughter Act required packing plants to slaughter animals humanely if they desired to sell products to federal agencies.

1959 Angus Herd Improvement Records (AHIR) established.

1960 AI studs began emphasis on beef AI.

Transportation shifted from rail to truck as new slaughter plants were built near feedlots rather than in large metropolitan areas.

Retail chain stores processed beef centrally.

Boxed beef processing began to influence the industry.

1964 USDA feeder cattle grades officially accepted.

American Hereford Association created the Total Performance Records (TPR) Program.

Guidelines program (performance records) established by American Polled Hereford Association.

1965 Carcass yield grades adopted.

1966 Simmentals reintroduced (see 1886). Limousins also introduced at this time.

1967 Wholesome Meat Act extended meat inspection to products in interstate commerce.

1968 Beef Improvement Federation formed. This organization was to coordinate the performance record programs of state and breed associations.

American Simmental Association formed.

1969 Surplus grassland produced by government cropland diversion programs as 56 million acres are diverted from crop production (1955–1969).

1970 Cattle-Fax created by the American National Cattlemen's Association (ANCA).

Consumer activism surfaced. Issues were high beef prices, nutrition, and health concerns.

1972 Simmental Association established the first National Sire Summary for beef cattle.

1974 Monensin (trade name is Rumensin®), a newly discovered feed additive which improved feed efficiency.

1975 All-time cattle inventory record of 131.8 million head.

1976 Revision of USDA carcass grade standards. Major changes were (1) all federally graded beef would be yield graded in addition to graded for quality, (2) conformation eliminated as a quality grade factor, (3) marbling requirement for Prime, Choice, and Standard reduced by one degree of marbling, and (4) range of marbling for Good grade narrowed.

Surplus grassland available to beef cattle with 16 million head dairy cow reduction (1945–1976) and 38 million head decrease in sheep numbers (1942–1976).

1977 NCA (National Cattlemen's Association) organized through the merger of the American National Cattlemen's Association and the National Livestock Feeder's Association.

A United States Senate committee published a controversial report which recommended a de-emphasis of meat in the American diet as a measure of health improvement.

1978 First calf in the United States from a frozen embryo.

Humane Methods of Slaughter Act amended the Meat Inspection Act of 1906 to require that meat inspected and approved be produced only from livestock slaughtered in accordance with humane methods.

1979 Revision of USDA feeder cattle grades (new grades based on frame size and thickness).

DES (diethylstilbestrol) banned from use in cattle feeding.

The first prostaglandin (Lutalyse) approved for use in estrous synchronization.

1980 Large number of forage-producing acres converted to other crops since early 1970s as grain export market increased from $7.7 billion to nearly $40 billion (1972–1980).

1981 Proposed carcass grade changes submitted by NCA to USDA.

First test tube calf in the world born in the United States (PA). Fertilization of this Holstein calf occurred in a laboratory dish the size of a dollar.

1982 First identical twin calves born in the United States (CO) as a result of microsurgery (splitting the genetic material of a single 5 to 6 day embryo). First identical twin calves from this process occurred in Great Britain in 1981.

Proposed carcass grade changes made in 1981 were denied by the USDA.

Lasalocid (trade name is Bovatec®) approved for use in feedlot rations. This antibiotic improves both gain and efficiency of gain.

THE INFLUENCE OF CATTLE ON MAN'S CULTURE

Early Man

Prehistoric men hunted cattle to meet some of their immediate needs. The flesh provided food, while hides supplied various forms of raiment. Bones were used for weapons and various forms of tools. Cattle's first contribution to early man was one of simply assisting him in his survival.

Later, cattle dung was found useful as fuel in wood-sparse areas, building material, and eventually its fertilizing capabilities were identified.

From a technological standpoint, the most important contribution of cattle was their serving as beasts of burden and later of traction. It is highly probable that development of wheeled transport was closely associated with domestication of cattle.

A Measure of Wealth

Long before man derived the monetary system, approximately 700 BC, trade was based on barter and exchange. Cattle were highly valued and became the basis for value comparisons. In Homer's Iliad, prizes for wrestlers were based on cattle value:

"A massy tripod for the victor lies,
Of twice six oxen its reputed price;
And next, the loser's spirits to
restore,
A female captive, value but at four."

The noun "cattle" apparently has the same origin as "chattel," which means possession. The Romans took the Latin word decus (cattle) to give origin to the word pecunia, meaning money. Our present word, pecuniary, refers to, or relating to, money.

The first Roman coins made approximately 350 BC were crude, shapeless lumps of bronze. About 300 BC the bronze was cast into metal bars, shaped like flat, thin bricks. These were called aes signatum, and were obviously meant to represent the value of cattle since both sides displayed the figure of a bull. Although the bars were inconvenient to handle, they represented the first step toward eliminating cattle as a medium of exchange.

Alphabet

The first letter of the alphabet took its shape from an Egyptian hieroglyphical (picture symbol) which was the head of an ox. The Semites, who lived in Syria and Palestine, changed the symbol slightly and called it aleph, which means ox. The ancient Greeks used a similar symbol and called it alpha (Figure 24.2).

Religion and Mythology

Evidence indicates that man has held cattle in a sacred, mysterious image from the beginning of his existence until the present time. Certain individuals believe that pictures in the Lascaux Caves of France reflect the deep reverence that man had for cattle some 15,000 to 30,000 years BC. It is believed that superstitious forces caused them to paint the pictures. By capturing the cattle in their pictures and other rites, the actual chase would be easier and more productive.

Cattle worship probably reached its peak in Egypt. Evidence shows that it was occurring prior to 3200 BC and that it continued for centuries. Orisis, the most popular of all Egyptian gods, was believed by some to have originated from a cow impregnated by lightning or moonshine. The sacred bulls, Apis and Mnevis, which were to represent the Soul of Orisis, were actual bulls that were maintained in Egypt to be worshipped by the people. If any harm came to the bulls, the individual was cruelly tortured and hanged. Egyptian worship was based on the bull's strength, fierceness, and enviable reproductive powers.

It is little wonder that the Israelites, at the base of Mount Sinai, resorted to worshipping the Golden Calf after being under the Egyptian, bull-worshipping influence for centuries. Even the Prophet Moses, attempting to convey to them a new God, could not sway them from some of their traditional influences. Even centuries later, worshipping calf idols was being practiced—". . . have made molten images of their silver . . . let the men that sacrifice kiss their calves" (Hosea 13:2).

Other people besides the Egyptians worshipped and revered cattle. Many people believed cattle had supernatural powers. Supposedly, the assassination of Caesar and the destruction of Pompeii were foretold by an ox.

Today this sacred reverence of cattle exists in several countries. In some African countries, cattle are used only as ceremonial beasts where slaughter is limited or prohibited. Cattle are sacred to the Hindus of India. Over 200 million head of cattle, the largest cattle population of any country in the world, roam the streets and countryside of India while some people starve. This is an interesting paradox; however, it vividly displays the deep reverence people have for cattle.

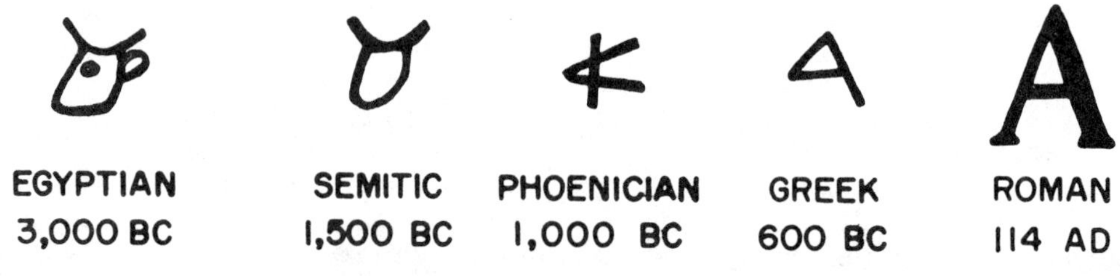

Figure 24.2. The letter A evolved from an ox head. Drawn by CSU Graphics.

Vocabulary

Reference has already been made to the origin of the word cattle and cattle's influence on the Roman alphabet. The early ancestors of cattle were given many different names. From Latin and Greek, we get bos, with the genetive bovis, from which are derived the words bovine, beeves, and beef. From the Teutonic, we get the German ochs, the Danish oxe, and the English ox. Combining this with the old sanskrit word ur, meaning forest or stony place, we get urochs or aurochs, which signifies wild cattle. The word ur was likely connected with cattle and used for centuries, eventually being Latinized into urus, the plural being uri. The latter word was used by Caesar (65 BC) in describing the big, wild cattle being hunted at that time.

Bos became the name of the genus in the zoological classification. Cattle producers using the words, "bossy," "so-o-o bos," or "come, bos" didn't realize they were speaking in such scientific terms.

The word, Brahman, as applied to our Zebu cattle in this country, is probably a misnomer. Brahman is the name of the highest, strictly vegetarian caste in the Hindu hierarchy.

The English talked and wrote about the word "neat." They borrowed the Icelandic word "naut," meaning ox, changed it to "neat" and associated it with cattle only. Colonists at Plymouth Rock referred to a shipment of "neats" from England. Even today, reference is made to neat's-foot-oil and neat's leather.

Entertainment

Man has pitted himself against the fighting bull for centuries. It appears that man's desire for blood-lust and courage, while popular today, is as old as civilization. There is clear evidence that as early as 2500 to 1500 BC the sports arena existed on the Island of Crete. Here "bull grappling" occurred which involved women as well as men, because some of these events had religious tones. Scenes from some of the world's first rodeos were cast in imperishable bronze which today gives us a clear vision of what occurred. A bull grappler would grab the charging bull by the horns, then when the bull flipped his head in the air, the acrobat would turn a somersault and land on the bull's back. The person would then vault to the ground or into the arms of an accompanying performer. Similar kinds of events spread into neighboring Greece and Italy.

In Rome wild cattle, along with lions, were turned loose on the unarmed Christians which resulted in mass slaughter of people. Also during this time, wild cattle battled gladiators who were armed with swords and spears.

The history of bull fighting in Spain, and eventually in Mexico, is older than the Spaniards themselves since the natives were known as Iberians. The exact origin of this sport in Spain is not known, but documented records show that during the Eleventh Century AD, men were competing against bulls. During the 1700s, bull fighting became a commercial enterprise in Spain with paid professionals doing the bull fighting. Breeding establishments came into existence where bulls were bred for superiority in their fighting instinct. Today thousands of four- and five-year-old bulls are required for bull fights in Spain, Mexico, and South America (Figure 24.3).

In the United States today, cattle are an integral part of man's sports activity and the entertainment of thousands. The rodeo would be far from complete without steer

Figure 24.3. Bull fighting is a popular sport in several Spanish-speaking countries. Courtesy of José Rafael Cortes, editor of "La Nación" newspaper, San Cristobal, Venezuela.

wrestling, team roping, calf roping, and bull riding (Figure 24.4). The latter is considered one of the most dangerous events in rodeo.

County, state, and national fairs are a source of pleasure and entertainment to many people. The vast majority of the United States population in urban centers enjoy the opportunity to see their steaks, roasts, and hamburger on-the-hoof.

Figure 24.4. Calf roping is a popular rodeo event. Courtesy of Larry Thomas.

Modern America

Cattle have had a tremendous influence on development of the United States. The original colonies burst at the seams as men set their eyes westward for more land, more grass, and more opportunities. Cattle was second only to gold as the stimulus to develop the West. Cattle left some everlasting marks on certain localities considering such present names as Beefhide, Kentucky; Bulltown, West Virginia; Greybull, Wyoming; Cowtown, New Jersey; Cowpens, South Carolina; Holstein, Iowa; Hereford, Texas.

Today's children continue to relive the historic past by playing "Cowboys and Indians."

The longevity of Americans today is extended because glands, organs, and other cattle parts yield products to help relieve serious health problems.

Above all else, Americans are a society of meat eaters where preference for beef is clearly demonstrated. Beef is king, from the glorified hamburger to the prime rib and T-bone steak. In a sense, beef is a luxury item in man's diet but is considered by many a necessity for a satisfying meal.

ZOOLOGICAL CLASSIFICATION

All animals have been classified by zoologists based on the differences and similarities in their structure and function. The basic unit of the classification system is the species. All members of the same species have a common gene pool and are capable of breeding with one another. There has been disagreement in placing the Zebu cattle in a different species since they do cross in the taurus species. Table 24.1 shows where cattle fit into the zoological classification.

THE FUTURE OF THE BEEF INDUSTRY

During the 1970s and early 1980s, the beef industry has experienced many volatile conditions and uncertain times. These periodic crisis situations have centered around such areas as consumer reactions to price, consumer health issues, volatile market prices sometimes widely fluctuating from one week to the next to change a profit to a loss, severe cost-price squeezes, and competition from other meats because of economic pressures on consumers.

Table 24.1. The Zoological Classification of Cattle

Kingdom—Animalia: Includes all animals in the animal kingdom.

Phylum—Chordata: Animals which have a backbone (vertebrae) or the rudiment of a backbone (chorda).

Class—Mammalia: Warm-blooded, hairy animals that suckle their young from a milk-producing mammary gland.

Order—Artiodactyla: Even-toed, hoofed mammals.

Family—Bovidae: Ruminants having hollow, up-branched horns and a placenta type of numerous cotyledons. This family not only includes the ox but the bison and buffalo.

Genus—Bos: Ruminant quadrupeds, including domestic and wild cattle having a stout body and curved, hollow horns that stand out laterally from skull.

Species—taurus: Includes most cattle found in the United States, including their European ancestors.
indicus: These are the Zebu (humped) cattle including the Brahman breed in the United States.

During the early 1980s, several conferences and reports have surfaced to assess the future of the beef industry. These activities reflect the concern that many individuals are considering as to what the future holds for the industry. Highlights of these conferences are given here reflecting the thoughts of individuals who are considered current beef industry leaders.

A special advisory committee for the NCA, after consulting with representatives of all segments of meat production and marketing, submitted a report, The Future of Beef. The report identified the following seven major trends expected to develop and continue through 1990:

1. Business Environment
   —An improving economy, with rapid real growth, declining inflation, and interest rates.
   —Ample supplies of energy but at higher prices.
   —An information explosion that will force cattlemen to change their business methods.
   —New and greater risks but tools to deal with the risks.
   —A relatively less important role for government in the economy and in agriculture.
2. Demand
   —A mature meat industry, with little or no growth in average per capita consumption of meat.
   —Improvement in the competitive relationships between beef, other meats, and the potential for beef's market share to be maintained or even grow.
   —Continuation of beef as consumers' preferred meat product.
   —A widening of the beef product line, providing consumers with more choices in terms of kinds and prices of beef products.
   —Increased demand for leaner beef; a change in beef grading standards or less use of grades in beef merchandising or both.
   —A leveling off and eventual decline in conern about beef in relation to health.
   —Limited but gradually improving opportunities to increase beef exports.
   —Development and implementation of a total beef marketing program—including research, product development, education and public information, and advertising and promotion.
3. Production
   —Ample financing for cattle production at competitive interest rates.
   —Increased difficulty in servicing debt (cash flow) because of higher interest rates.
   —Ample supplies of grain for feedings.
   —Potential for greatly improved range and pastures.
   —Water is becoming a more costly and potentially limiting factor.
   —Limited growth in total cattle numbers, with the herd remaining below the peak level of 1976. A "leveling" of cyclical changes in total herd size.
   —Continued large numbers of "part-time" cattle operators, decline in number of "medium" size operations, and slight increase in number of large, specialized operations.
   —No significant increase in concentration of ownership in cattle and beef businesses.
   —Slow rate of change in numbers of cattle by region.
   —Further separation of land and cattle ownership.

—Improvements in efficiency, through science, technology, and management—improving beef's competitive position and potential for profit among cattlemen.
—More market orientation—producing products that more readily satisfy the demands of different market segments.
—Improved marketing techniques restraining cost increases from producer to retailer. More coordination of production and marketing.

4. Efficiency
—Increased efficiencies in reproductive performance, nutrition (including forage utilization), disease control, and carcass quality and yield.
—Increased efficiencies in processing and marketing.
—More potential for further advances in beef production technology than in competitive meat production.
—Greater use by innovative operators of currently available technology and management techniques.
5. Business Management
—More use by successful operators of new communications tools, advisory services, and risk management techniques.
—More market orientation among individual cattlemen and adoption of new methods of marketing cattle and beef.
—Opportunities for profitable operations—with profits going largely to better informed and more able planners and managers.
6. Industry Unity
—Better coordination of beef industry programs.
—Increased funding for research and development in key areas.
—Support and growth of a strong national industry organization.
7. General Conditions
—Collectively, trends suggesting a rapidly changing environment for the cattle industry.
—A profitable future for cattlemen willing and able to meet consumers' needs.
—A complex future, not likely to mirror the past.

Boehlje, et al. (1981) reached the following conclusion after reviewing many topics associated with the beef industry's future: "Livestock will continue to be an important source of high-quality protein in the world's food supply. About 60% of the world's grazing land is unsuitable for cultivation. Ruminants convert various feeds, wastes, and by-products not directly consumable by humans into high-quality food, so they will continue to be the best use of those resources. Those animals fed high-energy grains that could be eaten directly by humans will be increasingly high-cost sources of protein, especially those with the least efficient conversion rates. Thus, while the highest income segments of the world population will continue to eat grain-fed beef, beef will have to become more competitive with other sources of animal protein for middle and low income consumers here and abroad, and that likely will require a greater emphasis on forage and by-product utilization in the ration.

"The U.S. beef industry must and will adjust to produce a competitively priced nutritious and palatable protein product. But these adjustments will require changes on the part of consumers, producers and government regulators. One possible change is in the grading system, but equally as important will be changes in production technology, specifically changes in the nutrition, health, breeding and processing technology to produce leaner beef at a lower cost. Changes in the grading system to compensate those who produce beef with the desired characteristics will facilitate the required adjustments, but additional research and information is required to demonstrate the private as well as social benefits to producers and processors of changes in production and processing techniques and procedures."

Dickerson (1981) identified the following ways in which lean beef could be produced more efficiently—both biologically and economically: "What biological objectives in beef improvement have the greatest potential? These three would reduce lean beef production costs most: (1) more calves marketed per cow annually, (2) faster growth rate—of market calves but not of replacement females, and (3) less waste fat in market animals.

"What would be the potential gains in productive efficiency? The total effects of a 20% improvement in all three, as an example, would approach 21% in either biological or economic efficiency. Biological efficiency is defined as food energy per unit of lean beef produced. Economic efficiency includes nonfeed inputs for labor, capital, housing, processing, and transport.

"A 20% genetic reduction in fat content of market animals apparently would improve biological efficiency more than comparable increases in number of calves marketed per cow or growth rate of market calves only.

"In terms of economic efficiency (dollar costs per kilogram of beef protein), the effects of 20% genetic change in calves marketed per cow, calf growth rate or fat deposition would be similar.

"On the first of the objectives—more calves marketed per cow annually—where would the potential savings come? More than 50% of feed energy, and nearly 50% of total cost per pound of beef protein produced, goes just to maintain and replace cows. So any net increase in calf crop would spread these major fixed cow-herd overhead costs over more calves (omitting extra cow costs for pregnancy and lactation, that would not be reduced).

"Each cow now weans 7/10 to 9/10 of a calf per year, on the average, and markets 5/10 to 7/10 of a calf. The difference between weaned and marketed is heifers retained as replacements.

"What about comparative costs of raising twins or single-born calves? Twin calves would be smaller at birth and weaning. So they would require (1) less preweaning feed, (2) more postweaning time and feed, but (3) little more total feed than single-born calves to reach the same market weight. The potential savings from more calves per cow would be reduced to the extent that twin births increase calf mortality and reduce subsequent fertility.

"What are the prospects for increasing incidence of twinning? Genetic twinning is lowly heritable. But it can be increased initially by total population screening, and steadily by using incidence of twinning among paternal half-sisters as a basis for selecting young sires (i.e. sire progeny test). A first step is the experimental evaluation of nutrition and management appropriate for twin-producing cows.

"Hormonally induced twinning has been erratic. Perfected ova-transfer twinning seems likely to be too expensive for commercial use.

"How would faster growth rate of market animals, as another breeding objective, increase efficiency of beef production? More than half of the feed energy—and nearly two-thirds of the total cost for growing the present market beef animal—goes just to maintain normal life processes. Net gain in body weight from protein synthesis and fat deposition used only about 45 percent of the feed and 35 percent of total growing costs.

"The proportion of total input used for maintenance alone can be reduced by increasing the animal's genetic rate of growth. Specifically, this means faster cell multiplication and increase in cell size, thus converting more of the feed intake into new body protein and fat. Total feed and nonfeed inputs per unit of gain would then be reduced.

"To make the case for rapid growth even stronger, 30% to 40% of the superiority of bulls or heifers selected on their own postweaning growth or yearling weight is genetic and is expected to appear in progeny of the selected parents.

"What limits faster growth rate as a breeding objective? The sobering limitation is the nearly proportional increase in cow size—and hence increase in the half of total production costs going for cow-herd maintenance and replacements.

"Mainly for this reason, numerous studies have shown little net reduction, from faster growth rate, in total cost of beef marketed. Cow replacements have the same genes as market animals in straight breeding or sire-breed rotation crossbreeding systems. So faster growth is associated with proportional increases in cow size in these systems.

"When can size in terminal sire breeds be a problem? Too great an excess in mature size of sire above dam breeds can increase birth difficulties, calf mortality and infertility. Thus, even in terminal sire breeds, the objective of faster growth rate should be accompanied by (1) some restriction on increased birth weight—and thus, indirectly, on increased gestation length calving difficulty and mature size as well—and (2) possibly by some changes in cow and calf conformation to minimize birth difficulty.

"So how can faster growth rate for market animals but not cows be achieved? The strongest case for faster growth rate as a biological objective is paternal or sire breeds—specifically, those used to produce crossbred market animals from older purebred or crossbred cows of smaller maternal breeds.

"Increasing growth rate of only the crossbred market calves directly reduces growing costs, per unit of gain, to the same slaughter weight. If these faster-growing market calves are carried to heavier slaughter weights, part of the faster growth savings is shifted from lower growing-maintenance costs to lower cow-herd costs per unit of slaughter weight.

"Does increased growth rate have an advantage in maternal breeds—those used to produce commercial cow replacements? Here, the advantage of faster growth is severely limited to changing the shape of the growth curve toward more rapid postnatal growth and earlier reproductive maturity, with little increase in cow size.

"Now, turning to the objective of leaner market animals: How are production costs affected? Above-maintenance costs of energy for depositing fat tissue is more than four times that of laying down lean tissue. Trim fat is also worth much less than lean retail cuts. Thus, leaner market animals would reduce production costs and especially increase the value of the market animal.

"For maintenance, on the other hand, feed per unit of body size is greater for lean animals than for fat ones.

"Some level of fat deposition is optimum for vital body functions, as well as for processor and consumer acceptability. Present market beef animals in this country contain 25% to 35% body fat—and the 10% to 15% fat content of turkeys and broilers is being criticized as too high.

"Can further progress be made toward leaner market animals? Differences in body fat content are rather highly heritable. Live animal measurements of subcutaneous fat could be used in selection programs. Also, existing breed differences and carcass progeny tests can be used in selecting sires of market animals. Higher liveweight prices for leaner market animals, however, are a needed incentive.

"What other biological objectives might be considered? Lower metabolic rate and activity would directly reduce the large maintenance costs for both the cow-herd and growing market animals. Some evidence shows breed and individual genetic variation in metabolic rate of cattle independent of fatness. But direct measurements of metabolic rate in individual animals would be very expensive, and a lower metabolic rate might have adverse effects on other performance traits.

"Optimum milk production is a biological objective that is heavily dependent upon the genetic, feed resource and marketing situation. Abundant inexpensive feed for cows would favor higher inherent milk levels, especially if calves were sold at weaning with no penalty for heavier calves. A higher milk level would be optimum if cows are producing twins, and both a higher nutrient level for cows and better calf-rearing management would be required."

Ritchie (1982) identified the major constraints on beef production efficiency and possible steps to correct these deficiencies:

Constraints

1. Long production cycle coupled with high interest rates.
2. Nearly 70% of dietary energy is for maintenance.
3. Transportation and associated costs: (a) Trucking, (b) Shrink, (c) Morbidity, and (d) Mortality.
4. High degree of fatness to ensure Choice grade.

Possible Steps to Improve Constraints

1. Dilute maintenance costs; sell more weight per cow.
2. If possible, retain ownership of calves to slaughter.
3. Reduce time on feed to minimum needed for acceptable palatability.
4. Reduce emphasis on marbling; stress lean growth.
5. Fine-tune the trade-off between lean growth and: (a) more energy to maintain fertility and (b) dystocia.

RESEARCH AND EDUCATIONAL PRIORITIES

Several groups have identified established research priorities for the beef industry. Research accomplished in the years ahead will have a marked effect on the continued shape and direction of the beef industry.

Animal Agriculture Conference

The conference was sponsored by several national associations of animal and poultry associations, 26 agricultural experiment stations, 10 colleges of veterinary medicine, and other livestock and government organizations. Pond, et al. (1980) prepared the executive summary of the conference, where research priorities were identified in the following areas: (1) human nutrition, (2) food processing and acceptability, (3) food safety, (4) animal nutrition and digestive physiology, (5) animal genetics and reproduction, (6) animal health, (7) feed production, (8) production, marketing, and distribution, (9) resources and environment, and (10) public policy.

National Cattlemen's Association

The NCA, through its Beef Cattle Research and Improvement Committee, recommended research needs in five major areas: (1) beef production, (2) pasture, range, and harvested forage, (3) industry economies, (4) product quality and safety, and (5) animal health. No attempt was made to rank the five major categories; however, with each of the five categories, several specific subject areas were defined and ranked.

Overall NCA Beef Cattle Research Priorities

1. Nutritive value of beef in the human diet
   a. Bioavailability of nutrients.
   b. Optimum consumption for sustaining good health in life cycle nutrition.
   c. Use in weight control and physical fitness diets.
2. Market analysis
   a. More precisely define what factors determine consumer purchasing preferences.
   b. Develop new or expanded domestic and foreign markets or both.
   c. Develop alternative marketing strategies for cattle and beef.
3. Consumer acceptance of beef
   a. Flavor (e.g. retention of natural flavor in precooked or reheated products or both).
   b. Tenderness enhancement through processing technology.
4. Processing and packaging
   a. Improve technology for restructuring of intermediate value products.
      1. Develop techniques to alter connective tissue.
      2. Develop improved binding agents.
   b. Improve quality, appearance and shelf life of fresh beef products.
5. Develop techniques to manipulate rumen microflora to improve cellulose and starch utilization
6. Profile plant morphology and physiology related to grazing management and animal performance
7. Implement integrated reproduction management (IRM) program
   a. Enhance female fertility.
      1. Improve efficiency of ovulation.
      2. Develop methods to reduce embryonic mortality.
      3. Diagnosis, treatment, and prevention of reproductive and nutritional disorders.
   b. Develop methods to reduce perinatal mortality (e.g., dystocia).
   c. Enhance male fertility.
      1. Disease.
      2. Libido.
   d. Develop techniques to enhance the onset of early puberty (male and female).

8. Improved beef by-product utilization
   a. Variety meat product development.
   b. Develop processing technology to improve competitive position of domestic hides and leather.
   c. Develop new or improved medical applications or both.
9. Identify pre-disposing management systems and infectious agents related to reducing losses from respiratory disease
10. Brucellosis
    a. Develop a vaccine maker.
    b. Develop improved methods of diagnosis and prevention.
11. Identify value determining carcass traits and develop instrumentation to measure them objectively.
12. Develop improved price discovery and reporting systems.

Overall NCA Beef Cattle Educational Priorities

1. Develop a computerized national data bank of beef cattle management and technical information that can be accessed by producers.
2. Extend factual information on nutritive value and preparation of beef.
3. Implement Grazing Lands and People project.
4. Implement Integrated Reproduction Management (IRM) program.
5. Increase application of accounting, financial and risk management information.
6. Implement forage management systems to optimize kinds, amounts and timing of grazing and inputs for maximum production efficiency.
7. Optimize utilization of heterosis and breed complementarity.
8. Effective utilization of performance and sire evaluation information.
9. Develop and apply computer systems modeling to facilitate management decisions.
10. Animal stress reduction.
11. Implement Residue Avoidance Program (RAP).
12. Predator management.

The NCA Research and Education Committee pointed out that their priorities were developed from a national scope reflecting the needs of the entire industry. They recognized that research and education needs of some states and regions might differ from the national list. States and regions were encouraged to draft their own specific needs and communicate them to those responsible for the research and educational programs in their area.

Carcass to Consumer Confab

The carcass to Consumer Confab, held September 1981, involved industry and university personnel and was sponsored by the Texas Cattle Feeders Association. The Confab was organized to: (1) Review current beef end-product research with end-product being defined as from carcass to table, (2) identify current research needs and prioritize them, and (3) assist in locating research dollars and possible universities interested in end-product research.

After identifying and prioritizing short-range and long-range goals, the participants narrowed the list into two major areas of need: No. 1 priority would be development of an intermediate value product. No. 2 priority would be a marketing survey of wants and needs of consumers, retailers, and HRI trade.

The first priority, that of an intermediate value product, could be defined as: providing the consumers and retailers a low-salt, uniform quality product between hamburger and steaks, with more convenience than a roast, at a price between hamburger and steaks and which tastes good. This goal would contain the following several research components: (1) pre-cooked, (2) packaged for extended shelf life, (3) utilization of meat extenders, flavor control, and (4) limitations on salt and phosphate content.

The second major priority, that of a market survey, should be done by a consortium effort of the beef industry. National Cattlemen's Association should take the lead and this should be an on-going project taking into consideration trends as they occur. Survey hopefully to begin in early 1982.

RECENT BEEF INDUSTRY INNOVATIONS

Following are relatively new innovations which recently have been made available to the industry or have been accepted on a relatively small scale.

Full-scale acceptance of some of these will no doubt cause significant changes in the beef industry over the next several years.

> Embryo splitting (induced) to produce identical twins; frozen embryos; embryo transfer; estrous synchronization; microcomputers; feeding bulls instead of steers (Seideman, et al. 1982); management systems; electronic marketing; methane gas production from manure; Ivermectin (the most potent broad-spectrum anthelmintic yet discovered—not yet released by FDA); electrical stimulation; blade tenderization; radurization of retail cuts; intermediate valued beef products; mechanical deboning and hot boning of carcasses.

FUTURE BEEF INDUSTRY GROWTH

Future beef industry growth will depend on: (1) population growth, which is expected to average about 0.8% per year in the 1980s, (2) expansion of exports, which now accounts for only 1% of the U.S. beef production, and (3) beef demand (development of new beef uses and beef products, improvement in the competitive relationship between beef and other meats improving beef's acceptance, and capturing a greater share of the available market for meat). These three items were identified by NCA Executive Vice-President, W. T. Berry, Jr., at the 1982 National Beef Profit Conference.

Numerous beef industry leaders in the early 1980s state that the beef industry is a mature industry. The rapid growth, experienced in the 1950s, 1960s, and early 1970s, is not to be expected in the 1980s and 1990s. Over the past two decades, the total per capita consumption of meat or animal protein foods has changed very little. Per capita consumption of the individual meat or protein foods comprising the total has changed significantly. In the future, the total is expected to remain approximately the same. The mixture, however, can be expected to change in direct proportion to the production and marketing skills behind each commodity.

Even in a mature industry, economic conditions can change sharply in a short period of time. High interest costs, high grain prices, decreased consumer buying power, and high energy costs could cause major changes in the beef industry (Figure 24.6). On

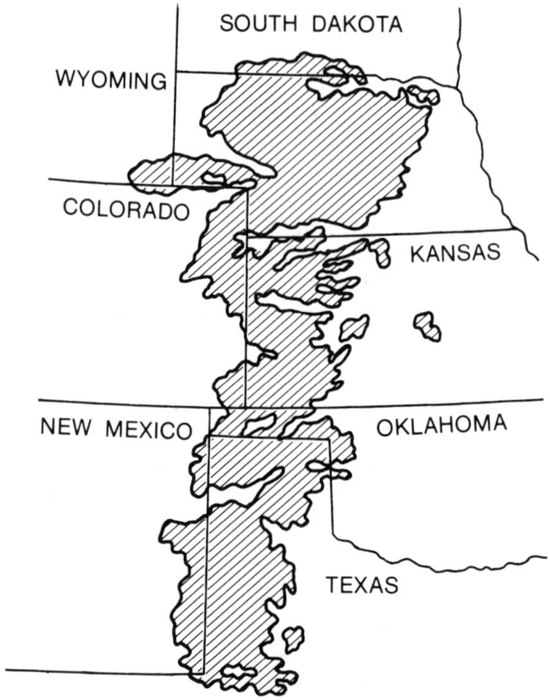

Figure 24.5. Irrigation water from the Ogallala Aquifer may be significantly reduced by the year 2000. High plains cattle feeding and slaughter would likely shift towards the Midwest, where a larger feed source exits. Drawn by CSU Graphics.

the other hand, a reverse of these negative economic conditions coupled with improved standards of living in several key foreign countries could well stimulate growth in the mature beef industry.

Even if the beef industry remains in a relatively mature status for several years, internal changes not only are likely but inevitable. There may well be shifts in geographical locations of efficient beef production. There are those who predict marked shifts in fed cattle production and slaughter from the Southern Great Plains States to the Corn Belt States. There is evidence that the water in the Ogallala Aquifer (Figure 24.5) is being depleted. In future years, this could have a marked reduction on the feed supply which became abundant as irrigation water was developed from this large aquifer over the past few decades.

While the depletion of the Ogallala Aquifer is a known possibility with significant effects on the beef cattle industry, there are no doubt numerous other future challenges which are not known at this time. Cattle producers and others in the beef industry will undoubtedly meet most of these challenges and create new opportunities. This is the way it has been in the past, and there is no reason to expect this to change in the future.

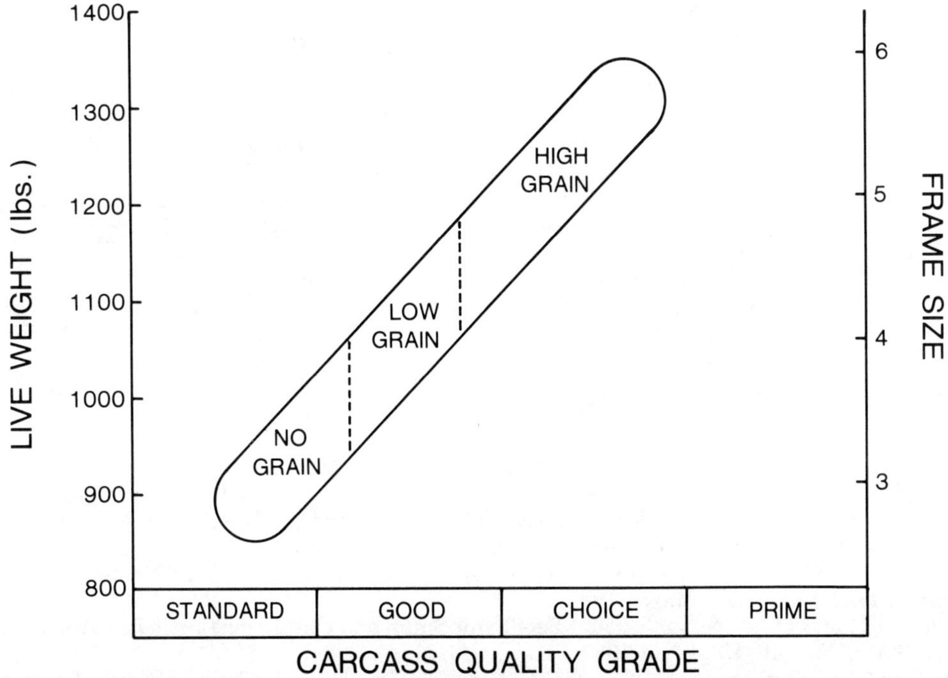

Figure 24.6. Future projections by many beef industry leaders state that feed grains will diminish in cattle rations. If this occurs, economics will likely dictate significant changes from the slaughter weights, frame sizes and quality grades currently in demand. Drawn by CSU Graphics.

SELECTED REFERENCES

Publications

A Brief History of the U.S. Meat and Livestock Industry, 1980. Meat Industry Magazine and the Natl. Livestock and Meat Board.

Baker, F. H., 1967. *History and development of beef and dairy performance programs in the United States.* J. Anim. Sci. 26:1261.

Boehlje, M., Rouse, G., Russell, J., Futrell, G., Jolly, B., and Hayenga, M. *The future of the beef industry.* Proceedings, National Beef Grading Conference. Iowa State Univ. 6–8 Jan. 1981.

Breidenstein, B. C., 1982. *Intermediate Value Beef Products.* Chicago: Natl. Live Stock and Meat Board.

Casteret, N., 1948. *Lascaux Cave, Cradle of World Art.* Natl. Geographic Magazine. 94:771.

Clemen, R. A., 1923. *The American Livestock and Meat Industry.* New York: The Ronald Press Co.

Dale, E. E., 1960. *The Range Cattle Industry. Ranching on the Great Plains from 1865 to 1924.* Norman: Univ. of Okla. Press.

Dickerson, G. E. *Reducing costs of lean beef production.* Better Beef Business. October to November, 1981.

Dobie, J. F., 1941. *The Longhorns.* Boston: Little, Brown and Co.
Field, R. A. New developments in beef processing. The Range Beef Cow Symp. VII. December 1981.
Koch, R. M. and Alego, J. W., 1983. *The beef cattle industry: Changes and challenges.* J. of Anim. Sci. 57:28.
Meat for the Multitudes. The Natl. Provisioner. July 4, 1981. Vol. I and II.
Park, E., 1969. *The World of the Bison.* New York: J. B. Lippincott Co.
Pond, W. G., Merkel, R. A., McGillard, L. D., and Rhodes, V. J., 1980. *Animal Agriculture. Research to Meet Human Needs in the 21st Century.* Boulder, CO: Westview Press, Inc.
Proceedings, Natl. Beef Profit Conference. NCA. Denver, CO. 28–29 June 1982.
Qeuner, F. E., 1963. *A History of Domesticated Animals.* London: Hutchison and Co. Ltd.
Ritchie, H. D. *Future and direction of the U.S. Beef Industry.* Proceedings from the National Association of Animal Breeders Conference. Denver, CO. 16 January 1982.
Research and educational priorities. Beef Business Bulletin (NCA). 15 July 1983.
Rouse, J. E., 1977. *The Criollo. Spanish Cattle in the Americas.* Norman: Univ. of Okla. Press.
Sanders, J. O., 1980. *History and development of the Zebu Cattle in the United States.* J. Anim. Sci. 50:1188.
Seidel, G. E., Jr., 1980. *Management of reproduction in cattle in the 1990's.* Ark. Agric. Expt. Sta. Spl. Rpt. 77.
Seideman, S. C., Cross, H. R., Oltjen, R. R., and Schanbacher, B. D., 1982. *Utilization of the intact male for red meat production: a review.* J. Anim. Sci. 55:826.
Simoons, F. J., 1968. *A Ceremonial Ox of India.* Madison: The Univ. of Wisc. Press.
The Future for Beef. NCA 5 March 1982.
The Future of the Beef Industry. National Beef Symposium and Oklahoma Cattle Feeders Seminar. Okla. State Univ. 10–12 March 1982.
Towne, C. W. and Wentworth, E. N., 1955. *Cattle and Men.* Norman: Univ. of Okla. Press.
Walters, F. E. *The fed beef industry: A new beginning.* Mountain Plains Annual Cattle Feeders Journal (Published by The Record Stockman). 25 June 1981.
Ward, G. M. 1984. *Energy impacts on future livestock production.* Boulder, Colo.: Westview Press.
Willham, R. L., 1982. *Genetic improvement of beef cattle in the United States: cattle, people and their interaction.* J. Anim. Sci. 54:659.

Visuals

"Cattlemen and Cowboys (A History)." An eleven projector, five screen presentation. Also available in slide-tape form. Contact NCA.
"What Makes a Cattleman." 16 mm film (approx. 10 min.). 1982. Ralston Purina, Checkerboard Square, St. Louis, MO 63164.

Appendix Contents

METRIC SYSTEM

The metric system has been accepted as the preferred system in nearly all countries of the world. The United States is the only major trading nation worldwide that does not presently use the metric system.

The National Bureau of Standards recommended in 1971 that the United States convert to the metric system. The president signed the Metric Conversion Act in 1975, establishing a national goal of converting, voluntarily, to predominantly metric measurements. Currently it is necessary to understand both measurement systems and be able to convert one to the other. Tables A.1, A.2, A.3, A.4, and A.5 presents these measurement systems and their conversions.

Table A.1. Customary (U.S.) Measurements

Length	
12 inches	= 1 foot (ft)
3 feet	= 1 yard (yd)
5½ yd or 16½ ft	= 1 rod
40 rods or 220 yds	= 1 furlong
5,280 ft 1,760 yds 8 furlongs	= 1 mile
3 miles	= 1 league
1 nautical mile	= 1.15 land miles
1 fathom	= 6.08 ft
1 hand (used in measuring horses)	= 4 inches
Area	
144 square (sq) inches	= 1 sq ft
9 sq ft	= 1 sq yd
43,560 sq ft 4,840 sq yds	= 1 sq mile
Surveyors or Chain Measure	
7.92 in.	= 1 link
25 links	= 1 rod
100 links 66 feet 4 rods	= 1 chain
80 chains	= 1 mile
160 sq rods 10 sq chain	= 1 acre
640 acres	= 1 sq mile
36 sq miles	= 1 township
Dry Measure	
2 pints	= 1 quart or 67.2 cubic in.
8 quarts	= 1 peck
4 pecks	= 1 bushel
Solid or Cubic Measure	
1728 cubic in.	= 1 cubic ft
46,656 cubic in. or 27 cubic ft	= 1 cubic yd

Table A.1. (*Continued*)

128 cubic ft (or 8 ft x 4 ft x 4 ft)	= 1 cord (wood)
144 cubic in. (or 1 ft x 1 ft x 1 in.)	= 1 board ft
2150.42 cubic in.	= 1 bushel
sq ft 4 in floor	= 0.0123 cu yds concrete
sq ft 6 in floor	= 0.0185 cu yds concrete
Avoirdupois Weight	
27$^{11}/_{22}$ grains	= 1 dram
16 drams	= 1 ounce or 437.5 grains
16 ounces	= 1 pound
2000 pounds	= 1 short ton
2240 pounds	= 1 long ton
Liquid Measure	
60 drops	= 1 teaspoon or ⅙ ounce
3 teaspoons	= 1 tablespoon or ½ ounce
16 tablespoons	= 1 cup
2 cups	= 1 pint or 16 fluid ounces
2 pints	= 1 quart
4 quarts	= 1 gallon
31½ gallons	= 1 barrel
7.48 gallons water	= 1 cubic foot
1 U.S. gallon	= 231 cubic in.
1 gallon of milk	= 8.6 lbs (approx)
1 gallon of water	= 8.32 lbs
1 acre foot	= 325.85 gallons

Table A.2. Prefixes in the Metric System

Prefixes	Symbol	Power and Meaning	
Tera	T	10^{12}	
Giga	G	10^{9}	
Mega	M	10^{6}	1,000,000 times base
kilo	k	10^{3}	1000 time base
hecto	h	10^{2}	100 times base
deca	da	10	10 times base
deci	d	10^{-1}	tenth
centi	c	10^{-2}	hundredth
milli	m	10^{-3}	thousandth
micro	μ	10^{-6}	millionth
nano	n	10^{-9}	
pico	p	10^{-12}	
femto	f	10^{-15}	
atto	a	10^{-18}	

Table A.3. Converting Customary (U.S.) Measurements to Metric

Length	
1 inch	= 25.40 millimeters (mm)
1 foot	= 0.30 meter (m)
1 yard	= 0.90 meter
1 mile	= 1.60 kilometers (km)
Area	
1 sq inch	= 6.50 sq centimeters (cm2)
1 sq foot	= 0.09 sq meter (m2)
1 sq yard	= 0.80 sq meter
1 acre	= 0.40 hectare (ha)
1 sq mile	= 2.60 sq kilometers (km2)
Mass	
1 grain	= 64.80 milligrams (mg)
1 ounce (dry)	= 28.30 grams (g)
1 pound	= 453.60 grams
1 pound	= 0.45 kilogram (kg)
1 ton (2,000 lbs)	= 907.20 kilograms
Volume	
1 cubic inch	= 16.40 cubic centimeters (cm3)
1 cubic foot	= 0.03 cubic meter (m3)
1 cubic yard	= 0.76 cubic meter
1 teaspoon	= 5.00 milliliters (ml)
1 tablespoon	= 15.00 milliliters
1 fl. ounce	= 29.60 milliliters
1 cup	= 0.24 liter (l)
1 pint (liq)	= 0.47 liter
1 quart (liq)	= 0.95 liter
1 quart (dry)	= 1.10 liters
1 gallon	= 3.78 liters
1 gallon (liq)	= 0.004 cubic meter
1 gallon (liq)	= 0.133 cubic feet
1 bushel	= 0.035 cubic meter
Rate Equivalents	
1 pound/acre	= 1.04 g/100 sq feet
1 pound/acre	= 1.121 kg/hectare
1 mile/hour	= 1.61 km/hour
Heat and Power	
1 British thermal unit (Btu)	= 1055.06 joules (J)
1 Btu/ft	= 11356.33 joules/sq meter (J/m2)
1 Btu/hr	= 0.29 watt
1 horsepower (550 ft lb force/sec	= 745.70 watts

Table A.4. Conversion Metric Measurements to Customary (U.S.)

Length		
1 millimeter (mm)	=	0.04 inch
1 meter (m)	=	3.28 feet
1 meter	=	1.09 yards
1 kilometer (km)	=	0.62 mile
Area		
1 sq centimeter (cm2)	=	0.16 sq inch
1 sq meter (m2)	=	10.76 sq feet
1 sq meter	=	1.20 sq yards
1 hectare	=	2.47 acres
1 sq kilometer	=	0.39 sq mile
1 sq kilometer	=	247.10 acres
Mass		
1 milligram(mg)	=	0.015 grain
1 gram (g)	=	0.035 ounce
1 gram	=	0.0022 pound
1 kilogram (kg)	=	2.205 pounds
1 metric ton	=	1.102 tons (short)
Volume		
1 cubic centimeter (cm3)	=	0.06 cubic inch
1 cubic meter (m3)	=	35.31 cubic feet
1 cubic meter	=	1.31 cubic yards
1 milliliter (ml)	=	0.2 teaspoon
1 milliliter	=	0.07 tablespoon
1 milliliter	=	0.03 ounce
1 liter (l)	=	4.23 cups (liq)
1 liter	=	2.11 pints (liq)
1 liter	=	1.06 quarts (liq)
1 liter	=	0.91 quart (dry)
1 liter	=	0.264 gallon
1 cubic meter	=	264.20 gallons (liq)
1 cubic meter	=	28.38
Temperature		
Degrees Celsius to degrees Fahrenheit (Multiply $9/5$ then add 32)		
Degrees Fahrenheit to degrees Celsius (Multiply $5/9$ after subtracting 32)		

Table A.5. Weight-Unit Conversion Factors

Units Given	Units Wanted	For Conversion Multiply by
lb	g	453.6
lb	kg	0.4536
oz	g	28.35
kg	lb	2.2046
kg	mg	1,000,000
kg	g	1,000
g	mg	1,000
g	μg	1,000,000
mg	μg	1,000
mg/g	mg/lb	453.6
mg/kg	mg/lb	0.4536
μg/kg	g/lb	0.4536
Mcal	kcal	1,000
kcal/kg	kcal/lb	0.4536
kcal/lb	kcal/kg	2.2046
ppm	μg/g	1
ppm	mg/kg	1
ppm	mg/lb	0.4536
mg/kg	%	0.0001
ppm	%	0.0001
mg/g	%	0.1
g/kg	%	0.1

Table A.6. Average Bushel Weights or Selected Grains and Seeds

Grain or Seed	Aver. Lbs per Bushel
Alfalfa	60
Barley	48
Bluegrass	14–30
Clover	60
Corn	56
Oats	32
Orchardgrass	14
Sorghum	56
Soybeans	60
Wheat	60

Table A.7. Determining the Amount of Corn Silage in Trench Silo. Tons = Average Width × Length × Average Depth of Silage (in Feet) × Tons per Cubic Foot of Depth

Depth of Settled Silage (Feet)	Tons per Cubic Foot	Depth of Settled Silage (Feet)	Tons per Cubic Foot
1	0.00925	9	0.01320
2	0.00985	10	0.01365
3	0.01040	11	0.01405
4	0.01090	12	0.01445
5	0.01140	13	0.01490
6	0.01190	14	0.01530
7	0.01235	15	0.01565
8	0.01280	16	0.01605

Example: A trench silo measures 8 feet wide at the top and is 50 feet long. The silage averages 8 feet deep. The calculation for total tons in the trench silo is as follows:

Solution:

$$\frac{8+12}{2} \times 50 \times 8 \times 0.0128 = 51.2 \text{ tons}$$

Table A.8. Silo Volumes per Foot and Pounds per 2″ layer

Silo Diameter	Volume per Foot of Depth (Cu Ft)	Pounds of Silage in 2″ Layer Based on 40 Lb/Cu Ft
12	113.1	755
14	153.9	1,025
16	201.1	1,340
18	254.5	1,696
20	314.2	2,094
22	380.1	2,534
24	452.4	3,015
26	530.9	3,539
28	615.8	4,105
30	706.9	4,712
34	908.0	6,053
38	1134.1	7,560
42	1385.4	9,235
50	1963.5	13,089
60	2827.4	18,847

Table A.9. Approximate Silo Capacity in Tons

Inside Diameter of Silo (Ft)	Silo height in feet					
	20	30	40	50	60	70
10	33	56	77			
12	48	80	110			
14	66	109	150	193		
16	86	143	196	252		
18		180	248	320	392	
20		223	307	394	483	574
22		270	371	477	585	694
24		321	442	570	697	827
26		377	520	668	818	970
28			600	773	947	1125
30			690	886	1087	1290

Table A.10. Nail Sizes and Descriptions

Size	Length Inches	Approx. No./Lb
Common nails		
2d	1	847
3d	1¼	543
4d	1½	294
5d	1¾	254
6d	2	167
7d	2¼	150
8d	2½	101
9d	2¾	92
10d	3	69
12d	3¼	63
16d	3½	49
20d	4	31
30d	4½	24
40d	5	18
50d	5½	14
60d	6	11
Spikes		
10d	3	32
12d	3¼	31
16d	3½	24
20d	4	19
30d	4½	14
40d	5	12
50d	5½	10
60d	6	9
5⁄16	7	6
3⁄8	8–12	5–3

VOLUMES AND WEIGHTS OF STACKED AND BALED HAY

Determining the volume and weight of hay is important whenever hay is sold or yields are determined. Volume of hay is expressed in cubic feet. Weight is expressed in pounds for individual bales, or tons for stacks and loads. Converting from volume, which is reasonably easy to measure, to weight requires that density (weight per cubic foot) of the hay be measured or estimated.

Stacks

The formula commonly used for estimating the volume of loose hay in stacks is:

$$V = \frac{(0 - 5.6\ W)}{2} \times W \times L$$

V = Stack volume in cubic feet
O = Average distance over stack in feet
W = Average stack width in feet
L = Stack length in feet

The over measurement (O) can be obtained using a tape or string with attached weight that is thrown over the stack. The stack should be checked in about four places, then those measurements averaged.

Bales

The volume of a stack of baled hay can be determined by measuring length, width and height; then multiplying these together. Another technique used with baled hay

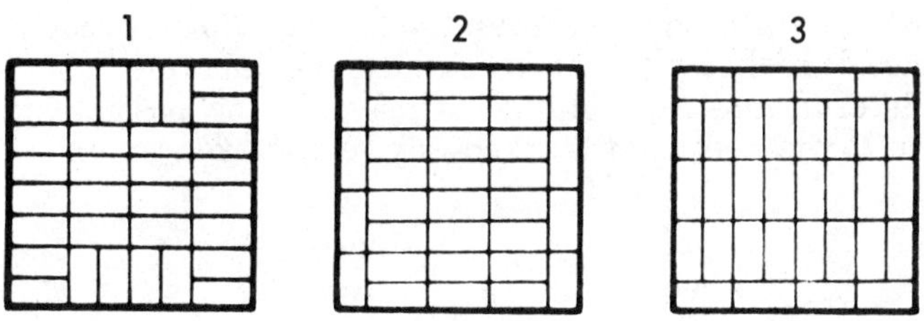

Figure A.1. Layers for stacking a small square stack (No. 1 for first and fourth layers; No. 2 for second and fifth layers; No. 3 for third and sixth layers). Courtesy of CSU.

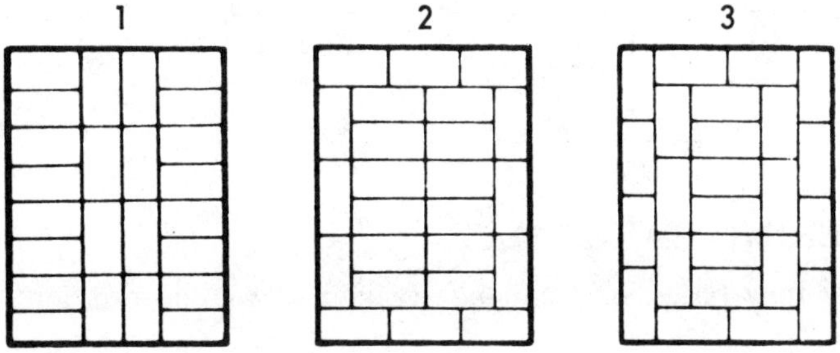

Figure A.2. Layers for stacking a small rectangular stack (No. 1, 2 and 3 represent the same layer as for illustration A). Courtesy of CSU.

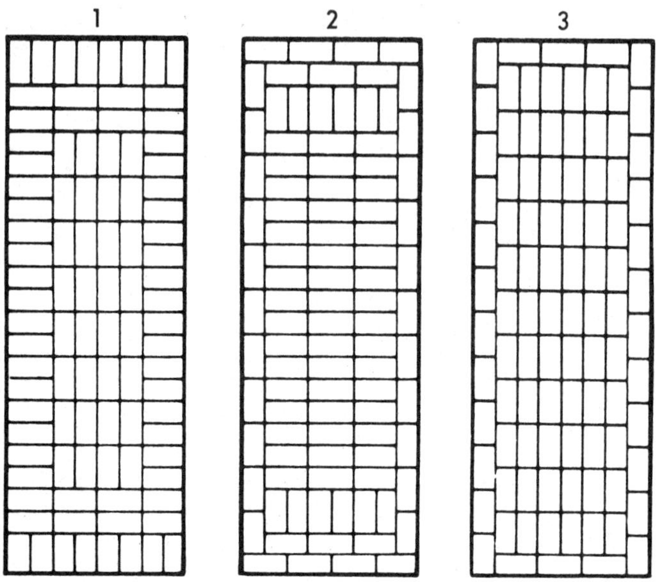

Figure A.3. Layers for a large rectangular stack. (No. 1 or first, fourth, seventh and tenth layers; No. 2 for second, fifth and eighth layers; No. 3 for third, sixth and ninth layers). Courtesy of CSU.

is to count the number of bales, then multiply by an estimated or determined weight per bale to determine total weight. The volume of the bales is occasionally needed for storage purposes. With round bales an estimate of total tonnage can be made by weighing a few bales and counting the number of bales involved.

WEIGHT OF HAY

The density, or pounds per cubic foot, of both stacked and baled hay varies greatly. The following table gives some guides. It is always better to actually calculate the density after weighing a few bales and determining the volume. It may also be possible to weigh a stack or portion of a stack. Remember that stacked hay settles over time.

	Weight of Loose Hay Pounds per Cubic Foot	Weight of Baled Hay Pounds per Cubic Foot
Alfalfa	4–5	8–14
Grass Hay	3–5	6–10
Straw	2–3	4–6

STACKING BALED HAY

Figures A.1, A.2, and A.3 show how baled hay can be stacked to reduce nutrient loss.

ROUND GRAIN BIN VOLUMES

This formula assumes that grain has fallen freely into the bin with the top forming a cone.

$$\text{Volume in bushels} = (.974 \times d^2) \times \left(h + \frac{d}{20}\right)$$

d = diameter (in feet)
h = height (in feet)

Example: $d = 18 \quad h = 12 \quad V = (.974 \times 18^2) \times \left(12 + \frac{18}{20}\right) \quad V = 4071$ bushels

MEASURING IRRIGATION WATER FLOW

Water measurement is a necessary part of water management. Measurement is also used to verify water rights.

Several devices can be installed for measuring water delivery in open ditch irrigation. Common devices are weirs and flumes. For accurate water flow measurements, weirs and flumes must be properly constructed, installed and maintained. Details for constructing and using these measuring devices are given in the publication: Irrigation Water Measurement, Wyo. Agric. Ext. Serv. Bull. 583R, April 1978.

Inches of Water Applied With Hand Move and Wheel Line Sprinklers

Factors that affect the inches of water applied are pressure, nozzle size, sprinkler spacing, and time. Pressure is measured with a gauge. The size is stamped on the nozzle, but older sprinklers should be verified using a drill bit as a measure. Spacing is the distance between sprinklers and line positions. Use Tables A.11, A.12, and A.13 in sequence, to determine inches of water applied.

Examples:
a. Pressure of 50 psi and 13/65″ nozzle from Table A.11 read the discharge as 8.71 gpm.
b. Sprinklers spaced 40′ apart and lines 60′ apart. From Table A.12 estimate the precipitation rate as 0.35 inches per hour.
c. Sprinklers are run for 12 hours. From Table A.13 estimate the amount of water entering the root zone as 3.15″.

Table A.11. Discharge in Gpm for Various Nozzle Sizes and Pressures

	Nozzle Diameter in Inches					
P.S.I.	9/64″	5/32″	11/64″	3/16″	13/64″	7/32″
30	3.26	4.01	4.83	5.75	6.80	7.85
40	3.74	4.61	5.54	6.64	7.80	9.02
50	4.18	5.15	6.19	7.41	8.71	10.10
60	4.50	5.65	6.80	8.12	9.56	11.05
70	4.96	6.10	7.34	8.78	10.32	11.95
80	5.29	6.52	7.84	9.39	11.02	12.74

Table A.12. Precipitation Rates in Inches per Hour for Various Sprinkler Discharges and Spacings

	Sprinkler Discharge in Gpm						
Spacing Feet	**3.0**	**4.0**	**5.0**	**6.0**	**8.0**	**10.0**	**12.0**
30x50	0.19	0.25	0.32	0.38	0.51	0.64	0.76
30x60	0.16	0.21	0.27	0.32	0.43	0.53	0.64
40x50	0.14	0.19	0.24	0.29	0.38	0.48	0.58
40x60	0.12	0.16	0.20	0.24	0.32	0.40	0.48

Table A.13. Water Applied per Set in Inches @ 75% Efficiency

	Precipitation Rate—Inches per Hour					
Hours	**0.20**	**0.30**	**0.40**	**0.50**	**0.60**	**0.70**
6.0	0.90	1.13	1.80	2.25	2.70	3.15
8.0	1.20	1.80	2.40	3.00	3.60	4.20
12.0	1.80	2.70	3.60	4.50	5.40	6.30
24.0	3.60	5.40				

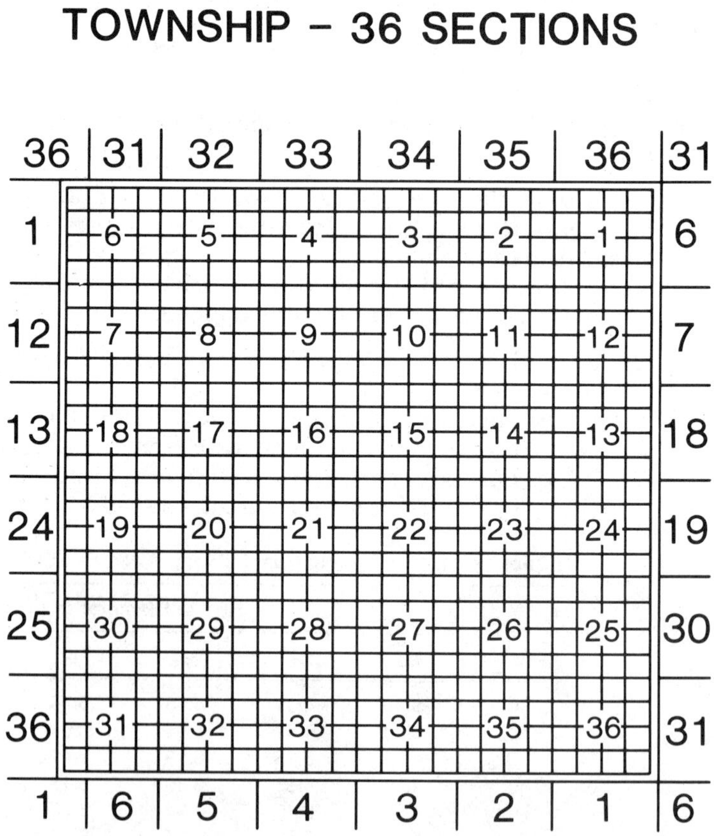

Figure A.4. A township divided into 36 sections. Adjoining sections are also shown. Drawn by CSU Graphics.

Land Description for Legal Purposes

The federal government has established land surveys which are used for the legal description of land. Townships, six miles square, are located north or south of standard parallels and east and west of prime meridians.

Each township is divided into 36 sections, each section being a mile square. The sections are numbered 1 through 36, starting with number 1 in the northeast corner of the township and ending with number 36 in the southeast corner (Figure A.4).

Sections are further divided into smaller units so that the location of every land parcel can be identified for legal purposes (Figure A.5). For example, the legal description of a certain 40 acres may read: The south half of the west half of the southwest quarter of section 6 in township 32, north of range 4 west.

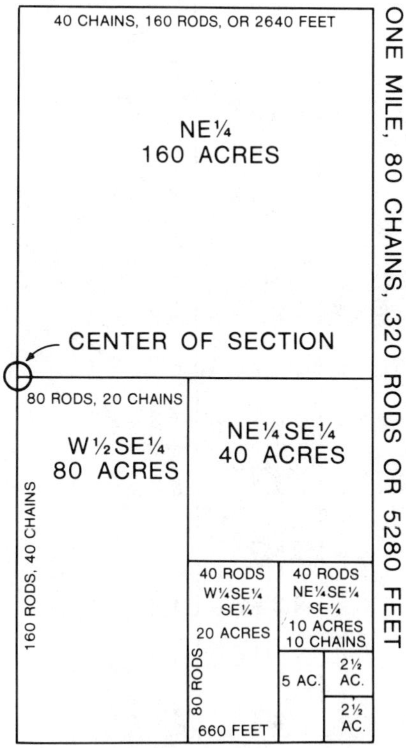

Figure A.5. A half-section divided into several component units. Drawn by CSU Graphics.

Table A.14. Age Measurement Table

To determine expected birth date of calf add 283 to the "day-number" of the date bred.

Day of Mo.	1 Jan.	2 Feb.	3 Mar.	4 Apr.	5 May	6 June	7 July	8 Aug.	9 Sept.	10 Oct.	11 Nov.	12 Dec.
1	1	32	60	91	121	152	182	213	244	274	305	335
2	2	33	61	92	122	153	183	214	245	275	306	336
3	3	34	62	93	123	154	184	215	246	276	307	337
4	4	35	63	94	124	155	185	216	247	277	308	338
5	5	36	64	95	125	156	186	217	248	278	309	339
6	6	37	65	96	126	157	187	218	249	279	310	340
7	7	38	66	97	127	158	188	219	250	280	311	341
8	8	39	67	98	128	159	189	220	251	281	312	342
9	9	40	68	99	129	160	190	221	252	282	313	343
10	10	41	69	100	130	161	191	222	253	283	314	344
11	11	42	70	101	131	162	192	223	254	284	315	345
12	12	43	71	102	132	163	193	224	255	285	316	346
13	13	44	72	103	133	164	194	225	256	286	317	347
14	14	45	73	104	134	165	195	226	257	287	318	348
15	15	46	74	105	135	166	196	227	258	288	319	349
16	16	47	75	106	136	167	197	228	259	289	320	350
17	17	48	76	107	137	168	198	229	260	290	321	351
18	18	49	77	108	138	169	199	230	261	291	322	352
19	19	50	78	109	139	170	200	231	262	292	323	353
20	20	51	79	110	140	171	201	232	263	293	324	354
21	21	52	80	111	141	172	202	233	264	294	325	355
22	22	53	81	112	142	173	203	234	265	295	326	356
23	23	54	82	113	143	174	204	235	266	296	327	357
24	24	55	83	114	144	175	205	236	267	297	328	358
25	25	56	84	115	145	176	206	237	268	298	329	359
26	26	57	85	116	146	177	207	238	269	299	330	360
27	27	58	86	117	147	178	208	239	270	300	331	361
28	28	59	87	118	148	179	209	240	271	301	332	362
29	29	—	88	119	149	180	210	241	272	302	333	363
30	30	—	89	120	150	181	211	242	273	303	334	364
31	31	—	90	—	151	—	212	243	—	304	—	365

To find the date on which a calf becomes 205 days of age locate the "day-number" of its birth date. Add 205 and locate the date he becomes 205 days old. For yearling measures at 365 days or 550 days follow the same procedure. The table applies to ordinary years only. For leap years, one day must be added after February 28.

Day of Mo.	1 Jan.	2 Feb.	3 Mar.	4 Apr.	5 May	6 June	7 July	8 Aug.	9 Sept.	10 Oct.	11 Nov.	12 Dec.
1	366	397	425	456	486	517	547	578	609	639	670	700
2	367	398	426	457	487	518	548	579	610	640	671	701
3	368	399	427	458	488	519	549	580	611	641	672	702
4	369	400	428	459	489	520	550	581	612	642	673	703
5	370	401	429	460	490	521	551	582	613	643	674	704
6	371	402	430	461	491	522	552	583	614	644	675	705
7	372	403	431	462	492	523	553	584	615	645	676	706
8	373	404	432	463	493	524	554	585	616	646	677	707
9	374	405	433	464	494	525	555	586	617	647	678	708
10	375	406	434	465	495	526	556	587	618	648	679	709
11	376	407	435	466	496	527	557	588	619	649	680	710
12	377	408	436	467	497	528	558	589	620	650	681	711
13	378	409	437	468	498	529	559	590	621	651	682	712
14	379	410	438	469	499	530	560	591	622	652	683	713
15	380	411	439	470	500	531	561	592	623	653	684	714
16	381	412	440	671	501	532	562	593	624	654	685	715
17	382	413	441	472	502	533	563	594	625	655	686	716
18	383	414	442	473	503	534	564	595	626	656	687	717
19	384	415	443	474	504	535	565	596	627	657	688	718
20	385	416	444	475	505	536	566	597	628	658	689	719
21	386	417	445	476	506	537	567	598	629	659	690	720
22	387	418	446	477	507	538	568	599	630	660	691	721
23	388	419	447	478	508	539	569	600	631	661	692	722
24	389	420	448	479	509	540	570	601	632	662	693	723
25	390	421	449	480	510	541	571	602	633	663	694	724
26	391	422	450	481	511	542	572	603	634	664	695	725
27	392	423	451	482	512	543	573	604	635	665	696	726
28	393	424	452	483	513	544	574	605	636	666	697	727
29	394	—	453	484	514	545	575	606	637	667	698	728
30	395	—	454	485	515	546	576	607	638	668	699	729
31	396	—	455	—	516	—	577	608	—	669	—	730

Index

Student Survey

Beef Production and the Beef Industry:
A Beef Producer's Perspective
by Robert E. Taylor

Students, send us your ideas!

The author and the publisher want to know how well this book served you and what can be done to improve it for those who will use it in the future. By completing and returning this questionnaire, you can help us develop better textbooks. We value your opinion and want to hear your comments. Thank you.

Your name (optional) ______________________________ School ______________

Your mailing address ____

City ______________________ State ____________________ ZIP ___________

Instructor's name (optional) _________________________ Course title _________

1. How does this book compare with other texts you have used? (Check one)

   ☐ Superior ☐ Better than most

   ☐ Comparable ☐ Not as good as most

2. Circle those chapters you especially liked:

   Chapters: 1 2 3 4 5 6 7 8 9 10 11 12 13 14 15 16 17 18 19 20 21 22 23 24

   Comments:

3. Circle those chapters you think could be improved:

   Chapters: 1 2 3 4 5 6 7 8 9 10 11 12 13 14 15 16 17 18 19 20 21 22 23 24

   Comments:

4. Please rate the following (check one for each):

	Excellent	Good	Average	Poor
Logical organization	( )	( )	( )	( )
Readability of text material	( )	( )	( )	( )
General layout and design	( )	( )	( )	( )
Match with instructor's course organization	( )	( )	( )	( )
Illustrations that clarify the text	( )	( )	( )	( )
Up-to-date treatment of subject	( )	( )	( )	( )
Explanation of difficult concepts	( )	( )	( )	( )
Selection of topics in the text	( )	( )	( )	( )

OVER, PLEASE

5. List any chapters that your instructor did not assign. ______________________

____

6. What additional topics did your instructor discuss that were not covered in the text? ______________________

____

7. Did you buy this book new or used? ☐ New ☐ Used
   Do you plan to keep the book or sell it? ☐ Keep it ☐ Sell it
   Do you think your instructor should continue to assign this book? ☐ Yes ☐ No

8. After taking the course, are you interested in taking more courses in this field? ☐ Yes ☐ No
   What is your major? ______________________

9. GENERAL COMMENTS:

May we quote you in our advertising? ☐ Yes ☐ No

Please remove this page and mail to:
Mary L. Paulson
Burgess Publishing Company
7108 Ohms Lane
Minneapolis, MN 55435

THANK YOU!